Discontinuous Phase Transitions in Condensed Matter

Symmetry Breaking in Bulk Martensite, Quasiperiodic and Low-Dimensional Nanostructures

Discontinuous Phase Transitions in Condensed Matter

Symmetry Breaking in Bulk Martensite, Quasiperiodic and Low-Dimensional Nanostructures

Vladimir Dmitriev

ESRF, France

World Scientific

NEW JERSEY • LONDON • SINGAPORE • BEIJING • SHANGHAI • HONG KONG • TAIPEI • CHENNAI • TOKYO

Published by

World Scientific Publishing Europe Ltd.

57 Shelton Street, Covent Garden, London WC2H 9HE

Head office: 5 Toh Tuck Link, Singapore 596224

USA office: 27 Warren Street, Suite 401-402, Hackensack, NJ 07601

Library of Congress Cataloging-in-Publication Data
Names: Dmitriev, Vladimir P., 1949- author.
Title: Discontinuous phase transitions in condensed matter : symmetry breaking in bulk martensite, quasiperiodic and low-dimensional nanostructures / Vladimir Dmitriev, ESRF, France.
Description: New Jersey : World Scientific, [2023] | Includes bibliographical references and index.
Identifiers: LCCN 2022025782 | ISBN 9781800612914 (hardcover) | ISBN 9781800612921 (ebook for institutions) | ISBN 9781800612938 (ebook for individuals)
Subjects: LCSH: Phase transformations (Statistical physics) | Condensed matter--Mathematical models.
Classification: LCC QC175.16.P5 D58 2023 | DDC 530.4/14--dc23/eng20220913
LC record available at https://lccn.loc.gov/2022025782

British Library Cataloguing-in-Publication Data
A catalogue record for this book is available from the British Library.

For any available supplementary material, please visit
https://www.worldscientific.com/worldscibooks/10.1142/Q0381#t=suppl

Desk Editors: Balasubramanian Shanmugam/Adam Binnie/Shi Ying Koe

Typeset by Stallion Press
Email: enquiries@stallionpress.com

To the Bandilet ~ Dmitriev ~ Friemann, my restless family

Preface

Although many of my colleagues regard me as more of a theorist, I see myself as primarily an experimental physicist. On the one hand, I have indeed published a number of theoretical works. Yet, on the other hand, I typically adhere to the hands-on problem-solving as an experimentalist at heart. This will be very apparent to the reader in the sections on fundamentals of the theory of phase transitions, where some subjects are omitted from the discussion. In my earlier work with Pierre Toledano (P. Toledano and V. Dmitriev, *Reconstructive Phase Transitions*, World Scientific, 1996), we offered examples of strongly first-order phase transitions and outlined a generalization of the phenomenological approach originally adopted only for continuous transformations. Here, with 24 years of continued research, we unify the currently available theory and experiments into advanced symmetry-based phenomenological descriptions.

This book continues the journey started by our work on reconstructive phase transitions. Pierre Toledano, due to other commitments, could not co-author this time, though his contribution to the present work is still significant. The first two chapters, reviewing the classic Landau theory, reiterate the first chapter of our earlier publication and a highly pedagogical review written mainly by Pierre back then. It is now repurposed to meet the new objectives and set a high bar for the rest of the book.

In brief, Chapters 1 and 2 offer the reader a collection of generic phenomenological models ready for practical use. The equations and phase diagrams can be employed even by those unfamiliar with the theory. Chapter 3 complements the introduction to the topic by the conventional terms' definitions.

Chapter 4 reveals the basic principles of the advanced phenomenological theory of phase transitions. The latter redefines the scope of the density-wave formalism, the order-parameter and order-parameter space. By that, it avoids the limitations of the classic Landau theory, accounts for non-critical background fluctuations, and extends beyond the limits of a mean-field theory.

The applications to realistic crystalline systems are demonstrated in Chapters 5 and 6. I give examples of classic structure reconstructions in bulk metals like Burger and Bain distortions, with special emphasis on their martensitic character. I illustrate step by step the symmetry-based method, identifying and accounting for its primary and secondary destabilizing mechanisms.

The density-wave approach to phase transitions results in a unified symmetry-based model-free theory of crystallization (Chapter 7). In the framework of a unified formalism, it considers the weak crystallization of molecular mixtures into liquid-crystalline mesophases. I also look into strongly discontinuous crystallization from molten metals and alloys into conventional fully segregated crystals and into aperiodic quasi-crystalline structures. I further cover here the assembly of aperiodic closed viral capsids with non-crystallographic symmetry.

Chapter 8 presents the symmetry-based approach for low-dimensional systems. It includes the stability comparison of different surface superstructures and monoatomic adlayers on single-crystal substrates. The approach permits fixing the atomic positions, identifying corresponding transformation mechanisms while relating them to the nanostructures' electrical conductivity. With the twisted bilayer graphene, I demonstrate how parametrization by the advanced phenomenological approach connects, in a symmetry-controlled and thus model-free manner, the geometrical parameters with the reconstruction of the Brillouin zone and energy bands.

Symmetry provides a straight path toward understanding condensed matter physics. I hope this work proves helpful to a wider readership from students to established researchers. This book offers new applications, where the theory reveals its strengths such as martensitic transformations, crystallization of the crystalline and aperiodic bulk materials, complex fluids and viruses. It further equips the reader with the necessary tools and recipes for new materials design.

Last but not least, I would like to express my deep gratitude to the colleagues and friends who encouraged the writing and generously shared their insights: Alexander Dmitriev, Valentin Gordeliy, Olga Konevtsova, Alexander Korzhenevskii, Sergey Rochal, Vladimir Shirokov and Pavel Timonin. I feel fortunate and deeply grateful collaborating with Elena Boldyreva and Leonid Dubrovinsky, who introduced me to specific materials physics and chemistry under extreme conditions. I am significantly indebted to Vadim Dyadkin for his kind assistance in the preparation of this book.

Vladimir Dmitriev
Grenoble
December 2021

About the Author

Vladimir Dmitriev is Emeritus at the European Synchrotron Radiation Facility (Grenoble, France), where he headed a Collaborative Research Group for the past 20 years. His research interests are both in experimental and in theoretical solid state physics. He obtained his PhD in experimental physics in 1979 from the Rostov State University (USSR, currently — Southern Federal University), and DSc in theoretical physics from the Institute for Low-Temperature Physics in Kharkiv (Ukraine). In the late 1990s, he worked as a professor at the University of Santa-Catarina (Brazil), later pursuing what he considers the best career to combine his experimental and theoretical skills, at the largest European synchrotron center in France. His research interests are in X-ray diffraction and Raman spectroscopy of materials under extreme conditions; theoretical models of structural phase transitions, strongly discontinuous phase transitions in crystals and quasicrystals, and phase transitions in complex fluids. He is the author, in collaboration with P. Tolédano, of the *Reconstructive Phase Transitions: In Crystals and Quasicrystals* (World Scientific Publishing, 1996).

Contents

Chapter 1

Phenomenological Theory of Weakly Discontinuous Phase Transitions

1.1. Introduction

Two main classes of phase transitions occur in crystals, which involve a modification of the crystal's atomic configuration. A *first class* is characterized by the fact that the structural changes which take place across the transition do not break the chemical bonds, but only slightly alter their lengths and orientations. When the transition can be described in terms of displacements, it means more concretely that the atomic shifts from one phase to another are generally of the order of 0.01–0.1 Å. At the experimental level, this category of structural transitions corresponds to latent heat smaller than a few *calories/gram* and to small discontinuities in the physical quantities variation at the transition point. Most importantly, a group–subgroup relationship between the symmetry invariance groups of the phases surrounding the transition can be inferred from the evolution of the X-ray and neutron diffraction spectrum across the transition. This type of transition is found in most of the insulators, in some alloys, and in a small number of crystals of pure elements. A minority of these structural transitions is found to be second order (continuous), and in most cases, they are slightly first-order (slightly discontinuous).

To a *second class* of structural transitions pertain most of the allotropic transformations of the elements, the majority of the transitions found in alloys, and also a few well-established examples of transitions which take place in insulating materials. These transitions, often denominated as *reconstructive*, since they involve breaking part of the chemical bonds of the initial structures, exhibit drastic changes at the transition point with large latent heat and thermal hysteresis. The corresponding atomic displacements are of the order of the lattice parameters, and *no group–subgroup* relationship is found between the symmetry of the phases. The transitions have generally a strongly first-order character.

The apparent simplicity of the symmetry connection between the structures in the first class of structural phase transitions has motivated, since the seminal work of Landau [1], a widespread effort at working out their general properties on the level of the phenomenological theories. The aim of this chapter is to make a general survey of this phenomenological approach to structural transitions in which the phases are group–subgroup related, as the theoretical methods used for their interpretation, and some of their general properties, will serve for the description of reconstructive phase transitions.

This chapter focuses on discontinuous phase transitions. This is not only because the subject of this book concerns the theory of a class of first-order transitions, but also because most of the extensions of the original Landau theory, valid only for the second-order ones, apply to the interpretation of discontinuous transitions, which constitute, by far, the more common situation among structural systems. Even when considering only the structural transitions, discontinuous transitions constitute a very heterogeneous family [2]. Obviously, the primary thermodynamic classification of phase transitions, which distinguishes only between the second- and first-order transitions, is insufficient, as it denominates similarly weakly discontinuous transitions and reconstructive ones. However, it should be possible to differentiate them also on a theoretical ground and not only quantitatively. It is one of the aims of this book to show that the *full* symmetry of the order-parameter, which includes *the intrinsic symmetry of the concrete transition mechanism*, may serve as a basis for distinguishing between the various categories of first-order transitions.

Before dealing with the first-order discontinuous transitions, we should introduce the basic concepts of the Landau theory of continuous phase transitions and the various general procedures which allow the phenomenological description of a phase transition.

1.2. Basic Concepts of the Landau Theory of Phase Transitions

1.2.1. *Aim of the Landau theory*

The Landau theory is a *phenomenological* theory. It does not aim to predict or establish that a phase transition exists in a given system. The *existence of a transition* is an *experimental fact* considered as a starting point of the theory. The explanatory power of the theory is to establish the overall consistency of the microscopic characteristics of the transition (space-symmetry and structural changes, anomalies in the phonon spectrum, etc.) and the results of the measurement of various relevant macroscopic quantities (thermal, dielectric, optical, elastic, etc.).

The *continuous* ("second-order") character of the transition plays an essential role in the working out of the general foundations of the Landau theory. However, though the *strict field of validity* of the theory is that of continuous transitions, the theory satisfactorily applies to most of the first-order transitions, for which a group–subgroup relationship exists between the phases surrounding the transition.

The Landau theory defines two basic concepts: the *order-parameter* and the *transition free energy.* From the symmetry properties of these two quantities, it is possible to infer, on the one hand, a certain number of observable symmetry characteristics of the system: degeneracy of the "low-symmetry" phase (i.e., the number of energetically equivalent domain orientations in this phase) and enumeration of the possible symmetries of the "low-symmetry" phase for a given symmetry of the "parent" high-symmetry phase. On the other hand, macroscopic quantities can be classified as functions of their symmetries with respect to the order-parameter from which can be deduced a classification of the type of anomalous behavior induced by the occurrence of the phase transition (see Chapter 3).

In order to give an intuitive approach to the basic arguments of the Landau theory, and to its use, we will first analyze a simplified working example of the structural phase transition.

1.2.2. *An introductory example*

1.2.2.1. *Crystallographic description and assumptions*

Figure 1.1 represents the unit cell of a speculative crystalline structure with primitive tetragonal Bravais lattice in which a phase transition is assumed to take place. Positive ions occupy the vertices of the tetragonal cell, which has the lattice constants $\boldsymbol{a} = \boldsymbol{b} \neq \boldsymbol{c}$. A negative ion M^- is at the center of the cell. This configuration is assumed to be the equilibrium state of the system *above* the temperature T_C of the transition. Below the transition, equilibrium is assumed to correspond to a structure which differs from the preceding *parent* structure by the fact that M^- lies out of the center of the cell in an unspecified direction. Hence, the latter equilibrium is characterized by the magnitude and direction of the displacement $\boldsymbol{v}_0 = (v_x^0, v_y^0, v_z^0)$ of the central ion. At high temperature, the equilibrium corresponds to $\boldsymbol{v}_0 = 0$.

Let us determine the displacement $\boldsymbol{v}_0$ as a function of temperature. This can be done by a *variational method.* One considers an *arbitrary* displacement $\boldsymbol{v}$ of the M^- ion. For a given temperature T and pressure P (or volume V), and specified values of the components of $\boldsymbol{v}_0$, there is a definite value $F(T, P, v_x, v_y, v_z)$ for the *free energy* of the system. This function is a *variational free energy* and not the equilibrium free energy of the system, since it is calculated for an arbitrary displacement. The equilibrium displacement $\boldsymbol{v}_0(T, P)$ is defined as the displacement which minimizes the variational free energy F. The equilibrium free energy of the system is F_eq $F_\mathrm{eq}(T, P) = F(T, P, \boldsymbol{v}_0)$. Note that, in the case of a change in the volume, one would have to consider a variational Gibbs function, $F + PV$, in order to determine the equilibrium of the system.

One cannot go further as long as one does not know the form of the variational free energy as a function of the components of $\boldsymbol{v}$. The second step of the Landau theory is to show that, given the general assumptions, one is able to determine simply the form of $F(T, P, \boldsymbol{v}_0)$ in the required range of values of the function arguments.

The *basic assumption* is that of *continuity of the phase transition* in two different meanings: on the one hand, one assumes that the equilibrium values $v^0{}_i(T, P)$ *vary continuously across the transition at* T_C. On the other hand, one assumes that F is a *continuous* and *derivable function* of the parameters T, P, $\boldsymbol{v}$, which can be expanded in the form of a Taylor expansion as a function of these arguments. Let us stress that, as on either side of T_C, $|\boldsymbol{v}_0|$ is small, one can restrict the determination of the functional form of $F(T, P, \boldsymbol{v})$ to *small values* of (v_x, v_y, v_z) and of $|T - T_C|$. F will then be equal to the *sum of the first relevant terms of a Taylor series in the preceding variables.*

In view of further applications of the theory, let us note that the above limitations to small values (displacements) appear in order to introduce and then operate with analytically expressible function F, and are not induced by any crystal-geometrical restriction.

1.2.2.2. *Symmetry constraints: Form of the free energy*

The central property of the variational free energy which allows to specify its form is a *symmetry property.* F is a function of (v_x, v_y, v_z), which is *invariant by the symmetry transformations of the high-temperature (parent) equilibrium structure.* In other terms, an arbitrary displacement $\boldsymbol{v}$ and the displacement $\boldsymbol{v}'$ obtained from $\boldsymbol{v}$ by applying one of the latter symmetry transformations correspond to the same value of the free energy. Indeed, both displacements determine an identical set of mutual distances between the positive and negative ions of the system, and the free energy only depends on this *internal* configuration of the ions.

In the example represented in Fig. 1.1, the set of symmetry transformations comprises, aside from the lattice transformations, four-fold rotations around the Z axis, mirror symmetries into planes as (m_x, m_y), as well as the products of these transformations. The set of rotations and reflections form a *group* G_0 of order 16, which is the crystallographic *point group* $4/mmm$. Note that the symmetry property of the free energy, to be invariant by the symmetry group of the parent structure, holds for each degree of the Taylor expansion of F. This property derives from the fact that the geometric transformations of G_0 act *linearly* on the components of $\boldsymbol{v}$. Hence, the terms of different degrees belonging to the expansion of F will not

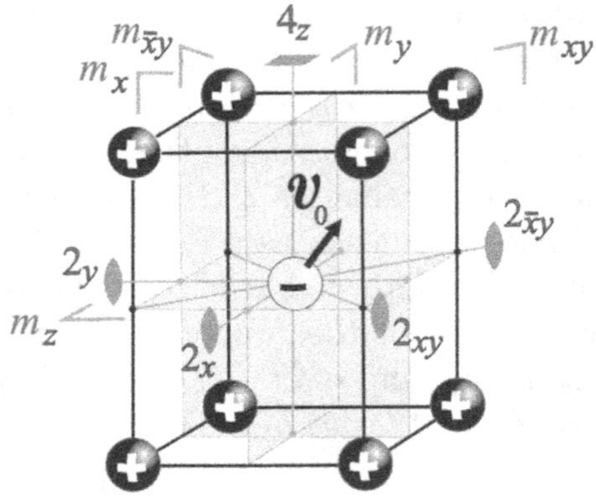

Fig. 1.1. Unit-cell reflecting the tetragonal $4/mmm$ symmetry of the structure considered in the introductory example. Vector $\boldsymbol{v}_0$ shows the displacement of the ion M^-.

mix and must be separately invariant. It is easy to check on Fig. 1.1 that by successive application of the mirror symmetries m_x, m_y, m_z to the components of $\boldsymbol{v}$, no linear combination of these components is invariant by G_0: each of the three former symmetry transformations reverses one component of $\boldsymbol{v}$ and preserves the other two. *Linear terms are, therefore, absent from the expansion.*

As for second-degree terms, the same symmetry transformations preclude the existence of configurations of bilinear products of the type $(v_x v_y)$. Actually, one finds that the four-fold symmetry imposes that the most general form of the second-degree contribution to the variational free energy is a linear combination of v_z^2 and $(v_x^2 + v_y^2)$. Hence, the Taylor expansion of F, restricted to its lowest degree terms in (v_x, v_y, v_z), is

$$F(T, P, v_x, v_y, v_z) = F_0(T, P) + \frac{\alpha_1(T, P)}{2} {v_z}^2 + \frac{\alpha_2(T, P)}{2} ({v_x}^2 + {v_y}^2). \tag{1.1}$$

1.2.2.3. *The order-parameter*

Let us now derive the key result of the Landau theory: *either the component v_z or the pair of components (v_x, v_y) will take non-zero values below T_C (but not both).* The derivations of this result rely on the fact that one and only one of the two coefficients α_i in Eq. (1.1) vanishes and changes sign at T_C, and that the other coefficients remain positive in the neighborhood of T_C. The demonstration is as follows: at T_C, the equilibrium values v_i^0 of the components of $\boldsymbol{v}_0$ are zero. Therefore, at this temperature, the variational free energy (1.1) is minimum for $v_i = 0$. Considering the form (1.1) of F, this property implies that we have $\alpha_i(T_C) \geq 0$ for $i = 1, 2$. These inequalities cannot be *strict* for *both* coefficients α_i, because their positiveness would hold on either side of T_C in the vicinity of this temperature. Consequently, the minimum of F would correspond to $v_i = 0$ on either side of the transition, while the situation assumed is only compatible with this result *above* T_C. Using the converse argument that the equilibrium values of the v_i are all not equal to zero *below* T_C leads easily to the conclusion that *one of the two coefficients* α_i, *at least vanishes at* T_C, *and becomes negative below this temperature*, as represented in Fig. 1.2(a).

The two coefficients α_i *cannot vanish simultaneously,* neither at T_C nor in the vicinity of this temperature. This result relies on the assumption that the two coefficients are *different functions* of temperature and pressure. Figure 1.2(b) shows, in the (T, P) plane, the lines corresponding to the vanishing of the two functions α_i. The simultaneous vanishing occurs only at an isolated point (T_0, P_0). Let us consider, for instance, the situation depicted in Fig. 1.2(b). For P above P_0, on lowering the temperature, α_1 vanishes at T_C', and α_2 remains positive in the neighborhood of T_C'. Hence, the equilibrium value of the set (v_x, v_y) remains equal to zero on either side of T_C'. A transition at this temperature will only concern a possible change in v_z^0. Likewise, for P below P_0, a transition at T_C'' only concerns

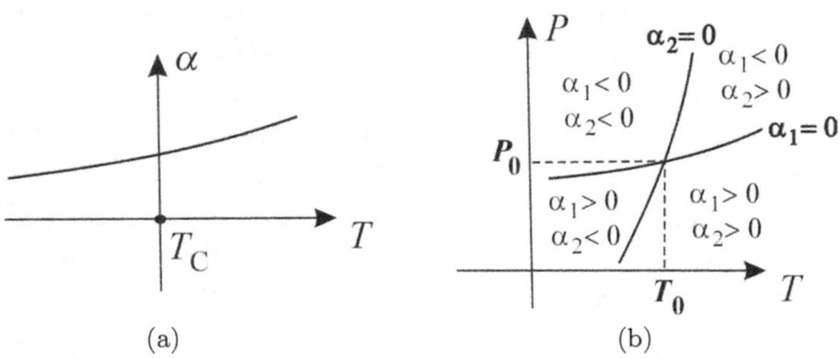

Fig. 1.2. (a) Almost linear dependence of $\alpha \cdot (T - T_C)$. (b) Region of the pressure–temperature plane corresponding to different signs of the coefficients α_1 and α_2.

a possible change of the set of components (v_x^0, v_y^0), the third component v_z^0 remaining equal to zero on either side of T''_C. Hence, an infinitesimal change of the pressure (for instance, a small fluctuation of the atmospheric pressure) from above P_0 to below P_0 will modify qualitatively the nature of the phase transformation. This is a singular situation not encountered in nature. Rather, the standard situation corresponds to pressures away from P_0, for which a slight change of the pressure does not modify significantly the direction of the displacement. In this case, *one coefficient* α_i *only vanishes and changes sign at the transition temperature.*

If one α_i coefficient is strictly positive, then the minimum of F with respect to the components v_i appearing in the second-degree term multiplying this coefficient *has zero equilibrium value in the vicinity of* T_C *above and below this temperature.* Thus, depending on the α_i which remains positive, either v_z or (v_x, v_y) can be omitted, in the first place, from the free energy expansion. *The remaining set of components constitutes the order-parameter of the transition.* At this stage, the order-parameter is defined as the set of degrees of freedom whose coefficient of the second-degree contribution to F *vanishes and changes sign at* T_C. The number of independent components of the order-parameter (one in the case of v_z, two in the case of (v_x, v_y)) is the *dimension of the order-parameter.* In reference to Fig. 1.1, it means that the displacement of the M^- ion below T_C cannot occur in the arbitrary direction of space. It is *either* along the Z axis *or* in the (X, Y) plane.

1.2.2.4. *Stable state below* T_C *and physical anomalies at* T_C

We have seen that either v_z or the couple (v_x, v_y) components of $\boldsymbol{v}$ constitute the order-parameter of the transition, and that the free energy needs only to be expanded as a function of the components of the order-parameter. Below the transition, the corresponding coefficients α_i are negative, and accordingly, a limitation of the expansion to the second degree is not sufficient to determine the equilibrium state of the system, which requires to add positive terms of higher degrees. We will first examine the simplest, one-dimensional situation for which the order-parameter coincides with the v_z component.

The same symmetry argument used to establish the form (1.1) of the Landau free energy allows straightforwardly to assert the absence of a third-degree term v_z^3 in the expansion of F as a function of v_z

and to check the possibility of a fourth-degree term. If we assume that this simplest form of expansion is sufficient to determine the equilibrium state of the system, the coefficient of the fourth-degree term must be positive in the neighborhood of T_C. Up to the latter degree, the form of the relevant contributions to the free energy is therefore

$$F(T, P, v_z) = F_0(T, P) + \frac{\alpha(T - T_C)}{2} {v^2}_z + \frac{\beta}{4} v_z^4. \tag{1.2}$$

In this expression, α_1, which is an odd function of $(T - T_C)$, has been expanded up to its *lowest* degree term, α being a positive constant. Likewise, the lowest degree expansion of $\beta(T - T_C)$ is a positive constant β in the vicinity of T_C. The function $F_0(T, P)$, which is the zero-degree term in the expansion, represents the normal "background" part of the free energy. It behaves smoothly across the transition since it does not depend on the order-parameter. A plot of $F(v_z) - F_0$ for three characteristic temperatures is indicated in Fig. 1.3.

The minima of F, determined by the set of conditions

$$\frac{\partial F}{\partial v_z} = 0, \quad \frac{\partial^2 F}{\partial v_z^2} > 0, \tag{1.3}$$

occur above T_C for $v_z^0 = 0$ as expected. For $T < T_C$, they occur for

$$v_z^0 = \pm \left[\frac{\alpha(T_c - T)}{\beta} \right]^{1/2}. \tag{1.4}$$

This behavior has a general validity: the order-parameter of a transition, in the framework of Landau's theory, should possess a

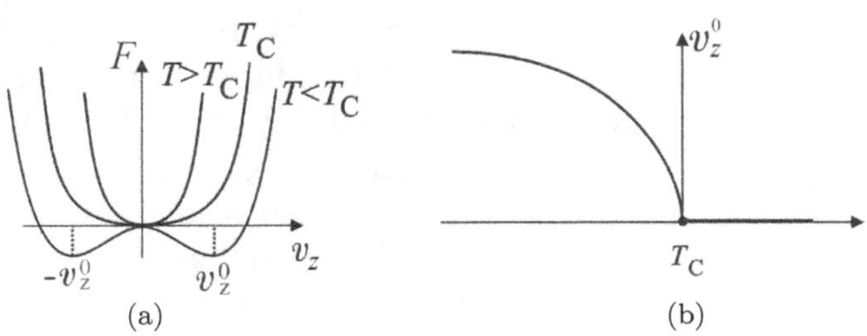

Fig. 1.3. (a) Plot of $F(v_z) - F_0$ for three characteristic temperatures. (b) Temperature dependence of v_z^0 following Eq. (1.4).

square-root dependence as a function of temperature. Note that one finds two minima corresponding to the same value of the free energy and opposite values of ${v^0}_z$. The corresponding upward and downward displacements of the M^- ion (Fig. 1.1) are distinct states of the system possessing the same stability.

Other physical consequences of the form (1.2) of the free energy can be drawn: *absence of latent heat* associated with the crossing of the transition, anomalous behavior of the *specific heat*, and behavior of the *susceptibility* related to the order-parameter.

The latent heat is $L = T_C \Delta S$, where ΔS is the difference of entropy between the two phases at T_C. We can derive S in each phase from the equilibrium free energy $F[T, P, v_z^0(T, P)]$ using the standard definition $S = -(\frac{\partial F_{eq}}{\partial T})_{P=\text{const}}$. We can write

$$\frac{\partial F_{eq}}{\partial T} = \left(\frac{\partial F}{\partial T}\right)_{v_Z} + \left(\frac{\partial F}{\partial v_z}\right)_T \left[\frac{d(v_z)}{dT}\right]_{v_Z^0}. \tag{1.5}$$

However, since F is minimum for $v_z = {v^0}_z$, the second contribution in Eq. (1.5) vanishes. Hence,

$$S = -\left(\frac{\partial F}{\partial T}\right)_{v_z^0} = -\frac{\alpha}{2}(v_z^0)^2 - \frac{\partial F_0(T, P)}{\partial T}. \tag{1.6}$$

Since both v_z^0 and $\frac{\partial F_0(T,P)}{\partial T}$ are continuous at T_C, there is no entropy jump ($\Delta S = 0$) at T_C, and thus, *there is no latent heat for the transition.*

Several specific heats can be considered for a system depending on the quantity, which is maintained constant. In the above example, the displacement v_z of a negative ion determines the occurrence of an electric dipole (or a macroscopic polarization P_z). The quantity ε_z, which is thermodynamically conjugated to v_z, is therefore proportional to an electric field (the conjugation between quantities X and Y is expressed by the fact that the infinitesimal work on the system has the form YdX). Let us show that the specific heat at constant electric field has a characteristic type of anomaly. The specific heat is expressed by $C_\varepsilon = T \cdot \left(\frac{\partial S}{\partial T}\right)_\varepsilon$. Using Eq. (1.6), we find the following expressions:

$$C_\varepsilon^0 = -\frac{\partial^2 F_0(T,P)}{\partial T^2} \quad \text{for } T > T_C,$$

and

$$C_\varepsilon = -\frac{\partial^2 F_0(T,P)}{\partial T^2} - \frac{\alpha}{2}\cdot\frac{d(v_z^0)^2}{dT} = C_\varepsilon^0 + \frac{\alpha^2}{2\beta} \quad \text{for } T < T_C. \tag{1.7}$$

Accordingly, above and below T_C, the specific heat is a different smoothly varying function of temperature determined by the background free energy $F_0(T,P)$ and by the smooth variation of the β coefficient. Figure 1.4(a) reproduces the anomaly of the specific heat, which has the form of an upward step on cooling through T_C.

The anomaly of the susceptibility χ, in the considered case, is proportional to the dielectric susceptibility of the material. It is defined as $\chi = \lim_{\varepsilon_z\to 0}(\frac{\partial v_z}{\partial \varepsilon_z})_{v_z^0}$. Its calculation requires examining the behavior of the system in the presence of a small field ε conjugated to the order-parameter. In this case, the appropriate thermodynamic potential, whose minimum determines the equilibrium of the system, is not F but $G = F - v_z\varepsilon_z$. Minimizing G with respect to v_z leads to

$$\varepsilon_z = v_z\left[\alpha(T - T_C) + \beta {v_z}^2\right]. \tag{1.8}$$

For small values of ε_z, the solution of this equation must tend toward the equilibrium values of v_z^0. Deriving these solutions with respect to

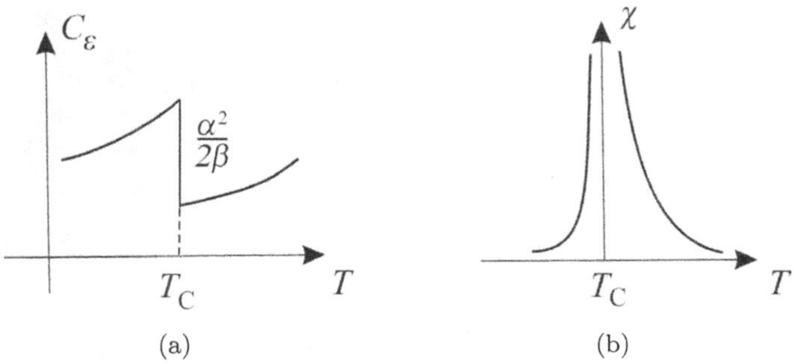

Fig. 1.4. Critical behavior of (a) the specific heat following Eq. (1.7) and (b) the susceptibility following Eq. (1.9).

ε_z, we obtain

$$\chi(T > T_C) = \frac{1}{\alpha(T - T_C)}; \quad \chi(T < T_C) = \frac{1}{2\alpha(T - T_C)}, \tag{1.9}$$

i.e., the susceptibility goes to infinity when $T \to T_C$ from either side of the transition (Fig. 1.4(b)).

The set of anomalies described in this paragraph represent the basic effects of temperature on quantities which are affected by a phase transition. They constitute the "canonical" signature of second-order phase transitions.

1.2.2.5. *Symmetry of the order-parameter*

In Sec. 1.2.2.3, we have defined the order-parameter as a set of degrees of freedom (either v_z or (v_x, v_y)), which has a second-degree contribution to the free energy, the coefficient of which has a specific temperature dependence $\alpha \cdot (T - T_C)$. Actually, the order-parameter can also be defined on the basis of its specific symmetry characteristics. With that goal, let us consider the manner by which the components (v_x, v_y, v_z) transform when we apply to the crystal each of the 16 symmetry operations of the point group $G_0 = 4/mmm$. Table 1.1 specifies the results of these transformations.

Let us first note that v_z is transformed either into itself or into $-v_z$. Considering this coordinate as the basis vector of a one-dimensional vector space, we can conclude that this vector space

Table 1.1. Transformation properties of the displacement components under the symmetry operations of $4/mmm$.

G_0	4	$4_2 = 2$	4_3	m_x	m_y
v_z	v_z	v_z	v_z	v_z	v_z
v_x	v_y	$-v_x$	$-v_y$	$-v_x$	v_x
v_y	$-v_x$	$-v_y$	v_x	v_y	$-v_y$
G_0	m_{xy}	$m_{\bar{x}y}$	I	$\bar{4}_3$	m_z
v_z	v_z	v_z	$-v_z$	$-v_z$	$-v_z$
v_x	v_y	$-v_y$	$-v_x$	$-v_y$	v_x
v_y	$-v_x$	v_x	$-v_y$	v_x	v_y
G_0	$\bar{4}$	2_x	2_y	2_{xy}	$2_{\bar{x}y}$
v_z	$-v_z$	$-v_z$	$-v_z$	$-v_z$	$-v_z$
v_x	v_y	v_x	$-v_x$	v_y	$-v_y$
v_y	$-v_x$	$-v_y$	v_y	v_x	$-v_x$

is invariant by all the transformations of G_0 (a vector space being defined as the set of vectors which are linear combinations of the basis). Such a space, containing no space of smaller dimension, is called *an irreducible invariant space relative to the group* G_0. On the other hand, each of the components (v_x, v_y) is not transformed into a proportional component by all the elements of G_0. Half of these elements transform v_x into $\pm\, v_y$, and vice versa. Hence, v_x and v_y are not bases for one-dimensional irreducible invariant spaces. However, their set generates a two-dimensional vector space which has the property to be invariant and irreducible by all the transformations of G_0. Note that the set of three components (v_x, v_y, v_z) carries a three-dimensional vector space which also has the property to be invariant by all the transformations of G_0. However, this vector space contains the two invariant spaces carried, respectively, by v_z and (v_x, v_y). Hence, it is *reducible.*

In summary, from a symmetry point of view, the order-parameter of a phase transition is a set of degrees of freedom which carries an irreducible vector space (an *irreducible representation*) with respect to the action of the group G_0, which is the symmetry group (point group) of the high-symmetry (parent) phase.

1.2.2.6. *Degeneracy of the low-symmetry phase*

We have noted above that in the stable state of the system, above T_C, the structure is invariant by G_0. When $\boldsymbol{v} \neq 0$, the structure becomes invariant by a smaller set of transformations. Let us enumerate these transformations for each possible stable state of the system below T_C.

When the order-parameter coincides with v_z, we determined below T_C two stable states $v_z^0 = \pm[\frac{\alpha(T_C-T)}{\beta}]^{1/2}$. The crystalline structures corresponding to the displacements of the M^- ion parallel to the Z axis are both invariant by the same set of eight symmetry transformations. These comprise the cyclic group of order 4 generated by the four-fold rotation around Z, and by the reflections in planes containing this axis. This set is the group $4mm$ *subgroup of* G_0. *The transition induces a lowering of the symmetry of the system.* Also note that the two states $\pm v_z^0$ are transformed into each other by some of the symmetry operations such as the mirror-symmetry m_z "lost" below T_C. These two states necessarily possess the same free energy; they are equally stable. Indeed, application to the system of m_z preserves the value of the free energy (which is invariant by all the transformations belonging to G_0), while the state $+v_z^0$ is not preserved and is transformed into $-v_z^0$.

We have not worked out explicitly the stable states of the system in the case of a two-dimensional order-parameter (v_x, v_y). A simple discussion along the line developed for the one-dimensional order-parameter v_z would show that the relevant form of the free energy is

$$F(T, P, v_x, v_y) = F_0(T, P) + \frac{\alpha \cdot (T - T_C)}{2} \cdot (v_x^2 + v_y^2) + \frac{\beta_1}{4}(v_x^4 + v_y^4) + \frac{\beta_1}{4}(v_x^2 v_y^2), \tag{1.10}$$

and the possible stable states below T_C are

$$\text{(I) } v_x = \pm \left[\frac{\alpha \cdot (T - T_C)}{\beta_1}\right]^{1/2}, v_y = 0;$$

$$\text{(II) } v_y = \pm \left[\frac{\alpha \cdot (T - T_C)}{\beta_1}\right]^{1/2}, v_x = 0;$$

$$\text{(III, IV) } v_x = \pm\, v_y = \pm \left[\frac{\alpha \cdot (T - T_C)}{\beta_1 + \beta_2}\right]^{1/2}.$$

Similarly to the case of v_z, there is a lowering of the crystal's symmetry below T_C. In the enumerated cases, one finds that the respective point groups of the structures are (I) $G_1 = mm2_x$, (II) $G'_1 = mm2_y$, (III) $G_2 = mm2_{xy}$, (IV) $G'_2 = mm2_{\bar{x}y}$. States I and II correspond to each other through one of the lost transformations of G_0 (e.g., a four-fold axis), and the pair of states III and IV corresponds to the same value of the equilibrium free energy. Note that the symmetry groups associated to equally stable states are *conjugate* relative to $G_0 = 4/mmm$, i.e., they satisfy the relationship $G'_i = g \cdot G_i \cdot g^{-1} (i = 1, 2)$, with g belonging to G_0.

1.2.3. *Generalization of the preceding example: Formulation of the Landau theory*

1.2.3.1. *Summary of the results obtained in the introductory example*

In the specific model described in the preceding section, we have shown that an order-parameter (e.g., v_z or (v_x, v_y)) is a set of scalar degrees of freedom which allows to describe the symmetry and physical changes accompanying the phase transition in a crystal. The equilibrium values of the n-components of the order-parameter are zero for $T \geq T_{\mathrm{C}}$, and *not all zero* for $T < T_{\mathrm{C}}$. The n components define a *vector space* ε_n which is an irreducible invariant space by the group G_0, which is the crystallographic symmetry group of the high-temperature parent phase. A variational free energy F associated with the transition can be expanded as a function of the components of the order-parameter. The second-degree term of this expansion has a coefficient which vanishes and changes sign at the transition temperature T_{C}. The form of F is determined by the symmetry properties of the order-parameter, i.e., by the mode of transformation of the components of the order-parameter into each other, when the operations of G_0 are applied to them. The specific form of F determines the symmetries of the potentially stable phases below T_{C}, the degeneracy of these phases, and the temperature dependences of relevant physical quantities.

In this section, we will briefly outline the arguments used in order to formulate the preceding approach in a general framework. These arguments rely in part on the properties of the irreducible representations of a group. We will consider here representations of a group G_0 carried by a set $(\eta_1, \eta_2, \ldots, \eta_n)$ of degrees of freedom, and use essentially the following properties:

(i) If the set is irreducible and not totally symmetric, there is no linear combination $f_1(\eta_i) = a_1\eta_1 + \cdots + a_n\eta_n$ of the η_i which is *invariant* by all the elements of G_0.

(ii) If the set is irreducible, there is a single homogeneous polynomial $f_2(\eta_i)$ of degree *two*, which is invariant by G_0. Its form is $f_2(\eta_i) = \sum_{i=1}^{n} \eta^2{}_i$.

(iii) If $g(\boldsymbol{r})$ is an arbitrary function of the space coordinates, one can always write

$$g(\boldsymbol{r}) = \sum_{\gamma} \sum_{j} \varphi_{j\gamma}(\boldsymbol{r}), \tag{1.11}$$

where each set of functions $\varphi_{j\gamma}(j = 1, 2, \ldots, p)$ defines an irreducible invariant space by the group G_0 corresponding to the irreducible representation Γ_γ of G_0.

1.2.3.2. *Description of a second-order phase transition*

In order to generalize the considerations developed in Sec. 1.2.2, we have to define independently from any specific structure the basic ingredients of the theory: definition of a variational set of degrees of freedom, construction of a free energy, determination of the stable states on either side of a transition temperature T_C. In a first step, we describe an arbitrary atomic configuration of a crystal by the set of densities $\rho_i(\boldsymbol{r})$ of the particles of type i. This set constitutes variational degrees of freedom, which can be used to construct the variational free energy of the crystal $F[T, \rho_i(\boldsymbol{r})]$. An equilibrium state of the crystal is defined by the set of functions $\rho_i^{\mathrm{eq}}(T\boldsymbol{r})$, which minimize the free energy.

The symmetry of the system at given temperature and pressure is defined as the set of geometrical transformations which leave all the ρ_i^{eq} invariant. This set forms a group. We are then in a position to define a second-order (continuous) transition at T_C by two conditions:

(1) The $\rho_i^{\mathrm{eq}}(T\boldsymbol{r})$ are the functions whose form changes continuously across T_C.
(2) The symmetry group of the system just above T_C is different from the symmetry just below T_C. Accordingly, the phase transition, though associated to a *continuous change of the spatial configuration of the atoms*, is associated to a *sudden change of the symmetry.*

1.2.3.3. *The transition order-parameter*

As compared to the set (v_x, v_y, v_z) used in the example of Sec. 1.2.2, the variational functions $\rho_i(\boldsymbol{r})$ have the drawback of *being not small and not scalar quantities*, thus making an expansion of the free energy questionable. Without loss of generality, let us restrict to a single type of particles. In order to remove the mentioned difficulty, we write

$$\rho(\boldsymbol{r}) = \rho_0(\boldsymbol{r}) + \sum_{\gamma}\sum_{j} \eta_{\gamma j}\varphi_{\gamma j}(\boldsymbol{r}), \tag{1.12}$$

where $\rho_0(\boldsymbol{r}) = \rho^{\mathrm{eq}}(T_C, \boldsymbol{r})$ is the equilibrium density at T_C. Let us denote by G_0 the symmetry group of this equilibrium density. The sum in the second member of Eq. (1.12) *is a small increment since the transition is continuous.* This increment has been expanded, as in Eq. (1.11), as a function of irreducible functions with respect to the group G_0. Moreover, each function has been expressed as the product of a *normalized function* φ and of a *small scalar parameter* η. It is easy to convince oneself that it is possible, at choice, to consider that under the action of G_0, either the φ_j transform into each other, the η_j being fixed coefficients, *or the* η_j *transform into each other, the functions* φ_j *being fixed.* We adopt the second convention.

In thermodynamics, a system is characterized by its *free energy* in any point of the phase space (in fact, at certain thermodynamic conditions):

$$F = \langle E \rangle - TS = -k_B T \cdot \ln Z = -k_B T \cdot \ln \int \exp\left\{-\frac{1}{T}E(q)\right\} d\Gamma,$$

where k_B denotes the Boltzmann constant, Z is the partition function, and the integral is taken over phase space. In the phenomenological approach suggested by L. Landau, degrees of freedom q are segregated into two groups: (a) few *critical* (non-equilibrium) degrees responsible for the symmetry breaking, and (b) remaining equilibrium variables inducing no singularities in energy. The free energy, therefore, reads

$$F = -k_\mathrm{B} T \cdot \ln \int \exp\left\{-\frac{1}{T}[E_0(q) + E_1(q,\eta)]\right\} d\Gamma = F_0(P,T,\ldots)$$
$$+ F_\mathrm{L}(P,T,\eta).$$

F_L is a variational free energy and not an equilibrium energy, since it is calculated for an arbitrary *order-parameter* η. The equilibrium free energy of the system is $F_\mathrm{L}^\mathrm{eq}(P,T) = F(P,T,\eta_0)$, where η_0 is found as a result of minimization of $\partial F_\mathrm{L}(\eta)/\partial\eta = 0$. The Landau free energy is, therefore, an incomplete (partly integrated) thermodynamic potential.

The variational free energy can now be written in the form $F[T, \eta_{\gamma j}, \varphi_{\gamma j}(\boldsymbol{r})]$. At each temperature, the characteristics of the system are specified by the following conditions: (i) The equilibrium states correspond to the values $\eta^0_{\gamma j}$, which make the free energy minimum. These values define through Eq. (1.12) the equilibrium density of particles in the system. (ii) The symmetry of the system is the group of invariance of the determined equilibrium density. Note that in $(\rho - \rho_0)$, the contribution of the degrees of freedom $\eta_{\gamma j}$ spanning a totally symmetric representation can be ignored in the first place, as one can show that such a contribution would have the same symmetry on either side of the transition. Therefore, it is not crucially associated with the symmetry change defining the phase transition.

Following the ideas introduced in Sec. 1.2.2, the free energy form is obtained as a Taylor expansion as a function of the small parameters $\eta_{\gamma j}$. Besides, each polynomial term of given degree in this expansion is invariant by G_0. Indeed, F being a scalar is unchanged by any rotation or reflection. Among the transformations, those belonging to G_0 have the additional property of leaving invariant the density ρ_0, which is, besides T and $\eta_{\gamma j}$, an argument of F.

The group theoretical rules recalled in Sec. 1.2.3.1 determine the absence of an invariant linear term in the expansion. Moreover, these rules specify the form of the second-degree contribution. We have

$$F[T, \eta_{\gamma j}, \varphi_{\gamma j}(\boldsymbol{r})] = F_0(T) + \sum_{\gamma} \alpha_\gamma \left(\sum_j {\eta^2}_{\gamma j} \right). \tag{1.13}$$

On the other hand, the increment of the equilibrium density at the transition is

$$\delta\rho^{\mathrm{eq}}(\boldsymbol{r}) = \rho^{\mathrm{eq}}(\boldsymbol{r}) - \rho_0(\boldsymbol{r}) = \sum_{\gamma} \sum_{j} \eta^0_{\gamma j} \varphi_{\gamma j}(\boldsymbol{r}). \tag{1.14}$$

Let us first consider the system at T_{C}. The equilibrium values of all the $\eta_{\gamma j}$ are zero. Hence, on the basis of Eq. (1.13), we conclude that all the α_γ satisfy the condition $\alpha_\gamma \geq 0$ (the second-degree expansion must be minimum at the origin). Note that *this condition cannot be strict for all* γ. Otherwise, these coefficients would also be positive in the vicinity of T_{C} on either side of this temperature. As a consequence, the equilibrium values of the η_j would all be zero, and the symmetry would be unchanged. Hence, *at least one of the* α_γ *has to vanish and change sign at* T_{C}. An argument already invoked in Sec. 1.2.2 allows to assert that *one* α_γ *coefficient only has this property*. The $\eta_{\gamma j}$ corresponding to the other indices γ keep zero equilibrium values in the vicinity of T_{C}, and can be ignored in the first place. In Eq. (1.13), we can, therefore, drop all degrees of freedom except the ones associated with the α_γ, which vanishes at T_{C}. This set

$(\eta_{\gamma j})$ $(j = 1, 2, \ldots, n)$ constitutes *the n-dimensional order-parameter of the considered transition*. As it comprises all the degrees of freedom contributing to a single second-degree term in the free energy, it necessarily constitutes a basis for an irreducible vector space with respect to G_0 following the group theoretical rules recalled above.

1.2.3.4. *Stable states and symmetry in the vicinity of T_C*

Above T_C, due to the positivity of α (we drop the γ index), the equilibrium values of the η_j are zero, and the symmetry of the system is G_0 identical to the symmetry at T_C. Below T_C, α is negative, and the minimum of F occurs away from the origin in the η_j space. The symmetry of the system comprises all the transformations leaving the density invariant:

$$\rho^{\text{eq}} = \rho_0 + \sum_j \eta_j^0 \varphi_j(\boldsymbol{r}). \tag{1.15}$$

Since the η_j^0 contribution to the second member *is small*, these transformations have to be selected among those belonging to the invariance group of ρ_0. As the space $\{\eta_j\}$ defines a non-trivial representation of the latter group, the increment of density in Eq. (1.15) cannot be invariant by all the transformations of G_0: *the symmetry group of the system below T_C is a subgroup G of G_0*. Note that the group–subgroup relationship $G \subset G_0$ requires to assume that the η_j^0 are small. For large η_j^0, Eq. (1.15) does not imply *necessarily* a group–subgroup relationship between the invariance groups of ρ^{eq} and ρ_0 (see Fig. 4.21).

As pointed out in Sec. 1.2.2, in order to determine the minimum of F below T_C, it is necessary to expand the free energy to degrees higher than two. The relevant expression of the free energy is then

$$F(T, P, \eta_j) = F_0(T, P, \rho_0) + \frac{\alpha(T - T_C)}{2} \cdot \left(\sum_j {\eta^2}_j\right) + f_3(\eta_j) + f_4(\eta_j) + \cdots, \tag{1.16}$$

where we have developed the coefficient of the second-degree term to the lowest degree in $(T - T_C)$. It can be shown that the existence of a *third-degree term* $f_3(\eta_j)$ depends exclusively on the nature of the representation Γ_n associated with the order-parameter. If the symmetry of the order-parameter is such that a third-degree term is not symmetry-forbidden, the transition will be *necessarily discontinuous*. This symmetry-predicted type of first-order transition will be discussed in Sec. 2.2.

For any symmetry of the order-parameter, fourth-degree terms will always be present in the free energy expansion (if only the square of the second-degree term). No further general statement can be made. Depending on the form and coefficients of this term, and eventually of higher-degree terms (see Sec. 1.3), a continuous (or discontinuous) transition will be possible toward one or several distinct low-symmetry phases. The form of the $f_4(\eta_j)$ term is determined by searching the most general fourth-degree polynomial invariant by the set of transformations belonging to G_0.

1.2.3.5. *Summary of the Landau scheme for the study of a phase transition*

In the light of the considerations developed in this section, the investigation of a phase transition, in the framework of the Landau theory, proceeds on the following lines:

(1) Search, as starting information on the system, the symmetry group G_0 of the more symmetric phase surrounding the transition, and the nature of the irreducible representation Γ_n associated with the order-parameter.
(2) Construct the free energy by determining the form of the invariant polynomials of the required degrees.
(3) Determine, as a function of the coefficients of the free energy expansion, the absolute minima of F.
(4) For each minimum, determine the invariance group of the density ρ^{eq}, i.e., the "low-symmetry" groups of the system.
(5) Derive the temperature dependence of the quantities related to the order-parameter.

A further step is to consider the coupling of the order-parameter to other relevant degrees of freedom and derive the temperature dependence of these "secondary" order-parameters.

1.2.4. *Secondary order-parameters*

1.2.4.1. *Primary and secondary order-parameters*

Starting from Eq. (1.13), we have considered that the free energy F of the system and the particle density $\rho(\boldsymbol{r})$ describing its structure were functions of a single irreducible set of degrees of freedom of the system: the n components η_j of the order-parameter. The rejection of other degrees of freedom corresponding to other irreducible representations of G_0 (equivalent or inequivalent) was based on the form of the second-degree expansion of F, which imposed that only the η_j can take non-zero equilibrium values below T_C. *This conclusion is no longer valid if the expansion of F to terms of degree higher than two is considered.*

In this section, we analyze the general conditions which preside over the onset, below T_C, of non-zero equilibrium values for degrees of freedom distinct from the order-parameter components η_j. These degrees of freedom will be termed *secondary order-parameters*, while the set of η_j can be termed the *primary order-parameter*. We will examine the form of the contribution of the secondary order-parameter to the free energy and show that the symmetry characteristics of the transition (e.g., the possible symmetries of the phases stable below T_C) are entirely determined by the primary order-parameter. Before establishing this result on a general ground (Sec. 1.2.4.4), we will examine the influence of a particular secondary order-parameter in the example discussed in Sec. 1.2.2.

1.2.4.2. *Return to the introductory example*

The only relevant degree of freedom assumed in the example of Sec. 1.2.2 is the set of three components of the displacement $\boldsymbol{v}$. Let us allow the possibility that the c-parameter in Fig. 1.1, equal to the distance along Z between the planes of the positive ions, can change with temperature and pressure. This change corresponds to a deformation of the unit cell along Z, which coincides with the component e_{zz} of the strain tensor. In the same manner as for the components of $\boldsymbol{v}$, the equilibrium value of e_{zz} can be determined by minimizing the variational free energy F assumed to be a function of e_{zz} as well as $F(T, P, \boldsymbol{v}, e_{zz})$.

The form of the contribution of e_{zz} to F is determined by noting that e_{zz} is invariant under the application of all the transformations of $G_0 = 4/mmm$. The part of the expansion of F, depending exclusively on e_{zz}, is therefore of the form

$$\lambda_1 e_{zz} + \frac{\lambda_2}{2} {e^2}_{zz} + \frac{\lambda_3}{3} {e^3}_{zz} + \cdots, \tag{1.17}$$

as terms of any degree in e_{zz} are allowed to be present in F. Besides, any power of e_{zz} multiplied by any invariant polynomial of the components of $\boldsymbol{v}$ are invariant by G_0 and may be included in F. The *lowest-degree* term of this type is

$$e_{zz}[\delta_1({v^2}_x + {v^2}_y) + \delta_2 {v^2}_z]. \tag{1.18}$$

In order to examine the effect of the presence of e_{zz}-dependent terms, one has to minimize F with respect to e_{zz}. It yields

$$\lambda_1 + \lambda_2 e_{zz} + \lambda_3 {e^2}_{zz} + \delta_1({v^2}_x + {v^2}_y) + \delta_2 {v^2}_z = 0. \tag{1.19}$$

As $\lambda_1(T, P)$ is a parameter which, in general, *is not* vanishingly small, e_{zz} cannot be vanishingly small. λ_1 is actually attached to the thermal expansion of the unit cell along Z, which is not related to the phase transition. We can eliminate the resulting component of e_{zz} by setting $e_{zz} = e^0_{zz} + e^c_{zz}$, where e^0_{zz} is the non-zero value of e_{zz} at T_C, defined by

$$\lambda_1 + \lambda_2 e^0_{zz} + \lambda_3 (e^0_{zz})^2 + \cdots = 0,$$

and e^c_{zz} is the *spontaneous* contribution to e_{zz} *resulting from the transition.* As both λ_1 and e^0_{zz} vary smoothly across the transition, Eq. (1.19) can be rewritten as

$$e^c_{zz} = -\frac{1}{C}[\delta_1(v_x^2 + v_y^2) + \delta_2 v_z^2], \tag{1.20}$$

where the coefficient C, which depends on e^0_{zz}, is the *effective* coefficient of the quadratic contribution in e^c_{zz} to the free energy.

If α_1 or α_2 vanishes at T_C, C necessarily remains strictly positive in the neighborhood of T_C. We can then deduce from Eq. (1.20) that e^c_{zz} vanishes above T_C and becomes non-zero below T_C. Hence, the order-parameter (v_z or (v_x, v_y)) *is not the only quantity possessing non-zero values for $T < T_C$ and zero value for $T > T_C$.*

Quantities such as e^c_{zz} also possess this property and are *secondary order-parameters.* They can be distinguished from the primary order-parameters by the fact that the α_i coefficient associated with it vanishes at T_C, while the coefficient C remains strictly positive near T_C. The onset of a non-zero value for e^c_{zz} is related to the occurrence of a *coupling term* in the free energy. If the coefficient (δ_1 or δ_2) of the coupling term is zero, Eq. (1.20) shows that e^c_{zz} will remain equal to zero.

1.2.4.3. *Influence of the secondary order-parameter on the properties of the system*

Assuming $\alpha_1(T_C, P_C) = 0$, i.e., that the primary order-parameter of the transition is v_z, we can eliminate, using Eqs. (1.2) and (1.20), $e^c{}_{zz}$ from the free energy and obtain an expansion of F, which depends only on v_z. This expression is

$$F = F_0(T, P) + \frac{(\alpha_1 - e^c_{zz}\delta_2)}{2} \cdot v_z^2 + \frac{(\beta - 2\delta_2^2)}{4} \cdot v_z^4$$
$$= F_0(T, P) + \frac{a_1}{2} \cdot v_z^2 + \frac{a_2}{4} \cdot v_z^4, \tag{1.21}$$

which shows that F has the same functional form as Eq. (1.2) *but different coefficients* $a_1 = (\alpha_1 - e^c_{zz}\delta_2)$ and $a_2 = (\beta - 2\delta_2^2)$. The effect of coupling between primary and secondary order-parameters is thus *to modify the coefficients of the free energy associated with the primary order-parameter* by amounts increasing with the magnitude δ_2 of the coupling. Equation (1.21) establishes that none of the conclusions reached in Sec. 1.2.2 is changed: the possibly stable phases and the form of the anomalies associated with the transition, which only depend on the functional form of F, are the same as in the absence of coupling. Only *the boundaries* of the phases *are displaced,* and *the amplitudes* of *the anomalies* are modified. For instance, the transition temperature corresponds to the vanishing of $(\alpha_1 - 2e_{zz}\delta_2)$ instead of α_1 and will take place at $(T_C + \frac{2e^c_{zz}\delta_2}{\alpha})$ instead of T_C. Furthermore, this point is not necessarily significant, as one can consider that Eq. (1.2) is the free energy obtained when the coupling to secondary degrees of freedom, such as $e^c{}_{zz}$, has been considered, and already eliminated through procedures similar to the one leading to Eq. (1.21).

Let us also emphasize that the *symmetry* of the various phases *is the same*, as found in Sec. 1.2.4, as e^c_{zz} *preserves* the symmetry G_0 and also the breaking of symmetry resulting from the onset of v_z: *the symmetry characteristics of the transition are exclusively determined by the primary order-parameter.*

1.2.4.4. *Physical anomalies related to secondary order-parameters*

Along the same line, we can show that the primary order-parameter is also essential for working out the anomalies related to the secondary order-parameters. This is a consequence of the fact that *the existence of a coupling* between the secondary degrees of freedom and the primary order-parameter is a *necessary* condition for the singular behavior of the additional quantities. Equation (1.20) can be used to deduce the temperature dependence of e^c_{zz}. One has $e^c_{zz} = 0$ above T_C, and as below T_C v_z is proportional to $(T_C - T)$, it yields

$$e^c_{zz} = -\frac{1}{C}(T_C - T), \tag{1.22}$$

corresponding to the *linear dependence* shown in Fig. 1.5(a), which differs from the square-root dependence of the primary order-parameter in function of $T_C - T$ represented in Fig. 1.3(b).

The quantity which is thermodynamically conjugated to e^c_{zz} is the stress-tensor component σ_{zz}. The susceptibility χ_e associated with e^c_{zz}, defined by $\chi_e \lim_{\sigma_{zz}\to 0} = (\frac{\partial e^c_{zz}}{\partial \sigma_{zz}})$, has the meaning of an *elastic compliance*. Minimizing the thermodynamic potential $F - \sigma_{zz} e_{zz}$ with

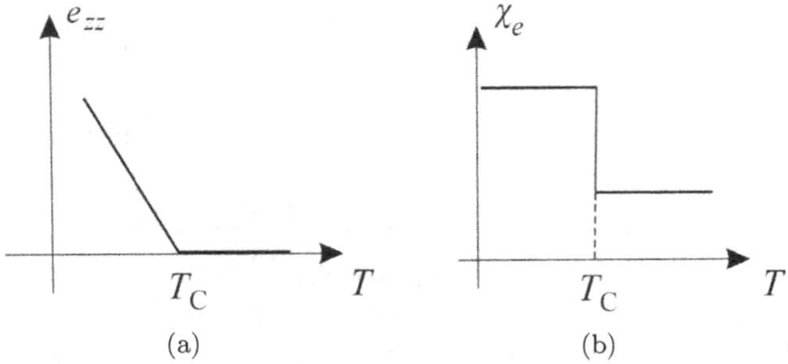

Fig. 1.5. Temperature dependence of (a) the secondary order-parameter $e_{zz} \sim (T_C - T)$ and (b) the elastic susceptibility χ_e following Eq. (1.28).

respect to σ_{zz}, one gets as a substitute to Eq. (1.22):

$$e^c_{zz} = \frac{\sigma_{zz}}{C} - \frac{\delta_2}{C}v_z^2. \tag{1.23}$$

Likewise, instead of Eq. (1.21), one gets the following expression:

$$F - \sigma_{zz}e_{zz} = \frac{1}{2}\left(\alpha_1 - 2\delta_2 e^0_{zz} + \frac{\sigma_{zz}\delta_2}{C}\right)\cdot v_z^2 + \frac{1}{4}\left(\beta - \frac{2\delta_2^2}{C}\right)\cdot v_z^4. \tag{1.24}$$

If we assume that the mechanical susceptibility χ_e corresponds to the conditions of zero applied stress σ_{zz}, we can write the minimum of Eq. (1.24) with respect to v_z as follows:

$$v_z \cdot \left[\left(\alpha_1 - 2\delta_2 e^0_{zz} + \frac{\sigma_{zz}\delta_2}{C}\right) + \left(\beta - \frac{2\delta_2^2}{C}\right)\cdot {v^2}_z\right] = 0. \tag{1.25}$$

From Eq. (1.23), we draw the expression of χ_e

$$\chi_e = \frac{1}{C}\left[1 - \delta_2\left(\frac{\partial v_z^2}{\partial \sigma_{zz}}\right)\right], \tag{1.26}$$

where $(\frac{\partial v_z^2}{\partial \sigma_{zz}})$ is obtained below T_{C} by minimizing the term into bracket in Eq. (1.25) with respect to σ_{zz}

$$\frac{\partial v_z^2}{\partial \sigma_{zz}} = \frac{-\delta_2}{\beta C - 2\delta_2^2}. \tag{1.27}$$

Replacing Eq. (1.27) into Eq. (1.26), we finally obtain

$$\chi_e = \frac{1}{C} \text{ for } T > T_{\mathrm{C}} \text{ and } \chi_e = \frac{1}{C} + \frac{\delta_2^2}{\beta C - 2\delta_2^2} \text{ for } T < T_{\mathrm{C}}. \tag{1.28}$$

The temperature dependence of χ_e is given in Fig. 1.5(b). χ_e experiences an *upward* jump at T_{C}. Note that α_1 is modified by the stress σ_{zz}: the temperature is shifted linearly by the application of the stress σ_{zz}. This shift, as well as the magnitude of the jump of $\chi_e(T)$ at T_{C}, is controlled by the magnitude δ_2 of the coupling between e^c_{zz} and by the *primary* order-parameter.

1.3. The Angular and Radial Problems in the Landau Theory of Structural Transitions

1.3.1. *Introduction*

The general considerations exposed in the preceding section apply to any type of phase transitions in which a group–subgroup relationship exists between the symmetries of the phases: structural transitions in crystals, but also magnetic and superconducting transitions, transitions in liquid crystals, etc. For each type of transition, one has to take into account the specific symmetries of the system and their irreducible degrees of freedom, which may imply particular procedures in applying the theory. For example, magnetic systems require the use of magnetic space groups and their co-representations [3], and to follow the approach described by Dzialoshinskii [4], which separates in the thermodynamic potential the terms representing the exchange and relativistic forces. In the normal-to-superconducting transition, the symmetry of the phases includes the rotations in real and spin space, the gauge symmetry, and time-reversal [5], whereas in liquid crystals, continuous groups in real space and their representations have to be used [6].

Let us examine the particular ingredients needed to apply the Landau theory to transitions between *crystalline* phases. In this case, the invariance group, G_0 or G, coincides with a *crystallographic space* group. Let us only recall here that a space group is a group of infinite order formed by the elements which can be written as $\{R|\boldsymbol{t}\}$, where R is a point-symmetry operation, and $\boldsymbol{t}$ is a translation. The symmetry operations R generate the point group of the crystal. On the other hand, among the translations $\boldsymbol{t}$, there is an infinite group $\boldsymbol{T}$ of "primitive" translations, generating the three-dimensional Bravais lattice of the crystal.

There is an infinite set of nonequivalent irreducible representations of a space group G_0. They cannot be tabulated in a systematic manner as the more known representations of finite groups, and have to be constructed starting from the simplest representations. Namely, each representation has a double index: the first index is a $\boldsymbol{k}$-vector in reciprocal space belonging to the first Brillouin zone of the space. The former vector defines a subgroup $G(\boldsymbol{k})$ of G_0, which is the set of elements $\{R|\boldsymbol{t}\}$ of G_0 whose elements R leave $\boldsymbol{k}$ unchanged or transform it into an equivalent vector, i.e., differing from $\boldsymbol{k}$ by a reciprocal lattice vector. This group $G(\boldsymbol{k})$ has irreducible representations $\hat{\tau}_i(\boldsymbol{k})$ of dimension n_i, which are tabulated [7–9]. A representation of G_0 can be denoted $\Gamma_{k\gamma}$. It is constructed according to the systematic rules using $\hat{\tau}_i(\boldsymbol{k})$. Its dimension is $n = n_i \times r$, where r is the number of vectors in *the star* $\boldsymbol{k}^*$ of $\boldsymbol{k}$. This star is the set of vectors nonequivalent to $\boldsymbol{k}$, having the same modulus as $\boldsymbol{k}$, and obtained from $\boldsymbol{k}$ by the application of all the point symmetry group elements R.

As shown in Sec. 1.2.3.6, the difference between the two stable phases surrounding a transition is specified by the increment of equilibrium density (1.12), which, following the preceding consideration, can be written for a structural transition as

$$\delta\rho^{\mathrm{eq}}(\boldsymbol{r}) = \rho(\boldsymbol{r}) - \rho_0(\boldsymbol{r}) = \sum_{kni} \eta^i_{kn}\varphi^i_{kn}(\boldsymbol{r}). \tag{1.29}$$

The symmetry properties of $\delta\rho^{\mathrm{eq}}(\boldsymbol{r})$ determine the form of the non-equilibrium variational free energy $F(T, \eta^i_{kn}\varphi^i_{kn})$. As underlined in Sec. 1.2.3.3, because the transformation properties of the η^i_{kn} and φ^i_{kn} are identical, F can be considered as a function only of the η^i_{kn}. When the transition occurs between strictly periodic structures, and is associated to a single irreducible representation, the indices k and i can be omitted, and the set $(\eta_j)\,(j = 1 - n)$ constitutes the order-parameter components. The equilibrium values (η^e_j) are determined from the condition that F is minimal with respect to the order-parameter components:

$$\frac{\partial F}{\partial \eta} = 0 \quad (j = 1 - n), \tag{1.30}$$

and

$$\left\| \frac{\partial^2 F}{\partial \eta_j \partial \eta_k} \right\| \geq 0 \quad (j, k = 1 - n). \tag{1.31}$$

Equation (1.31) expresses the positiveness of the eigenvalues of the $n \times n$ matrix of the second derivatives of F with respect to the order-parameter components.

In the example discussed in Sec. 1.2.2, the number of components (η_j) is one (v_z) or two (v_x, v_y), and the expansion $F(T, \eta_j)$ was restricted to the minimal fourth power, which may give rise to a stable low-symmetry structure across a *second-order* transition. When the number n of η_j components increases (and there exist many examples of structural transitions with $n = 3, 4, 6, 8, \ldots$), or when F is expanded to higher powers (and this will be shown to be necessary in the description of some first-order transitions, or for disclosing low-symmetry structures), the minimization of F presents mathematical difficulties. Systematic investigations of the order-parameter symmetries of structural transitions [10–12], and the theoretical methods considering the more essential symmetries involved in the Landau approach [11,13], have allowed to overcome most of these difficulties. The corresponding methods, mainly proposed and applied by Gufan and his collaborators [11], have shown that the problems related to the description of structural transitions break into two categories: the *angular problem* and the *radial problem.*

The angular problem involves the working out of the characteristics of the system which are determined only by the transformation properties (the symmetry) of the order-parameter but not by the specific form of the thermodynamic potential: classification of the solutions of the equations of states (1.30), enumeration of the possible low-symmetry phases, number of domains in each phase, symmetry of the external interactions which act on the domains, etc. In this part, phenomenological theory is *model-free* and *geometrically correct* in solving the crystal-geometric and symmetry-related problems. The introduction of simplifying the concepts as the ε-space, the image-group, the complete rational basis of invariants, etc. has made possible to obtain a geometrically exact solution of the angular problem.

The radial problem consists in finding the equilibrium values of the order-parameter and subsequently of the other physical quantities inducing anomalies at the transition and related to the order-parameter. It includes also the construction of the phase diagram, with the description of the boundaries between the phases, the coexistence regions, etc. In this subdomain, the problem cannot be solved without model assumptions, namely without knowing the specific form of the thermodynamic potential and the power at which it is truncated. Much progress has also been made in this field using the results obtained in the theory of singularities [14].

In this section, we will present successively the main results obtained in the solutions of the angular problem (Sec. 1.3.2) and of the radial problem (Sec. 1.3.3).

1.3.2. *The angular problem and its solutions*

1.3.2.1. *Image of G_0 in the order-parameter representation*

Consider the set of matrices $M(g_i)$ defining a representation Γ of G_0 in a given space ε (irreducible or not). These matrices are not necessarily distinct. For instance, in Table 1.1, the matrices of $4/mmm$ spanned by the one-dimensional irreducible subspace z are equal to 1 for half of the g_i and -1 for the other half. In the two-dimensional irreducible subspace (x, y), the matrices $M(E)$ and $M(m_z)$ are both equal to $\left[\begin{smallmatrix}1&0\\0&1\end{smallmatrix}\right]$. Γ is a *faithful* representation of G_0, when all its matrices are distinct. In this case, the correspondence between the g_i and the $M(g_i)$ is an *isomorphism*. If Γ is not faithful, the correspondence is a *homomorphism*.

The set of distinct matrices $M(g_i)$ forms a *group*, which is called the image or image-group (denoted I_0) of G_0 in the homomorphism. The *kernel* K of the homomorphism is the set $g_i \in G_0$, such as the matrices $M(g_i)$ coincide with the identity matrix in the space ε. K is an *invariant* subgroup of G_0 (i.e., $g_iK = Kg_i$ for all $g_i \in G_0$), and G_0 can be expressed as a sum of left or right *cosets* associated to K:

$$G_0 = K + g_2K + g_3K + \cdots , \tag{1.32}$$

where the g_i *do not belong to* K. The set of cosets forms the *quotient group of* G_0 *by* K, which is denoted G_0/K. This group is isomorphous to the image G_0, and we can write $G_0 = I_0 \times K$. For a *finite* group G_0, the orders (n_G) of the three groups G_0, I_0 and K satisfy the equality $n_{G_0} = n_{I_0} \times n_K$. Reciprocally, *the order of* I_0 *can be finite, even if* G_0 *is an infinite group.*

The preceding considerations apply to the n-dimensional irreducible representations carried by the n-components of the transition order-parameter. The set of distinct matrices of this irreducible representation constitutes a group I_0, which is the image of the group G_0 of invariance of the high-symmetry phase. Each matrix in I_0 is *orthogonal,* since it is real and unitary. Its action on the components (η_j) of the order-parameter can, therefore, be considered as a geometrical transformation in the n-dimensional space ε_n of the order-parameter, which has the property to preserve the origin in this space *as well as the length of vectors.* In the three-dimensional space, such transformations comprise rotations about an axis passing through the origin, reflections about a plane passing through the origin as well as the products of these transformations. The rotations are the orthogonal transformations whose matrices have a determinant Δ equal to $+1$. The set of all rotations about any axis passing through the origin forms a group denoted $SO(3)$, while the set of all orthogonal transformations is a group denoted $O(3)$. More generally, in the n-dimensional space, the set of all orthogonal transformations with determinant $\Delta = +1$ forms a group denoted $SO(n)$, while the set of all orthogonal transformations is the group denoted $O(n)$.

The image I_0 of G_0 in the order-parameter representation is isomorphous to a subgroup of $O(n)$ since it is composed of $n \times n$ orthogonal matrices. I_0 also has an additional peculiarity, which is related to the irreducibility of the order-parameter space, under the action of I_0: it can only be a subgroup of $O(n)$ which has no invariant subspace of the n-dimensional space. The subgroups of $O(n)$ possessing this property are termed *irreducible subgroups of* $O(n)$. Another possible terminology is based on the remark that the n-dimensional space ε_n is spanned by the components of a vector $(\eta_1, \ldots, \eta_n)$. Thus, ε_n spans the "*vector representation*" of the n-dimensional space. The preceding requirement can be expressed by stating that the vector representation of the group I_0 is *irreducible*. For instance, if $n = 3$, the group $4/mmm$ in Table 1.1 leaves the $\overleftrightarrow{z}$ direction and the (x, y) plane separately invariants. This group *is not* an irreducible subgroup of $O(3)$. By contrast, the restriction of $4/mmm$ to $\overleftrightarrow{z}$ or to (x, y) are irreducible subgroups of $O(1)$ or $O(2)$, respectively.

1.3.2.2. *Image-groups in the application of the Landau theory*

The action of G_0 in the order-parameter space is affected by means of the elements of its image-group I_0. Hence, once the order-parameter symmetry and the related image I_0 are known, it is possible, on the basis of this sole knowledge to determine the form of the free energy, and to work out the possible symmetries of the low-symmetry phase.

Each degree in the free-energy expansion will be constituted by the most general polynomial of this degree in the η_j components, which is invariant by I_0. As I_0 is generally a simpler group than G_0 (e.g., a finite group instead of an infinite one), the consideration of I_0 will simplify the procedure of determination of the successive terms of the Landau free energy. Besides, consideration of the images shows that the determination of the forms of the free energies, which arise for any phase transition associated with an n-component order-parameter, is a mathematically well-defined and solvable task. This task consists in enumerating the irreducible subgroups of $O(n)$, and for each subgroup, in working out the form of the invariant Pth-degree polynomials constructed from the set η_j of coordinates in the n-dimensional space.

A similar simplification will occur in the derivation of the symmetry group of the low-symmetry phase, which is a subgroup of G_0 leaving invariant a given vector in the space ε of the order-parameter. In this space, G is represented by its image I_G, which is the subgroup of I_0 leaving invariant the considered vector. If I_G is known, G can be determined as its reciprocal image: if $M(g_i)$ is an element of I_G, all the elements $g_i K$ belong to G. Hence, in order to determine G, it is sufficient to determine the simpler group I_G and use the relation $G = I_G \times K$. Examples of this procedure are treated in Sec. 1.3.2.6.

Many authors contributed to making a list of the irreducible subgroups of $O(n)$, and the results were summarized in tables [12]. Thus, today, with the tables of Ref. [12], one has access to a full list of the order-parameter symmetries for order-parameter dimensions up to $n = 24$.

1.3.2.3. *Integrity bases*

For a *finite group*, any invariant polynomial can be obtained as a linear combination of the powers of a limited number of basic polynomials forming the *integrity basis of the group* [11,13,15]. This mathematical property has generated the methods of finding the integrity bases for certain types of groups [16]. We will restrict ourselves to summarizing the corresponding results.

In the case of the two-dimensional groups, $n\,(C_n)$ and $nm(C_{nv})$, the integrity bases are, for n:

$$I_1 = \rho^2, \quad I_2 = \rho^n \cdot \cos(n\theta), \quad I_3 = \rho^n \cdot \sin(n\theta), \tag{1.33}$$

and for nm:

$$I_1 = \rho^2, \quad I_2 = \rho^n \cdot \cos(n\theta). \tag{1.34}$$

For instance, for $n = 3$, the basis comprises ρ^2 and $\rho^3 \cdot \cos(3\theta)$ for $3m$, while for 3, it contains the additional polynomial $\rho^3 \cdot \sin(3\theta)$.

Table 1.2. Integrity bases for invariant polynomials corresponding to five irreducible subgroups of $O(3)$.

23	$I_1 = \eta_1^2 + \eta_2^2 + \eta_3^2; I_2 = \eta_1\eta_2\eta_3;$ $I_3 = \eta_1^4 + \eta_2^4 + \eta_3^4;$ $I_4 = (\eta_1^2 - \eta_2^2)(\eta_2^2 - \eta_3^2)(\eta_3^2 - \eta_1^2)$
$m\overline{3}$	$I_1,\ I_3,\ I_4,\ I_5 = \eta_1^2\eta_2^2\eta_3^2$
$\overline{4}3m$	$I_1,\ I_2,\ I_3$
432	$I_1,\ I_3,\ I_5,\ I_2 \times I_4$
$m\overline{3}m$	$I_1,\ I_3,\ I_5$

The corresponding free energy will be written, for example, for $3m$ as follows:

$$F = a_1 I_1 + b_1 I_2 + a_2 I_1^2 + \gamma_{12} I_1 I_2 + \cdots . \tag{1.35}$$

Analogously, one has for $4m$, ρ^2 and $\rho^4 \cdot \cos(4\theta)$ and

$$F = a_1 I_1 + a_2 I_1^2 + b_1 I_2 + b_2 I_2^2 + \gamma_{12} I_1 I_2 + \cdots . \tag{1.36}$$

The irreducible subgroups of three-dimensional $O(3)$ form an infinite set, and as in the case $n = 2$, we can identify the subgroups, which are conjugates to each other as representing equivalent physical situations. One then finds [16] that, besides $SO(3)$, there are seven irreducible groups among the subgroups of $O(3)$: the three subgroups of $SO(3)$, denoted 23, 432 and 532, and the three subgroups of $O(3)$, denoted $m\overline{3}$, $m\overline{3}m$ and $\overline{532}/m$, which are obtained by adjoining to (23, 432, 532) the symmetry about the origin. The seventh group, $\overline{4}3m$, results from the combination with 23 of the reflections about a symmetry plane of the regular tetrahedron.

The two continuous groups $O(3)$ and $SO(3)$ give rise again to the free energy (1.35)–(1.36), where $\rho^2 = \eta^2{}_1 + \eta^2{}_2 + \eta^2{}_3$. Table 1.2 lists the integrity bases of the invariant polynomials corresponding to the five finite subgroups 23, $m\overline{3}$, $\overline{4}3m$, 432 and $m\overline{3}m$. The integrity basis of the invariant polynomials for the irreducible groups 532 and $\overline{5}$ $\overline{3}2/m$ can be found in Ref. [16].

1.3.2.4. *Isotropy groups and other simplifying mathematical concepts*

In order to determine the absolute minima of the Landau free energy, one has to search the values of the order-parameter components η_j, which satisfy Eqs. (1.30) and (1.31). The standard minimization process presents no major difficulty for small dimensionalities of the order-parameter and for the order-parameter expansions truncated to the lowest powers of the η_j. For a large number of order-parameter components or (and) when higher-degree expansions are taken into account, there may exist strong technical difficulties. A number of methods, more elaborate than the standard one, have been described, which take advantage of the symmetry properties of the minima η_j^{eq} satisfying Eqs. (1.30) and (1.31), and use some symmetry concepts, which will be summarized in this section.

(a) Isotropy group

In Sec. 1.3.2.2, we have seen that the space $\varepsilon_n = (\eta_1, \eta_2, \ldots, \eta_n)$ is irreducible under the action of the image $I_0 \in O(n)$. No subspace of ε_n is invariant by I_0. However, every vector direction η in ε_n is left invariant by some subgroup I_η of I_0. I_η, which is defined as the complete set of elements of I_0 leaving η unchanged, is called the *isotropy group* of η (or *little group* of η). Along this line, the symmetry group G of the phase stable below T_{C} is related to the isotropy group $I_{\eta^{\mathrm{eq}}}$ of the vector $\boldsymbol{\eta}^{\mathrm{eq}} = (\eta_1^{\mathrm{eq}}, \eta_2^{\mathrm{eq}}, \ldots, \eta_n^{\mathrm{eq}})$ corresponding to the absolute minimum of the free energy by $G = I_{\eta^{\mathrm{eq}}} \times K$, where K is the kernel of the homomorphism $G_0 \rightarrow I_0$.

(b) Orbit, stratum

Since η is not invariant by all the elements of I_0, when applying the elements of the group I_0 to a given vector η, we generate a set of distinct vectors $\eta\,(p)$ which forms the *orbit* (or the *star*) of η. The isotropy groups $I_\eta^{(p)}$ and $I_\eta^{(q)}$ of any two vectors $\eta(p)$ and $\eta(q)$ of the orbit are *conjugated* in I_0. This result applies to the orbit of the absolute minimum η^{eq} of F and has the consequence that the groups $G^{(p)}$, such as $G^{(p)}/K = I_\eta^{(q)}$ are conjugated in G_0.

Two vectors, whose isotropy groups are conjugated in I_0, do not necessarily belong to the same orbit. If they belong to different orbits, these distinct orbits are said to be of the *same type*. The set of all orbits of a given type (i.e., characterized by the isotropy groups which are conjugated in I_0) is called *a stratum.* Distinct orbits of a stratum contain the same number of vectors, i.e., they are associated with the same number of minima of the free energy. However, the distinct orbits do not correspond to the same *value* of the free energy.

(c) Minimal and maximal isotropy groups

The isotropy group I_η of the vector η can leave invariant the other vectors in the ε-space. These vectors and η generate a subspace ε_η of ε, each vector of which is invariant by I_η. This subspace is termed the *isotropy manifold* of I_η. If I_0 is finite, ε_η belongs to the stratum of η. The isotropy groups I_η associated to all the vector directions in ε possess a common subgroup I_k, possibly reduced to the identical transformation, which is an invariant subgroup of I_0 and of all the I_η. Consequently (as it is a self-conjugate), it is the isotropy group of all the vectors of a stratum. This stratum contains a continuous set of distinct vector directions. If a vector $\boldsymbol{\eta} \in \varepsilon$ does not belong to this stratum, it constitutes a *symmetry direction* in ε: Its isotropy group will contain I_k as a *strict subgroup.* The isotropy manifold of I_η will be a *symmetry manifold.*

A partial order exists between the isotropy groups I_η up to a conjugation in I_0. Thus, for two directions $\boldsymbol{\eta}$ and $\boldsymbol{\eta}'$, the isotropy groups I_η and $I_{\eta'}$ can satisfy an inclusion relation $I_\eta \subset I_{\eta'}$ up to a conjugation (i.e., I_η belongs to a group conjugated to $I_{\eta'}$). If the direction $\boldsymbol{\eta}$ is the subgroup of no other isotropy group I_η, I_η is called a *maximal isotropy group*, and $\boldsymbol{\eta}$ can be termed a *maximal symmetry direction* in ε.

1.3.2.5. *Illustration of the preceding concepts*

Let us translate the above rather abstract mathematical definitions to the crystal-geometrical picture applying the preceding mathematical concepts to the example treated in Sec. 1.2.2 considering only the two-dimensional order-parameter space $\varepsilon_2 = (v_x, v_y) = (\eta_1, \eta_2)$ of the group $G_0 = 4/mmm$. The set of distinct matrices of $4/mmm$ in the order-parameter representation $(v_x, v_y) = (\eta_1, \eta_2)$, which expresses the action of $4/mmm$, is shown in Table 1.3. This is the vector representation E_u of the $4/mmm$ group. One can verify, on the first row of Table 1.3, that it corresponds to the image $I_0 = 4mm$

Table 1.3. Symmetry elements and the corresponding matrices of $4/mmm$, and symmetry elements of I_0 ($4mm$).

$G_0 = P4/mmm$	$E; m_z$	$4; \bar{4}$	$2; \bar{1}$	$4_3; \bar{4}_3$
E_u	$\begin{bmatrix} 1 & 0 \\ 0 & 1 \end{bmatrix}$	$\begin{bmatrix} 0 & 1 \\ -1 & 0 \end{bmatrix}$	$\begin{bmatrix} -1 & 0 \\ 0 & -1 \end{bmatrix}$	$\begin{bmatrix} 0 & -1 \\ 1 & 0 \end{bmatrix}$
$I_0(E_u)$	1	4_1	4_2	4_3
$G_0 = P4/mmm$	$m_x; 2_y$	$m_y; 2_x$	$m_{xy};\ 2_{\bar{x}y}$	$m_{\bar{x}y}; 2_{xy}$
E_u	$\begin{bmatrix} -1 & 0 \\ 0 & 1 \end{bmatrix}$	$\begin{bmatrix} 1 & 0 \\ 0 & -1 \end{bmatrix}$	$\begin{bmatrix} 0 & -1 \\ -1 & 0 \end{bmatrix}$	$\begin{bmatrix} 0 & 1 \\ 1 & 0 \end{bmatrix}$
$I_0(E_u)$	m_x	m_y	m_{xy}	$m_{\bar{x}y}$

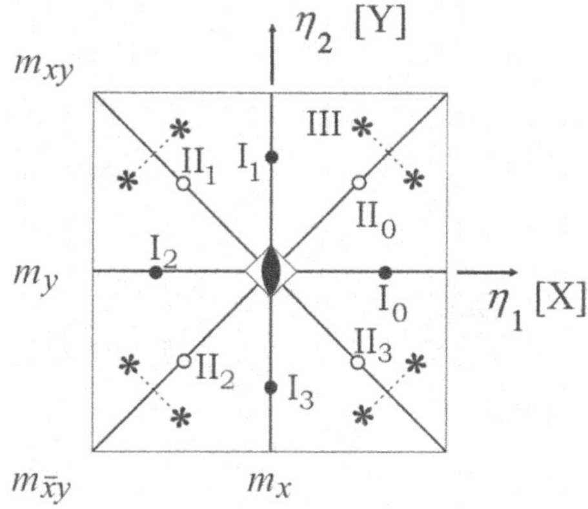

Fig. 1.6. The three types of orbits corresponding to the image $4mm$: $[\pm(v_x, 0); \pm(0, v_x)]$ (orbit I), $[\pm(v_x, \pm v_x)]$ (orbit II), and $[\pm v_x, \pm\, v_y)]$ (orbit III).

(last row in Table 1.3). As $4/mmm$ and $4mm$ are groups of order 16 and 8, respectively, the kernel of the homomorphism $G_0 \to I_0$ is a group K of order 2 comprising the identity E and the reflection m_z. The isotropy groups of the directions (η_1, η_2), subgroups of I_0, can be derived by taking the quotient groups (G/K). It is straightforward to work them out directly by simple inspection of Fig. 1.6.

However, such a straightforward "visual" method can be easily realized only for I_0 groups in low-dimensional spaces ($d \leq 3$) but becomes doubtful in case of order-parameters spanning irreducible representations of higher dimensionality. This is why, in the general case, we should follow more universal geometric algorithms in n-dimensional order-parameter space. Another reason is all existing related software packages (ISOTROPY [12, 17] is a typical example) compute the matrix procedure operating in multidimensional space where I_0 image-groups are their vector representations.

The procedure starts by finding n-dimensional vectors ($n = 2$ in the case) by actions of each symmetry element (matrix) of the ε_2 point group, i.e., the orbits of $\boldsymbol{\eta}$:

$$
\begin{aligned}
&E\colon \begin{pmatrix} 1 & \\ & 1 \end{pmatrix} \cdot \begin{pmatrix} \eta_1 \\ \eta_2 \end{pmatrix} = \begin{pmatrix} \eta_1 \\ \eta_2 \end{pmatrix} \to \begin{matrix} \eta_1 = \eta_1 \\ \eta_2 = \eta_2 \end{matrix} \to \begin{pmatrix} \eta_1 \\ \eta_2 \end{pmatrix};\\
&4\colon \begin{pmatrix} & 1 \\ -1 & \end{pmatrix} \cdot \begin{pmatrix} \eta_1 \\ \eta_2 \end{pmatrix} = \begin{pmatrix} \eta_1 \\ \eta_2 \end{pmatrix} \to \begin{matrix} \eta_1 = \eta_2 \\ -\eta_1 = \eta_2 \end{matrix} \to \eta_1 = \eta_2 = 0 \to \begin{pmatrix} 0 \\ 0 \end{pmatrix};\\
&2: \begin{pmatrix} -1 & \\ & -1 \end{pmatrix} \cdot \begin{pmatrix} \eta_1 \\ \eta_2 \end{pmatrix} = \begin{pmatrix} \eta_1 \\ \eta_2 \end{pmatrix} \to \begin{matrix} -\eta_1 = \eta_1 \\ -\eta_2 = \eta_2 \end{matrix} \to \eta_1 = \eta_2 = 0 \to \begin{pmatrix} 0 \\ 0 \end{pmatrix};\\
&4_3: \begin{pmatrix} & -1 \\ 1 & \end{pmatrix} \cdot \begin{pmatrix} \eta_1 \\ \eta_2 \end{pmatrix} = \begin{pmatrix} \eta_1 \\ \eta_2 \end{pmatrix} \to \begin{matrix} -\eta_2 = \eta_1 \\ \eta_1 = \eta_2 \end{matrix} \to \eta_1 = \eta_2 = 0 \to \begin{pmatrix} 0 \\ 0 \end{pmatrix};\\
&m_x\colon \begin{pmatrix} -1 & \\ & 1 \end{pmatrix} \cdot \begin{pmatrix} \eta_1 \\ \eta_2 \end{pmatrix} = \begin{pmatrix} \eta_1 \\ \eta_2 \end{pmatrix} \to \begin{matrix} -\eta_1 = \eta_1 \\ \eta_2 = \eta_2 \end{matrix} \to \begin{matrix} \eta_1 = 0 \\ \eta_2 = \eta \neq 0 \end{matrix}\\
&\quad \to \begin{pmatrix} 0 \\ \eta \end{pmatrix} \text{ or } \begin{pmatrix} 0 \\ -\eta \end{pmatrix};\\
&m_y: \begin{pmatrix} 1 & \\ & -1 \end{pmatrix} \cdot \begin{pmatrix} \eta_1 \\ \eta_2 \end{pmatrix} = \begin{pmatrix} \eta_1 \\ \eta_2 \end{pmatrix} \to \begin{matrix} \eta_1 = \eta_1 \\ -\eta_2 = \eta_2 \end{matrix} \to \begin{matrix} \eta_1 = \eta \neq 0 \\ \eta_2 = 0 \end{matrix}\\
&\quad \to \begin{pmatrix} \eta \\ 0 \end{pmatrix} \text{ or } \begin{pmatrix} -\eta \\ 0 \end{pmatrix};\\
&m_{xy}\colon \begin{pmatrix} & -1 \\ -1 & \end{pmatrix} \cdot \begin{pmatrix} \eta_1 \\ \eta_2 \end{pmatrix} = \begin{pmatrix} \eta_1 \\ \eta_2 \end{pmatrix} \to \begin{matrix} -\eta_2 = \eta_1 \\ -\eta_1 = \eta_2 \end{matrix} \to \eta_1 = -\eta_2 = \eta \neq 0\\
&\quad \to \begin{pmatrix} \eta \\ -\eta \end{pmatrix} \text{ or } \begin{pmatrix} -\eta \\ \eta \end{pmatrix};\\
&m_{\bar{x}y}: \begin{pmatrix} & 1 \\ 1 & \end{pmatrix} \cdot \begin{pmatrix} \eta_1 \\ \eta_2 \end{pmatrix} = \begin{pmatrix} \eta_1 \\ \eta_2 \end{pmatrix} \to \begin{matrix} \eta_2 = \eta_1 \\ \eta_1 = \eta_2 \end{matrix} \to \eta_1 = \eta_2 = \eta \neq 0\\
&\quad \to \begin{pmatrix} \eta \\ \eta \end{pmatrix} \text{ or } \begin{pmatrix} -\eta \\ -\eta \end{pmatrix}.
\end{aligned}
\tag{1.37}
$$

Three types of orbits can be distinguished (i.e., three strata):

(I) The orbit of the vector $(\eta_1 \neq 0, \eta_2 = 0) = (\eta, 0) = (v_x, 0)$, which comprises three additional vectors $(-v_x, 0), (0, v_x)$ and $(0, -v_x)$. Their isotropy group is, up to a conjugation, the group of order 2: $I_1 = (E, m_y)$. The isotropy manifold of I_1 is reduced to the $\pm(v_x, 0)$ direction: it is a line which corresponds to a *maximal symmetry direction* in ε_2, i.e., the corresponding isotropy group is *maximal.*

(II) The orbit of the vectors $(\eta_1 = \eta_2 \neq 0) = (\eta, \eta) = (v_x, v_x)$ comprises the additional vectors $(v_x, -v_x), (-v_x, -v_x)$ and $(-v_x, v_x)$. The isotropy group is also of the order 2, $I_2 = (E, m_{\overline{x}y})$. Here, again, the isotropy manifold is a line coinciding with one of the bisectors, which is a maximal symmetry direction in ε_2 corresponding to a maximal isotropy group.

(III) The orbit of the vector $(\eta_1 \neq \eta_2 \neq 0) = (\eta_1, \eta_2) = (v_x, v_y)$ with $v_x \neq v_y \neq 0$. There are eight distinct vectors in the orbit: $(\pm v_x, \pm v_y)$ and $(\pm v_y, \pm v_x)$. The isotropy group of any vector of the orbit is 1, which is obtained for all the orbits whose representative vector (v_x, v_y) is in a general direction of the (v_x, v_y) plane. The stratum corresponding to the isotropy group is composed of all the directions in the plane, except the directions corresponding to the two first orbits. Such directions correspond to the *minimal isotropy group.*

Evidently, the vector $(0) = (\eta_1 = \eta_2 = 0) = (v_x = v_y = 0)$ represents the parent state, and its $I_0 = 4mm$.

In the considered example, we have determined the isotropy groups of every direction in the two-dimensional space, by checking in each case which ones of the eight elements of $I_0 = 4mm$ leave the considered vector direction invariant.

The orbits of vectors (I) and (II) have similar characters: their vectors transform to each other within the sets forming, therefore, two independent vector stars. It means that the vectors correspond to two different minima of the free energy, i.e., the invariant vectors (I) and (II) define different phases of the system. Forming m-component stars corresponds to m-fold energy degeneration of the corresponding distorted state that, in turn, indicates the existence of m equivalent domains in the low-symmetry structure.

Summarizing in crystal-geometric terms of the introductory example (Sec. 1.2.2), we conclude that if, in a crystal of symmetry $G_0 = P4/mmm$, the order-parameter components span two-dimensional irreducible representation E_u, its phase diagram can contain four phases

$$0:\begin{pmatrix}0\\0\end{pmatrix};\quad \mathrm{I}:\begin{pmatrix}\eta\\0\end{pmatrix}_0,\begin{pmatrix}0\\\eta\end{pmatrix}_1,\begin{pmatrix}-\eta\\0\end{pmatrix}_2,\begin{pmatrix}0\\-\eta\end{pmatrix}_3;$$

$$\mathrm{II}:\begin{pmatrix}\eta\\\eta\end{pmatrix}_0,\begin{pmatrix}-\eta\\\eta\end{pmatrix}_1,\begin{pmatrix}-\eta\\-\eta\end{pmatrix}_2,\begin{pmatrix}\eta\\-\eta\end{pmatrix}_3;\quad \mathrm{III}:\begin{pmatrix}\eta_1\\\eta_2\end{pmatrix}.$$

It is worth underlining that the above prediction is model-free: no free energy approximation has been used.

1.3.2.6. *Determination of the low-symmetry space groups*

If the absolute minimum of the order-parameter expansion below T_C corresponds to the set of order-parameter components $(\eta_1^0, \eta_2^0, \ldots, \eta_n^0)$, the invariance group G of the low-symmetry phase is the largest subgroup of G_0, which leaves invariant the vector $\eta_0 = (\eta_1^0, \eta_2^0, \ldots, \eta_n^0)$ of the order-parameter space. $G = I_G \times K$, where I_G is the isotropy group of the η_0 direction, and K is the kernel of the homomorphism $G_0 \to I_0$. In the case of structural transitions, the advantage of making use of the image I_0 to determine G is clear: I_0 is a finite group, which will be simpler to handle than the infinite group G_0. G is one of the 230 space groups. Accordingly, once the vector η_0 is specified, the determination of G can be performed in three main steps: (1) determination of K *and* I_0; (2) determination of the *isotropy group*; (3) determination of the *space group* G. The last step requires the identification of the point group $\hat{G}$, the *Bravais lattice* of G, and the screw-axes and glide-planes contained in G.

To illustrate this procedure, we shall consider the irreducible representation, denoted $\Gamma_1(\boldsymbol{k}_8)$ [6], of the $Fm\bar{3}m$ space group, associated with the wave-vector $\boldsymbol{k}_8 = \frac{3}{4}\boldsymbol{a}^* + \frac{1}{2}\boldsymbol{b}^* + \frac{1}{4}\boldsymbol{c}^*$, where $\boldsymbol{a}^*$, $\boldsymbol{b}^*$ and $\boldsymbol{c}^*$ are the reciprocal lattice vectors of the face-centered cubic (FCC) Brillouin zone.

(*a*) *Image of* $\Gamma_1(\boldsymbol{k}_8^*)$

The wave-vector $\boldsymbol{k}_8$ ends at the W-point of the FCC Brillouin zone. As $G_0 = Fm\bar{3}m$ is a symmorphic space group, the *small* representations $\hat{\boldsymbol{\tau}}_i(\boldsymbol{k}_8)$ coincide with the ordinary irreducible representations of the point group $\hat{G}(\boldsymbol{k}_8) = \overline{4}2m$ [6, 7]. Among the five irreducible representations $\hat{\boldsymbol{\tau}}_i$ of $\overline{4}2m$, we will consider the *totally symmetric* representation $\hat{\boldsymbol{\tau}}_1$. As the ratio of the orders of G_0 (48) and $\hat{G}(\boldsymbol{k}_8)$ (8) is *six*, the *star* $\boldsymbol{k}_8^*$ has six arms: $\boldsymbol{k}_8^{(1)}$, $\mathbf{k}_8^{(2)} = -\boldsymbol{k}_8^{(1)}$, $\boldsymbol{k}_8^{(3)} = 3_1 \cdot \boldsymbol{k}_8^{(1)}$, $\boldsymbol{k}_8^{(4)} = -\boldsymbol{k}_8^{(3)}$, $\boldsymbol{k}_8^{(5)} = 3_2^2 \cdot \boldsymbol{k}_8^{(1)}$, $\boldsymbol{k}_8^{(6)} = -\boldsymbol{k}_8^{(5)}$. Thus, the irreducible representation $\Gamma_1(\boldsymbol{k}_8^*)$ constructed from $\hat{\boldsymbol{\tau}}_1$ is six-dimensional. On the other hand, it is isomorphous to the product $\hat{G}_0 \times \boldsymbol{T}$, where $\hat{G}_0$ is the point group associated to G_0, and $\boldsymbol{T}$ is the group of primitive translations. Consequently, the image I_0 of G_0 for the representation $\Gamma_1(\boldsymbol{k}_8^*)$ can be written $I_0 = I_T \times I_{\hat{G}_0}$, where I_T is the image of $\boldsymbol{T}$, and $I_{\hat{G}_0}$ is the image of the group of elements $\{R|0\}$ isomorphous to $\hat{G}_0$. I_T is a group of order 32 generated by the following three 6×6 matrices:

$$\boldsymbol{a} = \left[\begin{array}{cc|cc|cc} 0 & 1 & 0 & 0 & 0 & 0 \\ -1 & 0 & 0 & 0 & 0 & 0 \\ \hline 0 & 0 & 0 & -1 & 0 & 0 \\ 0 & 0 & 1 & 0 & 0 & 0 \\ \hline 0 & 0 & 0 & 0 & -1 & 0 \\ 0 & 0 & 0 & 0 & 0 & -1 \end{array}\right], \quad \boldsymbol{b} = \left[\begin{array}{cc|cc|cc} 0 & -1 & 0 & 0 & 0 & 0 \\ 1 & 0 & 0 & 0 & 0 & 0 \\ \hline 0 & 0 & -1 & 0 & 0 & 0 \\ 0 & 0 & 0 & -1 & 0 & 0 \\ \hline 0 & 0 & 0 & 0 & 0 & 1 \\ 0 & 0 & 0 & 0 & -1 & 0 \end{array}\right],$$

$$\boldsymbol{a} + \boldsymbol{b} + \boldsymbol{c} = \left[\begin{array}{cc|cc|cc} -1 & 0 & 0 & 0 & 0 & 0 \\ 0 & -1 & 0 & 0 & 0 & 0 \\ \hline 0 & 0 & -1 & 0 & 0 & 0 \\ 0 & 0 & 0 & -1 & 0 & 0 \\ \hline 0 & 0 & 0 & 0 & -1 & 0 \\ 0 & 0 & 0 & 0 & 0 & -1 \end{array}\right]. \tag{1.38}$$

I_2 is a group of order 48, isomorphous to $\hat{G}_0 = m\overline{3}m$, generated by the following matrices:

$$\mathrm{A} = \left[\begin{array}{cc|cc|cc} 0 & 0 & 0 & 0 & 1 & 0 \\ 0 & 0 & 0 & 0 & 0 & 1 \\ \hline 1 & 0 & 0 & 0 & 0 & 0 \\ 0 & 1 & 0 & 0 & 0 & 0 \\ \hline 0 & 0 & 1 & 0 & 0 & 0 \\ 0 & 0 & 0 & 1 & 0 & 0 \end{array}\right], \quad \mathrm{B} = \left[\begin{array}{cc|cc|cc} 1 & 0 & 0 & 0 & 0 & 0 \\ 0 & -1 & 0 & 0 & 0 & 0 \\ \hline 0 & 0 & 0 & 0 & 1 & 0 \\ 0 & 0 & 0 & 0 & 0 & 1 \\ \hline 0 & 0 & 1 & 0 & 0 & 0 \\ 0 & 0 & 0 & -1 & 0 & 0 \end{array}\right],$$

$$\mathrm{C} = \left[\begin{array}{cc|cc|cc} 1 & 0 & 0 & 0 & 0 & 0 \\ 0 & 1 & 0 & 0 & 0 & 0 \\ \hline 0 & 0 & 0 & 0 & 1 & 0 \\ 0 & 0 & 0 & 0 & 0 & 1 \\ \hline 0 & 0 & 1 & 0 & 0 & 0 \\ 0 & 0 & 0 & 1 & 0 & 0 \end{array}\right], \quad \mathrm{D} = \left[\begin{array}{cc|cc|cc} 1 & 0 & 0 & 0 & 0 & 0 \\ 0 & -1 & 0 & 0 & 0 & 0 \\ \hline 0 & 0 & 1 & 0 & 0 & 0 \\ 0 & 0 & 0 & -1 & 0 & 0 \\ \hline 0 & 0 & 0 & 0 & 1 & 0 \\ 0 & 0 & 0 & 0 & 0 & -1 \end{array}\right]. \tag{1.39}$$

So, the image I_0 possesses $32 \times 48 = 1536$ elements.

(*b*) *Order-parameter expansion*

Let us calculate the integrity basis corresponding to I_0 using a simple procedure adapted to the specific *form* of I_0. With that goal, it can be noted that all the matrices (1.38) and (1.39) consist of three *diagonal* blocks acting separately on the three sets of order-parameter components (η_1, η_2), (η_3, η_4) and (η_5, η_6) spanning the six-dimensional space ε_6 of $\hat{\boldsymbol{\tau}}_i(\boldsymbol{k}_8)$. Accordingly, we can form the required invariant polynomials of two variables (η_i, η_{i+1}) $(i = 1, 3, 5)$, which are left invariant by the following set of 2×2 matrices:

$$\begin{pmatrix} & 1 \\ -1 & \end{pmatrix}, \quad \begin{pmatrix} 1 & \\ & -1 \end{pmatrix}, \quad \begin{pmatrix} -1 & \\ & -1 \end{pmatrix}.$$

These matrices generate a group identical to the group of matrices in Table 1.3, i.e., isomorphous to $4mm$, the basic invariants of which are ρ_i^2 and $\rho_i^4 \cdot \cos(4\theta_\iota)$. Any function of these invariants (with $i = 1, 3, 5$) will be invariant by the subgroup of I_0 considered. We are interested in the polynomials of nth degree, which are also invariant by the matrices (1.39). Clearly, the set generated by these matrices operates a permutation of the invariants of index i. Hence, the solutions to the investigated problem are polynomials of ρ_i^2 and $\rho_i^4 \cdot \cos(4\theta_\iota)$, which are invariant by permutation of the i-indices. Polynomials of the third degree are absent, and polynomials of fourth degree are easily determined allowing to write the order-parameter expansion, related to I_0, as

$$F(T,\rho_i,\theta_i) = F_0(T) + \frac{\alpha}{2}\left(\sum_{i=1}^{3}\rho_i^2\right) + \frac{\beta_1}{2}\left(\sum_{i=1}^{3}\rho^4{}_i\right) + \frac{\beta_2}{2}\left(\sum_{i=1,i\neq j}^{3}\rho_i^2\rho_j^2\right) + \frac{\beta_3}{2}\left(\sum_{i=1}^{3}\rho_i^4\cos(4\theta_i)\right). \quad (1.40)$$

(c) Kernel of the homomorphism $G_0 \to I_0$ and isotropy group I_G

As stressed above, the image I_0 contains a subgroup $I_{\hat{G}_0}$ isomorphous to $\hat{G}_0$. Hence, the kernel of the homomorphism $G_0 \to I_0$ necessarily coincides with a subgroup of the translation group $\boldsymbol{T}$ of G_0, as G_0 is symmorphic. The image of $\boldsymbol{T}$ being of order 32, the kernel K is a subgroup of $\boldsymbol{T}$ of index 32. In other words, the unit-cell of the infinite translation group $\boldsymbol{T}$ is 32 times larger than the unit-cell ($\boldsymbol{a}$, $\boldsymbol{b}$, $\boldsymbol{c}$) of the FCC lattice. The corresponding basic vectors are

$$\boldsymbol{a}_k = 2(\boldsymbol{a} - \boldsymbol{b} + \boldsymbol{c}) = 2a(100), \boldsymbol{b}_k = 2(\boldsymbol{a} + \boldsymbol{b} - \boldsymbol{c}) = 2a(010),$$
$$\boldsymbol{c}_k = 2(-\boldsymbol{a} + \boldsymbol{b} + \boldsymbol{c}) = 2a(001), \quad (1.41)$$

where the basic translations of the FCC lattice are

$$\boldsymbol{a} = \frac{a}{2}(110), \quad \boldsymbol{b} = \frac{a}{2}(011), \quad \boldsymbol{c} = \frac{a}{2}(101). \quad (1.42)$$

One can check that $(\boldsymbol{a}_k \cdot (\boldsymbol{b}_k \times \boldsymbol{c}_k)) = 32(\boldsymbol{a} \cdot (\boldsymbol{b} \times \boldsymbol{c}))$.

In the six-dimensional space spanning the matrices (1.38) and (1.39), let us assume that for a certain range of the coefficients, the absolute minimum of the free energy corresponds to the direction $(\eta - \eta 0000)$ of the order-parameter space (using Eqs. (1.30) and (1.31), it is straightforward to show that this range is defined by $\beta_3 > 0$, $\beta_2 > \beta_1 - \beta_3 > 0$). The isotropy group I_G is composed of the elements of I_0, which preserve the manifold (η_1, η_2) of the order-parameter space, and which, in the manifold, leave the first bisector invariant. As I_0 acts, within this manifold, as the group $4mm$, among the elements of this group only two elements (E and $m_{\overline{x}y}$) leave the [11]-direction unchanged, as they are represented by the 2×2 matrices $\left(\begin{smallmatrix}1&0\\0&1\end{smallmatrix}\right)$ and $\left(\begin{smallmatrix}0&1\\1&0\end{smallmatrix}\right)$ (see Table 1.3). Inspection of the matrices (1.38) and (1.39) shows that I_G is generated by the matrices which possess as first diagonal block one of the two former matrices. Thus, only the matrices denoted C in Eq. (1.39), or C × B and C × D, belong to I_G. Hence, I_G is generated by the subgroup of order 16 of I_0, formed by the matrices B, C, D, and by the subgroups of order 8 of I_T, whose first diagonal block is $\left(\begin{smallmatrix}1&0\\0&1\end{smallmatrix}\right)$, I_G is constituted by the product of these two sets: its order is $8 \times 16 = 128$, i.e., it is a subgroup of order 12 of G_0.

(d) Point group $\hat{G}$ and Bravais lattice of G

In the isomorphism between the groups generated by the matrices (1.39) and $m\bar{3}m$, the matrices A, B, C, D correspond, respectively, to the three-fold rotation 3_{xyz}, the four-fold rotation 4_x, the two-fold rotation $2_{\bar{y}z}$, and the inversion I. The isotropic group I_G of the considered direction in the order-parameter space contains the matrices denoted C, C × B and C × D. Accordingly, the low-symmetry group contains the reciprocal images of these matrices, namely $\{2_{\bar{y}z}|0\} \times K$, $\{4_x|\boldsymbol{a}+\boldsymbol{b}+\boldsymbol{c}\}$ and $\{\bar{1}|\boldsymbol{a}+\boldsymbol{b}+\boldsymbol{c}\}$. Since K is the group of primitive translations (1.41), and since the reciprocal images of the other generators of I_G are also primitive translations of G_0, we can infer that the *point group* $\hat{G}$ is generated by the elements I, 4_x and $2_{\bar{y}z}$. It is the crystallographic point group $4_x/mmm$ of order 16, the four-fold axis of which is along the x-axis.

The group $4_x/mmm$ belongs to the tetragonal system, and the translation group $\boldsymbol{T}_G$ of the crystal in the low-symmetry phase of space group G defines one of the Bravais lattices (primitive P or body-centered I of this system. The specification of the relevant Bravais lattice involves two steps:

(i) The *first step* consists in working out the basic translations of $\boldsymbol{T}_G$. Taking into account the fact that the index of $\boldsymbol{T}_G$ in $\boldsymbol{T}$ is 4 (since the index of G in G_0 is 3), the basic translations of $\boldsymbol{T}_G$ must define a unit-cell four times larger than ($\boldsymbol{a}$, $\boldsymbol{b}$, $\boldsymbol{c}$). One set of basic translations fulfilling this property is

$$\boldsymbol{a}'_G = -\boldsymbol{a}+\boldsymbol{b}+\boldsymbol{c} = a(001), \quad \boldsymbol{b}'_G = \boldsymbol{a}+\boldsymbol{b}-\boldsymbol{c} = a(010),$$
$$\boldsymbol{c}'_G = \boldsymbol{a}+\boldsymbol{c} = a\left(1\frac{1}{2}\frac{1}{2}\right). \tag{1.43}$$

(ii) The *second step* is the identification of the set (1.43) of basic translations with the standard set associated with the relevant

Bravais lattice, as given, for example in tables [7, 8]. When the four-fold axis is along X, the two Bravais lattices of the tetragonal system are defined by the following basic translations:

$$P\text{-lattice} : C(100), A(010), A(001),$$
$$I\text{-lattice} : \left(\frac{C}{2}\frac{A}{2}\frac{A}{2}\right), \left(-\frac{C}{2}\frac{A}{2}\frac{A}{2}\right), \left(\frac{C}{2}\frac{A}{2} - \frac{A}{2}\right). \quad (1.44)$$

In order to identify the Bravais lattice generated by the translations $\boldsymbol{a}'_G$, $\boldsymbol{b}'_G$, $\boldsymbol{c}'_G$ given by Eq. (1.43), we must find a combination of vectors (1.43), which defines a unit-cell of the same volume as $\boldsymbol{a}'_G$, $\boldsymbol{b}'_G$, $\boldsymbol{c}'_G$, and displays the same geometrical configuration as one of the sets of vectors (1.44). One can see that the combinations

$$\boldsymbol{a}_G = \boldsymbol{c}'_G = \left(a\frac{a}{2}\frac{a}{2}\right), \quad \boldsymbol{b}_G = \boldsymbol{a}'_G + \boldsymbol{b}'_G - \boldsymbol{c}'_G = \left(-a\frac{a}{2}\frac{a}{2}\right),$$
$$\boldsymbol{c}_G = \boldsymbol{b}'_G - \boldsymbol{c}'_G = \left(-a\frac{a}{2} - \frac{a}{2}\right), \quad (1.45)$$

form a set of translations identical to the standard translations of the I-Bravais lattice in Eq. (1.44) provided C $= 2a$ and A $= a$. Thus, the translations $\boldsymbol{a}_G$, $\boldsymbol{b}_G$, $\boldsymbol{c}_G$ are the three basic translations of the I-lattice of the low-symmetry phase.

(*e*) *Identification of the space group G*

G contains the elements $\{2_{\overline{y}z}|0\}$, $\{4_x|\boldsymbol{c}\} = \{4_x|\frac{C}{4}0\frac{A}{2}\}$, $\{\overline{1}|\boldsymbol{c}\} = \{\overline{1}|\frac{C}{4}0\frac{A}{2}\}$, as well as the product of these elements. Finding the space group G (its symbol) with $\hat{G} = 4/mmm$ requires specifying the nature of the four-fold axis (pure or helicoidal rotation) and the nature of the three types of reflection-planes m_x, m_y and $m_{\overline{y}z}$ (pure or glide reflection). One finds here the following elements:

$$\left\{4_x|\frac{C}{4}0\frac{A}{2}\right\}, \quad \left\{m_x|\frac{C}{4}0\frac{A}{2}\right\}, \quad \{m_y|0\}, \quad \left\{m_{\overline{y}z}|\frac{C}{4}0\frac{A}{2}\right\},$$

which are written in a nonstandard setting. For a standard setting, one has to perform the transformation $(x \to z, z \to y, y \to x)$, which gives the elements

$$\left\{4_z|\frac{A}{2}0\frac{C}{4}\right\}, \quad \left\{m_z|\frac{A}{2}0\frac{C}{4}\right\}, \quad \{m_x|0\}, \quad \left\{m_{\overline{x}y}|\frac{A}{2}0\frac{C}{4}\right\},$$

showing that the four-fold axis is a *screw-axis* whose non-primitive translation is one-fourth of the period C. m_z is a *glide-plane* with a translation A/2, m_x is a pure reflection, and $m_{\overline{x}y}$ is a glide-plane with a translation C/4. This allows, using the space-group tables [7–9], to find that $G = I4_1/amd$.

1.3.3. *The radial problem and its solutions*

1.3.3.1. *Specific form of the Landau potential*

Solving the angular problem allows us to find, in a model-free procedure, the number of possible low-symmetry states of the system which can be stabilized with a given order-parameter (irreducible or reducible) and to disclose the structure of solutions minimizing the corresponding free energy, that means finding the number of non-zero components of the order-parameter in each equilibrium state and determining some additional functional ratios between them.

The solution of the radial problem consists, therefore, in finding the dependence of the equilibrium values of the order-parameter components (η_j) on the phenomenological coefficients of the free energy (Landau potential), which depend on the external conditions (temperature, pressure and eventually concentration) [11]. The equilibrium values (η_j^{eq}) allow to evaluate the critical behavior (anomalies) of various relevant physical quantities: susceptibilities, specific heat, etc. They also lead to the construction of the phase diagram in the space of the external parameters, which includes the transition lines bounding the phases, the shape and width of the regions of coexistence of the phases (for first-order transitions), and possibly, various types of singular points. One can also deduce the effect of external (conjugated) fields on the phase diagram and on the transition anomalies, as well as the dynamics of the order-parameter, using equations of motion such as the Landau–Khalatnikov [18] and the Landau–Lifshitz [19] equations.

Although the *general form* of the Landau potential $F(T, P, \eta_j)$ has been shown in Sec. 1.3.2. to be entirely determined by the symmetry of the order-parameter, the equilibrium values of the order-parameter components, as functions of the phenomenological coefficients, depend on the *specific model* chosen for F, namely: (1) on the maximal power of the (η_j) at which F is truncated; (2) on the sign and dependence on the external parameters assumed for the phenomenological coefficients; (3) on the influence of the secondary order-parameters, and (4) on the maximal degree considered for the corresponding coupling terms, which, as shown in Sec. 1.2.4, lead to a renormalization of the coefficients of the primary order-parameter expansion.

Usually the choice of the model of Landau potential is made with some *arbitrariness*, and is essentially determined by empirical considerations: one takes the *minimal "necessary"* power for the η_j and the coupling terms which allow to describe the more realistically full set of available experimental data, namely the observed sequence of phases and the critical anomalies. Along the same line, the simplest (linear or constant) dependence on the external variables is assumed for the phenomenological coefficients. Such a procedure is coherent in a phenomenological approach, since the most "economic" model, which contains the smallest number of phenomenological variables, is indeed the most predictive one. In the examples discussed in this section, we will show that the degree of arbitrariness in the choice of the Landau potential can be reduced using a number of simple rules.

In the present section, we present the main results, which have been obtained in working out the solutions of the radial problem for a number of typical models containing only the even-degree invariants, i.e., those allowing the second-order phase transitions. The free-energy models operating with odd-degree invariants are considered in Chapter 2. Since the features of these models are chiefly determined by the number n of the order-parameter components and by the corresponding image-groups, we shall order our presentation as before in Secs. 1.3.2.2 and 1.3.2.3, i.e., following the increasing values of n.

1.3.3.2. *Solutions of the radial problem for one-component order-parameter expansions*

(a) Canonical model of first-order phase transition

In Sec. 1.2.2.4, the simplest one-component order-parameter expansion (1.2) truncated to the fourth degree was shown to describe a single second-order phase transition between a high- and low-symmetry phase at T_C. Another canonical example is the sixth-degree expansion:

$$F_L(P,T,\eta) = F_0(P,T) + a_1\eta^2 + a_2\eta^4 + a_3\eta^6, \tag{1.46}$$

which describes, when assuming $a_1 = a_{10}(T - T_C)$, $a_2 < 0$ and $a_3 > 0$, a first-order transition between the high- and low-symmetry phases. Expansion (1.46) corresponds to the one-component order-parameter spanning the image-group $\overline{1}$. It allows, despite its simple form, describing the main features as continuous (second-order) and discontinuous (first-order) phase transitions.

The equations minimizing F are

$$\frac{dF}{d\eta} = 2\eta \cdot (a_1 + 2a_2\eta^2 + 3a_3\eta^4) = 0, \tag{1.47}$$

$$\frac{d^2F}{d\eta^2} = a_1 + 6a_2\eta^2 15a_3\eta^4 \geq 0. \tag{1.48}$$

The discussion of the conditions (1.47) and (1.48) is visualized in Fig. 1.7. It can be summarized as follows: For $T > T_2 = T_C + a_2^2/3a_3$,

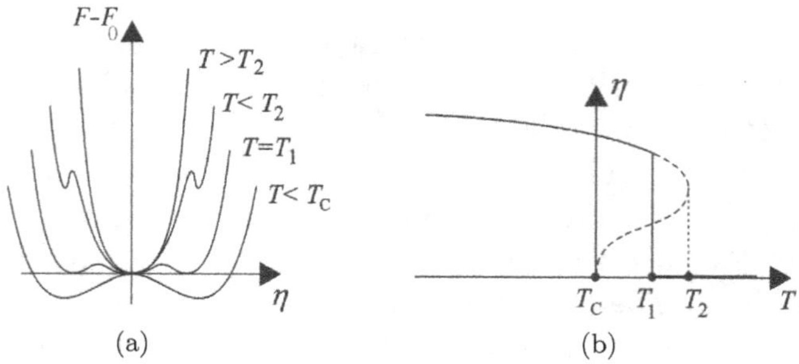

Fig. 1.7. (a) Variation of the non-equilibrium potential $F - F_0$ given by Eq. (1.46) as a function of temperature. (b) Temperature dependence of the order-parameter $\eta(T)$.

only one phase is stable, namely the parent phase I corresponding to $\eta_{\mathrm{I}}^{\mathrm{eq}} = 0$. At $T = T_2$, phase II, corresponding to

$$\eta_{\mathrm{II}}^2 = \frac{-a_2 \pm \sqrt{a_2^2 - 3a_1a_3}}{3a_3}, \tag{1.49}$$

appears as a metastable state (i.e., corresponding to the secondary minima in the non-equilibrium curve $[F(\eta)-F_0])$. As the temperature is lowered, the stability of phase II increases. Phases I and II are equally stable when the additional condition $F(\eta) - F_0$ is fulfilled at

$$T_1 = T_{\mathrm{C}} + \frac{a_2^2}{4a_{10}a_3}. \tag{1.50}$$

Below T_1, phase I becomes less stable than phase II and remains as a metastable state until a_1 changes sign for $T = T_{\mathrm{C}}$.

From the preceding discussion, the main distinctive properties of the first-order transitions can be inferred. These transitions are characterized by the following:

(i) By *a region of coexistence of the two phases* I and II within the interval:

$$\Delta T = T_2 - T_{\mathrm{C}} = a_2^2/3a_3, \tag{1.51}$$

the two phases being alternately metastable for $T > T_1$ and $T < T_1$. Although the transition may in reality take place at any temperature within the interval ΔT, the most probable transition temperature is T_1, which corresponds to an equal stability and equal energy for the two phases. As a consequence, a *thermal hysteresis* will generally be observed, since the measured transition temperature will not be the same on heating and cooling.

(ii) By a discontinuous jump of the order-parameter at the transition, which is given by:

$$\Delta\eta(T_1) = -a_2/6a_3. \tag{1.52}$$

Below T_1, η increases with decreasing temperature following the law (1.49).

(iii) By a *latent heat* L, which results from the discontinuity of $\eta(T)$ at the transition:

Transformation latent heat L is the change in entropy of the system at the fixed phase transition temperature point T_1 (or T_C in the case of continuous transformation), when it goes from phase I to II, $\Delta S = S_I - S_{II}$: $L = T_1 \cdot \Delta S(T_1)$. Entropy of each contacting phase can be expressed as follows:

$$S = -\left(\frac{\partial F_{\text{eq}}}{\partial T}\right)_P = -\left(\frac{\partial F_0}{\partial T}\right)_P - \left(\frac{\partial F_L^{\text{eq}}}{\partial T}\right)_P = S_0 + S_L,$$

where F_0 and S_0 relate to non-critical degrees of freedom corresponding to the background energy. Extra entropy ΔS_L, therefore, reads

$$S_L = -\frac{\partial F_{\text{L}}}{\partial T} = -\frac{\partial F_{\text{L}}}{\partial a_1} \cdot \frac{\partial a_1}{\partial T} = -\eta^2 \cdot \frac{\partial a_1}{\partial T} = -\alpha\eta^2.$$

Typically for the phenomenological approach, only a_1 coefficient in the energy expansion (1.46) depends on temperature. The latter equation shows that at the continuous I–II phase transition ($a_2 > 0$) $\Delta S_{\text{I-II}} = 0$, thus no latent heat is needed to bring the system from state I to II.

The change of sign of quadratic term in Eq. (1.46) to negative makes the corresponding phase transition discontinuous, then L is not equal to zero:

$$\Delta S_{\text{I-II}} = -\frac{a_{10}a_2}{6a_3}.$$

It is worth noting that the convenient phenomenological hypothesis that only a_1 coefficient in the free energy expansion is temperature dependent finally results in the strong conclusion that the extra entropy S_{L} is a harmonic function of the order-parameter value: $S_{\text{L}} \sim \eta^2$. Thus, it is precisely the reverse hypothesis on harmonicity of S_{L} that restricts the set of temperature-dependent coefficients, in the Landau potential F_{L}, to a single one $a_1(T)$.

One has, therefore:

$$L = T_1 \cdot \Delta S_{\text{I-II}}(T_1) = a_{10} \cdot T_1 \cdot \eta^2(T_1) = T_1 \cdot \frac{a_{10}a_2}{6a_3}. \qquad (1.53)$$

In the same way, a discontinuous variation should be found at T_1 for all the physical quantities proportional to the first derivatives of the thermodynamic potential, e.g., the volume, the lattice constants, the induced strains or polarization, etc.

(iv) By specific discontinuities at T_1 for the physical quantities related to the second derivatives of the thermodynamic potential: For example, if ξ is the field conjugated to η, defined as $\chi_\eta = \lim_{\xi_\eta \to 0}(\frac{\partial \eta}{\partial \xi_\eta})_{\eta^0}$, minimization of $F(\eta) - \eta\xi$ with respect to η and ξ provides the following form for the susceptibility relative to the order-parameter:

$$\chi_{\mathrm{I}}^{-1} = 2a_1 \text{ for } T > T_1, \quad \chi_{\mathrm{II}}^{-1} = -8(a_1 + a_2\eta^2) \text{ for } T < T_1.$$

χ_η^{-1} is represented in Fig. 1.8. It differs from the standard Curie–Weiss law (Fig. 1.4(b) and Eq. (1.9)) obtained for a second-order transition, as no divergence of χ_η occurs at T_1 but a *finite jump.*

Although at T_1 the specific heat curve $c_P(T)$

$$c_P = T\left(\frac{\partial S}{\partial T}\right)_P = -T\left(\frac{\partial^2 F_{\mathrm{eq}}}{\partial T^2}\right)_P,$$

undergoes an upward step on cooling proportional to $a_{10}^2/2a_2$ as for a second-order transition (Fig. 1.4(a) and Eq. (1.7)), the influence of higher-degree terms in the thermodynamic potential will manifest itself in the temperature dependence of $c_P(T)$ below T_1. In this region, $c_P(T)$ varies as

$$\Delta c_P(a_2 < 0) = \frac{\alpha^2}{a_2}\left(\frac{a_2^2}{4\alpha a_3} + T_{\mathrm{C}}\right), \tag{1.54}$$

and accordingly, the decrease of $c_P(T)$ below T_1 will be sharper for a first-order transition than for a second-order one as shown in Fig. 1.8. Table 1.4 and Fig. 1.8 summarize the results of the above consideration.

In the preceding discussion, only the temperature has been considered as an external variable. Assuming a dependence of the phenomenological coefficients a_1 and a_2 in Eq. (1.46) on *two* external variables (e.g., temperature and pressure) and a constant value for a_3, one gets the phase diagram in the (a_1, a_2) plane shown in Fig. 1.9. In this diagram, one can see that the line of first-order transitions

Table 1.4. Latent and specific heat for phase transitions represented in Fig. 1.8.

Transition I–II	Transition temperature T_1	Latent heat L	Specific heat jump Δc_P	Coexistence range
$a_2 > 0$ (IInd order)	T_C	0	$T_C \cdot \frac{a_{10}^2}{a_2}$	0
$a_2 < 0$ (Ist order)	$T_C + \dfrac{a_2^2}{4a_{10}a_3}$	$-T_1 \cdot \dfrac{a_{10}a_2}{6a_3}$	$\dfrac{a_{10}^2}{a_2}\left(\dfrac{a_2^2}{4a_{10}a_3} + T_C\right)$	$a_1 = \dfrac{a_2^2}{3a_3}$

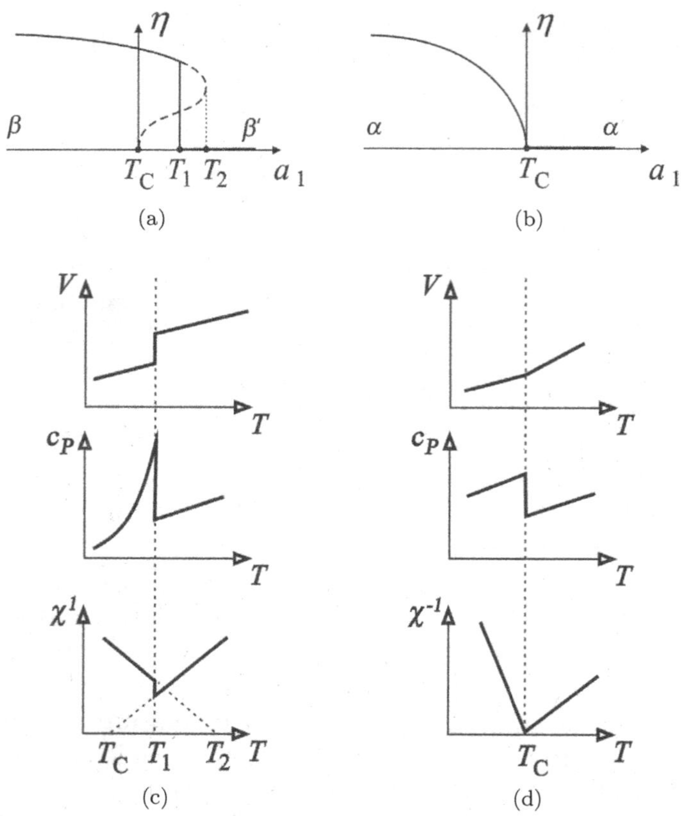

Fig. 1.8. Temperature dependence of the order-parameter η, volume V, the specific heat c_P, and susceptibility χ^{-1} at $a_2 < 0$ (a, c) and $a_2 > 0$ (b, d) corresponding to the potential (1.46).

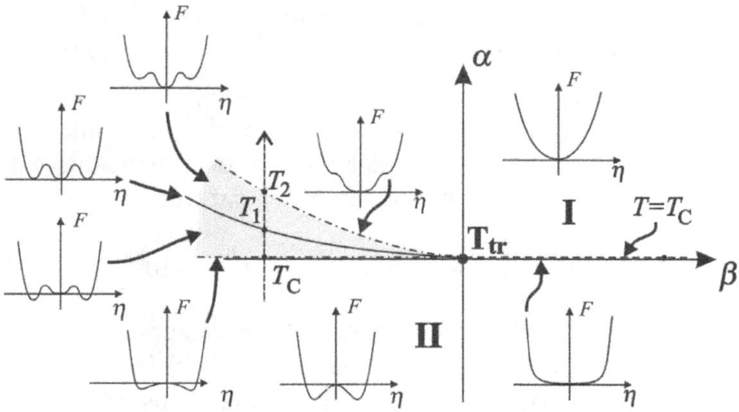

Fig. 1.9. Phase diagram corresponding to the potential (1.46). The tricritical point T_{tr} separates the second-order transition line $T = T_C$ from the first-order transition line $T = T_1$. The curves $(F - F_0)(\eta)$ are shown in some significant regions of the phase diagram. Shadowed is the coexistence region.

defined by

$$a_1 = \frac{a_2^2}{4a_3} \quad (a_2 < 0), \tag{1.55}$$

goes over the second-order transition line

$$a_1 = 0 \quad (a_2 > 0), \tag{1.56}$$

at the tricritical point T_{tr} located at the origin ($a_1 = a_2 = 0$). Close to this point, for $a_2 < 0$, a_2 can be neglected in Eq. (1.47), and one has the approximate equilibrium value for η:

$$\eta^{eq} = \left(\frac{a_1^2}{3a_3}\right)^{1/4}. \tag{1.57}$$

For $a_2 > 0$, close to the tricritical point, one can assume a power law $a_2 = a_{10} \cdot A^u$, where A is a positive constant, and $0 \leq u \leq 1/2$. Thus, the order-parameter close to the second-order transition line $T = T_C$ has the form

$$\eta^{eq} \cong \left(-\frac{a_1}{2a_2}\right)^{\frac{1}{2}} \cong \left(-\frac{a_1}{A}\right)^{\frac{1}{2}-u}. \tag{1.58}$$

When approaching the tricritical point, the critical exponent associated with the order-parameter varies in the interval $\left[\frac{1}{4}, \frac{1}{2}\right]$.

(b) General features of the phase diagrams associated with one-component order-parameter expansions with the image-group $\bar{1}$

The considerations developed for the sixth-degree potential (1.46), and previously for the fourth-degree potential (1.2), can be generalized to any one-component order-parameter expansion of higher degree containing even powers of the order-parameter η:

$$F(T,P,\eta) = F_0(T,P) + a_1\eta^2 + a_2\eta^4 + a_3\eta^6 + a_4\eta^8 + \cdots . \quad (1.59)$$

As indicated in Sec. 1.3.2.2, such potentials correspond to the image-group $\bar{1}$ with only one basic polynomial ($I = \eta^2$) in their integrity basis. The equation of state and stability conditions are

$$\frac{\partial F}{\partial \eta} = 2\eta(a_1 + 2a_2\eta^2 + 3a_3\eta^4 + 4a_4\eta^6 + \cdots) = 0, \quad (1.60)$$

$$\frac{\partial^2 F}{\partial \eta^2} = 2(a_1 + 2a_2\eta^2 + 3a_3\eta^4 + 4a_4\eta^8 + \cdots) + 4\eta^2(2a_2 + 6a_3\eta^2 + 12a_4\eta^4 + \cdots). \quad (1.61)$$

If $2m$ is the highest degree assumed for Eq. (1.59), i.e., if m is maximal number of non-zero coefficients a_i, Eq. (1.60) possesses at the utmost $(m-1)$ positive roots for the variable η^2, which are associated with $(m-1)$ stable low-temperature phase *having identical symmetries.* Hence, if m is even, the phase diagram will display a maximal number of $m/2$ successive isostructural transitions, whereas the maximal number of such transitions is $(m-1)/2$ for m odd. In Fig. 1.10, one can verify that for m odd *all the transitions are first-order*, while for m even, *an initial second-order transition is followed by* $(m/2-1)$ *first-order transitions.* We will now discuss in more detail the phase diagrams and physical anomalies corresponding to the potentials with $2m = 8$ and $2m = 10$, which have been studied in Refs. [11, 20, 21].

(c) Phase diagram for an eighth-degree expansion ($2m = 8$)

As one can see in Fig. 1.10, the eighth-degree expansion is the lowest degree which allows to describe *two* low-symmetry phases. These phases have identical symmetries but differ by the magnitude of the order-parameter η. The properties of the phase diagram, which

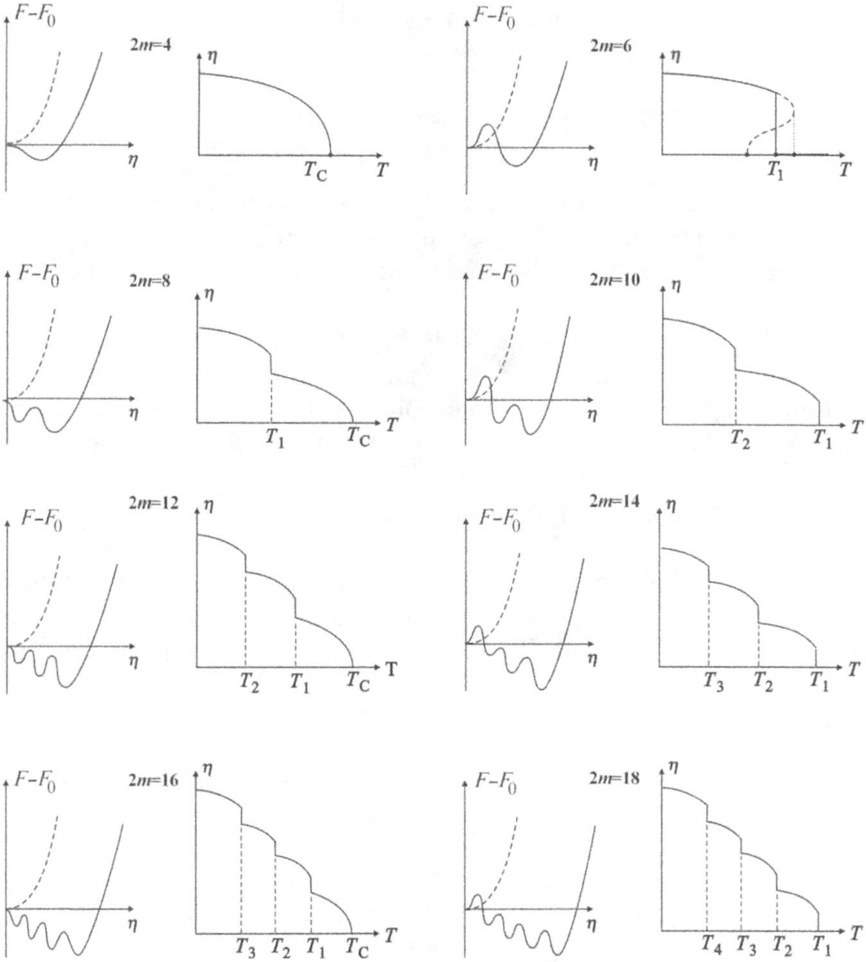

Fig. 1.10. Order-parameter dependence of the non-equilibrium thermodynamic potential $F(\eta)-F_0$, and temperature dependence of the order-parameter $\eta(T)$, for different maximal degrees $2m$ of the order-parameter expansion. Figures on the left-hand side correspond to m even, whereas the right-hand side figures show cases with m odd.

include a second-order transition line followed by a first-order transition line (or a single first-order transition line), can be worked out from the equation of state:

$$\eta \cdot (a_1 + 2a_2\eta^2 + 3a_3\eta^4 + 4a_4\eta^6) = 0, \qquad (1.62)$$

and from the limit of stability lines, which are $a_1 = 0$ for the parent phase I, obtained for $\eta = 0$, and:

$$\eta^2 \cdot (2a_2 + 6a_3\eta^2 + 12a_4\eta^4) = 0, \tag{1.63}$$

for the low-temperature phases denoted II and III. The detailed mathematical procedure for solving Eqs. (1.62) and (1.63) can be found in Ref. [11]. We will restrict ourselves, hereafter, to show its essential steps.

In addition to the solution $\eta = 0$, Eq. (1.62) possesses either one or three positive solutions for η^2 depending on the sign of a_3. As shown in Fig. 1.11(a) in the three-dimensional coefficient space (a_1, a_2, a_3), one has for $a_3 > 0$ only one low-temperature stable phase (as in the case $2m = 6$), while for $a_3 < 0$ and $a_4 > 0$, one finds two distinct regions of stability for the low-temperature phase. Assuming these latter conditions for the signs of a_3 and a_4, one can represent the complete set of phases for $2m = 8$, using as variables the coefficients a_1 and a_2, which are supposed to vary linearly as functions of external parameters (e.g., T and P). Figure 1.11(b) represents the lines of transitions bounding the various phases in the plane $(\tilde{a}_1, \tilde{a}_2)$, where $\tilde{a}_1 = a_1 - a_1^K$, $\tilde{a}_2 = a_2 - a_2^K$:

$$\tilde{a}_1 = a_1 - \frac{a_3^3}{16a_4^2}, \quad \tilde{a}_2 = a_1 - \frac{3a_3^2}{8a_4}.$$

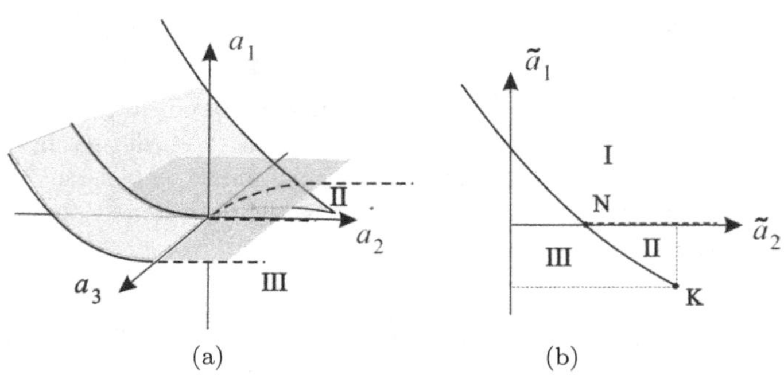

Fig. 1.11. Topology of the phase diagrams corresponding to an eighth-degree potential (1.59): (a) in the space (a_1, a_2, a_3), (b) in the $(\tilde{a}_1, \tilde{a}_2)$ plane for $a_3 > 0$. Full and dashed lines in (b) represent the first- and second-order transition lines, respectively.

Thus, the isostructural transition line II–III is a straight-line (N–K) segment of equation:

$$\tilde{a}_1 = \frac{a_3}{2a_4}\tilde{a}_2 \quad (a_3 < 0), \tag{1.64}$$

at which the jump of the order-parameter, given by $F(\eta_{\mathrm{II}}) = F(\eta_{\mathrm{III}})$, is evaluated by

$$[\eta^2_{\mathrm{III}} - \eta^2_{\mathrm{II}}]_{\mathrm{II-III}} = -\frac{a_3}{2a_4}. \tag{1.65}$$

The first-order transition segment (1.64) starts at a three-phase point N of coordinates $a_1^N = 0, a^N{}_2 = \frac{a_3^2}{4a_4}$, where two other transition lines converge: the second-order transition line I–II $\tilde{a}_1 = 0$ $(a_2^N > \frac{a_3^2}{2a_4})$ and the line of first-order transitions I–III, of equation:

$$a_1^E = \frac{a_3}{a_4}a_2^E + \left(-\frac{2^2}{3^3}\frac{(a_2^E)^2}{a_4}\right)^{1/2}, \tag{1.66}$$

where $a_1^E = a_1 - \frac{a_3}{3^3 a_4^2}$, $a_2^E = a_2 - \frac{a_3^2}{3a_4}$. The line (1.66) is tangent to the II–III line at the N-point. The order-parameter jump on the I–III line defined by $F(\eta_{\mathrm{II}}) = F(\eta_{\mathrm{III}}) = 0$ corresponds to

$$\left(\eta^2_{\mathrm{III}}\right)_{\mathrm{I-III}} = \frac{1}{3a_4}\left(-a_3 + \sqrt{a_3^2 - 3a_2a_4}\right). \tag{1.67}$$

At the N-point, the order-parameter discontinuities (1.65) and (1.67) coincide, and one has

$$\left(\eta^N_{\mathrm{III}}\right)^2 - \left(\eta^N_{\mathrm{II}}\right)^2 = -\frac{a_3}{2a_4}(a_3 < 0). \tag{1.68}$$

The isostructural transition segment II–III ends at a critical point K of coordinates:

$$a_1^K = \frac{a_3^3}{27a_4^2}, \quad a_2^K = \frac{a_3^2}{3a_4}, \tag{1.69}$$

at which the values of the order-parameters in phases II and III coincide:

$$\left(\eta^K_{\mathrm{III}}\right)^2 = \left(\eta^K_{\mathrm{II}}\right)^2 = -\frac{a_3}{3a_4}. \tag{1.70}$$

It can be noted that when going on a thermodynamic path between the I–II and II–III transition lines, close to the critical point K, the temperature dependence of the order-parameter varies as $(T_C - T)^{\beta}$, where β goes from $\beta = 1/2$ to $\beta = 1/3$ close to the K-point. Figures 1.12(a) and 1.12(b) summarize the form of the non-equilibrium potential (1.59), up to the eighth degree, as a function of the order-parameter in different regions of the $(\tilde{a}_1, \tilde{a}_2)$ plane (Fig. 1.12(a)) and along the lines of equal energies (Fig. 1.12(b)).

Using Eqs. (1.64)–(1.70), one can determine the form of the physical anomalies which take place at the various transitions shown in Figs. 1.11 and 1.12. For the second-order I–II transition and for the first-order I–III transition, the anomalies have the same shape as represented in Figs. 1.5 and 1.8, respectively. The region of the phase diagram corresponding to $\frac{a_3^2}{4a_4} \leq \tilde{a}_2 \leq \frac{a_3^2}{3a_4}$ (N-K projection) displays an interesting succession of physical anomalies, which are represented in Fig. 1.13 for the susceptibility χ_η. Thus, for this interval, a Curie–Weiss law (as in Fig. 1.5(b)) is followed by a first-order transition type anomaly (as in Fig. 1.8(a)), whereas at the critical point K $(\frac{a_3^2}{3a_4})$, one will observe a succession of two Curie–Weiss divergences of χ_η. In the vicinity of the K-point $(\tilde{a}_2 \geq \frac{a_3^2}{3a_4})$, the jump in $\Delta\chi$ should decrease progressively with the increasing values of $\tilde{a}_2$.

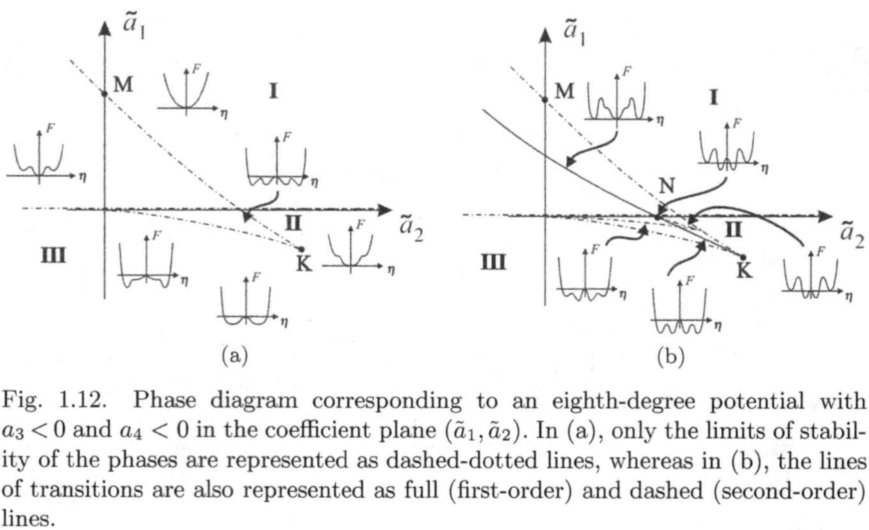

Fig. 1.12. Phase diagram corresponding to an eighth-degree potential with $a_3 < 0$ and $a_4 < 0$ in the coefficient plane $(\tilde{a}_1, \tilde{a}_2)$. In (a), only the limits of stability of the phases are represented as dashed-dotted lines, whereas in (b), the lines of transitions are also represented as full (first-order) and dashed (second-order) lines.

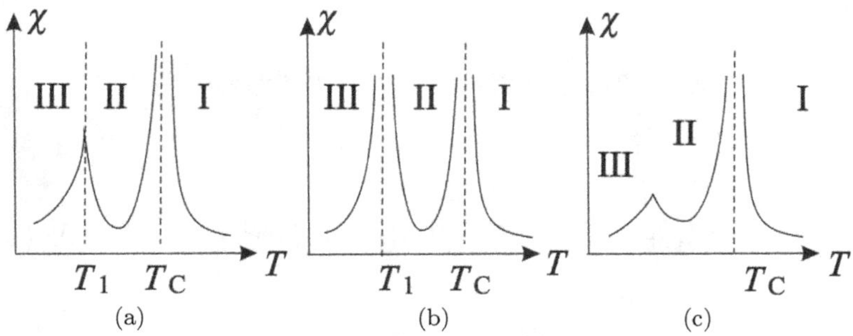

Fig. 1.13. Temperature dependence of the susceptibility χ_η associated with the order-parameter for a succession of one second-order transition followed by an isostructural transition. (a) $\frac{a_3^2}{4a_4} \leq \tilde{a}_2 \leq \frac{a_3^2}{3a_4}$, (b) $\tilde{a}_2 = \frac{a_3^2}{3a_4}$ (at the critical point K), (c) $\tilde{a}_2 > \frac{a_3^2}{3a_4}$ in the vicinity of the K-point.

1.3.3.3. *Phase diagrams for two-component order-parameter expansions*

As mentioned in Sec. 1.3.2.3, the image-groups associated with the two-component *irreducible* order-parameters are nm or n with $n = 3, 4, 6, 8$ and 12, and their integrity bases are given by Equations (1.33) and (1.34) respectively. The phase diagrams corresponding to the groups generated by reflections $3m$, $4mm$, $6mm$, and to the cyclic groups 3, 4, have been discussed first in Refs. [10, 21, 22], and later in a more general way in the unpublished dissertations of E. Kut'in [23] and V. Lorman [24]. In the present section, we will successively describe the phase diagrams related to the image-groups $4mm$ and 4 before summarizing the general results which hold for any nmm and n (see Sec. 2.2.1.3).

(a) Phase diagrams associated with the image-groups $4mm$ *and* 4

The basic invariants forming the integrity basis for the image-group $4mm$ are

$$I_1 = \eta_1^2 + \eta_2^2, \quad I_2 = \eta_1^2\eta_2^2. \tag{1.71}$$

Accordingly, the order-parameter expansion invariant by the $4mm$ group and restricted to the fourth degree contains the monomials $I_1, I_1I^2{}_1$ and I_2. One can use equivalently the order-parameter expansion

$$F(\eta_1, \eta_2) = a_1(\eta_1^2 + \eta_2^2) + b_1(\eta_1^4 + \eta_2^4) + b_2\eta_1^2\eta_2^2. \quad (1.72)$$

Since $\eta_1^4 + \eta_2^4 = I_1^2 - 2I_2$ is also invariant by $4mm$, the equations minimizing (1.72) are

$$2\eta_1(a_1 + 2b_1{\eta^2}_1 + b_2{\eta^2}_2) = 0, \quad 2\eta_2(a_1 + 2b_1\eta_2^2 + b_2{\eta^2}_1) = 0, \quad (1.73)$$

and

$$\frac{\partial^2 F}{\partial \eta_1^2} \geq 0, \quad \frac{\partial^2 F}{\partial \eta_1^2} \cdot \frac{\partial^2 F}{\partial \eta_2^2} - \left(\frac{\partial^2 F}{\partial \eta_1 \partial \eta_2}\right)^2 \geq 0, \quad (1.74)$$

with

$$\frac{\partial^2 F}{\partial \eta_1^2} = 2\left(a_1 + 2b_1\eta_1^2 + b_2\eta_2^2\right) + 8b_1\eta_1^2,$$

$$\frac{\partial^2 F}{\partial \eta_2^2} = 2\left(a_1 + 2b_1\eta_2^2 + b_2\eta_1^2\right) + 8b_1\eta_2^2,$$

$$\frac{\partial^2 F}{\partial \eta_1 \partial \eta_2} = 4b_2\eta_1\eta_2. \quad (1.75)$$

Simple calculations show that in addition to the trivial solution $(I)\,\eta_1^e = \eta_2^e = 0$, corresponding to the parent phase, which is stable for $a_1 > 0$, one has eight solutions for Eqs. (1.73) and (1.74), which are as follows:

$$(\mathrm{II}_1, \mathrm{II}_2) \quad \text{for } \eta_1^e = 0, \quad \eta_2^e = \pm\sqrt{-a_1/2b_1};$$

$$(\mathrm{II}_3, \mathrm{II}_4) \quad \text{for } {\eta^e}_2 = 0, \quad \eta_1^e = \pm\sqrt{-a_1/2b_1};$$

$$(\mathrm{III}_1, \mathrm{III}_2, \mathrm{III}_3, \mathrm{III}_4) \quad \text{for } \eta_1^e = \pm{\eta^e}_2 = \pm\sqrt{-a_1/(2b_1 + b_2)}.$$

However, as all the solutions of the type $\mathrm{II}_i\,(i = 1\text{–}4)$ coincide with the same stability conditions:

$$a_1 < 0, b_1 > 0, 2b_1 - b_2 < 0, \quad (1.76)$$

whereas the solutions $\mathrm{III}_i\,(i = 1\text{–}4)$ are all stable for

$$a_1 < 0, b_1 > 0, 2b_1 - b_2 > 0, 2b_1 + b_2 > 0, \quad (1.77)$$

it is obvious that the four states II_i and III_i constitute, respectively, four different domains of the same states (denoted hereafter

II and III). This can also be foreseen from the fact that Eqs. (1.73) and (1.74) are invariant by the permutations $(\pm\eta_1) \leftrightarrow (\pm\eta_2)$.

Using Eqs. (1.76) and (1.77), one can construct the phase diagram corresponding to expansion (1.72) in which the conditions $b_1 > 0$ and $(4b_1^2 - b_2^2) > 0$ (which are necessary for the stabilization of the phases II and III) are assumed. Figures 1.14(a) and 1.14(b) show the phase diagrams in the three-dimensional space (a_1, b_1, b_2) and in the plane (a_1, b_1) with $b_1 = \text{const} \neq 0$. One can see in these phase diagrams that the transition I–II, taking place for $a_1 = 0$, is always second-order, while the transition II–III, which occurs on the line $b_2 = 2b_1$ for $a_1 > 0$, is first-order, and corresponds to a jump of the order-parameter given by

$$\eta_{\text{II}}^2 - \eta_{\text{III}}^2 = -\frac{a_1}{4b_1}. \tag{1.78}$$

This jump vanishes for $a_1 = 0$, and the three phases are contiguous on the line $a_1 = 0$, $b_2 = 2b_1$ (i.e., at the N-point: $a_1 = 0, b_1 = 0$ in Fig. 1.14(b)).

The preceding description clearly reveals the incompleteness of the model using a fourth-degree expansion (1.72), as it predicts no region of coexistence between phases II and III, since the limit of stability line ($b_2 = 2b_1$) coincides with the first-order transition line. On the other hand, no first-order transition regime is predicted from phase I to the phases II or III. Expanding $F(\eta_1, \eta_2)$ to the sixth

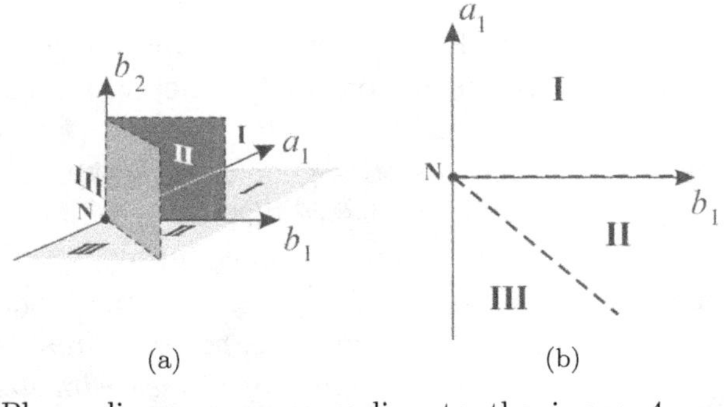

Fig. 1.14. Phase diagrams corresponding to the image $4mm$ and to the fourth-degree order-parameter expansion (1.72). (a) Phase diagram in the space (a_1, b_1, b_2). Gray vertical planes are the second-order and first-order transition surfaces. (b) Phase diagram in the plane $(a_1, b_1), b_2 > 0$. Dashed and solid lines are the second-order and first-order transition lines.

degree, with the two additional invariants

$$c_1\left(\eta_1^6+\eta_2^6\right) \quad \text{and} \quad c_2\left(\eta_1^4\eta_2^2+\eta_1^2\eta_2^4\right), \tag{1.79}$$

yields the following equations of state:

$$\begin{aligned}
2\eta_1\left(a_1+2b_1\eta_1^2+b_2\eta_2^2+3c_1\eta_1^4+2c_2\eta_1^2\eta_2^2\right)&=0,\\
2\eta_2\left(a_1+2b_1\eta_2^2+b_2\eta_1^2+3c_1\eta_2^4+2c_2\eta_1^2\eta_2^2\right)&=0,
\end{aligned} \tag{1.80}$$

which possess, in addition to the solutions of Eq. (1.73), the additional solutions $\eta_1^e \neq 0$, $\eta_2^e \neq 0$, $\eta_1^e \neq \eta_2^e$, with:

$$(\eta_1^e)^2+(\eta_2^e)^2=-\frac{2b_1-b_2}{3c_1},$$

$$(\eta_1^e)^2-(\eta_2^e)^2=\pm\sqrt{\frac{1}{12c_1}\cdot\left[\frac{(2b_1-b_2)^2}{3c_1}-4\frac{3a_1 3c_1-b_2(2b_1-b_2)}{3c_1-2c_2}\right]}.$$

However, the corresponding state (phase IV) is unstable, since it coincides with a *maximum* of the sixth-degree expansion. Phase IV possesses a domain of stability in the phase diagram only if the additional eighth-degree invariants are considered in the order-parameter expansion, namely:

$$d_1(\eta_1^8+\eta_2^8), \quad d_2(\eta_1^6\eta_2^2+\eta_1^2\eta_2^6), \quad d_3\eta_1^4\eta_2^4. \tag{1.81}$$

Accordingly, the lowest degree at which $F(\eta_1,\eta_2)$ should be truncated in order to describe the set of low-symmetry phases II, III and IV is eight.

Since phase IV corresponds to a general direction in the order-parameter space ε_2, no other phase of distinct symmetry (no other type of orbit) can be stabilized for the image-group $4mm$. Thus, the preceding considerations illustrate the general property that *the maximal degree at which the order-parameter expansion must be truncated in order to obtain the maximal number of phases with distinct symmetries is 2N, where N is the highest degree of the invariants forming the integrity basis* (i.e., $N = 4$ for the image). Lower-degree expansions *will not* provide a consistent ("*structurally stable*") model (see Ref. [14]).

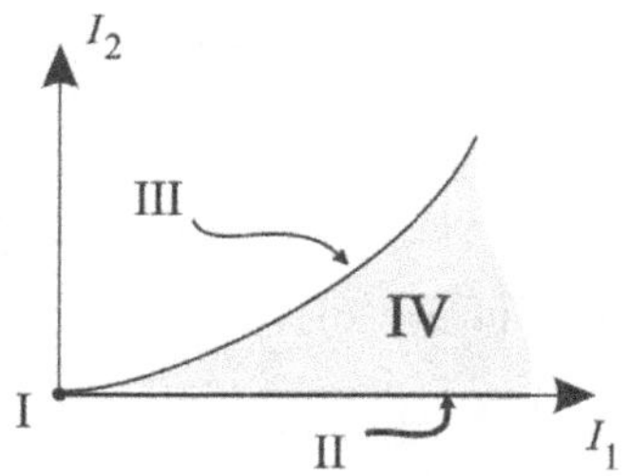

Fig. 1.15. Phase diagram in the orbit space ("manifold of orbits"), corresponding to the image-group $4mm$. Stability domain of phase IV ($\eta_1 \neq \eta_2 \neq 0$), is the portion of the plane (I_1, I_2) located between phases II and III.

A simple way to verify the structural stability of a model is to construct the phase diagram in the *orbit space*, which is the space of the integrity basis. Figure 1.15 shows such a phase diagram for the image-group $4mm$. The diagram has been constructed using expressions (1.71) of I_1 and I_2 and taking into account the equilibrium relationship between the order-parameter components in each phase. Thus, one can see in Fig. 1.15 that phase I coincides with the origin of the orbit space (I_1, I_2), since $(\eta_1 = \eta_2 = 0)$ yields $I_1 = I_2 = 0$. Phase II $(\eta_1 \neq 0, \eta_2 = 0)$ corresponds to one of the axes $(I_2 = 0, I_1 \neq 0)$, whereas phase III $(\eta_1 = \eta_2 \neq 0)$ is the parabolic branch $I_2 = \frac{1}{4}I_1^2 = \eta^4$. The region of stability of phase IV $(\eta_1 \neq \eta_2 \neq 0)$ is the portion of the plane (I_1, I_2) located between phases II and III.

Phase diagrams in orbit spaces reveal the full set of possibly stable states for the corresponding image-groups, but also provide indications on the topological properties of the phases in the real (e.g., temperature–pressure) phase diagram. From Fig. 1.15, one can deduce for example that phase IV will always be located *between* phases II and III, or that phase IV can touch phase I only at an isolated point of the phase diagram (at least when the second-order transitions are assumed).

In order to work out the phase diagram associated with an eighth-degree expansion $F(\eta_1, \eta_2)$, it is convenient to use the polar coordinates $(\eta_1 = \rho \cdot \cos\theta, \eta_2 = \rho \cdot \sin\theta)$ in the ε_2-space and the invariant basis

$$I_1 = \eta_1^2 + \eta_2^2 = \rho^2, \quad I_2' = I_1^2 - 4I_2 = \rho^4 \cos 4\theta. \tag{1.82}$$

The order-parameter expansion $(F(\eta_1, \eta_2) = F'(\rho, \theta) = F''(I_1 I'_2))$ has the form

$$F''(I_1 I'_2) = a_1 I_1 + a_2 I_1^2 + a_3 I_1^3 + a_4 I_1^4 + b_1 I'_2 + b_2 I_2'^2 + c_{12} I_1 I'_2 + c_{112} I_1^2 I'_2. \tag{1.83}$$

Taking $\Phi_1 = \frac{\partial F''}{\partial I_1}, \Phi_2 = \frac{\partial F''}{\partial I'_2}$, the equations of state are

$$\frac{\partial F'}{\partial \rho} = 2\rho(\Phi_1 + 2\rho^2 \Phi_2 \cos 4\theta) = 0,$$
$$\frac{\partial F'}{\partial \theta} = -4\rho^4 \Phi_2 \sin 4\theta = 0. \tag{1.84}$$

The equilibrium values of the order-parameter components are as follows:

$$\text{for phases II and III:} \quad \cos 4\theta = \pm 1 \text{ and } \Phi_1 \pm 2\rho^2 \Phi_2 = 0, \tag{1.85}$$

$$\text{and for phase IV:} \quad \Phi_1 = 0, \Phi_2 = 0. \tag{1.86}$$

Equation (1.85) has the explicit form

$$a_1 + 2(a_2 + b_1)\rho^2 + 3(a_3 \pm c_{12})\rho^4 + 4(a_4 + b_2 \pm c_{112})\rho^6 = 0. \tag{1.87}$$

The stability condition $\frac{\partial^2 F}{\partial \rho^2} \geq 0$ is

$$a_2 \pm b_1 + 3(a_3 \pm c_{12})\rho^2 + 6(a_4 + b_2 \pm c_{112})\rho^4 \geq 0. \tag{1.88}$$

Upto normalization of the coefficients, Eqs. (1.87) and (1.88), which account for the stabilization of phases II and III, are identical to the equations solved in Sec. 1.3.3.2 for an eighth-degree potential with a one-component order-parameter. Using the corresponding results, one obtains the phase diagrams shown in Fig. 1.16, for which the conditions $a_3 = c_{112} = 0$ and $a_2 > 0$ have been assumed. Figure 1.16(a) corresponds to the case $(c_{12}^2/4b_2) < a_2 < (c_{12}^2/32b_2)$ and $a_4 < 8b_2$. One can see that the transitions I–II and I–III can be either the second- or first-order. The first-order transition line I–II becomes a second-order transition line at the tricritical point T_{tc}, whereas the first-order transition line I–III, which separates the two isostructural phases III^+ and III^- for $a_1 < 0$, ends at a critical point K of the liquid-vapor type. Phase IV merges with phases I, II and III at a

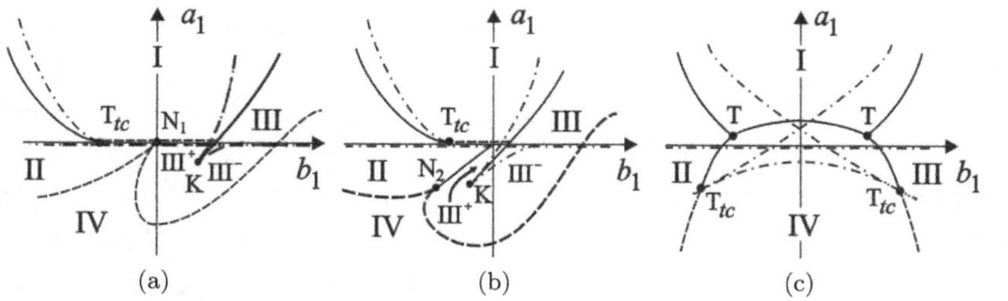

Fig. 1.16. Phase diagrams corresponding to the image $4mm$ and to the eighth-degree expansion (1.83) for $a_3 = c_{112} = 0$. The diagrams are obtained for the following conditions: (a) $a_4 < 8b_2$, $c_{12} > 0$, $(c_{12}^2/4b_2) < a_2(c_{12}^2/32b_2)$; (b) $0 < a_2 < (c^2{}_{12}/4b_2)$, $c_{12} > 0$; (c) $a_2 < 0$, $4a_2b_2 < c_{12}^2$. Dashed, full and dashed-dotted lines are, respectively, the second-order, first-order and limit of stability lines.

four-phase point N_1, and is separated from phases II and III by the second-order transition lines. Let us note that for large values of b_1 in Fig. 1.16(a), the transition line II–III crosses the b_1 axis, i.e., the second-order transition II–III takes place within the metastability region of phase I.

Figure 1.16(b) represents the phase diagram corresponding to $0 < a_2 < (c^2{}_{12}/4b_2)$. In this case, phase IV cannot be reached directly from phase I, and phases II, III and IV merge at a three-phase point N_2. The II–III transition may occur directly across a first-order transition line N–0.

In the preceding discussion, the condition $c_{12} > 0$ has been assumed. For $c_{12} < 0$, the left- and right-hand sides of the phase diagrams of Figs. 1.16(a) and 1.16(b) are exchanged symmetrically, i.e., the critical point K occurs within phase II, and the transition lines II–IV and III–IV meet in the $a_1 < 0$, $b_1 > 0$ portion of the phase diagram. On the other hand, for $a_2 > 0$ and $4a_2b_2 - c_{12}^2 < 0$, one has the phase diagram represented in Fig. 1.16(c) in which phase IV can be reached directly across a line of first-order transitions, and the II–IV and III–IV transition lines can be either the second- or first-order, the two regimes being separated by tricritical points. The T-points are triple points.

For the image-group 4, the order-parameter expansion depends on I_1, I_2' and on the additional invariant $I_3 = \rho^4 \sin 4\theta$. The equations of state, which correspond to the potential $F(I_1, I'_2, I_3)$, become

$$\rho[\Phi_1 + 4\rho^2(\Phi_2 \cos 4\theta + \Phi_3 \sin 4\theta)] = 0,$$

$$4\rho^4(-\Phi_2 \sin 4\theta + \Phi_3 \cos 4\theta) = 0, \qquad (1.89)$$

with $\Phi_i = \frac{\partial F}{\partial I_i}$ (i = 1–3). In addition to the trivial solution $\rho = 0$ (the parent phase I), there exists only one possible stable solution with $\rho \neq 0$ to the system of equations (1.89), namely $\eta_1 \neq 0$, $\eta_2 \neq 0$, $\eta_1 \neq \eta_2$ (phase II).

Between the invariants forming the integrity basis, one has the relationship $I^4{}_1 = I_2'^2 + I^2{}_3$, which determines a circle of radius $I^2{}_1$ in the plane $(I_2'^2 I_3)$. Accordingly, in the orbit-space, phase II is represented by a cone-type surface, shown in Fig. 1.17(a), the apex of which coincides with phase I.

As stressed for the $4mm$ image-group, a consistent model requires to consider at least an eighth-degree expansion. However, due to the complexity of the complete model, which contains not less than 13 invariants, only the simplified models have been solved till now.

A consistent model was studied in Ref. [14]. It contains the eighth-degree invariant $b_2 I'^2{}_2$:

$$F(I_1, I_2', I_3) = a_1 I_1 + a_2 I^2{}_1 + b_1 I'_2 + c_1 I_3 + b_2 I_2'^2. \tag{1.90}$$

The model corresponds to the phase diagram represented in Fig. 1.17(b) in the space of the phenomenological coefficients (a_1, b_1, c_1). One can see in this figure that a second-order transition I–II is possible across a disk-like surface, located in the plane $a_1 = 0$, which closes a curved surface of the first-order transitions between

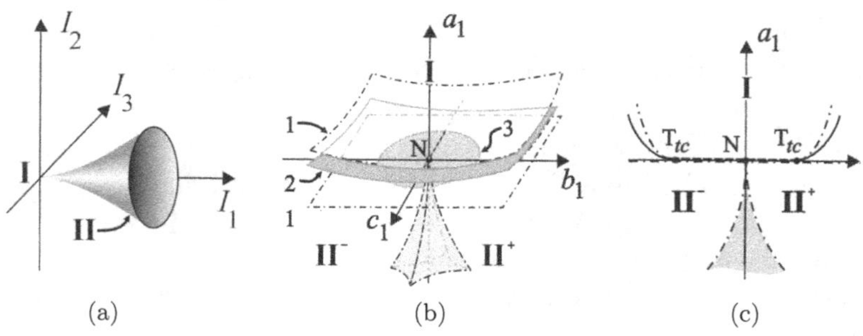

Fig. 1.17. Phase diagrams corresponding to the image group 4. (a) Location of the phases in the orbit space. (b) Three-dimensional phase diagram corresponding to an order-parameter expansion with fourth-degree invariants, and a single ($I_2'^2$) eighth-degree invariant. 1 and 2 denote the limit of stability and first-order transition surfaces. The hatched disc 3 is a second-order transition surface. (c) Intersection of the phase diagram (b) by a plane parallel to the axes a_1, b_1. Dashed-dotted, full and dashed lines are, respectively, the limit of stability, first-order and second-order transition lines. N is a three-phase point.

phases I and II. For $a_1 < 0$, one finds a region of metastability into which the two isostructural phases II^+ and II^- coexist, separated by the plane joining the two opposite edges of the pyramid. These edges are the lines of critical points of the liquid-vapor type. A simpler view of the phase diagram is given in Fig. 1.17(c), which represents a cross-section of Fig. 1.17(b) parallel to the (a_1, b_1) plane.

1.3.3.4. *Multiphase points on phase diagram: Gibbs's rule*

Preparing our further analysis of phase diagrams, let us analyze the restrictions stemming from Gibbs's phase rule relating to multiphase points. Gibbs's phase rule provides, in general, the number of degrees of freedom (control variables: pressure, temperature, concentration, etc.) for a system, which can be arbitrarily varied without destroying the equilibrium. The Gibbs rule forbids, in particular, the coexistence of more than three phases of a given compound at one (P, T) point.

The phase rule stems, however, from a pure mathematical restriction when one considers a system consisting of different π phases in contact. In general, each phase contains all the substances in which all the phases are independent: no structure links are considered. The number of independent components in the system is C. Then each phase is described by its pressure, temperature and chemical potential μ. In the general case, if P and T are the common pressure and temperature, the phase equilibrium condition will be that all chemical potentials are equal to each other:

$$\mu_i^I = \mu_i^{II} = \cdots = \mu_i^{\pi}(i = 1 - C). \tag{1.91}$$

Each of these potentials ${\mu^{\pi}}_i$ is a function of $(C + 1)$ independent variable: P, T and $(C - 1)$ concentrations. Hence, the number of unknowns is $2 + \pi(C - 1)$. To have solutions, the number of equations must not be greater than the number of unknowns, i.e., $C(\pi - 1) \leq 2 + \pi(C - 1)$ or $\pi \leq C + 2$, which is called the *phase rule*. The number of variables which can be arbitrarily varied is called the number of *thermodynamic degrees of freedom* (f) of the system. The phase rule may be written as

$$f = C + 2 - \pi. \tag{1.92}$$

Thus, when a one-component system $(C = 1)$ is considered, the number of *independent* phases, π, that exist in contact on the phase diagram cannot exceed three.

However, Gibbs's phase rule still holds when all phases merging at the triple point are assumed to be *independent*, i.e., *no parent phase* and *no order-parameters* are defined. The phenomenological approach increases the number of equations by adding the equations of state minimizing free energy (for instance, Eq. (1.94)). This increases, in turn, the number of variables that can be arbitrarily varied, and allows, therefore, the existence of N-phase point ($N > 3$) on phase diagrams.

1.3.3.5. *Phase diagrams for three-component order-parameter expansions: The image-group $m\bar{3}m$*

As indicated in Sec. 1.3.2.2, there are five image-groups corresponding to the structural phase transitions associated with an irreducible three-component order-parameter. These are the cubic point groups of which the vector representation is irreducible, namely $m\bar{3}m$, $m\bar{3}$, 432, $\bar{4}3m$ and 23. Their integrity bases have been listed in Table 1.2. The discussion of the corresponding phase diagrams can be found in a number of works by Gufan and his coworkers [11, 14, 23–26]. We will summarize the main results obtained in these studies starting from the image-group $m\bar{3}m$.

The integrity basis associated with the image-group $m\bar{3}m$ is (see Table 1.2)

$$I_1 = \eta_1^2 + \eta_2^2 + \eta_3^2, \quad I_2 = \eta_1^4 + \eta_2^4 + \eta_3^4, \quad I_3 = \eta_1^2\eta_2^2\eta_3^2. \qquad (1.93)$$

Since $m\bar{3}m$ is a group generated by reflections, it has the property that the dimension m of its basis of integrity is equal to the dimension n of the order-parameter space ε_2. As indicated in the preceding section, one can expect to have $2^n = 8$ distinct phases, differing by their symmetries, in the phase diagram. Using the notations $\Phi_i = \frac{\partial F'}{\partial I_i}$, $\Phi_{ij} = \frac{\partial^2 F'}{\partial I_i \partial I_j}$ $(i, j = 1\text{–}3)$, where $F'(I_1, I_2, I_3) = F(\eta_1, \eta_2, \eta_3)$, one can write the equations of state in the form

$$\frac{\partial F}{\partial \eta_j} = 2\eta_j(\Phi_1 + 2\eta^2{}_i\Phi_2 + \eta^2{}_j\eta_k^2\Phi_3)(i, j, k = 1-3; j \neq k \neq i). \quad (1.94)$$

The stability conditions can be calculated using the second derivatives as follows:

$$\frac{\partial^2 F}{\partial \eta_j^2} = 2\Phi_1 + 12{\eta^2}_i\Phi_2 + 2{\eta^2}_j{\eta^2}_k\Phi_3 + 2\eta_i\Phi_{11} + 4{\eta^3}_i\Phi_{21} \quad (1.95)$$

$$+ 2\eta_i{\eta^2}_j{\eta^2}_k\Phi_{31},$$

and

$$\frac{\partial^2 F}{\partial \eta_i \partial \eta_j} = 2\eta_i\Phi_{ij} + 4{\eta^3}_i\Phi_{jk} + 4\eta_i\eta_j{\eta^2}_k\Phi_{jk}$$

$$(i, j, k = 1 - 3; j \neq k \neq i). \quad (1.96)$$

Table 1.5 summarizes the equilibrium relations characterizing the eight stable phases which may be found in the phase diagram of the $m\overline{3}m$-image group, as deduced from Eqs. (1.94)–(1.96). In particular, column (f) of Table 1.5 allows to represent the $m\overline{3}m$-phase diagram in the orbit space (I_1, I_2, I_3), which is shown in Fig. 1.18.

Since the maximal degree for the invariants I_i is six (I_3), a structurally stable model describing the full set of phases listed in Table 1.5 requires to consider an expansion $F(\eta_1, \eta_2, \eta_3)$ of not less than the 12th degree in the η_i. Let us first consider the sixth-degree expansion:

$$F'(I_1, I_2, I_3) = a_1I_1 + a_2I_1^2 + a_3I_1^3 + b_1I_2 + d_1I_3 + c_{12}I_1I_2, \quad (1.97)$$

which has been first discussed by Devonshire [27] in order to describe the structural transitions in $BaTiO_3$. In this case, it is easy to deduce from the stability conditions given in column (e) that only the low-symmetry phases II, III and IV are stable (for example, the condition of stability of phase VI becomes $-\Phi_{12}^2 > 0$, and for phase V, one gets the contradictory constraints $\eta_1^2 > \eta_3^2$, $\eta_1^2 < \eta_3^2$, $\eta_1 \neq \eta_3$). On the other hand, from the equilibrium conditions listed in column (d) of Table 1.5, one can conclude that a vanishingly small solution of η may occur only along the line of equation $\Phi_1(P, T, \eta)$. For phases II and III, the stability conditions do not impose any restriction for a second-order transition along the preceding line. By contrast, for

Table 1.5. Characteristic features of the phase diagrams associated with the image-group $m\bar{3}m$. The columns give, for each stable phase, the following information: (a) labeling of the phase in reference to the text; (b) equilibrium value ("direction") of the order-parameter in the ε_3 space; (c) corresponding isotropy group I_η; (d) equation of state; (e) stability conditions; (f) specific form of the invariants I_1, I_2, I_3.

(a)	(b)	(c)	(d)	(e)	(f)
I	000	$m\bar{3}m$	$\eta_1 = \eta_2 = \eta_3 = 0$	$a_1 > 0$	$(0\ 0\ 0)$
II	$\eta 0 0$	$4mm$	$\Phi_1 + 2\eta^2\Phi_2 = 0$	$\Phi_2 \geq 0, 2\Phi_2 + \Phi_{11} \geq 0$	$(\eta^2\ 0\ 0)$
III	$\eta\eta\eta$	$3mm$	$\Phi_1 + 2\eta^2\Phi_2 + \eta^4\Phi_3 = 0$	$2\Phi_2 + \Phi_{11} \geq 0, 4\Phi_2^2 +$ $4\Phi_2\Phi_{11} - 2\eta^2\Phi_2\Phi_{11} \geq 0$	$(3\eta^2\ 3\eta^4\ \eta^6)$
IV	$\eta\eta 0$	$mm2$	$\Phi_1 + 2\eta^2\Phi_2 = 0$	$\Phi_2 \geq 0, \Phi_{11} \geq$ $0 - 2\Phi_2 + \eta^2\Phi_3 \geq 0$	$(2\eta^2\ \eta^4\ 0)$
V	$\eta_1\eta_1\eta_3$	m	$\Phi_1 + 2\eta^2\Phi_2 + \eta^4\Phi_3 =$ $0\Phi_1 + 2\eta^2\Phi_2 + \eta_1^2\eta_3^2\Phi_3 = 0$	$\Phi_{11} \geq 0, (\eta_1^2 - \eta_3^2)\Phi_3 \geq$ $0(\eta_3^2 - \eta_1^2) \geq$ $0[\Phi_3\Phi_{11} + \eta_1^2\Phi_3^2 + \eta_3^2\Phi_3\Phi_{12} +$ $4(\eta_1^2 - \eta_3^2)] \cdot (\eta_3^2 - \eta_1^2) \geq 0$	$(2\eta_1^2 + \eta_3^2,\ 2\eta_4^2 + \eta_3^2,\ \eta_1^2\eta_3^4)$
VI	$\eta_1\,\eta_2\,0$	m	$\Phi_1 + 2\eta_1^2\Phi_2 = 0\Phi_1 + 2\eta_2^2\Phi_2 = 0$	$\Phi_2 \geq 0, \Phi_{11} \geq$ $0\Phi_{11}\Phi_{22} - \Phi_{12}^2 \geq 0$	$(\eta_1^2 + \eta_2^2,\ \eta_1^4 + \eta_2^4,\ \eta_1^4\eta_2^2)$
VII	$\eta_3\,\eta_3\,\eta_1$	m	$\Phi_1 + 2\eta_3^2\Phi_2 + \Phi_1^4\Phi_3 =$ $0\Phi_1 + 2\eta_3^2\Phi_2 + \eta_1^2\eta_3^2\Phi_3 = 0$	$\Phi_{11} \geq 0, (\eta_1^2 - \eta_3^2)\Phi_3 \geq$ $0[\Phi_3\Phi_{11} + \eta_3^2\Phi_3^2 + \eta_1^2\Phi_3\Phi_{12} +$ $4(\eta_3^2 - \eta_1^2)] \cdot (\eta_3^2 - \eta_1^2) \geq 0$	$(\eta_1^2 + 2\eta_3^2,\ \eta_1^4 + 2\eta_3^4,\ \eta_1^2\eta_3^4)$
VIII	$\eta_1\,\eta_2\,\eta_3$	1	$\Phi_1 = \Phi_2 = \Phi_3 = 0$	General	$(I_1\ I_2\ I_3)$

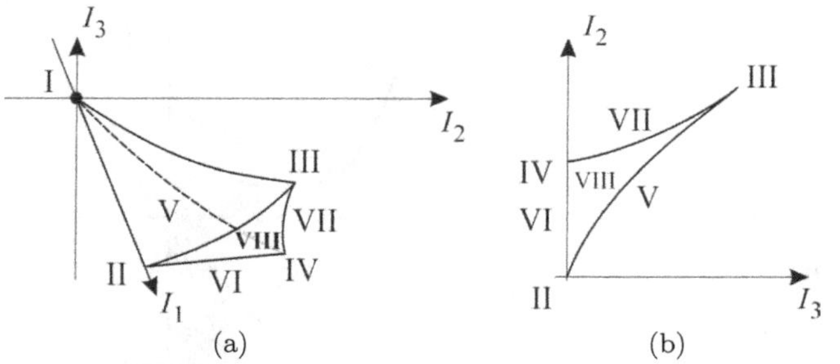

Fig. 1.18. Topology of the phases corresponding to the phase diagram associated with the image-group $m\bar{3}m$ (a) in the orbit space (I_1, I_2, I_3), (b) in a cross-section (I_2, I_3). The positions of the phases are deduced from column (f) of Table 1.5.

phase IV, the inequalities

$$\Phi_{11} \geq 0, \quad \Phi_2 \geq 0 \quad \text{and} \quad 2\Phi_2 + \eta^2\Phi_3 \geq 0,$$

are inconsistent at finite $\Phi_2(\eta = 0)$ and small η^2. It means that a second-order transition I–II may occur only at one point of the phase diagram given by $\Phi_1(P, T, \eta = 0) = \Phi_2(PT\eta = 0)$. Simple calculations [28] give the following additional features for the sixth-degree model, which are illustrated in Fig. 1.19.

(1) The second-order transition I–II occurs for $a_1 = 0$ and $b_1 \geq 0$. Phase II is stable for $a_1 < 0$ when $a_2 > 0$ and $b_1 + a_2 > 0$, or for $a_1 < -(b_1 + a_2)^2/3(a_3 + c_{12})$ when $b_1 + a_2 < 0$. The I–III transition can be either the second- or first-order, the two regimes being separated by a tricritical point.
(2) Phase IV exists only when $d_1 > 0$ and is always located between phases II and III (Fig. 1.19(a)). For $d_1 > 0$, phase IV is stable, and the phase diagram contains a direct first-order transition between phases II and III (Fig. 1.18(b)).
(3) For $d_1 > 0$, the limit of stability of phase IV with respect to phase III is given by

$$a_1 = \frac{2a_2b_1}{c_{12}}, \tag{1.98}$$

and with respect to phase II it is $a_1 = 8a_2b_1/(4c_{12} - d_1)$. On the other hand, phase II possesses a limit of stability within phase IV given by $a_1 = 12a_2b_1/(6c_{12} - d_1)$. The transitions III–IV are always first-order. This can be foreseen by the discontinuities of

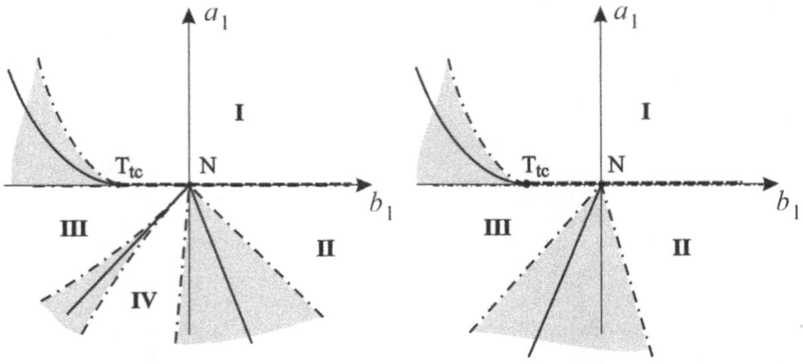

Fig. 1.19. Phase diagrams corresponding to the image-group $m\bar{3}m$ for the sixth-degree expansion (1.97) assuming $a_2 > 0$, $\Delta = 4a_2b_2 - c_{12}^2 < 0$ and (a) $d_1 > 0$, (b) $d_1 < 0$. Full, dashed and dashed-dotted lines are, respectively, the first-order, second-order and limit of stability lines. Shadowed are the coexistence regions.

the equilibrium values of the order-parameter components in the three phases. One finds, respectively:

$$\eta_{\mathrm{III}}^2 = -\frac{a_1}{2a_2} + \frac{a_1 b_1}{2a_2} - \frac{3(a_3 + c_{12})}{8a_2^3} a_1^2,$$

$$\eta_{\mathrm{II}}^2 = -\frac{a_1}{6a_2} + \frac{a_1 b_1}{18a_2} - \frac{27(a_3 + 9c_{12} + d_1)}{(6a_2)^3} a_1^2,$$

$$\eta_{\mathrm{IV}}^2 = -\frac{a_1}{2a_2} + \frac{a_1 b_1}{8a_2} - \frac{3(2a_3 + c_{12})}{32a_2^3} a_1^2. \tag{1.99}$$

Expressions (1.99) yield the equilibrium values of the potential (1.97) in each phase:

$$F(\eta_{\mathrm{III}}) = -\frac{a_1^2}{4a_2} + \frac{a_1^2 b_1}{4a_2^2} - \frac{(a_3 + c_{12})}{8a_2^3} a_1^3,$$

$$F(\eta_{\mathrm{II}}) = -\frac{a_1^2}{4a_2} + \frac{a_1^2 b_1}{12a_2^2} - \frac{(27a_3 + 9c_{12} + d_1)}{(6a_2)^3} a_1^3,$$

$$F(\eta_{\mathrm{IV}}) = -\frac{a_1^2}{4a_2} + \frac{a_1^2 b_1}{4a_2^2} - \frac{(2a_3 + c_{12})}{16a_2^3} a_1^3,$$

from which one can deduce the equations of the first-order transition lines between phases III and IV ($F_{\rm III} = F_{\rm IV}$):

$$b_1 = \frac{c_{12}}{2a_2} a_1, \tag{1.100}$$

and between phases II and IV ($F_{\rm II} = F_{\rm IV}$):

$$b_1 = \left(c_{12} - \frac{2d_1}{9} \right) \frac{a_1}{2a_2}.$$

Comparing Eqs. (1.98) and (1.100) shows that *the boundary between phases II and IV confounds with the limit of stability between the two phases*, i.e., there is no metastability region between the two phases. This result is indicative of the fact that the sixth-degree expansion (1.97) does not provide a consistent description of phase IV. One needs to consider, for this purpose, an eighth-degree expansion:

$$\begin{aligned} F'(I_1, I_2, I_3) = {} & a_1 I_1 + a_2 I_1^2 + a_3 I_1^3 + a_4 I_1^4 + b_1 I_2 + b_2 I_2^2 + d_1 I_3 \\ & + c_{12} I_1 I_2 + c_{112} I_1^2 I_2 + c_{13} I_1 I_3, \end{aligned} \tag{1.101}$$

which also describes the three additional phases listed in Table 1.5, namely phases V, VI and VII.

The equations of state (column (d) of Table 1.5) corresponding to phases V–VII can have solutions only at the point $\Phi_1(P,T) = \Phi_2(P,T) = 0$. Stability conditions for phase VI:

$$\Phi_{11} \geq 0, \quad \Phi_3 \geq 0, \quad \Phi_{11}\Phi_{22} - \Phi_{12}^2 \geq 0, \tag{1.102}$$

do not contain any further restriction on the possibility of a second-order transition at the preceding point. One finds for the equilibrium values of the order-parameter components in phase VI

$$\begin{aligned} I_1 = \eta_1^2 + \eta_2^2 &= \frac{b_1 c_{12} - a_1 b_2}{\Delta} + \frac{d_1 c_{12} - 6a_3 b_2}{\Delta^3} (b_1 c_{12} - 2a_1 b_2)^2, \\ I_2 = \eta_1^4 + \eta_2^4 &= \frac{a_1 c_{12} - 2a_2 b_1}{\Delta} + \frac{3a_3 c_{12} - 2a_2 d_1}{\Delta^3} (b_1 c_{12} - 2a_1 b_2)^2. \end{aligned} \tag{1.103}$$

In order that Eqs. (1.103) possess real solutions for η_1^2 and η_2^2, it is necessary that Eq. (1.102) is fulfilled. This is the case for the region determined by the inequalities $a_2 > 0$, $\Delta > 0$, $d_1 > 0$, which do not impose any restriction on the nature of the boundaries with the other low-symmetry phases. Actually, the boundaries of existence of Eq. (1.102) coincide exactly with the boundaries of phases III and IV, in agreement with Fig. 1.18, which shows that phase VI should be located between the preceding phases. Therefore, the transition between phase VI and phase III or IV may be second-order as shown in Fig. 1.20(a).

Let us stress that when the situation shown in Fig. 1.20(a) is realized, the origin ($a_1 = 0, b_1 = 0$), which is determined by the double condition $\Phi_1(P,T) = \Phi_2(P,T) = 0$, corresponds to a *five-phase point* N, the sequence of phases around N being I–III–VI–IV–II–I. The existence of five stable phases at the same point does not contradict Gibbs's rule, which forbids the coexistence of more than three phases at one point (triple point) in the plane of the external variables. Gibbs's rule holds when all phases merging at the triple point are assumed to be independent, i.e., their thermodynamic potentials, which are equal at the triple point, give rise to independent relationships between the external variables (see Sec. 1.3.3.4).

In the case shown in Fig. 1.20(a), the thermodynamic potentials correspond to phases which are also separated by the second-order transitions, and thus consist in extrema, related to each other of the

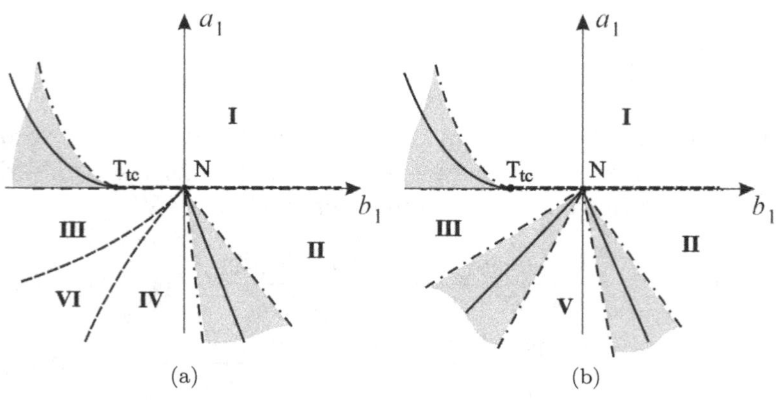

Fig. 1.20. Phase diagrams for the image-group $m\bar{3}m$ assuming an eighth-degree expansion (1.101): (a) $a_2 > 0$, $\Delta > 0$, $d_1 > 0$; (b) $a_2 < 0$, $\Delta < 0$, $d_1 < 0$. Full, dashed and dashed-dotted lines are, respectively, the first-order, second-order and limit of stability lines. Shadowed are the coexistence regions.

same non-equilibrium potential, and are thus expressed in terms of the same phenomenological coefficients.

For phase V (or phase VI), which is topologically close to phase III (see Fig. 1.18) near the line $\Phi_1(P,T)=0$, the contradictory stability conditions $\pm(\eta_1^2-\eta_3^2)\Phi_3>0$, which hold for small (η_1,η_3), disappear when taking into account the higher-order terms in η, i.e., in the region where $\eta_1^2-\eta_3^2\propto\eta^4$. Thus, the region of existence of phase V shown in Fig. 1.20(b) corresponds to the solutions which vanish on the line $\Phi_1=0$ and become narrower in proportion to η^4, as the point $(\Phi_1=\Phi_2=0)$ is approached. Such a region can only be near the boundary of the region of stability of phase II where the potentials of II and V (or VII) differ by an amount of I_1^3. Furthermore, phase V is metastable with respect to phase IV, the potential of which differs from the potential of phase II only by an amount of the order of I_1^2.

Solution VIII in Table 1.5 describes a phase with three independent components and with the lowest symmetry. The solution of the corresponding system of equations of state $(\Phi_1=\Phi_2=\Phi_3=0)$ allows a contact with phase I only across a first-order transition. A consistent model describing phase VIII requires taking into account a 12th-degree potential at least. In this case, illustrated by the phase diagrams from Ref. [23] represented in Fig. 1.21(a), all the eight phases listed in Table 1.5 possess a domain of stability. Additional

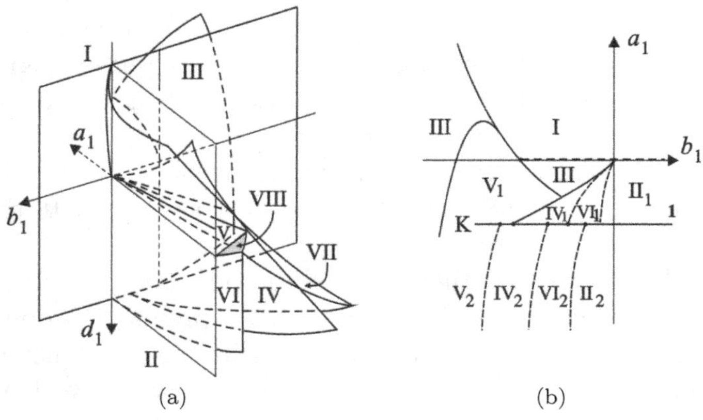

Fig. 1.21. (a) Location of the eight phases listed in Table 1.5 in the (a_1,b_1,d_1) phase diagram corresponding to the image-group $m\bar{3}m$. (b) Phase diagram associated with the image-group $m\bar{3}m$ involving an isostructural transition line (line 1) ending at a critical point of the liquid-vapor type.

isostructural transitions may also be stabilized for high-degree expansion models as shown in the phase diagram of Fig. 1.21(b) worked out in Ref. [24].

1.3.3.6. *First-order phase transitions: Outline of a theory of phase diagrams*

From the examples of solutions of the radial problem discussed in this section, a number of rules can be deduced and used as a guide in the choice of a phenomenological model. The inherent arbitrariness of this choice should be limited by the following procedure:

(1) The order-parameter expansion associated with a given image-group has to take into account the full set of invariants (I_i) forming the integrity basis of the image-group. The minimal degree of the expansion, which corresponds to a "structurally stable" model, i.e., a model describing consistently the whole set of phases, is $2N$, where N is the highest degree found among the invariants (I_i). For describing the first-order transitions from the parent phase instead of eventual second-order transitions, one has to consider $(2N + 2)$ degrees in the order-parameter components (η_i), and even higher degrees when isostructural transitions exist in the phase diagram. Structurally, stable models do not require to keep in the order-parameter expansion the whole set of coupled invariants I_iI_j, $I_iI_jI_k$ $(i \neq j \neq k)$ as their effect may be secondary for the essential topology of the phase diagram. On the other hand, one can reach a knowledge of the complete phase diagram by successive approximations, starting from the "zero approximation" (the fourth-degree expansion), which allows to describe the transitions to the phases corresponding to the maximal isotropy subgroups of the parent phase.

(2) One should start by the determination of the phase diagram in the orbit space (I_i), as it provides a useful preliminary description of the topology of the phases, i.e., their contact or neighboring in the real phase diagram, and is generally easier to work out than the phase diagram in the space of the phenomenological coefficients. Note that when the dimension m of the basis of integrity is equal to the dimension n of the order-parameter space, the number of distinct phases is 2^n, while for $m > n$, the number of distinct phases is smaller than 2^n.

(3) The correspondence between the theoretical phase diagram in the space of the phenomenological coefficients and the phase diagram in the space of the external variables ($X_e = T, P, x$) imposes the assumption of a dependence on the X_e of *all* the coefficients $a_1, b_1, c_1, \ldots$, of the invariants I_i, while the coefficients of the higher-power invariants $I_i^2, I_i^3, I_i I_j, \ldots$, can be assumed to be constant. Generally, a *linear* dependence is assumed for $a_1(X_e), b_1(X_e), c_1(X_e), \ldots$, in such a way that the real phase diagram ($T - P, T - x$, etc.) can be deduced by a linear transformation from the theoretical phase diagram, and thus displays the same configuration. However, close to singularities (e.g., N-phase points) or when strongly curved transition lines take place, a nonlinear dependence of $a_1(X_e)$, $b_1(X_e)$, $c_1(X_e), \ldots$, may be necessary.

Two main categories of the first-order transitions can be distinguished. A first category is constituted by the transitions whose discontinuous character depends on the values of the coefficients in the order-parameter expansion, i.e., on the values of the external parameters and correlatively on the highest necessary degree at which the order-parameter expansion needs to be truncated. However, a second category should be singled out, it contains transitions which can be predicted to be *always discontinuous from symmetry considerations.* This is for example the case of phase transitions between the low-symmetry phases, which possess no group–subgroup relationship between their symmetries. It is also the case for isostructural transitions and for transitions which possess a cubic invariant in their order-parameter expansion. The next chapter focuses in more detail on the classes of symmetry-predicted first-order transitions.

References

[1] L. D. Landau, *Zh. Eksp. Teor. Fiziki* **7** (1937) 19; **7** (1937) 627. Transl. in *Collected Papers of L. D. Landau*, ed. D. Ter Haar (Pergamon Press, Oxford, 1965), p. 193, 216.

[2] P. Toledano and V. Dmitriev, *Reconstructive Phase Transitions in Crystals and Quasicrystals* (World Scientific, Singapore, 1996).

[3] E. P. Wigner, *Group Theory and its Application to the Quantum Mechanics of Atomic Spectra* (Academic Press, New York, 1959).

[4] I. E. Dzialoshinskii, *Sov. Phys.: JETP* **19** (1964) 960; **20** (1965) 223; **20** (1965) 665.
[5] L. P. Pitaevskii, *Sov. Phys.: JETP* **37** (1960) 1267.
[6] V. L. Indenbom, S. A. Pikin and E. B. Loginov, *Sov. Phys. Crystallogr.* **21** (1976) 632.
[7] O. V. Kovalev, *Irreducible Representations of the Space Groups* (Gordon and Breach, New York, 1965).
[8] J. Zak, A. Casher, H. Glück and Y. Gur, *The Irreducible Representations of Space Groups* (Benjamin, New York, 1969).
[9] S. C. Miller and N. V. Love, *Tables of Irreducible Representations of Space Groups and Magnetic Corepresentations of Magnetic Space Groups* (Pruett, Boulder, 1964).
[10] (a) P. Toledano and J. C. Toledano, *Phys. Rev. B* **14** (1976) 3097.
(b) P. Toledano and J. C. Toledano, *Phys. Rev. B* **16** (1977) 386.
(c) J. C. Toledano and P. Toledano, *Phys. Rev. B* **21** (1980) 1139.
(d) P. Toledano and J. C. Toledano, *Phys. Rev. B* **25** (1982) 1946.
[11] Yu. M. Gufan, *Structural Phase Transitions* (Nauka, Moscow, 1982).
[12] H. T. Stokes and D. M. Hatch, *Isotropy Subgroups of the 230 Crystallographic Space Groups* (World Scientific, Singapore, 1988).
[13] Yu. A. Izyumov and V. N. Syromyatnikov, *Phase Transitions and Crystal Symmetry* (Kluwer, Dordrecht, 1990).
[14] E. I. Kut'in, V. L. Lorman and S. V. Pavlov, *Sov. Phys. Usp.* **34** (1991) 497.
[15] L. Michel, in *Group Theoretical Methods in Physics*, ed. R. Sharp and B. Kolman (Academic Press, New York, 1977), p. 75.
[16] J. C. Toledano and P. Toledano, *The Landau Theory of Phase Transitions* (World Scientific, Singapore, 1987).
[17] H. T. Stokes, D. M. Hatch and B. J. Campbell, *ISOTROPY Software Suite* (http://stokes.byu.edu/isotropy.html).
[18] L. D. Landau and I. M. Khalatnikov, *Dokl. Akad. Nauk SSSR* **96** (1954) 469. Transl. in *Collected Papers of L. D. Landau*, ed. D. Ter Haar (Pergamon Press, Oxford, 1965), p. 444.
[19] L. D. Landau and E. M. Lifshitz, *Phys. Z. Sowjet.* **8** (1935) 153. Transl. in *Collected Papers of L. D. Landau*, ed. D. Ter Haar (Pergamon Press, Oxford, 1965), p. 101.
[20] Yu. M. Gufan and E. S. Larin, *Sov. Phys. Dokl.* **23** (1978) 754.
[21] Yu. M. Gufan and E. S. Larin, *Izv. Akad. Nauk SSSR, Ser. Fiz.* **43** (1979) 14.
[22] Yu. M. Gufan and V. P. Sakhnenko, *Sov. Phys.-Solid State* **16** (1974) 1034.
[23] E. I. Kut'in, *Candidate Dissertation* (University of Rostov on the Don, 1988).

[24] V. L. Lorman, *Candidate Dissertation* (University of Rostov on the Don, 1988).
[25] Yu. M. Gufan and V. P. Sakhnenko, *Sov. Phys.: JETP* **42** (1976) 728.
[26] Yu. M. Gufan and E. S. Larin, *Sov. Phys.-Solid State* **20** (1978) 998.
[27] A. F. Devonshire, *Philos. Mag.* **40** (1949) 1040.
[28] Yu. M. Gufan and V. P. Sakhnenko, *Sov. Phys.: JETP* **36** (1973) 1009.

Chapter 2

Symmetry-Predicted Discontinuous Phase Transitions

2.1. Introduction

Different classes of transitions can be predicted to be discontinuous on the sole basis of symmetry considerations, i.e., by considering the symmetry of the phases surrounding the transition.

(1) The most obvious class of symmetry-predicted first-order transitions are the transitions between the phases which possess no group–subgroup relationship between their symmetries. One should distinguish, in this respect, two subclasses corresponding to very different phenomena:

(a) Transition between two phases, the structures of which are associated with different subgroups of a *common parent structure*, from which they can be deduced by slight distortions. In other words, the two phases, although they have no mutual group–subgroup relationship, correspond to different equilibrium values of the same set of order-parameter components, which transform as a single multidimensional irreducible representation of the symmetry group of the parent phase. Such type of transitions have been met in the preceding chapter for phase diagrams associated with the image-groups with $n \geq 2$ such as the $4mm$ or $m\bar{3}m$ phase diagram, and correspond to a large number of experimental examples of ferroelectric and ferroelastic transitions [1].

A textbook example is the orthorhombic-to-rhombohedral ($mm2$–$3m$) transition which occurs at 183 K in $BaTiO_3$ [2].

(b) Reconstructive-type phase transitions for which the loss of group–subgroup relationship between the phases results entirely from the drastic character of the transition mechanism, which involves large atomic displacements, specific ordering processes, etc. [3]. Here, the two phases do not possess, in general, a common parent structure within their phase diagram.

(2) Another important class of symmetry-predicted discontinuous transitions can be found in the phase diagrams associated with *more than one irreducible (symmetry-breaking) order-parameter*, i.e., in which some phases are induced by a *reducible* representation of the parent phase. This property is reflected by the low symmetry of the structure of these phases which cannot be induced by a single irreducible representation of the parent phase, by their domain structure, and by the form of the transition anomalies. Examples of phase diagrams involving coupled order-parameters will be given in Sec. 2.4.

(3) *Isostructural and anti-isostructural* phase transitions also represent a clearly recognizable class of first-order transitions, as the space groups of the phases possess the same symmetry (space-groups). A number of examples of isostructural transitions have been given in Sec. 1.3.3. It has been noted that their distinctive property is to display a critical point of the liquid–gas type. Anti-isostructural transitions were also found in the description of the $3m$ phase diagram. We address the intrinsic properties of these two types of first-order transitions in the following sections, and give additional illustrative examples of their occurrence.

(4) When a *cubic invariant* is allowed by symmetry in the order-parameter expansion, the transitions from the parent phase are necessarily first-order, except for an isolated point in the phase diagram (at which the coefficient of the cubic term vanishes). This is the formulation *a contrario* of the so-called Landau condition [5, 6], and it will be illustrated in Sec. 2.3 for the phase diagrams associated with the image 1, $3m$, $\bar{4}3m$ and 23. For the image-group 3, to which correspond *two* cubic invariants, a

second-order transition has been shown to be completely forbidden, even for an isolated point of the phase diagram.

(5) *Lock-in phases* occur generally in structurally modulated crystals below the range of stability of the incommensurate structure [4]. They may also occur directly from the parent phase across a first-order transition. Such situations have been described on a phenomenological basis by the competing influence of two types of invariants in the order-parameter expansion: (i) Invariants including spatial derivatives of the order-parameter components, which express the tendency of the system to stabilize an incommensurately modulated structure. The existence of these so-called "Lifshitz invariants" can be predicted by symmetry on the basis of the criterion introduced by Lifshitz [4]. (ii) High-degree invariants which favor the onset of lock-in phases.

According to the preceding scheme, one can identify (and predict the existence of) first-order "lock-in" transitions when the irreducible representation involved in the transition allows the existence of a Lifshitz invariant. In Sec. 2.3, we discuss the actuality of this class of first-order transitions.

2.2. The Landau Condition

Let us first come back on the elementary reason which forbids a cubic term in the order-parameter expansion of a second-order transition, and consider again the general form (1.16) of an expansion transforming as a single n-dimensional irreducible representation Γ of a space-group G_0. In Eq. (1.16), f_3, f_4, etc. are the successive terms of the Taylor expansion of F in the function of the order-parameter components (η_i). For $T = T_C$, the minimum of F corresponds to $\eta_i = 0$ for all j. Hence, for any infinitesimal values of η_i, the expansion $F - F_0$ must be positive. For such infinitesimal values, it is sufficient to consider the term of lowest degree, which at T_C is $f_3(\eta_i)$. Clearly, this term cannot be positive for any set of infinitesimal η_i values, as if f_3 is positive for a given direction in the order-parameter space, the set $(-\eta_i)$ corresponding to the opposite direction makes f_3 negative. Thus, in the presence of a third-degree term, the value $\eta_i = 0$ does not correspond to a minimum of the order-parameter expansion but to a saddle point of this function (Fig. 2.1(a)). Consequently, a

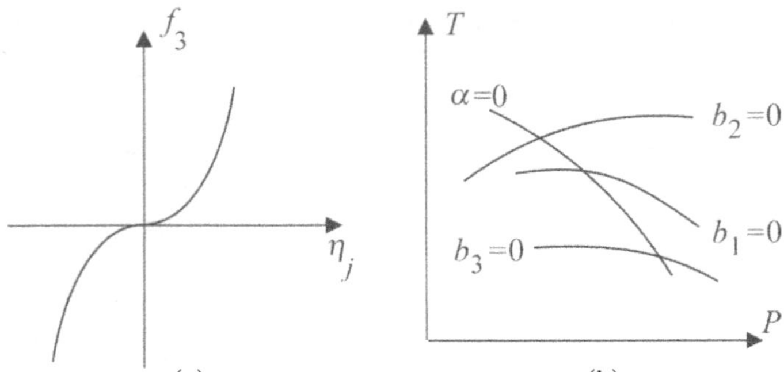

Fig. 2.1. Figures illustrating the facts: (a) that at $\eta_i = 0$, a third-degree term $f_3(\eta_i)$ corresponds to a saddle point and not to a minimum of the order-parameter expansion; (b) that when there are more than one independent cubic invariants, the simultaneous vanishing of the corresponding coefficients $b_i(T, P)$ with $\alpha\,(T, P)$ is impossible.

second-order transition, for which the state $\eta_i = 0$ is necessarily an equilibrium state, requires that no third-degree invariant exists in the expansion F.

The absence of $f_3(\eta_i)$ can be due to symmetry reasons, namely the impossibility to find a third-degree polynomial invariant by the symmetry group G_0. Another "accidental" reason can be the vanishing at T_C of the temperature-dependent coefficients in $f_3(\eta_i)$. Writing $f_3(\eta_j) = b_1(T, P)P_1(\eta_j) + b_2(T, P)P_2(\eta_j) + \cdots$, where P_i are linearly independent polynomials of the third degree separately invariant by G_0, one can see that the occurrence of a continuous transition at $T_C, P_C)$ requires, in addition to $\alpha(T_C, P_C) = 0$, the fulfillment of the conditions $b_1(T_C, P_C) = b_2(T_C, P_C) = \cdots = 0$, the transition point being at the intersection of the lines $\alpha(T_C,P_C) = 0$ and $b_i(T, P) = 0$ (Fig. 2.1(b)). If there are several independent $P_i(\eta_j)$, the simultaneous vanishing of the b_i at (T_C, P_C) will not be possible, and no continuous transition exists. If $f_3(\eta_j)$ includes a single $P_1(\eta_j)$ with the coefficient $b(T, P)$, a second-order transition will be possible at the isolated point $\alpha(T_C, P_C) = b(T_C,\ P_C) = 0$. If one excludes this singular type of transition, *the existence of a continuous transition requires that no third-degree polynomials of the order-parameter components are invariant by* G_0 (Landau condition [5, 6]).

The $P_1(\eta_j)$ polynomials of the third degree in the order-parameter components are the distinct invariant polynomials contained in the symmetrized third power of the irreducible representation Γ of G_0. Hence, the number of independent P_1 is equal to the number of totally symmetric irreducible representations Γ_1 contained in $[\Gamma^3]$. Accordingly, a necessary and sufficient condition for the absence of third-degree term in the order-parameter expansion can be expressed as

$$\Gamma_1 \not\subset [\Gamma^3], \tag{2.1}$$

which is the group-theoretical form of the Landau condition.

The third-degree invariants are found for the one-dimensional image-group 1, the two-dimensional images 3 and $3m$, and the three-dimensional images 23 and $\overline{4}3m$. A systematic investigation of the experimental examples of transitions associated with the irreducible representations which do not fulfill the Landau condition shows that this condition is a very reliable criterion to predict the first-order character of a transition [7]. All the observed transitions are indeed first-order, sometimes very weakly denoting the vicinity of an isolated "Landau" critical point of second-order transition, which has been evidenced in some structural systems. In ferroelastic phase transitions in which the order-parameter identifies to spontaneous strain components, the coefficient of the cubic term corresponds to a combination of third-order elastic constants [7]. On the other hand, for certain cubic-to-tetragonal ferroelastic transitions, the ratio $c/a - 1$, which expresses the tetragonality of the low-symmetry phase, coincides with the third-degree invariant [8].

2.2.1. *Order-parameters not fulfilling the Landau condition*

2.2.1.1. *One-component order-parameter expansions with the identity image-group: Isostructural phase transitions*

In one-dimensional order-parameter space (see Sec. 1.3.2.2), the group $O(1)$ is reduced to the identity or to the inversion. Thus, there are two possible symmetries for $n = 1$. The first is $O(1) = \overline{1}$, and we analyzed it in Sec. 1.3.3.2. Another one is the identity group. It transforms as the identity representation of any of the 230 space groups (at the Γ-point of the Brillouin zone) and contains all powers (even and odd) of the order-parameter η. The corresponding integrity basis is thus reduced to the linear invariant $I = \eta$, and the free-energy reads

$$F = a_0\eta + a_1\eta^2 + b_1\eta^3 + a_2\eta^4 + b_2\eta^5 + \cdots . \tag{2.2}$$

The equation of state and stability condition associated with the expansion (2.2) are

$$\frac{\partial F}{\partial \eta} = a_0 + \eta(2a_1 + 3b_1\eta + 4a_2\eta^2 + 5b_2\eta^3 + \cdots) = 0, \tag{2.3}$$

$$\frac{\partial^2 F}{\partial \eta^2} = a_1 + \eta(3b_1 + 6a2\eta + 10b_2\eta^2 + \cdots) \geq 0, \tag{2.4}$$

which clearly show that a stable state with $\eta = 0$ cannot exist if the coefficient a_0 of the linear term in Eq. (2.2) is different from zero. This is related to the fact that the identity representation of a space-group induces no symmetry change, η being a non-symmetry-breaking quantity. In other words, the potential (2.2) describes a transition (or a sequence of transitions depending on the degree to which it is truncated) between isostructural phases without definite parent phase.

The free energy associated with the simplest model of an isostructural phase transition has the form

$$F(\eta) = a_0\eta + a_1\eta^2 + a_2\eta^3 + \eta^4, \tag{2.5}$$

where $\eta = v_1 - v_2$ is the totally symmetrical (non-symmetry-breaking) order parameter, and a_i are the phenomenological parameters. It is worth reminding that exactly difference between two non-small parameters v_i (volume, valence, etc.) corresponds to the classic definition of the order-parameter. The change of variables $\eta = \tilde{a}_0\tilde{\eta} - a_2/4$ and the following elimination of the equilibrium part of the energy transforms Eq. (2.5) into the normal form [9]

$$F(\tilde{\eta}) = \tilde{a}_0\tilde{\eta} + \tilde{a}_1\tilde{\eta}^2 + \tilde{\eta}^4, \tag{2.6}$$

where $\tilde{a}_0 = a_0 - a_1a_2/2 + a_2^3/8$ and $\tilde{a}_1 = a_1 - 3a_2^3/8$. Minimization of the potential $F(\tilde{\eta})$ with respect to the order parameter $\tilde{\eta}$ yields an equation of state as follows:

$$\frac{\partial F}{\partial \tilde{\eta}} = \tilde{a}_0 + 2\tilde{a}_1\tilde{\eta} + 4\tilde{\eta}^3. \tag{2.7}$$

The above model is studied in detail since the form of the Landau potential (2.6) coincides with one of the elementary catastrophes (namely, the cusp catastrophe), and was studied comprehensively in the framework of the mathematical catastrophe theory [10]. The equilibrium values of the order parameter $\tilde{\eta}$ are the solutions of the equation of state (2.7), and because this is a cubic equation, it has either one or three real roots and only those have a physical meaning. The number of roots is determined by the determinant $D = 8\tilde{a}_1^3 + 27\tilde{a}_0^2$. If $D \leq 0$, there are three real roots:

$$\tilde{\eta}_1 = -2\rho\cos\frac{\varphi}{3}, \quad \tilde{\eta}_{2,3} = 2\rho\cos\frac{\pi \pm \varphi}{3};$$
$$\rho = \operatorname{sgn}\left\{\tilde{a}_0\sqrt{|\tilde{a}_1/6|}\right\}, \quad \cos\varphi = \tilde{a}_0/8\rho^3, \tag{2.8}$$

two solutions, $\tilde{\eta}_{2,3}$, corresponding to the minima of the Landau potential and one $\tilde{\eta}_1$ corresponding to a maximum; otherwise, there is only one ($\tilde{\eta}_1$). Figures 2.2(a) and 2.2(b) illustrate this by a three-dimensional diagram of the equilibrium surface (Fig. 2.2(a)) and its projection on the $(\tilde{a}_0, \tilde{a}_1)$ plane (Fig. 2.2(b)). One can see that smooth variations in $\tilde{a}_0$ and $\tilde{a}_1$ almost always produce smooth variation in $\tilde{\eta}$. The only exceptions occur when the system crosses the energy equality (isostructural first-order phase transition) line $F(\tilde{\eta}_2) = F(\tilde{\eta}_3)$, i.e., $\tilde{a}_0 = 0$. In this point, it jumps to the other sheet corresponding to the other energy minimum ($\tilde{\eta}_2 \leftrightarrow \tilde{\eta}_3$). This brings about a sudden change in $\tilde{\eta}$. The projections of the folds of the equilibrium surface onto the $(\tilde{a}_0, \tilde{a}_1)$ plane ($D = 0, \tilde{a}_1 = \tilde{a}_0^{2/3}/2$) are the stability limit (spinodal) lines (Fig. 2.2(b)). The variation regime in $\tilde{\eta}$ can be changed smoothly from discontinuous to continuous by varying the $\tilde{a}_1$ parameter from negative to positive sign. The change of regime occurs at $\tilde{a}_1 = 0$, i.e., at the critical point, E_{cr}, of the liquid–gas type (Fig. 2.2).

Figure 2.3 shows the phase diagram corresponding to Eq. (2.5). It should be especially underlined that the smooth variation regime of

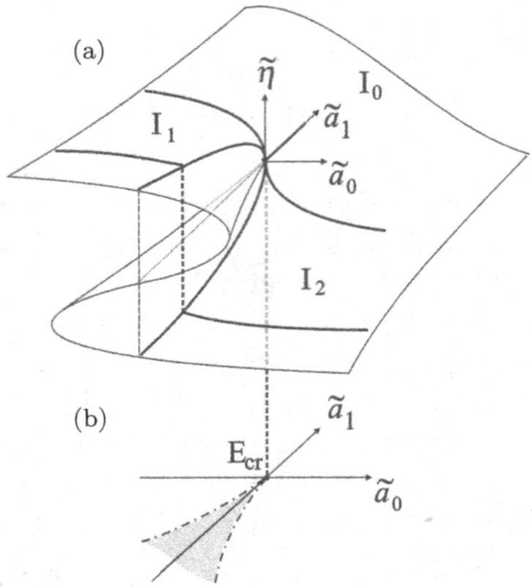

Fig. 2.2. (a) Equilibrium surface corresponding to Eq. (2.7). (b) Projection onto the $(\tilde{a}_0, \tilde{a}_1)$ plane. Dashed-dotted lines are the limit of stability lines, and the solid line is the first-order isostructural phase transition line. E_{cr} is the critical end-point. Shadowed is the coexistence region.

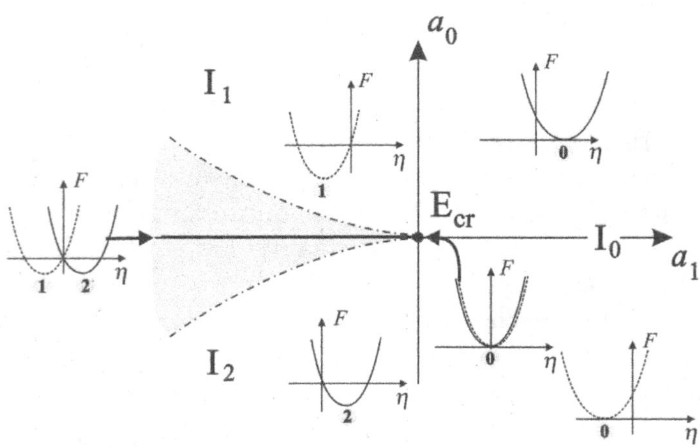

Fig. 2.3. Phase diagram corresponding to the potential (2.5). The point E_{cr} is the critical end-point. Dashed-dotted lines are the stability limit lines, and the solid line is the first-order isostructural phase transition line. Shadowed is the coexistence region.

η beyond the critical end-point E_{cr} ($\tilde{a}_1 > 0$) is a cross-over from one non-zero to another non-zero value of the non-symmetry-breaking order-parameter. The process evidently cannot be classified as a continuous phase transition.

Since isostructural transitions are present quite seldom in real systems, or can be described using non-trivial phenomenological models (see Sec. 2.2.2), the phase diagrams associated with one-component order-parameter models displaying odd-degree terms have not been investigated thoroughly with the exception of the following truncated potential:

$$F(T, P, \eta) = F_0(T, P) + a_1\eta^2 + b_1\eta^3 + a_2\eta^4, \tag{2.9}$$

which describes a single first-order transition. Let us stress that the form of expansion (2.9) does not correspond to any one-dimensional image-group. However, it will be shown in the next chapters (see also Sec. 2.2.1.2 in this chapter) to correspond, in some cases, to the *effective form of multicomponent order-parameter expansions*, namely to the *reduced form* which is obtained when the symmetry relationships between the equilibrium order-parameter components are taken into account.

The equations minimizing the potential (2.9) are

$$\eta[2a_1 + 3b_1\eta + 4a_2\eta^2] = 0, \tag{2.10}$$

$$a_1 + 3b_1\eta + 6a_2\eta^2 \geq 0. \tag{2.11}$$

Assuming $b_1 < 0, a_2 > 0$ and $a_1 = a_{10}(T - T_C)$, the discussion of these equations is visualized in Fig. 2.4(a). Thus, for $T_1 = T_C + b_1^2/4a_2a_{10}$, the low-temperature phase appears as a metastable state. The high- and low-temperature phases are equally stable for $T_2 = T_C + b_1^2/2a_2a_{10}$. T_2 coincides with the actual first-order transition temperature, at which the order-parameter undergoes the jump $\eta(T_2) = -b_1/2a_2$ and then increases following the law

$$\eta(T) = \frac{-3b_1 + \sqrt{9b_1^2 - 32a_1a_2}}{8a_2}, \tag{2.12}$$

which is represented in Fig. 2.4(b). The preceding description is made assuming that $b_1 > 0$ does not modify and has only the result of exchanging the left- and right-hand sides of Fig. 2.4(a).

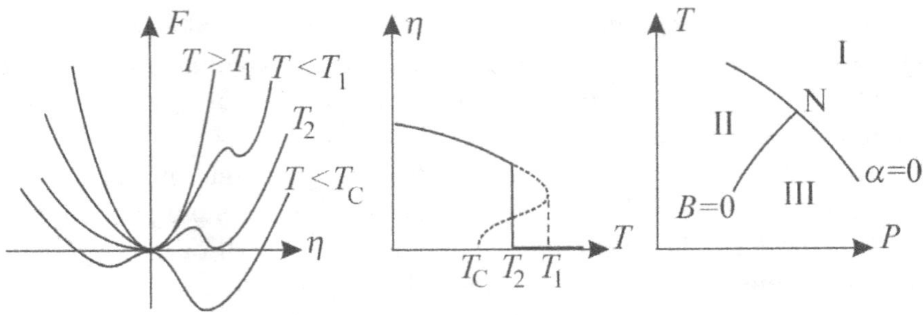

(a) (b) (c)

Fig. 2.4. (a) Order-parameter dependence of the potential (2.9) with $b_1 < 0$ for different temperatures. (b) Temperature dependence of the order-parameter given by Eq. (2.12) for $T < T_2$. (c) Neighborhood of the Landau critical point N. Solid lines are the first-order transition lines.

The preceding discussion resembles the one performed for the potential (1.46), and most of the features characterizing the first-order transitions, as described in Sec. 1.3.3.2a, are also found here: region of coexistence for the high- and low-temperature phases, discontinuous jump at the transition for the order-parameter, thermal hysteresis, latent heat, etc. However, the phase diagram corresponding to the potential (2.9) involves only a first-order transition regime and no tricritical point. This is due to the existence of the cubic invariant. As stressed by Landau [5], the vanishing of the coefficient of the preceding invariant determines on the line of first-order transitions $a_1(T_2, P) = 0$ a *three-phase point* N (so-called Landau point) given by

$$a_1(T_2, P) = 0 \text{ and } b_1(T_2, P) = 0. \tag{2.13}$$

As shown in Fig. 2.4(c), N is an isolated point of continuous transition ending a line of first-order transitions of equation $b_1(T, P)$, which separates two *anti-isostructural* phases II and III, i.e., two phases of identical symmetries having opposite values for the order-parameter, as follows:

$$\eta_{\mathrm{II,III}} = \pm\sqrt{\frac{-a_1}{2a_2}}. \tag{2.14}$$

As demonstrated by Landau [4], the latent heat on the anti-isostructural transition line II–III is equal to

$$L = \frac{2}{3} T \frac{\partial b_1}{\partial T} (\Delta \eta)^3_{\text{II-III}}, \tag{2.15}$$

where $(\Delta\eta)_{\text{II-III}} = 2\sqrt{-a_1}/2a_2$ is the jump of the order-parameter across the $b_1(T, P) = 0$ line. Hence, L increases with the "distance" $(-a_1/2a_2)$ to the N-point with a power law 3/2.

2.2.1.2. *Phase diagrams associated with the image-groups 3m and 3*

The phase diagrams associated with the image-group $3m$ have been thoroughly investigated since the corresponding integrity basis contains only two basic invariants:

$$I_1 = \eta_1^2 + \eta_2^2 = \rho^2, \quad I_2 = \eta_1^3 - 3\eta_1\eta_2^2 = \rho^3 \cos 3\theta, \tag{2.16}$$

having a maximal degree $N = 3$, so structurally stable models involve order-parameter expansions which can be truncated at the sixth ($2N$) degree, and are thus algebraically solvable. Let us first show, following Ref. [10], that a fourth-degree expansion is indeed insufficient. Up to the fourth degree, one has

$$F(\rho, \theta) = F'(I_1, I_2) = a_1 I_1 + b_1 I_2 + a_2 I_1^2, \tag{2.17}$$

which yields the equations of state

$$\frac{\partial F}{\partial \rho} = \rho(2a_1 + 3b_1\rho\cos 3\theta + 4a_2\rho^2) = 0, \tag{2.18}$$

and

$$\frac{\partial F}{\partial \theta} = -3b_1\rho^3 \sin 3\theta = 0. \tag{2.19}$$

Equation (2.19) determines, in addition to the parent phase I ($\rho = 0, \eta_1 = \eta_2 = 0$), two low-symmetry phases with identical symmetries corresponding, respectively, to $\cos 3\theta = +1$ (phase II) and $\cos 3\theta = -1$ (phase III). They are associated with the opposite values of the equilibrium order-parameter components (phase II) and

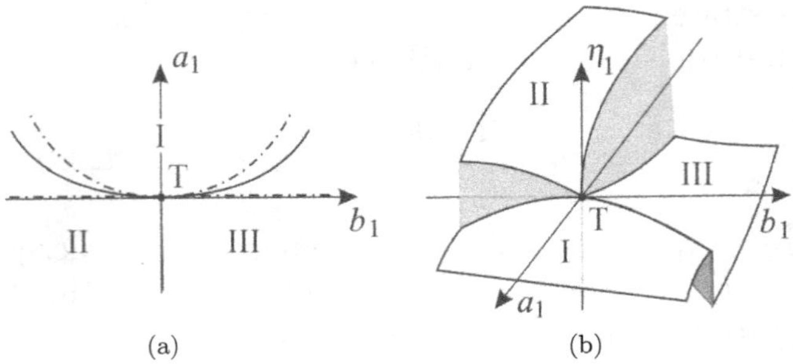

Fig. 2.5. Phase diagrams corresponding to the thermodynamic potential (2.17): (a) In the (a_1, b_1) plane. Dashed-dotted and full lines are, respectively, the limit of stability and first-order transition lines. T is a triple point. (b) In the (η_1, a_1, b_1) space. Hatched surfaces represent the jumps of the order-parameter $(\eta_1, 0)$ at the transitions.

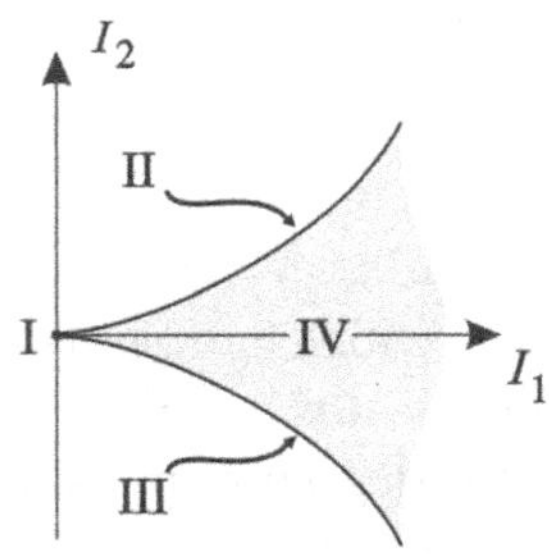

Fig. 2.6. Topology of the phases corresponding to the $3m$ image-group in the orbit-space (I_2, I_1).

$\eta_1 = -\eta \neq 0$, $\eta_2 = 0$ (phase III), i.e., phases II and III are *anti-isostructural* phases.

The phase diagrams represented in Fig. 2.5 show that, in addition to the first-order transitions I–II and I–III, the region of coexistence of phases II and III is reduced to the anti-isostructural first-order transition line II–III. In other words, even though the component η undergoes a finite jump at the preceding transition, the boundaries between these latter phases coincide. Another inconsistency of the fourth-degree expansion model can be foreseen from the topology of the phases in the orbit space (Fig. 2.6). Thus, the invariants I_1 and I_2 possess in phases II and III the following connections: $I_2 = \eta_{\text{II}}^3 = I_1^{3/2}$ and $I_2 = -\eta_{\text{III}}^3 = -I_1^{3/2}$, i.e., in the (I_2, I_1) plane, the two phases only coincide at the origin (phase I) and for $\eta_1 \neq 0$ are separated by a

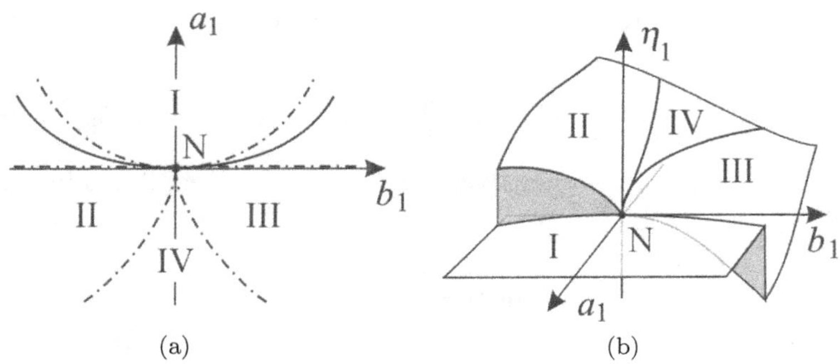

Fig. 2.7. Phase diagrams corresponding to the thermodynamic potential (2.20): (a) In the (a_1, b_1) plane. Dashed, full and dashed-dotted lines are, respectively, the second-order, first-order and limit of stability lines. N is a four-phase "Landau" point. (b) In the (η_1, a_1, b_1) space.

region in which an additional phase (IV) corresponding to $-I_1^{3/2} < I_2 < I_1^{3/2}$ or equivalently to $-1 < \cos 3\theta < +1$ can be stabilized. The construction of the phase diagram associated with the truncated sixth-degree order-parameter expansion:

$$F'(I_1, I_2) = a_1 I_1 + b_1 I_2 + a_2 I_1^2 + b_2 I_2^2, \tag{2.20}$$

clarifies the shortcomings of the fourth-degree model. The equation of state (1.140) becomes

$$\frac{\partial F}{\partial \theta} = -3b_1 \rho^3 \sin 3\theta \cdot (b_1 + 2b_2 \cos 3\theta) = 0, \tag{2.21}$$

which admits the additional solution $(b_1 + 2b_2 \cos 3\theta) = 0$ corresponding to the stable state IV (η_1, η_2) with $\eta_1 \neq 0$, $\eta_2 \neq 0$, $\eta_1 \neq \eta_2$. The symmetry of phase IV is the maximal common subgroup (of index 2) to the symmetry groups of phases II and III. The phase diagram associated with the expansion (2.20) is shown in Fig. 2.7. Hence, phases II and III are separated by two lines of second-order transitions bounding phase IV. These phases merge with phase I at a four-phase point N of the Landau type (see Fig. 2.4), as it is an isolated point of second-order transition on a line of first-order transitions.

The model based on the order-parameter expansion (2.20) is structurally stable, but overlooks a number of important features of the phase diagrams which may be constructed for the $3m$ image-group. These features are obtained when considering the full set of invariants, up to the sixth-degree, in the order-parameter expansion:

$$F'(I_1, I_2) = a_1 I_1 + a_2 I_1^2 + a_3 I_1^3 + b_1 I_2 + b_2 I_2^2 + c_{12} I_1 I_2, \qquad (2.22)$$

which has been comprehensively discussed in Ref. [9]. The corresponding equations of state are

$$\frac{\partial F}{\partial \rho} = 2\rho(a_1 + a_2 I_1 + a_3 I_1^2 + c_{12} I_2) + 3\rho^2 \cos 3\theta (b_1 + 2b_2 I_2 + c_{12} I_1) = 0, \qquad (2.23)$$

$$\frac{\partial F}{\partial \theta} = -3\rho^3 \sin 3\theta \cdot (b_1 + 2b_2 I_2 + c_{12} I_1) = 0. \qquad (2.24)$$

In phase II ($\rho \neq 0, \cos 3\theta = +1$), the equilibrium value of ρ is given by

$$2a_1 + 3b_1\rho + 4a_2\rho^2 + 5c_{12}\rho^3 + 6(a_3 + b_2)\rho^4 = 0,$$

and the stability conditions are

$$\frac{\partial^2 F}{\partial \rho^2} = \rho[3b_1 + 8a_2\rho + 15c_{12}\rho^2 + 24(a_3 + b_2)\rho^3] \geq 0, \qquad (2.25)$$

and

$$\frac{\partial^2 F}{\partial \theta^2} = -9\rho^3(-b_1 + c_{12}\rho^2 + 2b_2\rho^3) = 0. \qquad (2.26)$$

Accordingly, the limit of stability of phase II with respect to phase I can be expressed under the parametric form

$$a_1 - 2a_2\rho^2 - 10c_{12}\rho^3 - 18(a_3 + b_2)\rho^4 = 0,$$

$$b_1 - \frac{8}{3}a_2\rho + 5c_{12}\rho^2 + 8(a_3 + b_2)\rho^3 = 0,$$

which provides the equation of the line of stability II–I

$$a_1 = \frac{9}{2^5 a_2} b_1^2 + \frac{3}{2^{10} a_2^3} b_1^3 - \frac{3^6}{2^{17} a_2^5}\left[\frac{2^5 a_2}{3}(a_3 + b_2) - 5^2 c_{12}^2\right] b_1^4. \qquad (2.27)$$

The equation of the first-order transition line I–II is determined by the equality of the potentials $F(\rho_{\mathrm{I}} = 0) = F(\rho_{\mathrm{II}})$, namely by

$$\rho^2[a_1 + b_1\rho + a_2\rho^2 + c_{12}\rho^3 + 9(a_3 + b_2)\rho^4] = 0,$$

in which ρ is given by Eq. (2.23). It yields

$$a_1 = \frac{1}{2^2 a_2} b_1^2 + \frac{c_{12}}{2^3 a_2^3} b_1^3 - \frac{3c_{12} - 2^2(a_2 + b_2)a_2}{2^6 a_3} b_1^4. \qquad (2.28)$$

The boundary of phase III ($\rho \neq 0, \cos 3\theta = -1$) into phase I is obtained by replacing in Eqs. (2.23)–(2.21) b_1 by $-b_1$ and c_{12} by $-c_{12}$. As Eqs. (2.27) and (2.28) are invariant by these changes in sign, the equation of the stability line of phase III with respect to phase I and of the first-order transition line I–III are deduced from the analogous lines for phase II by a mirror symmetry with respect to $b_1 = 0$.

The limit of stability line between phases II and III can be worked out by introducing the corresponding equilibrium values ρ_{II} and ρ_{III} in Eq. (2.25). One gets

$$a_1 = \frac{2a_2}{c_{12}} b_1 \pm \Delta \cdot (-b1)^{3/2} + \frac{3(\Delta \cdot b_2 - a_3 c_{12}^2)}{c_{12}^4} b_1, \qquad (2.29)$$

where $\Delta = 4a_2 b_2 - c_{12}^2$. It is easy to show [11] that the mutual limits of stability lines of phases II and III have a common tangent $a_1 = 2a_2 b_2 / c_{12}$ at the origin ($a_1 = b_1 = 0$), and that depending on the sign of Δ, one has between the two phases either two lines of second-order phase transitions bounding phase IV ($\rho \neq 0, -1 < \cos 3\theta < +1$) for $\Delta < 0$ or a direct first-order (anti-isostructural) transition line II–III for $\Delta < 0$. Figures 2.8(a) and 2.8(b) represent the two preceding situations in the case $a_2 > 0$. As shown in Ref. [9], the topology of the phase diagrams changes drastically when $a_2 < 0$. In this case, represented in Figs. 2.8(c), one has, instead of a four-phase point N, two three-phase points located on a direct first-order transition line between phases I and IV.

Another important change in the topology of the $3m$ phase diagram occurs when the coefficients c_{12}, b_2 or a_3 become negative. Higher-degree invariants have to be considered in this case in order to ensure the stability of the phases. Figures 2.9(a)–2.9(c) represent the phase diagrams associated with an eighth-degree expansion in which the two additional invariants ($c_{112} I_1^2 I_2$ and $a_4 I_1^4$) have been added to the expansion (2.22). One can see that further complexifications of the phase diagrams take place. In Figs. 2.9(a) and 2.9(b), the transition II–III possesses two different, first- and second-order regimes,

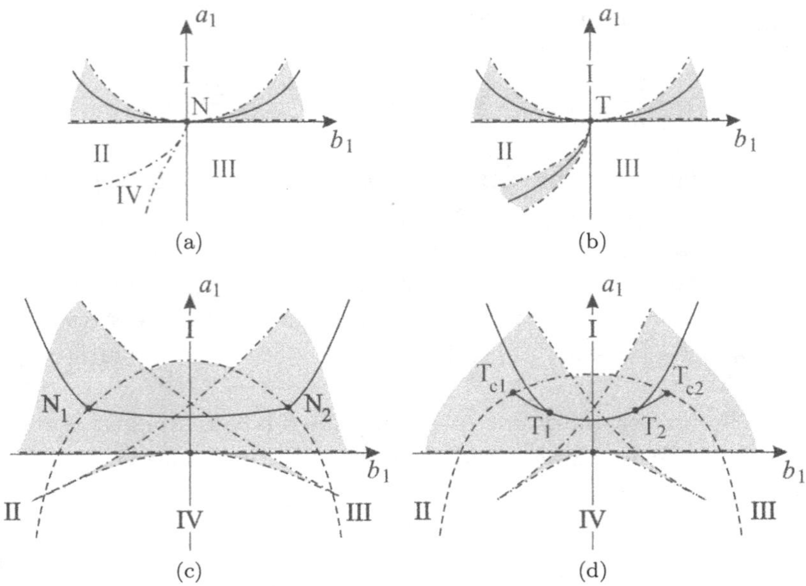

Fig. 2.8. Phase diagrams corresponding to the image-group $3m$ for a sixth-degree expansion (2.22). The diagrams are obtained for the following conditions: (a) $a_2 > 0$, $a_3 > 0$, $c_{12} > 0$, and $\Delta = 4a_2b_2 - c_{12}^2 > 0$; (b) $a_2 > 0$, $a_3 > 0$, $c_{12} > 0$, $\Delta < 0$; (c) $a_2 < 0$, $c_{12} = 0, 0 < b_2 < a_3$; (d) $a_2 < 0, c_{12} = 0, 0 < a_3 < b_2$. Full, dashed and dotted-dashed lines represent the first-order, second-order and limit of stability lines, respectively. Hatched surfaces are the metastability regions surrounding the first-order transition lines. N is a four-phase point. N_1 and N_2 are the three-phase points. T, T_1 and T_2 are the triple points. T_{tc1} and T_{tc2} are the tricritical points.

with either one or two tricritical points. In Fig. 2.9(c), phase II has two different regions of stability separated by a line of isostructural transitions ending at a critical point of the liquid-vapor type [9].

For the image-group 3, one has, with respect to $3m$, the additional invariant $I_3 = \eta_2^3 - 3\eta_1^2\eta_2 = \rho^3 \sin 3\theta$ in the integrity basis. Using for the order-parameter expansion $F(\rho, \theta) = F'(I_1, I_2, I_3)$ the notation $\Phi_i = \partial F'/\partial I_i$ and $\Phi_{ij} = \partial^2 F'/\partial I_i \partial I_j \, (i, j = 1 - 3)$, one gets the following equilibrium conditions:

$$\frac{\partial F}{\partial \rho} = \rho[\Phi_1 + 3\rho(\Phi_2 \cos 3\theta + \Phi_3 \sin 3\theta)] = 0, \tag{2.30}$$

$$\frac{\partial F}{\partial \theta} = 3\rho^3[-\Phi_2 \sin 3\theta + \Phi_3 \cos 3\theta] = 0. \tag{2.31}$$

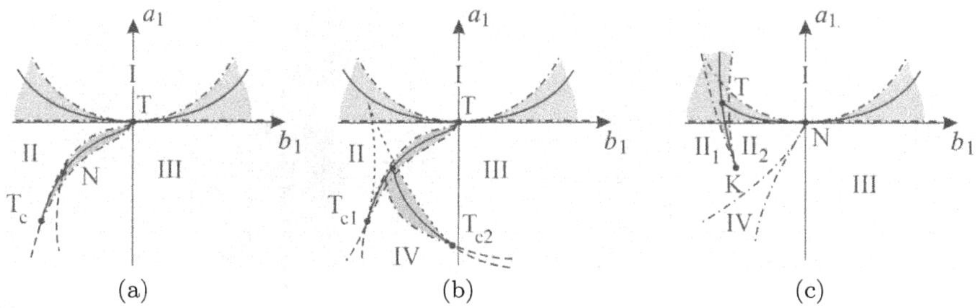

Fig. 2.9. Phase diagrams corresponding to the image-group $3m$ for an eighth-degree expansion (see text). The diagrams are obtained for the following conditions: (a) $a_2 > 0$, $a_3 > 0$, $c_{12} < 0$, $b_1 < 0$, $b_2 > 0$, $\Delta > 0$; (b) $a_2 > 0$, $a_3 < 0$, $b_2 < 0$, $a_4 > 0$, $\Delta < 0$; (c) $a_2 > 0$, $a_3 < 0$, $b_2 > 0$, $\Delta > 0$. Full, dashed and dotted-dashed lines represent the first-order, second-order and limit of stability lines, respectively. T_{tc1} and T_{tc2} are the tricritical points. N is a three-phase point in (a), and is a four-phase point in (c). T is a triple point. K is a critical point of the liquid-vapor type. II_1 and II_2 are two isostructural phases.

The corresponding stability conditions involve the second derivatives:

$$\frac{\partial^2 F}{\partial \rho^2} = 4\rho^2 \Phi_{11} + \rho(\Phi_2 \cos 3\theta + \Phi_3 \sin 3\theta) + \cdots,$$
$$\frac{\partial^2 F}{\partial \theta^2} = -9\rho^2(\Phi_2 \cos 3\theta + \Phi_3 \sin 3\theta) + \cdots,$$
$$\frac{\partial^2 F}{\partial \rho \partial \theta} = 6\rho^2(-\Phi_{21} \sin 3\theta + \Phi_{21} \cos 3\theta) + \cdots, \tag{2.32}$$

in which only the lowest degree terms in ρ have been written. Equations (2.31) and (2.32) show that, as for the image-group 3, only one low-symmetry phase (phase II) can be stabilized. It corresponds in the order-parameter space ε_2 to the "angle" θ^e given by

$$\tan 3\theta^e = \frac{\Phi_2(\rho^e, \theta^e)}{\Phi_3(\rho^e, \theta^e)},$$

and to the "modulus"

$$\rho^e = -\frac{\Phi_2(\rho^e, \theta^e)}{3[\Phi_2(\rho^e, \theta^e) \cos 3\theta^e + \Phi_3(\rho^e, \theta^e) \sin 3\theta^e]},$$

which yields the order-parameter components $\eta_1 = \rho^e \cos \theta^e \neq \eta_2 = \rho^e \sin \theta^e$. Figure 2.10(a) shows that, in the space (I_1, I_2, I_3), phase II

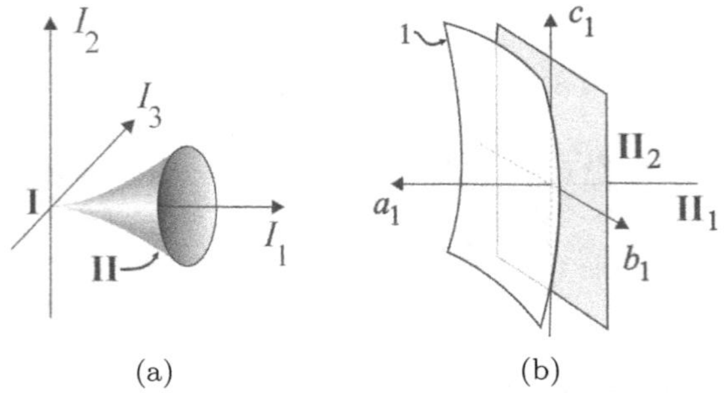

Fig. 2.10. (a) Topology of the phases in the orbit space for the image-group 3. (b) Phase diagram of 3 for the fourth-degree expansion (2.33). 1 denotes the surface of first-order transitions I–II.

is represented by a cone-type surface of equation $I_2^2 + I_3^2 = I_1^3$, which resembles (although the equation is different) the surface found for the image-group 4 (Fig. 1.17(a)).

From Eqs. (2.26)–(2.32), one can easily deduce that the stability conditions for phase II, namely $\frac{\partial^2 F}{\partial \rho^2} \geq 0$, $\frac{\partial^2 F}{\partial \rho^2} \cdot \frac{\partial^2 F}{\partial \theta^2} \geq (\frac{\partial^2 F}{\partial \rho \partial \theta})^2$, can be fulfilled only when $\Phi_1 = \Phi_2 = \Phi_3 = 0$, i.e., *for a finite value of* ρ^e *at the transition point*. In other words, *no second-order transition (even at an isolated point) is possible between phases I and II*. This is a general property that was already pointed out by Landau [5], of the order-parameter expansions involving *two* cubic invariants.

A discussion of the phase diagrams associated with the image-group 3 [11] provides an interesting insight of the respective influence of high-degree invariants on the topology of a phase diagram. Thus, using the (structurally unstable) model based on a fourth-degree expansion:

$$F'(I_1, I_2, I_3) = a_1 I_1 + a_2 I_1^2 + b_1 I_2 + c_1 I_3, \tag{2.33}$$

leads, as shown in Fig. 2.10(b), to the stabilization of two isostructural low-symmetry phases (II_1 and II_2) with a region of coexistence reduced to a single line ($b_1 = c_1 = 0, a_3 < 0$) in the (a_1, b_1, c_1) space. In a two-dimensional (e.g., pressure–temperature) phase diagram, it would correspond to an isolated point of first-order – transition. This unphysical result can be removed by adding to Eq. (2.29) the sixth-degree invariants $b_2 I_2^2$ and $c_2 I_3^2$. It yields, for $a_2 > 0$, the phase diagram represented in Fig. 2.11(a) in which the isostructural

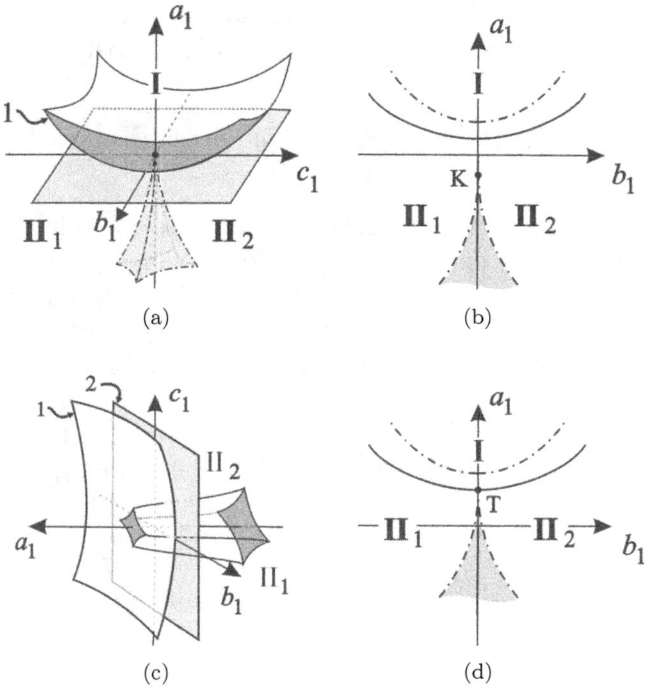

Fig. 2.11. Phase diagrams corresponding to the image-group 3 for a sixth-degree expansion of the type (2.33) in which the sixth-degree invariants ${b_2 I^2}_2$ and ${c_2 I^2}_3$ are added ((a) and (b)). In (c) and (d), the sixth-degree ${a_3 I^3}_1$ invariant is also taken into account (with $a_2 < 0$ and $a_3 > 0$). The surfaces denoted 1 and 2 in the three-dimensional diagrams (a) and (c) represent, respectively, the I–II transition and the isostructural $\mathrm{II}_1 - \mathrm{II}_2$ transition. These surfaces become full lines in the two-dimensional diagrams of (b) and (d), which show cross-sections $c_1 = \text{const.}$ c_1 is assumed to be small for the cross-section (d). The edges limiting surface 2 are the lines of critical points. They correspond to the K-point in (b). T is a triple point. For large values of c_1, T is replaced by a critical point K.

transition plane is surrounded by a pyramidal-type volume of coexistence. The two-dimensional phase diagram of Fig. 2.11(b) shows no contact between the I–II and – isostructural transition lines.

In contrast, when adding also the sixth-degree invariant $a_3 I_1^3$ to Eq. (2.33) with $a_2 < 0$ and $a_3 > 0$, one has a surface of contact between the $(\mathrm{II}_1 + \mathrm{II}_2)$ region and the stability region of phase I in the (a_1, b_1, c_1) space (Fig. 2.11(c)), i.e., the *critical point* K of

Fig. 2.11(b) *becomes* (for not too large values of c_1) a triple point as shown in the two-dimensional diagram of Fig. 2.11(d).

2.2.1.3. *General features of phase diagrams associated with the image-groups nmm and n*

The results obtained in Sec. 1.3.3.3 for $4mm$ and 4 and in Sec. 2.2.1.2 for $3m$ and 3 can be extended to the other image groups nmm and $n\,(n = 6, 8, 12)$. Clearly, one has to distinguish the groups generated by reflections (nmm) and the cyclic groups (n), which possess different specific features. For nmm image-groups, for which the number of basic invariants forming their integrity basis ($I_1 = \rho^2, I_2 = \rho^n \cos n\theta$) is equal to the number of order-parameter components ($\eta_1 = \rho\cos\theta, \eta_2 = \rho\sin\theta$), the equations determining the equilibrium values of η_1 and η_2 are [11]

$$\frac{\partial F}{\partial \rho} = \rho(2\Phi_1 + n\rho^{n-2}\Phi_2 \cos n\theta) = 0, \tag{2.34}$$

$$\frac{\partial F}{\partial \theta} = -n\rho^n \Phi_2 \sin n\theta = 0, \tag{2.35}$$

using the notation $\Phi_i = \partial F'/\partial I_i$ with $F\prime(I_1, I_2) = F(\rho, \theta)$.

In addition to the solution $\rho = 0$ (phase I), one has two types of solutions corresponding to different classes of low-symmetry phases: (1) Solutions coinciding with minima of F lying in symmetry planes of the order-parameter space ε_2. They are determined by the equation sin $n\theta = 0$, which yields two low-symmetry phases denoted II ($\cos n\theta = +1$) and III ($\cos n\theta = -1$) corresponding to the equilibrium values $(\eta, 0)$ and (η, η), respectively. The equilibrium value of the order-parameter "modulus" ρ is given by

$$\Phi_1 \pm n\rho^{n-2}\Phi_2 = 0, \tag{2.36}$$

and thus depends on the phenomenological coefficients, i.e., on the external variables, while the "direction" θ in ε_2 remains constant.

(2) Solutions corresponding to minima of F which do not lie in the symmetry planes of ε_2, ρ and θ are determined by the system of

(nonlinear) equations

$$\Phi_1 = 0, \Phi_2 = 0, \tag{2.37}$$

which means that both ρ and θ vary with temperature. There exists actually only one solution for Eq. (2.37) denoted IV and corresponding to the equilibrium values (η_1, η_2) with $\eta_1 \neq \eta_2 \neq 0$.

Accordingly, the maximal number of phases which can be found in the phase diagram associated with the image-group nmm is 4. This number is consistent with a theorem by Chevaley, which states that for groups generated by reflections, it should be 2^N (when N is the number of order-parameter components). Denoting G_{I}–G_{IV}, the symmetry groups of phases I–IV, one can establish the following results: G_{II} and G_{III} correspond to different symmetries for $n \geq 4$, as their respective "directions" θ in ε_2 belong to different classes of conjugated elements. In contrast, for $3mm$, all the reflections belong to the same class, and consequently, G_{II} and G_{III} have the same symmetry elements. G_{IV} is the maximal common subgroup (of index 2) of G_{II} and G_{III}.

The stability conditions for phases II and III are

$$\mp\Phi_2 \geq 0, \text{ and}$$
$$4\rho^2\Phi_{11} \pm (n-2)n\rho^{n-2}\Phi_2 \pm 4n\rho^4\Phi_{21} + n^2\rho^{n-2}\Phi_{22} \geq 0, \tag{2.38}$$

where the upper sign corresponds to phase II. From Eqs. (2.34) and (2.38), one can deduce that the boundary between phases II and III always corresponds to a discontinuous change in ρ, i.e., the II–III transition is always first-order in agreement with the absence of group–subgroup relationship between the symmetries of the phases. On the other hand, phases II and III are always contiguous to phase I, and from the fact that $\lim_{\rho\to 0}\Phi_1 = a_1 = 0$, one can deduce that the transitions I–II and I–III can be of the second-order, in agreement with the group–subgroup relationship existing between phase I and, respectively, phases II and III. For $n \geq 4$, one has a *line* of second-order transitions, while for $n = 3$, a second-order transition can take place only at a point which is determined by the two conditions $a_1(P,T) = 0$ and $\lim_{\rho\to 0}\Phi_2(P,T,\rho) = 0$. In contrast, no second-order transition is possible from phase I to phase IV, whereas the II–IV and III–IV transitions can be second-order.

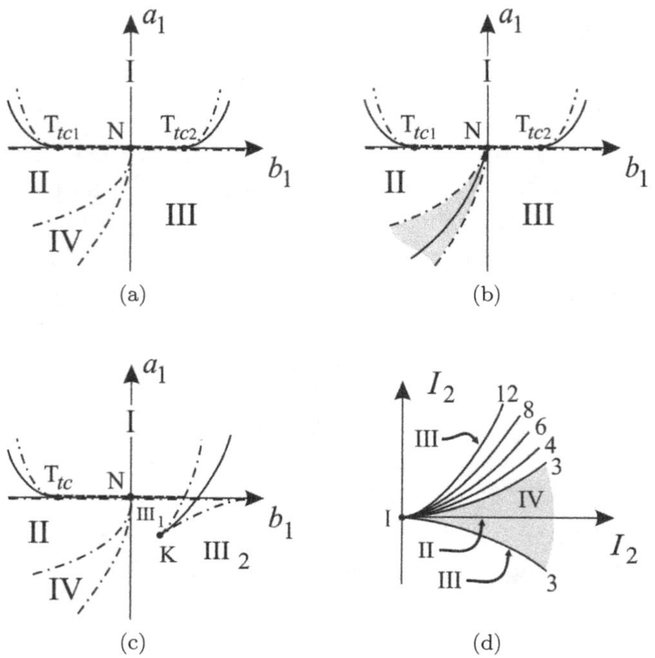

Fig. 2.12. (a)–(c) Phase diagrams associated with the image $6mm$ considering the 12th-degree expansion $F'(I_1, I_2) = a_1 I_1 + a_2 I_1^2 + a_3 I_1^3 + a_4 I_1^4 + b_1 I_2 + b_2 I_2^2 + c_{12} I_1 I_2$ with $I_1 = \rho^2, I_2 = \rho^6 \cos 6\theta$. The diagrams correspond to the following conditions: (a) $\Delta = 4a_2 b_2 - c_{12}^2 > 0$, $a_2 > 0$, $a_3 > 0$, (b) $\Delta < 0$, $a_2 > 0$, $a_3 > a_2$, (c) $\Delta > 0$, $a_2 > 0$, $a_3 < 0$. Full, dashed and dotted-dashed lines are the first-order, second-order,and limit of stability lines, respectively. N is a four- or three-phase point. N′ is a Landau point. $\mathrm{T_{tc1}}$ and $\mathrm{T_{tc2}}$ are the tricritical points. K is a critical point of the liquid-vapor type. (d) Phase diagram for the nmm image-groups in the orbit space. The numbers 3, 4, 6, 8, 12 denote the exponent $I_2 = I_2^{n/2}$ corresponding to phase III ($\eta_1 = \eta_2 = \eta \neq 0$).

The preceding properties are well illustrated by the phase diagrams obtained for $4mm$ (Fig. 1.16) and $3m$ (Fig. 2.9). For $n \geq 6$, due to the absence of cubic invariants, the phase diagrams will be similar to those established for $4mm$. Figure 2.12 shows, for instance, the phase diagrams corresponding to $6mm$ [11], for which 12-degree invariants have to be taken into account for a structurally stable model. Along the same line, one needs to consider 16- and 24-degree terms to work out consistently the phase diagrams associated with $8mm$ and $12mm$, respectively.

Figures 1.15 and 2.6 show that the general topology of the phase diagrams of nmm groups in the orbit space (I_1, I_2) is independent of the number n. Only the shape of the $I_2 = f(I_1)$ curve changes for phase III ($\eta_1 = \eta_2 = \eta = 0$), as one has $I_2 \propto \eta^n \propto I_1^{n/2}$. Figure 2.12(d) summarizes the phase diagrams (I_1, I_2) for the nmm image-groups.

Image-groups n arise in ε_2 for physically irreducible two-dimensional representations. For a given n, group n is a subgroup of index 2 of nmm, and it possesses a larger basis of integrity with the additional invariant $I_3 = \rho^n \sin n\theta$. Thus, the dimension of the basis $m = 3$ is larger than the number of order-parameter components. This implies a relationship between the invariants, which is for any n

$$I_2^2 + I_3^2 = I_1^n, \tag{2.39}$$

corresponding in the orbit-space (I_1, I_2, I_3) to the same cone-type surface representing phase II in Figs. 1.17(a) and 2.10(a). This phase, which is stabilized for $\eta_1 = \rho \cdot \cos\theta \neq \eta_2 = \rho \cdot \sin\theta$, is the only distinct low-symmetry phase. The associated equilibrium values of ρ and θ are determined by the following system of equations:

$$\Phi_1 + n\rho^{n-2}(\Phi_2 \cos n\theta + \Phi_3 \sin n\theta) = 0, \tag{2.40}$$

$$\Phi_2 \sin n\theta + \Phi_3 \cos n\theta = 0. \tag{2.41}$$

The stability conditions

$$\frac{\partial^2 F}{\partial \rho^2} \geq 0 \quad \text{and} \quad \left(\frac{\partial^2 F}{\partial \rho^2}\right) \cdot \left(\frac{\partial^2 F}{\partial \theta^2}\right) - \left(\frac{\partial^2 F}{\partial \rho \partial \theta}\right) \geq 0,$$

have the form

$$\frac{\partial^2 F}{\partial \rho^2} = 4\rho^2 \Phi_{11} + (n-2)\rho^{n-2}(\Phi_2 \cos n\theta + \Phi_3 \sin n\theta) + \cdots \geq 0, \tag{2.42}$$

and

$$\begin{aligned}\frac{\partial^2 F}{\partial \theta^2} &= [-n^2\rho^n(\Phi_2 \cos n\theta + \Phi_3 \sin n\theta) + \cdots] \\ &\quad - [2n\rho^{n-1}(-\Phi_{21} \sin n\theta + \Phi_{31} \cos n\theta) + \cdots]^2 \geq 0.\end{aligned} \tag{2.43}$$

For $n = 3$, it has already been stressed that these conditions lead to a first-order I–II transition. By contrast, for $n \geq 4$, conditions (2.42) and (2.43) are always fulfilled for small ρ, whatever value Φ_1 takes, and consequently the I–II transition may take place as a second-order transition. Accordingly, phase diagrams analogous to the one found for the image-group 4 (Figs. 1.17(b) and 1.17(c)) can be obtained for 6, 8, 12.

2.2.1.4. *Three-component order-parameters: Phase diagrams associated with the image-groups $\overline{4}3m, m\overline{3}$, 432, 23*

The possible stable phases corresponding to the other irreducible three-dimensional images $\overline{4}3m$, $m\overline{3}$, 432 and 23 are listed in Table 2.1. One can see that the image-group $\overline{4}3m$, which is also generated by reflection, displays (as $m\overline{3}m$) $2^3 = 8$ stable states, whereas the three other image-groups have only five possibly stable phases in their phase diagrams. Phases II ($\eta 00$) and III ($\eta\eta\eta$), which are common to all the three-dimensional image-groups, can be reached

Table 2.1. Equilibrium values of the order-parameter components (column (a)) and corresponding isotropy groups (column (b)) for the stable phases associated with the three-dimensional image-groups $\overline{4}3m$, $m\overline{3}$, 432 and 23.

$\overline{4}3m$				$m\overline{3}$	
(a)	(b)	(a)	(b)	(a)	(b)
I. 0 0 0	$\overline{4}3m$	V. $\eta_2\ \eta_2\ \eta_1$ $\eta_1 > \eta_2$	$mm2$	I. 0 0 0	$m\overline{3}$
II. η 0 0	$mm2$	VI. $-\eta\ \eta\ \eta$	$3m$	II. η 0 0	$mm2$
III. $\eta\ \eta\ \eta$	$3m$	VII. $-\eta_1\ \eta_1\ \eta_2$ $\eta_1 > \eta_2$	m	III. $\eta\ \eta\ \eta$	3
IV. $\eta_1\ \eta_1\ \eta_2$ $\eta_1 > \eta_2$	m			IV. $\eta_1\ \eta_2 0$	m
		VIII. $\eta_1\ \eta_2\ \eta_3$	1	V. $\eta_1\ \eta_2\ \eta_3$	1

432		23	
(a)	(b)	(a)	(b)
I. 0 0 0	432	I. 0 0 0	23
II. η 0 0	4	II. η 0 0	2
III. $\eta\eta\eta$	3	III. $\eta\eta\eta$	3
IV. $\eta\eta 0$	2	IV. $-\eta\ \eta\ \eta$	3
V. $\eta_1\ \eta_2\ \eta_3$	1	V. $\eta_1\ \eta_2\ \eta_3$	1

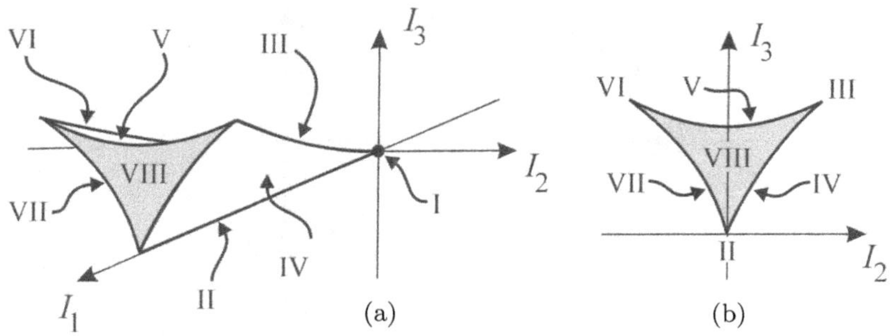

Fig. 2.13. Phase diagram of the $\bar{4}3m$ image-group: (a) In the orbit-space (I_1, I_2, I_3). (b) In a cross-section I_1 =const of the orbit-space.

from phase I (as for $m\bar{3}m$), across the lines of second-order transitions, for the image-groups 432 and $m\bar{3}$. In contrast, due to the existence of a cubic invariant ($I_2 = \eta_1\eta_2\eta_3$) in the integrity basis of the image-groups $\bar{4}3m$ and 23, the preceding phases are separated from phase I by the first-order transition lines, which become second-order at only one point N defined by $\Phi_1(P,T)_{\eta=0} = \Phi_2(P,T)_{\eta=0} = 0$. This isolated second-order critical point corresponds to the point of tangency of the first-order transition lines bounding phases I and II.

The invariants forming the basis of integrity of $\bar{4}3m$ are related to the invariants of $m\bar{3}m$ (see Table 1.2) by $I_1(\bar{4}3m) = I_1(m\bar{3}m)$, $I_2^2(\bar{4}3m) = I_3(m\bar{3}m)$, and $I_3(\bar{4}3m) = I_2(m\bar{3}m)$.

The location of the phases pertaining to the $\bar{4}3m$ phase diagram in the orbit-space (I_1, I_2, I_3) is shown in Fig. 2.13. Since the highest degree of the invariants is *four* (I_3), a consistent description of the phase diagram in the space of the phenomenological coefficients requires an eighth-degree order-parameter expansion model. Using the same notations as for the $m\bar{3}m$ image-group, one gets the equilibrium and stability conditions for the phases coexisting at the critical point N:

(1) for phase II: $\Phi_1 + 2\eta^2\Phi_3 = 0$, and $4\Phi_1^2 - 4\eta^2\Phi_2^2 > 0$, $2\Phi_1 + 4\eta^2(\Phi_{11} + 3\Phi_3) \geq 0, \Phi_3 < 0, \Phi_1 > 0$;
(2) for phases III and VI: $\Phi_1 \pm \eta\Phi_2 + 4\eta^2\Phi_3 = 0$, $\Phi_{11} + 3\Phi_3 > 0$ and $\pm 2\eta(-\Phi_2 \pm 4\eta\Phi_3) > 0$, $-2\Phi_1 + 4\eta^2(3\Phi_{11} + \Phi_3) \geq 0$, $2\Phi_1 + 4\eta^2(\Phi_{11} + 3\Phi_3) > 0$.

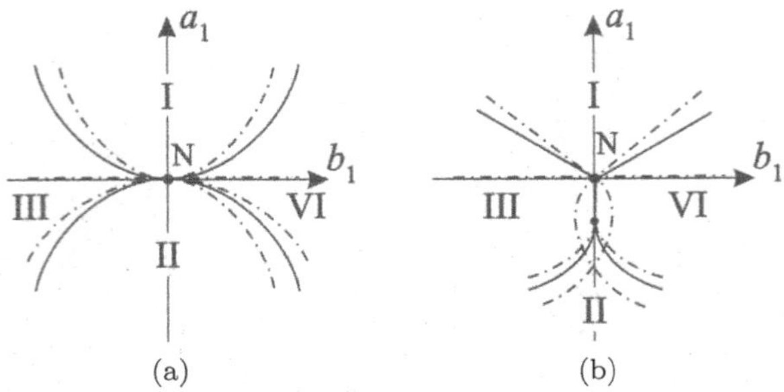

Fig. 2.14. Phase diagrams associated with the image-group $\bar{4}3m$ [11, 12]: (a) corresponds to a fourth-degree expansion with $a_2(I_1^2) > 0$, $a_2(I_1^2) > c_1(I_3) < 0$; (b) an eighth-degree expansion with $a_3(I_1^3) < 0$, $c_2(I_3^2) < 0$. Solid, dashed and dashed-dotted lines are the first-order, second-order and limit of stability lines, respectively.

It leads to the phase diagram represented in Fig. 2.14(a). Figure 2.14(b) shows the region of stability of the phases in an eighth-degree expansion. Figures 2.14(a) and (b) show cross-sections (a_1, b_1) in the vicinity of the four-phase point N.

2.2.1.5. *Multicomponent order-parameters: Six-dimensional image group*

When the number of order-parameter components is larger than $n = 3$, the corresponding number of image-groups strikingly increases. The images of all irreducible representations of 230 space groups, fulfilling the Lifshitz criterion (only commensurate sublattices, Sec. 2.4.3), have been identified and classified in tables of Ref. [13]. The list contains 131 image-groups' dimensionalities from 1 to 24. However, one finds, in literature, not so many examples of phase diagrams corresponding to the relevant order-parameters. This can be understood, since the dimension of the corresponding integrity bases, the degrees of the basis invariants, and the number of possible stable states become too large to solve algebraically and structurally stable models.

Let us briefly consider an example [14] of the six-dimensional order-parameter (E192d in notations of Ref. [12]) not fulfilling the Landau condition. It continues the series of the phenomenological models A1a(point group 1) – B6a($3m$) – C24b($\bar{4}3m$) – E192d – $\cdots$ relevant to the crystallization process of structures with increasing complexity (see Chapter 6). The corresponding image-group E192d consists of 192 matrices, and its basis of invariants includes 13 monomials [14]:

$$\begin{gathered} I_1 = \eta_1^2 + \eta_2^2 + \cdots ; \quad I_2 = \eta_1\eta_3\eta_5 + \cdots ; \quad I_3 = \eta_1^2\eta_2^2 + \cdots ; \\ I_4 = \eta_1^2\eta_3^2 + \cdots ; \quad I_5 = \eta_1^2\eta_4^2 + \cdots ; \quad I_6 = \eta_1\eta_2^2\eta_3\eta_5 + \cdots ; \\ I_7 = \eta_1^2\eta_2^2\eta_3^2 + \cdots ; \quad I_8 = \eta_1^2\eta_3^2\eta_6^2 + \cdots ; \quad I_9 = \eta_1\eta_2\eta_3\eta_4\eta_5\eta_6; \\ I_{10} = \eta_1\eta_2^2\eta_3\eta_4^2\eta_5 + \cdots ; \quad I_{11} = \eta_1^2\eta_2^2\eta_3^2\eta_4^2 + \cdots ; \\ I_{12} = \eta_1^2\eta_2^2\eta_3^2\eta_6^2 + \cdots ; \quad I_{13} = \eta_1\eta_2^2\eta_3\eta_4^2\eta_5^2 + \cdots . \end{gathered} \tag{2.44}$$

Only the first characteristic terms are shown in Eq. (2.44), the other can be produced by permuting subscript indexes. The structurally stable potential is limited by the sixth-degree terms $I_1 - I_9$ and reads

$$\begin{aligned} F = {} & a_1 I_1 + b_2 I_2 + c_{11} I_1^2 + c_3 I_3 + c_4 I_4 + c_5 I_5 + d_{12} I_1 I_2 + d_6 I_6 \\ & + f_{111} I_1^3 + f_{13} I_1 I_3 + f_{14} I_1 I_4 + f_{15} I_1 I_5 + f_{22} I_2^2 \\ & + f_7 I_7 + f_8 I_8 + f_9 I_9. \end{aligned} \tag{2.45}$$

The image-group method described in Chapter 1 (Sec. 1.3.2) provides the set of 17 low-symmetry phases – solutions of the corresponding equations of state. Five among them are one-parametric:

0: (0 0 0 0 0 0), I: (0 0 0 0 0 η), II$^{\pm}$: (0 η 0 η 0 $\pm\eta$), III: (0 0 η η 0 0), IV: ($\eta\,\eta\,\eta\,\eta\,\eta\,\eta$). They appear on the generic phase diagrams of the model (2.45), which is shown in Fig. 2.15, in the vicinity of the multicritical point L ($a_1 = b_1 = c_{11} = 0$). It is easy to see its remarkable similarity in the topology to the phase diagrams corresponding to $3m$ (Fig. 2.8), $\bar{4}3m$ (Fig. 2.14) image-groups.

2.2.2. *Isostructural and anti-isostructural phase transitions*

In the preceding sections, we disclosed and mathematically described two types of specific, isosymmetric phase transitions. These are non-symmetry-breaking isostructural (Secs. 1.3.3.2c and 2.2.1.1) and anti-isostructural transformations. Numerous examples of such transitions can be found in different materials.

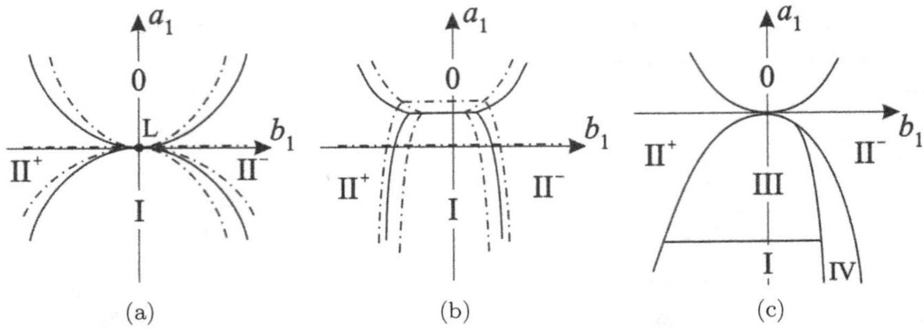

Fig. 2.15. Plane ($a_1 = b_1$) section of multidimensional phase diagram corresponding to the free-energy expansion (2.45) [14]: (a) $c_{11} > 0$, (b) $c_{11} < 0$, (c) $c_3 < 0$, $c_{11} > 0$; remaining coefficients in Eq. (2.45) are positive. Dashed-dotted line is the limit of stability line, solid line is the first-order transition line. L is the Landau point.

The liquid–gas phase transition is the most commonly encountered example of isostructural transition. The discontinuous character of this transition derives from the fact that the two phases adjacent to the transition line are "qualitatively" identical but "quantitatively" distinct, following the formulation of Landau [5]. In other words, the two phases cannot be distinguished by their symmetry, which is isotropic, but by the fluid's density, or the specific volume, which varies by several orders of magnitude between the two phases. This is true far from the *critical point*, which is the end-point of the line of discontinuous transition in the pressure–temperature plane. Beyond the critical point, the two phases are essentially indistinguishable (Figs. 2.2 and 2.3).

Another known example is the metal–insulator transition, where the change concerns the number of electrons localized near a metallic ion [15]. Lyotropic liquid crystals also provide examples of isostructural transitions related to the shape of the micellar aggregates [16]. Among the crystalline solids, there exist a small number of well-established isostructural transitions, such as the transition or cross-over between the FCC phases found in cerium [17, 18]], praseodymium [19], cesium [20] and chalcogenide lanthanides [21], or the transition in $(V_{1-x}Cr_x)_2O_3$ [22]. The isostructural transition is revealed by discontinuities in the cell parameter and by the existence of a critical point of the liquid–gas type, which is found, e.g., at 1.75 GPa and 550 K in Ce [17,23] or for $x = 0.17$ and 580 K (Fig. 2.16)

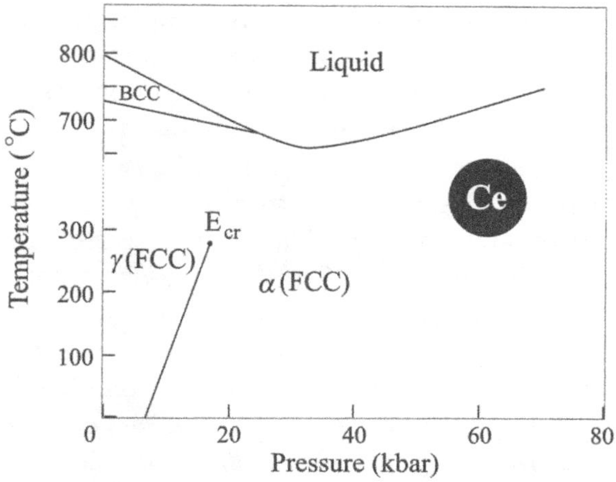

Fig. 2.16. Experimental P–T phase diagram of Ce [17, 23]. E_{cr} is the critical end-point.

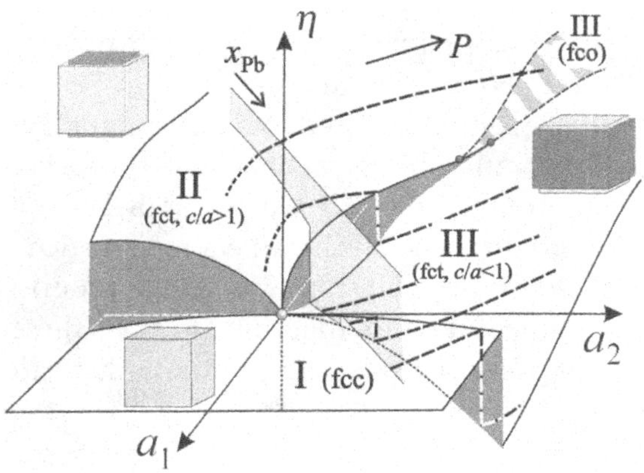

Fig. 2.17. Isothermal–isoconcentration sections of the theoretical phase diagram (Fig. 2.7) for $In_{1-x}Pb_x$ alloy. The grayish plane is an ambient isobaric plane. Vertical planes are (η, P) isothermal sections corresponding to $In_{1-x}Pb_x$ with different Pb content. Dashed lines in the planes show order parameter $\eta = (c/a - 1)$ as a function of pressure.

in $Sm_{1-x}Gd_xS$ [24]. Smooth anomalies of the dielectric and elastic constants as well as in the conductivity are generally observed at isostructural transitions.

The symmetry of the order-parameter (i.e., the equilibrium relationship between the order-parameter components) remains unchanged between two isostructural phases. In contrast, for *anti-isostructural transitions*, though the space groups of the phases are the same, the *sign* of the order-parameter components changes from one phase to another. Thus, the two phases can be distinguished qualitatively, and no critical end-point will be found on a first-order transition line of anti-isostructural transition. For instance, in the cubic–tetragonal transformations found in the In-based alloys $In_{1-x}Pb_x$, $In_{1-x}Sn_x$ and $In_{1-x}Cd_x$ [25], the two tetragonal low-symmetry phases correspond to the opposite signs of the quantity $(c/a - 1)$, where a and c are the tetragonal parameters. In terms of the spontaneous strain tensor components, one has in one phase $e_{zz} > e_{xx} = e_{yy}$, whereas in the other phase $e_{zz} < e_{xx} = e_{yy}$. Accordingly, the two anti-isostructural phases are associated with the opposite values of the order-parameter components $(\eta_1 0)$ and $(-\eta_1 0)$ in the framework of the $3m$-type model assumed for the phase diagram [7,11], the *cubic* invariant changing sign from one phase to the other (Fig. 2.17).

The solid-state solutions $Ni_{1-x}Cu_xCr_2O_4$ provide another example of phase diagram containing anti-isostructural phase transitions. A cooperative Jahn–Teller effect reduces the symmetry of the spinel-type structure from cubic to tetragonal and orthorhombic [26–28]. The $T{-}P{-}x$ diagram of the Ni–Cu–Cr–spinel system fully corresponds to the predictions of the phenomenological theory with the order-parameter possessing the $3m$ image-group (Sec. 2.2.1.2, Fig. 2.7). Its experimental phase diagram (Fig. 2.18) contains tetragonal phases showing both $c/a > 1$ and $c/a < 1$ distortions [26–28].

2.3. "Lock-in" Transitions and the Lifshitz Criterion

In Sec. 1.2.3.3, we have shown that the irreducible order-parameter associated with a transition can be distinguished from the other irreducible degrees of freedom by the fact that the coefficient α_{γ_0} multiplying the quadratic polynomial invariant $\sum_j \eta_{\gamma j}^2$ of the order-parameter components vanishes at T_C, while the similar coefficients $\alpha_{\gamma \neq \gamma_0}$ corresponding to other degrees of freedom are strictly positive at T_C. Hence, in the neighborhood of T_C, α_{γ_0} is the smallest of the coefficients α_γ in Eq. (1.13). In the case of a structural transition, Eq. (1.13) must be expressed differently since the high-symmetry

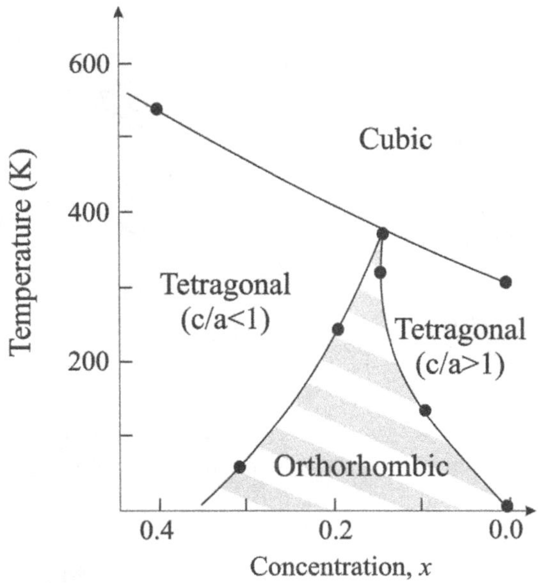

Fig. 2.18. Schematic phase diagram of $Ni_{1-x}Cu_xCr_2O_4$ in the vicinity of the Landau point. Experimental points on the transition lines are taken from [26–28].

group G_0 is a crystallographic space group, i.e., the set of its irreducible representations $\Gamma_{k\gamma}$ *is a continuous set*, since $\boldsymbol{k}$ is any vector of the Brillouin zone of G_0. Consequently, the discrete sum (1.13) over the irreducible representations of G_0 must be replaced by a continuous sum over the volume V^* of the Brillouin zone and also by a discrete sum over the small representations $\hat{\tau}_\gamma(\boldsymbol{k})$ of the "little" group $\hat{G}(\boldsymbol{k})$. One has

$$F = F_0 + \int_{V^*} \left\{ \sum_\gamma \alpha_\gamma[\boldsymbol{k}, T, P, \rho_0(\boldsymbol{r}), \phi_{k'j'}(\boldsymbol{r})] \sum_j \eta_{kj}^2 \right\} d\boldsymbol{k}, \quad (2.46)$$

where the normalized functions $\phi_{kj}(\boldsymbol{r})$ carry the n-dimensional irreducible representation $\Gamma_{k\gamma}$ of G_0. The coefficients $\alpha_\gamma(\boldsymbol{k})$ constitute "branches", with each branch being associated with a given index γ and a continuously varying vector $\boldsymbol{k}$.

Within the preceding continuous description, the order-parameter is the irreducible set η_{kj} corresponding to the minimum with respect to $\boldsymbol{k}$ of the lowest branch $\alpha_\gamma(\boldsymbol{k})$ in the vicinity of T_C. This minimum

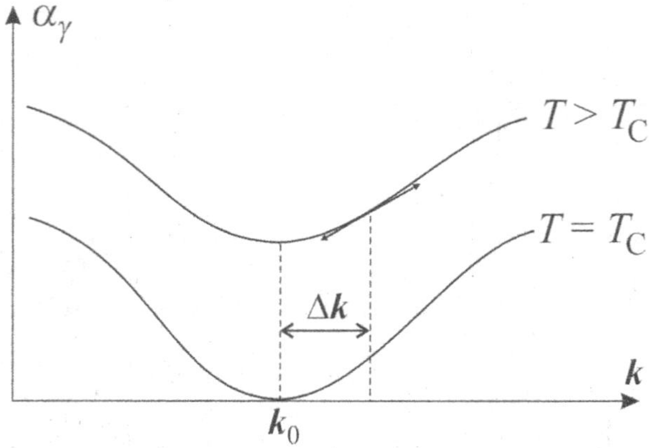

Fig. 2.19. Minimum at $\boldsymbol{k}_0$ for the lowest branch $\alpha_\gamma[\boldsymbol{k}(T)]$. At $\boldsymbol{k} = \boldsymbol{k}_0 + \Delta\boldsymbol{k}$, $\alpha_\gamma(\boldsymbol{k})$ possesses an oblique tangent that corresponds to a linear term in the expansion (2.47).

is equal to zero at T_C (Fig. 2.19). Symmetry considerations impose on the various branches $\alpha_\gamma(\boldsymbol{k})$ the occurrence of extrema at certain points of the corresponding Brillouin zone. If the minimum of $\alpha_\gamma(\boldsymbol{k})$ coincides with one of the extrema, the order-parameter will possess a symmetry-determined $\boldsymbol{k}$ vector whose location in the Brillouin zone will not be submitted to variations when the temperature or pressure is varied: the *entire line* of transitions $T_C(P)$ will be described by the *same irreducible* order-parameter. By contrast, if the minimum of $\alpha_\gamma(\boldsymbol{k})$ does not coincide with a symmetry extremum, one can have a continuous variation of the order-parameter symmetry with T and P.

The symmetry requirements imposed on the order-parameter which imply the occurrence of a symmetry-determined extremum for the relevant coefficient α_{γ_0} are expressed by a criterion initially proposed by Lifshitz [4] and clarified by Dzialoshinskii [29]. We will restrict ourselves to stress its origin: Let $\Gamma_{k_0\gamma_0}$ in Eq. (2.47) be the irreducible representation associated with the coefficient $\alpha_{\gamma_0}(\boldsymbol{k}_0)$. Equation (2.47) contains the degrees of freedom corresponding to $\boldsymbol{k} = \boldsymbol{k}_0 + \Delta\boldsymbol{k}$ whose coefficients $\alpha_\gamma(\boldsymbol{k})$ tend toward $\alpha_{\gamma_0}(\boldsymbol{k}_0)$ when $\Delta\boldsymbol{k} \to 0$. If $\boldsymbol{k}_0$ is a general vector of the Brillouin zone, or a vector ending at a point of high symmetry, there can be one or several branches meeting at $\boldsymbol{k}_0$ for a given direction $\Delta\boldsymbol{k}$. For each branch i

and direction $\Delta \boldsymbol{k}$, one can write the expansion

$$\alpha_\gamma^i(\boldsymbol{k}_0) = \alpha_{\gamma_0}(\boldsymbol{k}_0) + \sum_m \left[\frac{\partial \alpha_\gamma^i}{\partial(\Delta \boldsymbol{k}_m)}\right]_{\Delta k=0} \cdot \Delta \boldsymbol{k}_m + \cdots, \qquad (2.47)$$

where the $\Delta \boldsymbol{k}_m$ are the components of $\Delta \boldsymbol{k}$ in the Brillouin zone. If $\alpha_{\gamma_0}(\boldsymbol{k}_0)$ is an extremum, the term in the second member of Eq. (2.47), which is linear in the $\Delta \boldsymbol{k}_m$, must vanish. In other words, *the presence of a linear term in the expansion* (2.47) *is incompatible with the symmetry of the order-parameter*. Reciprocally, *a necessary condition for a second-order transition to take place toward a (periodic) structure corresponding to* $\boldsymbol{k}_0$ *is the absence of linear terms in the* $\alpha_\gamma(\boldsymbol{k})$ *expansion.* This is the *Lifshitz criterion* [4, 29], which can be formulated in the group-theoretical vocabulary [30]: the antisymmetric square $\{\Gamma^2_{k_0\gamma_0}\}$ of the irreducible representation spanning the order-parameter has no representation in common with the vector representation of the symmetry group G_0 of the parent phase. One can also show that the absence of linear terms in Eq. (2.47) is equivalent to the absence in the order-parameter expansion, which transforms as $\Gamma_{k_0\gamma_0}$, of antisymmetric invariants of the form $\eta_i \frac{\partial \eta_j}{\partial u} - \eta_j \frac{\partial \eta_i}{\partial u}$ (so-called Lifshitz invariants), which couple bilinearly the order-parameter components with their spatial derivatives other than those which are full differentials [30].

The main consequence of the Lifshitz criterion is that it selects in the Brillouin zones associated with the 14 Bravais lattices a small number of high-symmetry points, listed in Refs. [4, 13], which allow the existence of a second-order transition between the crystalline (strictly periodic) structures. Reciprocally, the non-fulfillment of the criterion corresponds to two different possible situations:

(1) A second-order (or slightly first-order) transition to an "inhomogeneous" phase corresponding to a non-strictly periodic structure associated with a temperature-dependent $\boldsymbol{k}$-vector varying closely to $\boldsymbol{k}_0$. The experimental examples [31] reveal that such *incommensurate* phases are stable within a limited interval of temperature, and are generally followed, at lower temperature, by a first-order "lock-in" transition at which the $\boldsymbol{k}$-vector "locks" at a value $\boldsymbol{k}_L$ (lock-in vector) corresponding to the definite multiple of the basic periods of the parent structure.

(2) A direct first-order transition to the lock-in phase characterized by the wave-vector $\boldsymbol{k}_L$.

The two preceding situations are described within the same unifying phenomenological scheme by the competing influence of the Lifshitz invariant $F_1(\eta_j, \frac{\partial \eta_j}{\partial u})$ and of higher-order invariants $F_2(\eta_j)$ of the order-parameter. For small values of η_j, the invariant F_1, which is proportional to the square of the η_j, predominates giving rise to the incommensurate structure, whereas for larger η_j values, the lock-in term F_2 becomes more significant and favors the stabilization of the lock-in phase. In a direct lock-in transition from the parent phase, the underlying incommensurate structure is "squeezed" by the jump of the order-parameter at the first-order transition. A comprehensive description of this phenomenological picture can be found in Ref. [7] (pp. 215–306).

In summary, the violation of the Lifshitz criterion leads to the first-order lock-in transitions, which appear either below an intermediate incommensurate phase or directly below the parent phase. In the first case, the lock-in transition is easily predictable. In the latter case, one may distinguish the lock-in transitions from the standard first-order transitions by the value of the lock-in vectors which often display unusual fractional coordinates associated with "non-Lifshitz" points of the Brillouin zone. Examples of materials undergoing direct or indirect lock-in transitions are given in Refs. [7, 31]. These transitions *always* possess a *first-order* character.

2.4. Phase Transitions Associated with Reducible Order-Parameters

2.4.1. *General features of the phase diagrams associated with a reducible order-parameter*

The transitions examined till now have been assumed to be related to a single irreducible order-parameter, in agreement with the essential statement of the Landau theory [5, 6], which associates to a second-order transition a single irreducible degree of freedom. This critical degree of freedom, which determines the symmetry of the low-temperature phase, has been shown (Sec. 1.2.4) to be possibly coupled to the secondary (non-critical) degrees of freedom, which modify in a non-essential manner the equilibrium and non-equilibrium properties of the system. There exist, however, a number of structural systems reported experimentally, the phase diagram of which must be partly or entirely connected to more than one order-parameter. In particular, we will show in Chapters 5 and 6 that a large fraction of the observed reconstructive transitions requires two and sometimes three symmetry-breaking order-parameters for their description.

When more than one order-parameter has to be used for the description of a phase transition, it means that the number of phenomenological variables which have to be introduced in the model is not minimal, as for one order-parameter, and consequently that the theory is less predictive regarding the properties of the transition. On the other hand, the linear or nonlinear coupling between the order-parameters has the result that the modifications accompanying a transition (change of symmetry, thermal, dielectric and elastic anomalies, etc.) are more complex making more difficult a phenomenological description. Such a description is nevertheless realizable at least when a reduced number of order-parameters are involved, and we will see that specific consequences can be deduced from the corresponding model regarding the topology of the phase diagrams and some physical properties of the systems.

Let us first examine the general symmetry aspects related to the existence of more than one order-parameter. If Γ_1, Γ_2, ..., Γ_j is a set of irreducible representations of a space-group G_0, which corresponds to the symmetry of the parent phase, the sum $\Gamma_1 \oplus \Gamma_2 \oplus \cdots \oplus \Gamma_j$ forms a reducible representation of G_0. If G^i_k are the space groups, respectively, induced by the Γ_i (where k denotes the various low-symmetry groups associated with the same irreducible representation of index i), one may expect, in the phase diagram, the stabilization of low-symmetry phases corresponding to the intersecting symmetries $G^i_k \cap G^i_{k'}$ under the condition that a sufficiently strong coupling exists between the order-parameters transforming as the irreducible representations Γ_i and Γ'_i. Let us note that the first-order character of the transition from the parent phase to a phase resulting from a strong coupling of two order-parameters is unquestionable, since it implies for the coefficients α_i and α'_i of the corresponding quadratic invariants to change the sign quasi-simultaneously in a narrow interval of temperature. As a consequence, part of the order-parameter components will exhibit a discontinuous jump at the transition. The fact that more than one order-parameter is involved for a given low-symmetry phase can be revealed by a variety of structural properties such as (i) the existence of a superstructure which has to be related to more than one wave-vector, (ii) an unusual lowering of the rotational symmetry as compared to the parent phase symmetry. Other properties, such as the domain texture, the structure of the regions of coexistence with neighboring phases associated with a single order-parameter, etc., should also be indicative of the contribution of more than one "primary" order-parameter to the transition mechanism. In the following subsections, we will discuss the most typical phase diagrams associated with different types of couplings between the two order-parameters, which have been mainly worked out in Refs. [32–38].

2.4.2. *Biquadratic coupling between two order-parameters*

The thermodynamic potential associated with a reducible representation $\Gamma_1 \oplus \Gamma_2$ contains invariants of the order-parameter components η_i and ξ_i corresponding, respectively, to Γ_1 and Γ_2, but also (for interacting order-parameters) mixed invariants which couple the two order-parameters. We will first discuss the most studied case of two one-component order-parameters η and ξ involving a biquadratic coupling $\eta^2\xi^2$, which transforms as the image-group $2/m$, i.e., the one-dimensional irreducible representations Γ_1 and Γ_2 span the reducible vector representation of the point-group $2/m$. The form of the order-parameter expansion is

$$F(\eta, \xi) = F_0 + a_1\eta^2 + a_2\eta^4 + \cdots + b_1\xi^2 + b_2\xi^4 + \cdots + \delta\eta^2\xi^2. \tag{2.48}$$

The integrity basis of $F(\eta, \xi)$ is formed by the two square invariants: $I_1 = \eta^2$, $I_2 = \xi^2$, and thus the minimal degree to which the expansion (2.48) has to be considered, in order to describe the full set of low-symmetry phases, is the fourth degree.

(a) The model corresponding to Eq. (2.48) with *two* fourth-degree invariants in η and ξ was first discussed in Ref. [39]. It contains five phenomenological parameters a_1, a_2, b_1, b_2 and δ, which are assumed to satisfy the conditions $a_2 > 0$, $b_2 > 0$, $\delta > -2(a_2b_2)^{1/2}$ ensuring the positive definiteness of $F(\eta, \xi)$ for large values of η and ξ. The corresponding equations of state are

$$\eta(a_1 + 2a_2\eta^2 + 2\delta\xi^2) = 0,$$
$$\xi(b_1 + 2b_2\xi^2 + 2\delta\eta^2) = 0. \tag{2.49}$$

In addition to the high-symmetry phase I, Eqs. (2.49) describe three low-symmetry phases with the following equilibrium values for η and ξ:

$$\text{Phase II}: \eta_{\text{II}}^2 = -\frac{a_1}{2a_2}, \quad \xi_{II} = 0;$$
$$\text{Phase III}: \eta_{\text{III}} = 0, \quad \xi_{\text{III}}^2 = -\frac{b_1}{2b_2};$$
$$\text{Phase IV}: \eta_{\text{IV}}^2 = -\frac{\delta b_1 - 2a_1b_2}{\Delta}, \quad \xi_{\text{IV}}^2 = -\frac{\delta a_1 - 2a_{2b_1}}{\Delta},$$
$$\Delta = 2a_2b_2 - \delta^2. \tag{2.50}$$

Assuming the usual linear dependence on temperature and pressure for a_1 and b_1: $a_1(P,T) = a_T(T - T_{\text{C1}}) + a_P(P - P_{\text{C1}})$, $b_1(P,T) = b_T(T - T_{\text{C1}}) + b_P(P - P_{\text{C1}})$, one can easily draw, from Eqs. (2.50), the following conclusions, which are illustrated in Fig. 2.20: (i) The transitions from the high-symmetry phase I to phases II and III are always second-order. (ii) A direct transition from phase I to phase

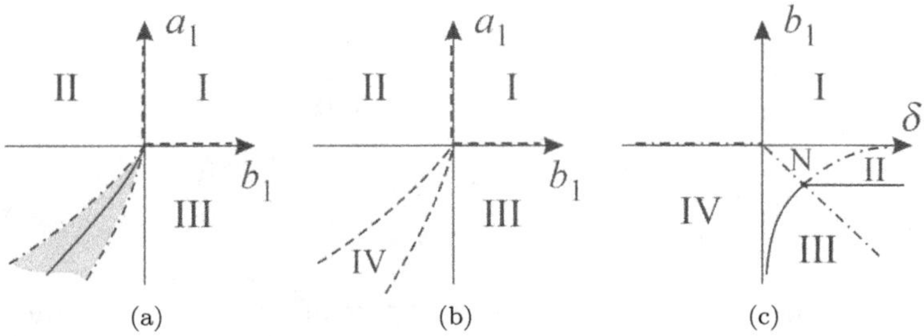

Fig. 2.20. Cross-sections of the phase diagram corresponding to a fourth-degree expansion (2.48) in η and ξ, for (a) $\Delta = 4a_2b_2 - \delta^2 < 0$, (b) $\Delta > 0$, (c) Δ varying from positive to negative values. Full, dashed and dashed-dotted lines are, respectively, the first-order, second-order and limit of stability lines.

IV can only take place at a single point of the phase diagram defined by $a_1(P,T) = b_1(P,T) = 0$. (iii) The transitions between phases II and III occur either as a first-order transition (Fig. 2.20(a)) or across two second-order transition lines enframing phase IV (Fig. 2.20(b)).

Two additional predictions reveal, however, the incompleteness of a fourth-degree expansion model, namely: (1) the stabilization of phase IV implies a negative interaction ($\delta > -2(a_2b_2)^{1/2}$) between the order-parameters, which contradicts the positive definiteness condition for $F(\eta,\xi)$. (2) The model predicts the existence of a three-phase point N, shown in Fig. 2.20(c), of coordinates $a_1 = b_1(a_2/b_2)^{1/2}$, $\delta = 2(a_2b_2)^{1/2}$, which corresponds to a *first-order transition point located at the meeting of two second-order transition lines*, i.e., at N, the second-order transition lines II–IV and III–IV and the first-order transition line II–III meet.

The preceding shortcomings show that a more realistic phase diagram would require to consider a sixth-degree term in $F(\eta,\xi)$ for at least one of the order-parameters.

(b) Let us take the sixth-degree asymmetric expansion first discussed in Refs. [33] and [40]:

$$F(\eta,\xi) = F_0 + a_1\eta^2 + a_2\eta^4 + a_3\eta^6 + b_1\xi^2 + b_2\xi^4 + \cdots + \delta\eta^2\xi^2, \tag{2.51}$$

which yields the following equilibrium values for the low-symmetry phases:

$$\text{Phase II}: \quad \eta_{\text{II}}^2 = \frac{-2a_1 + \sqrt{4a_2 - 12a_1a_3}}{6a_3}, \qquad \xi_{\text{II}} = 0;$$

$$\text{Phase III}: \quad \eta_{\text{III}} = 0, \quad \xi_{\text{III}}^2 = -\frac{b_1}{2b_2};$$

$$\text{Phase IV}: \quad \begin{cases} a_1 + 2a_2\eta_{\text{IV}}^2 + 3a_2\eta_{\text{IV}}^4 + \delta\xi_{\text{IV}}^2 = 0 \\ b_1 + 2b_2\xi_{\text{IV}}^2 + \delta\eta_{\text{IV}}^2 = 0. \end{cases} \tag{2.52}$$

Different phase diagrams are obtained from Eqs. (2.52), and from the corresponding stability conditions, depending on the value of $\Delta = 4a_2b_2 - \delta^2$. For $\Delta > 0$, one gets phase diagrams which are similar to the ones represented in Fig. 2.20. New topologies appear for $\Delta < 0$, i.e., for the value $\eta^2 = -\frac{\Delta}{12}a_3b_2$ which corresponds to the appearance of a new transition line. The results of the minimization of $F(\eta, \xi)$ can be summarized as follows.

(1) For $2(a_2 b_2)^{1/2} > \delta > 0$ and $a_2 > 0$, the III–IV transition is always first-order and takes place on the line of equation

$$a_1 = \frac{\delta b_1}{2b_2} + \frac{\Delta^2}{64 a_3 b_2^2}, \tag{2.53}$$

which is surrounded by the loss of stability line of phases IV and III, respectively, given by

$$a_1 = \frac{\delta b_1}{2b_2} + \frac{\Delta^2}{48 a_3 b_2^2} \text{ and } a_1 = \frac{\delta b_1}{2b_2}.$$

The II–IV second-order transition occurs on the line

$$a_1 = \frac{2a_2 b_1}{\delta} - \frac{3a_3 b_1^2}{\delta^2}. \tag{2.54}$$

At the three-phase point N_2 ($a_1^N = \frac{\Delta\delta}{6a_3 b_2}, b_1^N = \frac{\Delta\delta^2}{12 a_3 b_2^2}$), the line (2.53) transforms into the line of first-order transition II–III. Figures 2.21(a) and 2.21(b) represent two different situations corresponding, respectively, to $a_2 > 0$ (the I–II transition is second-order) and $a_2 < 0$. For such negative values of a_2, the I–II transition becomes first-order, and the corresponding line of transitions goes over continuously to the first-order transition line II–III at $a_1 = a_2^2/4a_3$.

(2) For $\delta < -2\sqrt{|a_2|b_2}$, which expresses a sufficiently large negative interaction between the two order-parameters, one has a different first-order transition line between phases I and IV following the "triggering" mechanism suggested by Holakovsky [40]. One has also two transition regimes between phases II and IV, which are separated by a tricritical point T_{tc} (Fig. 2.21(c)). Let us also note the second three-phase point N_2 which lies either on the b_1 axis (Fig. 2.21(c)) or above this axis (Fig. 2.21(d)).

(c) At last, let us consider the symmetric sixth-degree expansion

$$F(\eta, \xi) = F_0 + a_1\eta^2 + a_2\eta^4 + a_3\eta^6 + b_1\xi^2 + b_2\xi^4 + b_3\xi^6 + \delta\eta^2\xi^2, \tag{2.55}$$

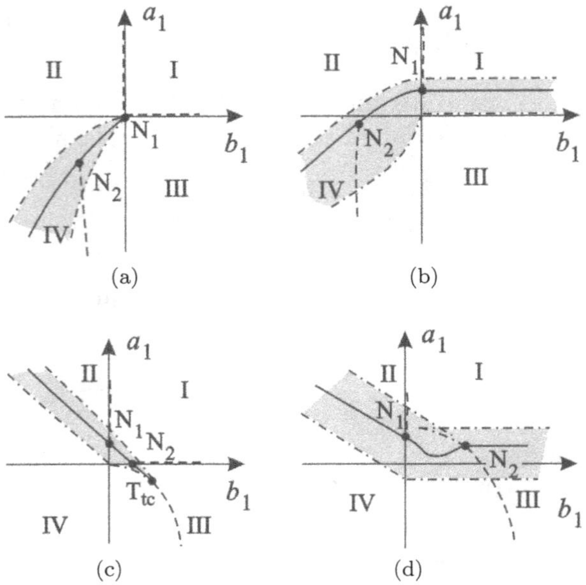

Fig. 2.21. Phase diagrams corresponding to the asymmetric sixth-degree expansion (2.51) for (a) $\delta > 2\sqrt{|a_2|b_2}$, $a_2 > 0$, (b) $\delta > 2\sqrt{|a_2|b_2}$, $a_2 < 0$, (c) $\delta < -2\sqrt{|a_2|b_2}$, $a_2 > 0$, (d) $\delta < -2\sqrt{|a_2|b_2}, a_2 < 0$. Full, dashed and dashed-dotted lines are, respectively, the first-order, second-order and limit of stability lines.

for which the conditions $a_3 > 0$, $b_3 > 0$ are assumed. Figures 2.22(a)–2.22(h) show various cross-sections (a_1, b_1) corresponding to different values of the phenomenological coefficients assuming either a positive (Figs. 2.22(a)–2.22(d)) or negative (Figs. 2.22(e)–2.22(h)) interaction between the two order-parameters. One can see that for $\Delta > 0$, $a_2 > 0$, one finds again the phase diagrams containing exclusively the second-order transitions (as in Fig. 2.20(b)). However, for $\Delta > 0$ and $a_2 < 0$, $b_2 < 0$, there is a region of coexistence of the phases II, III and IV, and in contrast to the asymmetric model, a first-order transition may occur directly between phases I and IV and between phases II and IV. Accordingly, tricritical points may appear both on the II–IV and III–IV transition lines. Furthermore, for $\Delta < 0$, one may have the "reentrant" sequences of phases II–IV–II and III–IV–III. Let us also note that the triple points may replace the three-phase points, and that the second-order phase transitions may disappear completely from the phase diagram.

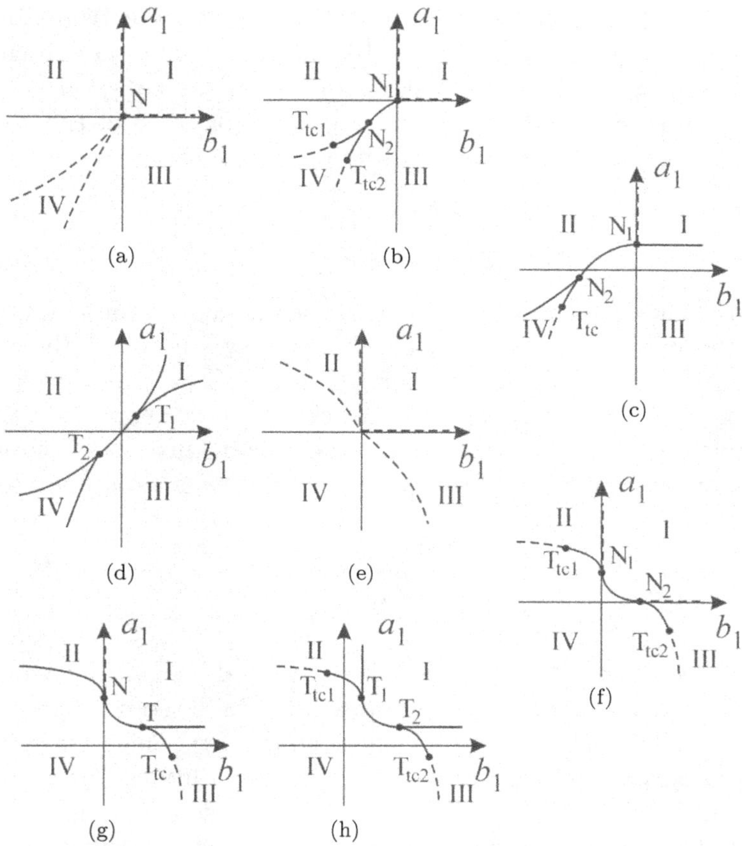

Fig. 2.22. Phase diagrams corresponding to the symmetric sixth-degree expansion (2.55) for: (a) $\delta > 0$, $a_2 > 0$, $b_2 > 0$, $\Delta > 0$; (b) $\delta > 0$, $a_2 < 0$, $b_2 < 0$, $\Delta > 0$; (c) $\delta > 0$, $a_2 < 0$, $b_2 > 0$, $\Delta > 0$; (d) $\delta > 0$, $a_2 < 0$, $b_2 < 0$, $\Delta < 0$; (e) $\delta < 0$, $a_2 > 0$, $b_2 > 0$, $\Delta > 0$; (f) $\delta < 0$, $a_2 > 0$, $b_2 > 0$, $\Delta < 0$; (g) $\delta < 0$, $a_2 < 0$, $b_2 > 0$, $\Delta < 0$; (h) $\delta < 0$, $a_2 < 0$, $b_2 < 0$, $\Delta > 0$. Full and dashed lines are, respectively, the first- and second-order transition lines. The T_{tc}, N and T points are the tricritical, N-phase and triple points, respectively.

(d) Models based on the expansion (2.48) can also be used to construct the phase diagram corresponding to biquadratically coupled multidimensional order-parameters, as this expansion represents the *effective form* of the thermodynamic potential in which there is only one equilibrium value for each of the moduli $|\eta_i|$ and $|\xi_j|$ (see Chapters 4–7). By contrast, when different non-zero equilibrium values are found for the order-parameter components, the resulting phase diagram may become more complex. This is illustrated by the model considered in Ref. [35] of a biquadratic coupling between two three-component order-parameters (η_1, η_2, η_3) and (ξ_1, ξ_2, ξ_3) with $\eta_3 = \xi_2 = \xi_3 = 0$ and $\xi_1 \neq 0$, $\eta_1 \neq \eta_2 \neq 0$ for some of the phases. The effective expansion discussed by Larin is

$$F(\eta_1, \eta_2, \xi) = a_1(\eta_1^2 + \eta_2^2) + a_2(\eta_1^2 + \eta_2^2)^2 + c(\eta_1^4 + \eta_2^4) + d(\eta_1^6 + \eta_2^6) \\ + b_1^2\xi + b_2^4\xi + b_3^6\xi + \delta(\eta_1^2 + \eta_2^2)\xi^2. \quad (2.56)$$

Let us note that the integrity basis for the image-group associated with the expansion (2.56), which has the symmetry of the three-dimensional reducible vector representation of the point-group $\overline{4}$, is $I_1 = \eta_1^2 + \eta_2^2$, $I_2 = \eta_1^2\eta_2^2$, $I_3 = \xi^2$. An eighth-degree expansion would thus be necessary for a structurally stable model instead of the sixth-degree expansion (2.56). The equations of state corresponding to this expansion are

$$\eta_i[a_1 + 2a_2(\eta_1^2 + \eta_2^2) + c\eta_i^2 + 3d\eta_i^4 + \delta\xi^2] = 0\,(i = 1, 2), \\ \xi[b_1 + 2b_2\xi^2 + 3b_3\xi^4 + \delta(\eta_1^2 + \eta_2^2)] = 0. \quad (2.57)$$

It leads to six possible stable states, denoted I–IV, which are listed in Table 2.2. This table provides for each phase the equilibrium values of the order-parameter components (column (b)), which are deduced from the respective equations of state and stability conditions given in columns (c) and (d) of Table 2.2. From these equations, one can deduce the phase diagrams shown in Fig. 2.23, which assume the conditions $c < 0$ and $a_2 + c > 0$. One can verify that the existence of two distinct components η_1 and η_2 modifies in many respects the one-component phase diagrams of Fig. 2.22.

A different biquadratic-coupled one-component order-parameter model containing a cubic invariant in the corresponding expansion:

$$F(\eta, \xi) = b_1\xi^2 + b_2\xi^3 + b_3\xi^4 + a_1\eta^2 + a_2\eta^4 + \delta\xi^2\eta^2, \quad (2.58)$$

has been studied in Ref. [41]. It will be discussed in Chapter 6.

2.4.3. *"Improper" coupling between two order-parameters*

Another typical coupling between two order-parameters is of the form $\delta\eta^n\xi$ with $n \geq 2$. This type of "improper" coupling, which has been

Table 2.2. Stable phases obtained by minimization of the order-parameter expansion (2.56). Column (b): equilibrium values of the order-parameter components. Column (c): equations of state. Column (d): stability conditions and equations of the limit of stability lines.

(a)	(b)	(c)
I	$\eta_1 = \eta_2 = \xi = 0$	–
II	$\eta_1 = \eta_2 = 0,\ \xi \neq 0$	$b_1 + 2b_2\xi^2 + 3b_3\xi^4 = 0$
III	$\eta_1 \neq 0,\ \eta_2 \neq 0,\ \xi \neq 0$	$\begin{cases} a_1 + 2(a_2 + c)\eta^2 + 3d\eta^4 + \delta\xi^2 = 0 \\ b_1 + 2b_2\xi^2 + 3b_3\xi^4 + \delta\eta^2 = 0 \end{cases}$
IV	$\eta_1 = \eta_2 \neq 0,\ \xi \neq 0$	$\begin{cases} a_1 + 2(2a_2 + c)\eta^2 + 3d\eta^4 + \delta\xi^2 = 0 \\ b_1 + 2b_2\xi^2 + 3b_3\xi^4 + 2\delta\eta^2 = 0 \end{cases}$
V	$\eta_1 \neq 0,\ \eta_2 = \xi = 0$	$a_1 + 2(a_2 + c)\eta^2 + 3d\eta^4 = 0$
VI	$\eta_1 = \eta_2 \neq 0,\ \xi = 0$	$a_1 + 2(2a_2 + c)\eta^2 + 3d\eta^4 = 0$

(a)	(b)	(d)
I	$\eta_1 = \eta_2 = \xi = 0$	$a_1 \geq 0, b_1 \geq 0$
II	$\eta_1 = \eta_2 0 = 0, \xi \neq 0$	$b_1 = 0, b_2 \geq 0$ $b_1 = \frac{b_2^2}{3b_3}, b_2 < 0, b_1 = \frac{2b_2}{\delta}a_1 - \frac{3b_2}{\delta^2}a_1^2$
III	$\eta_1 \neq 0, \eta_2 \neq 0, \xi \neq 0$	$a_1 = \frac{\delta b_1}{2b_2}, \Delta_1 = 4(a_2 + c)b_2 - \delta^2 \geq 0,$ $a_1 = \frac{\delta b_1}{2b_2} + \frac{\Delta_1}{48b_2^2}, \Delta_1 > 0,$ $a_1 = \frac{\delta b_1}{2b_2} + \frac{\Delta_0}{3b_2 d}, c \leq 0$ $(\Delta_0 = 4a_2b_2 - \delta^2 \geq 0)$
IV	$\eta_1 = \eta_2 \neq 0, \xi \neq 0$	$a_1 = \frac{\delta b_1}{2b_2}, \Delta_2 = 4(2a_2 + c)b_2 - \delta^2 \geq 0,$ $a_1 = \frac{\delta b_1}{2b_2} + \frac{\Delta_2^2}{12b_2 d}, \Delta_2 < 0,$ $a_1 = \frac{\delta b_1}{2b_2} + \frac{\Delta_0 c}{3b_2 d} + \frac{c^2}{d}$
V	$\eta_1 \neq 0, \eta_2 = \xi = 0$	$a_1 = 0, a_3 + c \geq 0, a_1 = \frac{4a_2c}{3d},$ $a_1 = \frac{(a_2 + c)^2}{3d}, a_2 + c < 0,$ $a_1 = \frac{2(a_2 + c)b_1}{d} - \frac{3b_1^2}{d^2}, b_1\delta < 0$
VI	$\eta_1 = \eta_2 \neq 0, \xi = 0$	$a_1 = 0, 2a_2 + c \geq 0, a_1 = \frac{4a_2c - c^2}{3d},$ $a_1 = \frac{(2a_3 + c)^2}{3d}, 2a_2 + c < 0,$ $a_1 = \frac{2(a_2+c)b_1}{d} - \frac{3b_1^2}{d^2}, b_1\delta < 0$

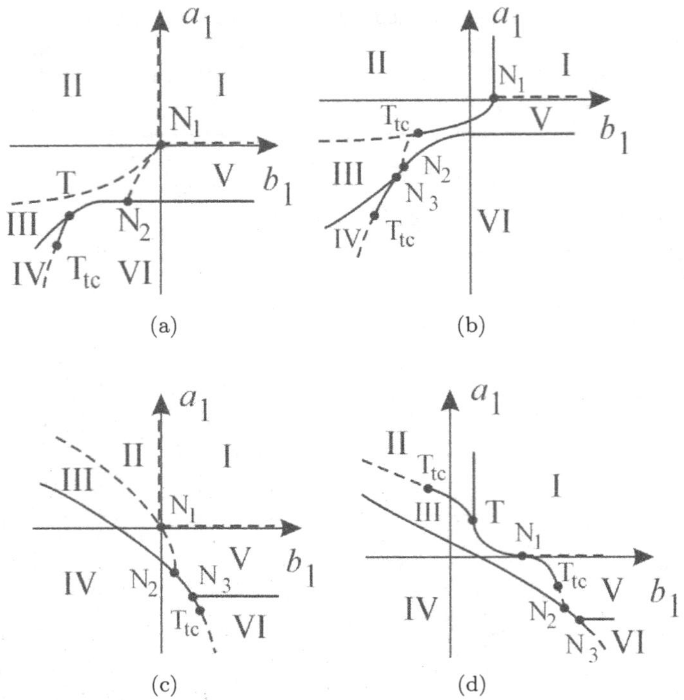

Fig. 2.23. Phase diagrams corresponding to the expansion (2.56) for $c < 0$, $a_2 + c > 0$, $d > 0$, $b_3 > 0$ and (a) $\Delta_0 = 4a_2b_2 - \delta^2 > 0$, (b) $\delta > 0$, $\Delta_0 < 0$, (c) $\delta < 0$, $\Delta_0 > 0$, (d) $\delta_0 < 0$. Full and dashed lines are, respectively, the first- and second-order transition lines. N_1, N_2 and N_3 are the N-phase points. T is a triple point. T_{tc}, T_{tc1} and T_{tc2} are the tricritical points.

discussed in Sec. 1.2.4, will be shown to result here in spontaneous values of ξ even in the absence of non-zero values for η. The lowest degree coupling ($n = 2$) has been considered in Refs. [36–38, 42].

Let us first discuss the order-parameter expansion

$$F(\eta, \xi) = F_0 + a_1\eta^2 + a_2\eta^4 + b_1\xi^2 + b_2\xi^4 + \delta\eta^2\xi, \qquad (2.59)$$

assuming $a_2 > 0$, $b_2 > 0$. In the absence of interactions between the two order-parameters ($\delta = 0$), one gets the same four stable phases found in the biquadratically coupled order-parameter model (2.48). The existence of the nonlinear interaction ($\delta \neq 0$) destabilizes the phase II ($\eta \neq 0, \xi = 0$) in the preceding model, and one has only three stable phases:

Phase I: $\eta = \xi = 0$; Phase II: $\eta = 0, \xi \neq 0$; Phase III: $\eta \neq 0, \xi \neq 0$.

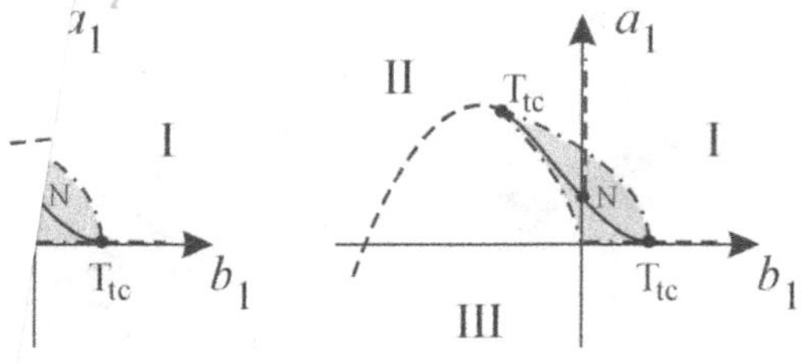

(a) (b)

Fig. agrams corresponding to: (a) the expansion (2.59) assuming
0; (b) to the expansion (2.59) with the additional term $\gamma\eta^2\xi^2$
shed and dashed-dotted lines are, respectively, the first-order,
a_2 mit of stability lines.

wi

d III, the non-zero equilibrium value of ξ is induced ng even in the absence of non-zero equilibrium valase II. This can be seen from the equations of state to (2.59):

2.60)

on-
at

$$\eta(a_1 + 2a_2\eta^2 + \delta\xi) = 0,$$
$$2b_1\xi + 4b_2\xi^3 + \delta\eta^2 = 0,$$

ch the condition $\eta = 0$ implies either $\xi = 0$ or $\xi^2 = -b_1/2b_2$. respect, let us stress that at variance with the improper cou- between the "primary" and "secondary" order-parameters dis- sed in Sec. 1.2.4, here ξ does not play the role of a secondary rder-parameter, since its equilibrium values influence the symmetry of phases II and III.

Figure 2.24(a) shows the phase diagram associated with expansion (2.59). It contains a second-order transition between phases I and III along the line $a_0 = 0$ for $b_1 > \delta^2/4$. At the tricritical point T_{tc1} of coordinates $(\delta^2/4a_2, 0)$, the preceding line becomes a first-order transition line, which transforms again, for negative values of b_1 into a second-order transition line, at a second tricritical point T_{tc2}. Between T_{tc1} and T_{tc2}, one has a three-phase point of coordinates $(0, \delta^2/8(a_2b_2)^{1/2})$.

A consistent model actually requires to take into account a biquadratic coupling $\gamma\eta^2\xi^2$ in Eq. (2.59), which is always allowed by symmetry. As one can see in Fig. 2.24(b), this adjunction does not modify the essential features of Fig. 2.24(a). By contrast, the topology of the phase diagram is drastically changed when a cubic

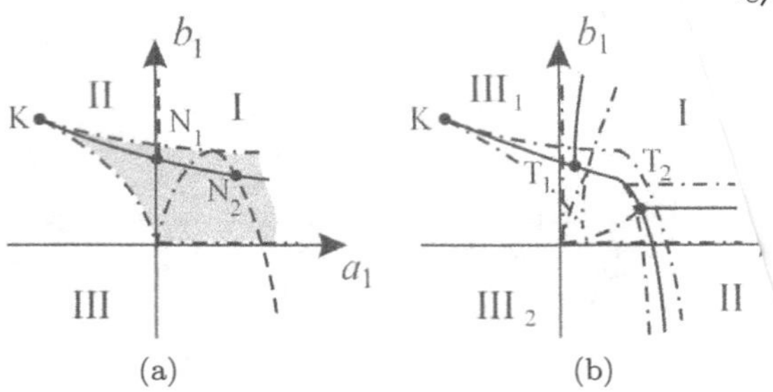

Fig. 2.25. Phase diagrams corresponding to the expansions (2.60) (
(b). Full, dashed and dashed-dotted lines are, respectively, the first-or
order and limit of stability lines.

invariant ξ^3 exists in the order-parameter expansion

$$F(\eta, \xi) = F_0 + a_1\eta^2 + a_2\eta^4 + b_1\xi^2 + b_2\xi^4 + b_3\xi^3 + \delta\eta^2\xi.$$

The same three phases I, II, III are found but with a different c figuration for their regions of stability. Thus, Fig. 2.25(a) shows t the first-order transition line between phases I and II, given by

$$b_1 = \frac{b_3^2}{4b_2}, \tag{2.61}$$

is prolongated for negative values of a_1 by a first-order transition line, separately for the two isostructural phases III_1 and III_2, of Eq. (2.62):

$$\tilde{a}_1 = -\frac{\delta a_1}{2a_2} - \frac{b_3}{2b_2}\tilde{b}_1 - \frac{b_3^2}{16b_2^2} = 0. \tag{2.62}$$

It meets the second-order transition line I–III_1 ($a_1 = 0$), and it meets the first-order transition line I–II (2.61) and the second-order transition line II–III_2 of equation $b_1 = b_3^2/4b_2$.

Because of the cubic invariant ξ^3, which, due to the nonlinear interaction $\eta^2\xi$, is proportional to η^6, one has to include a sixth-degree invariant in Eq. (2.59) for a better consistency of the model:

$$F(\eta, \xi) = F_0 + a_1\eta^2 + a_2\eta^4 + a_3\eta^6 + b_1\xi^2 + b_2\xi^4 + b_3\xi^3 + \delta\eta^2\xi. \tag{2.63}$$

It yields the phase diagram represented in Fig. 2.25(b). One can see that although the stability region of phases I and II remains almost unchanged, the I–III transition becomes first-order, and the corresponding transition line ends at a triple point T_1. A second triple point T_2 takes place at the convergence point of the first-order transition lines I–II and I–III.

2.4.4. *"Pseudo-proper" coupling between two order-parameters*

Another type of coupling may occur when two order-parameters representing different physical quantities transform as the same irreducible representation. Their "pseudo-proper" coupling, mentioned in Sec. 1.2.4.4, is expressed by a bilinear invariant of the form $\eta_i \xi_j$ in the order-parameter expansion. A model involving one-component order-parameters η, ξ which transform as the reducible two-dimensional image-group 2 was discussed in Ref. [9]. The corresponding basis of integrity is formed by three invariants:

$$I_1 = \eta^2, I_2 = \xi^2, I_3 = \eta\xi. \tag{2.64}$$

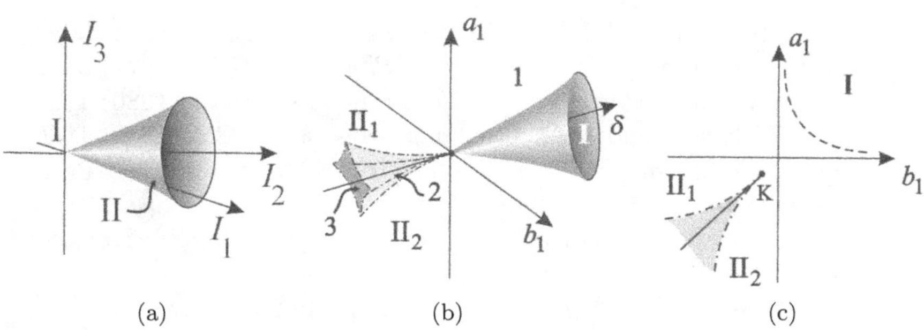

Fig. 2.26. Phase diagrams corresponding to the pseudo-proper coupling model based on expansion (2.65). (a) Configuration of the phases in the orbit-space. (b) Phase diagram in the (a_1, b_1, δ) space. The numbers denote the second-order transition surface I–II (1), the surface limiting the region of coexistence of the isostructural phases II_1 and II_2 (2), and the isostructural transition plane (3). The edges limiting this plane are the critical lines (4), which become a critical point in the cross-section (a_1, b_1). (c) Phase diagram in the (a_1, b_1) plane. Dashed, full and dashed-dotted lines are, respectively, the second-order, first-order, and limit of stability lines.

The order-parameter expansion has the form

$$F(\eta, \xi) = a_1\eta^2 + a_2\eta^4 + b_1\xi^2 + b_2\xi^4 + \delta\eta\xi + \gamma\eta^2\xi^2, \tag{2.65}$$

where the invariant I_3 expresses the interaction between the two order-parameters. The equations of state minimizing F are

$$\begin{aligned} &2\eta(a_1 + 2a_2\eta^2 + \gamma\xi^2) + \delta\xi = 0, \\ &2\xi(b_1 + 2b_2\xi^2 + \gamma\eta^2) + \delta\eta = 0, \end{aligned} \tag{2.66}$$

which give rise to *two* stable phases: phase I $-\ \eta = \xi = 0$ and phase II $-\ \eta = \xi \neq 0$. Figure 2.26(a) represents the configuration of the two phases in the orbit-space, whereas Figs. 2.26(b) and 2.26(c) show the phase diagram in the phenomenological coefficients space. Thus, the transition between phases I and II is always second-order, and one has an isostructural transition between the two variants of phase II. The metastability region in which those two variants coexist ends at a critical point of the liquid–gas type. In order to have a first-order regime between phases I and II, one needs to add a sixth-degree term in η or ξ to Eq. (2.65).

References

[1] (a) P. Toledano and J. C. Toledano, *Phys. Rev. B* **14** (1976) 3097.
(b) P. Toledano and J. C. Toledano, *Phys. Rev. B* **16** (1977) 386.
(c) J. C. Toledano and P. Toledano, *Phys. Rev. B* **21** (1980) 1139.
(d) P. Toledano and J. C. Toledano, *Phys. Rev. B* **25** (1982) 1946.

[2] *Landolt-Börnstein, Ferro- and Antiferroelectric Substances*, Band 3 (Springer Verlag, Berlin, 1969).

[3] P. Toledano and V. Dmitriev, *Reconstructive Phase Transitions* (World Scientific, Singapore, 1996).

[4] E. M. Lifshitz, *Zh. Eksp. Teor. Fiz.* **11** (1941) 255.

[5] L. D. Landau, *Zh. Eksp. Teor. Fiziki* **7** (1937) 19; **7** (1937) 627. Transl. in *Collected Papers of L. D. Landau*, ed. D. Ter Haar (Pergamon Press, Oxford, 1965), p. 193; p. 216.

[6] L. D. Landau and E. M. Lifshitz, *Statistical Physics. Part 3.* 3rd edition (Elsevier, Amsterdam, 1980).

[7] J. C. Toledano and P. Toledano, *The Landau Theory of Phase Transitions* (World Scientific, Singapore, 1987).

[8] V. P. Sakhnenko and V. M. Talanov, *Sov. Phys.-Solid State* **21** (1980) 1401.

[9] E. I. Kut'in, V. L. Lorman and S. V. Pavlov, *Sov. Phys. Usp.* **34** (1991) 497.
[10] T. Poston and I. N. Stewart, *Catastroph Theory and its Application* (Pitman, London, 1978).
[11] Yu. M. Gufan, *Structural Phase Transitions* (Nauka, Moscow, 1982).
[12] A. A. Mukovin and V. M. Talanov, *Eur. Phys. J. B* **87** (2014) 34.
[13] H. T. Stokes and D. M. Hatch, *Isotropy Subgroups of the 230 Crystallographic Space Groups* (World Scientific, Singapore, 1988).
[14] V. M. Talanov, M. V. Talanov and V. B. Shirokov, *Crystallogr. Rep.* **61** (2016) 159.
[15] N. F. Mott and E. A. Davis, *Electronic Processes in Non-Crystalline Materials* (Clarendon, Oxford, 1971).
[16] *Phase Transitions in Complex Fluids*, eds. P. Toledano and A. M. Figueiredo Neto (World Scientific, Singapore, 1998).
[17] A. Jayaraman, *Phys. Rev.* **137** (1965) 179.
[18] A. K. Singh, *High Temp.-High Press.* **12** (1980) 47.
[19] A. Yu. Kuznetsov, V. P. Dmitriev, O. I. Bandilet and H.-P. Weber, *Phys. Rev. B* **68** (2003) 064109.
[20] H. T. Hall, L. Merrill and J. D. Barnett, *Science* **146** (1964) 1297.
[21] A. Chatterjee, A. K. Sing and A. Jayaraman, *Phys. Rev. B* **6** (1972) 2275.
[22] A. A. Bugaev, B. P. Zakharchenya and F. A. Chudnovskii, *Metal-Insulator Phase Transition and Its Applications* (Nauka, Leningrad, 1979).
[23] D. A. Young, *Phase Diagrams of Elements* (University of California Press, Berkley, 1991).
[24] I. L. Aptekar' and E. Yu. Tonkov, *Sov. Phys.-Solid State* **21** (1979) 110.
[25] V. P. Dmitriev, D. Chernyshov, Y. E. Filinchuk and V. F. Degtyareva, *Phys. Rev. B* **75** (2007) 024111.
[26] Y. Kino and S. Miyahara, *J. Phys. Soc. Japan* **21** (1966) 2732.
[27] M. Tovar, R. Tobari, C. Welker and F. Fleischer, *Physica B: Condens. Matter* **385** (2006) 196.
[28] A. S. Mikheykin, V. I. Torgashev, V. M. Talanov *et al.*, *J. Phys.: Condens. Matter* **26** (2014) 505401.
[29] I. E. Dzialoshinskii, *Sov. Phys.: JETP* **19** (1964) 960.
[30] S. Goshen, D. Mukamel and S. Shtrikman, *Int. J. Magn.* **6** (1974) 221.
[31] *Incommensurate Phases in Dielectrics*, eds. R. Blinc and A. P. Levanyuk (North Holland, Amsterdam, 1986).
[32] Yu. M. Gufan, V. P. Dmitriev, *Sov. Phys. Crystallogr.* **25** (1980) 6.
[33] Yu. M. Gufan and E. S. Larin, *Sov. Phys.-Solid State* **22** (1980) 270.
[34] V. P. Dmitriev, *Sov. Phys. Crystallogr.* **25** (1981) 683.

[35] E. S. Larin, *Sov. Phys.-Solid State* **26** (1984) 1820.
[36] Yu. M. Gufan and V. I. Torgashev, *Sov. Phys.-Solid State* **22** (1980) 951.
[37] Yu. M. Gufan and V. I. Torgashev, *Sov. Phys.-Solid State* **23** (1981) 1129.
[38] E. I. Kut'in, V. L. Lorman and Ya. E. Cherner, *Sov. Phys.-Solid State* **29** (1988) 1819.
[39] E. M. Lifshitz, *Zh. Eksp. Teor. Fiz.* **14** (1944) 353.
[40] J. Holakovsky, *Phys. Stat. Sol. (b)* **56** (1973) 615.
[41] V. P. Dmitriev and S. B. Rochal, *J. Low Temp. Phys.* **14** (1988) 456.
[42] Yu. M. Gufan, E. I. Kut'in, V. L. Lorman and E. N. Sidorenko, *Sov. Phys.-Solid State* **29** (1987) 432.

Chapter 3

Phase Transitions in Crystals: Classes and Types

3.1. Classification of Phase Transitions

The revolutionary idea of the Landau approach to phase transitions presented in Chapter 1 consists in using symmetry: space groups, their irreducible representations, and the corresponding basis functions which directly identify the transformation mechanism. In the symmetry-based phenomenological schema, a phase transition breaks symmetry of the system, and the transformation is controlled by its order-parameter that is zero in the parent phase and takes non-zero value in the low-symmetry phase (see Sec. 1.2). From the symmetry properties of the order-parameter, it is possible to infer, on the one hand, a certain number of observable symmetry characteristics of the system. On the other hand, macroscopic quantities can be classified as a function of their symmetries with respect to the order-parameter. From this can be deduced a classification of the type of anomalous behavior induced by the occurrence of the phase transition. The order-parameter, $\eta_{\boldsymbol{k}_i}$, in the phenomenological approach is introduced via the increment of probability density in a crystal at a phase transition (see Eqs. (1.12) and (1.15)). It is worth underlying that these equations are an expansion in a *basis* but not in *series* that allows $\eta_{\boldsymbol{k}_i}$ to be non-small, as it primarily characterizes just a contribution ("weight") of each symmetry coordinate in the total structure distortion.

Expansion of Eqs. (1.12) and (1.15) allows assigning a phase transition to a certain class that typifies the transformation. We consider in the following sections such classes of phase transitions in order to summarize the vocabulary used in the preceding chapters and establish a unified terminology we use in further analysis.

3.1.1. *Thermodynamic: Ehrenfest classification*

The most general characteristic of a phase transition, which does not come, however, from the symmetry-based phenomenological approach, is its "order": this basic thermodynamic classification considers the thermodynamic potential $F(T,P)$ (free energy) and its derivative at the transition. If the first derivations of $F(T,P)$ exhibit jump singularities there, the transition is classified *first-order* (*discontinuous*). If the first derivatives are continuous, but the second ones are singular, we have a *second-order* (*continuous*) transition. Although the Ehrenfest scheme is an inaccurate method of classifying the phase transitions, it is still widely used. Its main criticism is its "mean-field" character – it is inaccurate in the vicinity of phase transitions, as it neglects the role of critical thermodynamic fluctuations. However, in general, at a first-order transition, there is no diverging correlation length, and therefore, no critical phenomena are to be expected. Moreover, for structural phase transitions, the crucial role of fluctuations has also been shown to be of importance in the extremely limited vicinity of the point of continuous phase transition (see Sec. 4.5.1).

3.1.2. *Symmetry criteria*

In Sec. 1.2.3, we showed that the increment of probability density in a crystal at a phase transition can be described in terms of basis functions $\varphi_{\boldsymbol{k}_i}(\boldsymbol{r})$ of certain "active" irreducible representations of the parent G_0 space group (see Eq. (1.12)). The basis functions $\varphi_{\boldsymbol{k}_i}(\boldsymbol{r})$ are Bloch-type functions consisting of a plane wave envelope function $\exp\{i\boldsymbol{kr}\}$ is multiplied by a periodic function $U_{\boldsymbol{k}_i}(\boldsymbol{r})$ as follows:

$$\varphi_{kj}(\boldsymbol{r}) = e^{i\boldsymbol{kr}} \cdot U_{kj}(\boldsymbol{r}), \tag{3.1}$$

where $U_{\boldsymbol{k}_i}(\boldsymbol{r})$ has the period $\boldsymbol{T}$ of the crystal lattice with $U_{\boldsymbol{k}_i}(\boldsymbol{r}) = U_{\boldsymbol{k}_i}(\boldsymbol{r}+\boldsymbol{T})$. The exponential functions from Eq. (3.1) constitute the basis for an irreducible representation of the *subgroup of translations* of the space group G_0. $U_{\boldsymbol{k}_i}(\boldsymbol{r})$ does the same for representations of the "little" group $G_{\boldsymbol{k}}$ of the relevant vector $\boldsymbol{k}$ of the reciprocal space.

Hence, the symmetry-related classification distinguishes (i) the translation symmetry breaking, and (ii) the change of the total symmetry characterized by a space group.

3.1.2.1. *Translation symmetry*

In analyzing the translation properties of a phase G_d that arises as a result of the $G_0 \to G_d$ phase transition, it suffices to consider an expansion of the G_d density function in the form

$$\rho(\boldsymbol{T}_j) = \sum_j \exp(i\boldsymbol{k}_j \boldsymbol{T}_j) \cdot \rho_j(0), \tag{3.2}$$

where the summation is performed over all the arms $\boldsymbol{k}_1, \ldots, \boldsymbol{k}_j$ of the reciprocal space vector star $\{\boldsymbol{k}\}$. This equation relates the density in the jth crystal unit cell to the density $\Sigma_j \rho_j(0)$ in the zero cell, and it is a direct consequence of Eq. (3.1). The unit cell vectors $\boldsymbol{T}_d$ of a distorted, low-symmetry structure are determined from the condition that the density function $\rho(\boldsymbol{T}_j)$ is invariant under all the translations of the new phase G_d. Using Eq. (3.2), one obtains

$$\exp(i\boldsymbol{k}_j \boldsymbol{T}_d) = 1. \tag{3.3}$$

Two possibilities exist: (i) we consider OPs with zero $\boldsymbol{k}$-vector (Γ-point of the corresponding Brillouin zone). Then, there is no change of translational symmetry at the transition. In this case, the transition is called *equitranslational* or *ferrodistortive.* (ii) If $\boldsymbol{k}_j \neq 0$, Eq. (3.3) defines new lattice primitive vectors which are linear combinations of the parent phase lattice basis vectors. In this case, we call the transition *non-equitranslational* or *antiferrodistortive.*

3.1.2.2. *Total symmetry*

The total symmetry change is considered mainly in a symmetry-based definition of the reconstructive phase transformation. The latter is identified as one breaking the group–subgroup relationship between the structures adjusting the transformation point. In contrast, it is worth recalling the transitions without changing the space group of the structure — *isosymmetric* transformations. Two transformation types can be distinguished for this case: (i) *isostructural* transformations, and (ii) *anti-isostructural* ones (see Sec. 2.2.2).

3.1.3. *Microscopic mechanisms*

A phase transition from the parent phase gives rise to a state which, at a microscopic level, may be specified by the occurrence of a spontaneous tensorial property on each atom position. In group-theoretical terms, it means that a representation relevant to the order-parameter (active irreducible representation) is the direct product of three representations: the representation of the subgroup of translations, a permutation representation, and the representation of the point group operating with the components of the tensor which specifies the *microscopic mechanism* of the phase transition. Variation of a scalar (zero-rank tensor) results in the *ordering*-type phase transition characterized by a change in the relative probability for different atoms of occupying certain crystallographic positions.

In the case of *displacive*-type phase transitions, a polar displacement vector (first-rank tensor) is assigned to an atom. It is worth underlying that the attribution, a transition to the displacive type, means confirming its *phonon-driven* character.

The assignment of an axial vector (pseudovector) yields a static magnetic moment in a *magnetic* phase transition.

3.1.4. *Macroscopic mechanisms*

General classification can be formulated as well in terms of the level of the point symmetry. In the general case, the point group of one phase (the ferroic phase) is a strict subgroup of the other one (the parent phase). The transition is then classified as *ferroic.* At a ferroic transition, components of certain macroscopic tensor acquire non-zero values in the low-symmetry phase. If two point groups are identical, the corresponding transition is classified as *non-ferroic.*

The lowering of the point-group symmetry at ferroic phase transitions is accompanied by an increased number of independent components of some property tensors, i.e., the appearance of spontaneous tensor components, which equal zero in the parent phase and non-zero in the ferroic phase. In the case of equitranslational (ferrodistortive) phase transitions, the symmetry of the corresponding tensor may be identical to the symmetry of the primary (symmetry-breaking) order parameter. Then, if the primary OP transforms as a vector, the equitranslational phase transition is called a *proper ferroelectric* transition. Similarly, if the primary OP transforms as components of a symmetric second-rank tensor, the corresponding principal tensor parameter is a spontaneous strain, and the equi-translational phase transition is called *proper ferroelastic*.

The property tensor components (secondary OP) may transform as a symmetric square $[\tau_i]^2$ of the primary order-parameter belonging to a non-central point of the Brillouin zone of the parent crystal lattice. The corresponding non-equitranslational phase transition is called *improper* (see Sec. 2.4.3). For example, if a symmetric square of the primary order-parameter contains vector representation, $[\tau_i]^2 \supset [V]$, we classify the transformation as improper ferroelectric. In case $[\tau_i]^2 \supset [e_{ij}]$, we attribute the transition to improper ferroelastic.

3.2. Martensitic Transformations

3.2.1. *Definition*

The body-centered cubic (BCC), face-centered cubic (FCC), and hexagonal close-packed (HCP) monoatomic structures and phase transitions between them appear in our book as the most expressive and transparent examples of structure reconstruction. On the one hand, we use, in the following chapters, transformations between these simple structures in order to demonstrate the main ideas constituting the basement of the theory of reconstructive phase transitions: transcendental form of the order-parameter and onset of new symmetry elements breaking the group–subgroup relationship. On the other hand, these three structures represent the basic structures of the majority of so-called martensites, which occur in alloy systems [1]. The transformations between three close-packed structures represent a very important class of *martensitic* transformations so far only mentioned in our book. We did not discuss yet these transformations although the comparison of two dominating approaches to phase transitions in metals and alloys is inevitable, as both of them are applied to the same materials and the same structural transitions.

There is a widespread and broad significance of martensitic transformations in materials science. The first-analyzed example was the transformation of a solid solution of carbon in FCC iron (austenite) to the one based on BCC iron (martensite), the latter giving the name to the whole class of such transformations.

Transitions between the three above structures, widely presented in most part of the metals from the periodic table, allow us briefly to compare the two approaches, phenomenological theory of phase transitions and theory of martensitic transformations, showing their strong and weak points, emphasizing difference in vocabularies, and finally, demonstrating their complementarity.

The most convenient and concise definition of the martensitic transformation reads today as "*a shear dominant, lattice distortive, diffusionless transformation occurring by nucleation and growth*" [2]. Using definitions from Sec. 3.1, one translates this definition as follows: (i) "shear dominant, diffusionless" means "displacive"; (ii) "lattice distortive" predicts the onset of spontaneous strains – "ferroelastic" (proper or improper); (iii) "transformation occurring by nucleation and growth" concludes on its strongly first-order character, i.e., the existence of a high interphase energy barrier which impedes bulk transformation.

To extend the definition, morphological, thermodynamic and kinetic features are considered in the literature. A dozen features identify "true" martensitic reaction: a plate-like *morphology* resulting from the transformation; certain *orientational relationship* between the lattices of the initial and final phases; a change in the form of the transformed region, which manifests in *characteristic relief* on the plane surface of the specimen; the high *growth rate* of the martensitic crystals; the athermal- or isothermal-operative *nucleation kinetics* of martensitic transformations, etc. [3].

3.2.2. *Martensitic vs structural*

Two decisive conclusions rise both from the definition and the list of characteristic details identifying the martensitic transformation: (i) the corresponding analysis does not speak in symmetry terms, and (ii) it does consider a transformation of a massive *bulk sample* but not a material crystal structure unit, constructing, therefore, a general pattern of the *real structure of crystalline body*. In fact, the term "martensitic transformation" has a much more general interpretation than the textbook definition of the "phase transition" in the field of structural transformations. Indeed, some of the metallurgists insist, e.g., that even deformation twinning should strictly be regarded as a special case of stress-induced martensitic transformation [4]: to classify twinning as a *transformation* is possible, but it is evidently not a *transition* since in a phase space both twins fix the same multidimensional "point", i.e., they are identical thermodynamic states of the material.

However, the approach is fully justified if we remember when and how the theory was created. First, the core ideas of the theory of martensitic transformations appeared in the 1920s, i.e., in the period of time when X-ray crystallography was only starting its glorious history, and it did not yet provide researchers with comprehensive structural information. Second, the corresponding studies were strongly motivated by the practical industrial challenges. The theory was raised in connection with the study of steel – the major *construction* material. Metallurgy requested explanations and predictions of strength, plasticity, endurance and durability of bulky constructions, i.e., *macroscopic* properties considerably controlled by the mesoscopic structure of sample: defects, its grain and twin structure, and structure of the walls separating them. Theory of martensitic transformations managed to successfully establish links between the mesoscale structure elements and the elastic properties of the real crystalline bodies.

By contrast, the crystal-geometrical symmetry-based approach to phase transitions appeared in the late 1930s and looked for primary *microscopic* symmetry-breaking instability; it considered the latter as a reason for a uniform transformation of a parent crystal structure and its lattice to a distorted one. However, it was limited to map the relevant atomic displacements as shifts of dimensionless points. The consideration evidently neglects the inter-atomic interactions which are responsible particularly for the general capability of the crystal lattice undergoing deformations. To account for the elastic subsystem of the crystal, the symmetry-based approach should consider not only the symmetry-breaking primary order-parameter but also the secondary, induced ones (see Sec. 1.2.4).

Roughly speaking, theory of martensites aims at integrating different components of the transformation mechanism into the unifying reaction of the whole material body to external effects, while the symmetry-based phenomenological approach dissects a whole transformation process into elementary components (symmetry coordinates) revealing primary instability factors of the crystal structure at an atomic level, and estimating the impact of each contribution. This leads to a difference in the theory vocabularies – the same terms sometimes do not have exactly the same meaning in two approaches (see, for instance, review paper [5]). To avoid misunderstanding when comparing the results and predictions by the two theories, let us consider an example demonstrating such language divergences.

3.2.3. *Displacive character of the transformation*

The term "phase transition mechanism" means, in symmetry-based theory, a pure lattice instability action (change in occupation probability, atom displacement, etc.), while in martensite-type theory, it is attributed to a mesoscopic event transforming a single crystal specimen to twinned, defected agglomeration of distorted crystallites. In particular, the theory of martensitic transformations considers atomic displacements in much more general aspect compared to the strict definition of the phenomenological Landau-type theory. Historically, the "metallurgical" definition intended to emphasize, first of all, the diffusionless character of the transformation, i.e., its principal distinction from the widely presented ordering process in alloys.

In the symmetry-based approach, as we demonstrated in Sec. 3.1.2, the symmetry of relevant functions (3.1) identifies transformation type. The term "displacive" unequivocally directs one to the phonon-driven mechanism – corresponding symmetrized functions form eigenvectors of the relevant acoustic or optic phonon modes.

On the one hand, by limiting the number of active components of the transformation mechanism, symmetry-based theory remains far from answering a simple but fundamental question of metallurgical theory: How do the atoms move? On the other hand, Landau-type approach allows distinguishing primary, critical for the phase, stability factors, and secondary, simply following the former. Anticipating the results of detailed analysis of the transitions between three close-packed structures developed in Chapter 5, we can already say that in the BCC–FCC transformation both acoustic phonons (displacements, share modes) and elastic effects (spontaneous strains) control the structure rearrangement and play the role of primary, symmetry-breaking, order-parameters. The conclusion directly stems from the fact that BCC–FCC transformation is equitranslational, and both mechanisms have identical symmetry: their $U_{\boldsymbol{k}_i}(\boldsymbol{r})$ functions in Eq. (3.1) transform as components of the symmetrical second-rank tensor (two-dimensional irreducible representation of the $m3m$ point group). The transformation, therefore, should be classified as (pseudo)proper ferroelastic (Sec. 3.1.4).

In variance, the BCC–HCP transformation breaks the translational symmetry doubling the number of atoms in the *primitive* unit cell of the structure: $\mathrm{Z_{BCC}} = 1$, while $\mathrm{Z_{HCP}} = 2$. Thus, symmetry-breaking phonon belongs to the Brillouin-zone (BZ) boundary (M-point), and all relevant anomalies in the BZ center (Γ-point), including elastic one, are induced by the primary displacive order-parameter. The BCC–HCP transformation is improper ferroelastic (see Sec. 3.1.4).

Distinguishing two mechanisms on the symmetry basis is not simply a formal procedure. It detects a difference in the temperature, pressure etc. dependence of two important contributions in the martensitic order-parameter. In the BCC–FCC mechanism, atom displacements and spontaneous strains are coupled bilinearly (see Eq. (2.65)), and their temperature behavior is identical: $e_{ij} \sim \eta \sim \sqrt{T}$. In the BCC–HCP mechanism, due to the secondary character of spontaneous strains, the coupling is linear-quadratic (see Eq. (2.59)), and the corresponding dependences are different: $\eta \sim \sqrt{T}, e_{ij} \sim \eta^2 \sim T$.

Summarizing, in the theory of martensitic transformation, due to its integrating complexity and macroscopic character, defining an order-parameter is an old and difficult problem. Moreover, one finds even very pessimistic but misleading statements that "there is no obvious order-parameter associated with martensitic transformation" [6]. By contrast, identifying the symmetry-breaking order-parameter in the Landau-type formalism and establishing the form of its coupling to the secondary mechanisms are routine procedures. Thus, it is obvious immediately how the symmetry-based analysis can contribute to the theory of martensitic transformation.

References

[1] G. B. Olson and M. Cohen, *Metall. Trans.* **7A** (1976) 1905.

[2] J. B. Christian, G. B. Olson and M. Cohen, *J. Phys. IV* **5** (1995) C8-3-10.
[3] A. L. Roitburd and G. V. Kurdjumov, *Mater. Eng.* **39** (1979) 141.
[4] J. B. Christian and J. Mahajan, *Progr. Mater. Sci.* **39** (1995) 1.
[5] C. Cayron, *Crystals* **8** (2018) 181.
[6] P. C. Clapp, *J. Phys. IV* **5** (1995) C8-11.

Chapter 4

Advanced Phenomenological Theory of Discontinuous Phase Transitions

4.1. Introduction

In Chapter 1 (Sec. 1.2.3.3), we have seen that the Landau theory uses the increment $\delta\rho(\boldsymbol{r})$ of the probability density, expressing the difference between the initial density in the high-symmetry parent phase, $\rho_0(\boldsymbol{r})$, and the final low-symmetry phase, $\rho_d(\boldsymbol{r})$, expanded as a function of the basis functions of an irreducible representation (IR) τ_{kj} of the space group G_0 of the parent phase [1, 2]. This expansion has the form

$$\delta\rho(\boldsymbol{r}) = \rho_d(\boldsymbol{r}) - \rho_0(\boldsymbol{r}) = \sum_{k,j,i} \eta^i_{kj}\varphi^i_{kj}(\boldsymbol{r}). \tag{4.1}$$

The wave vector $\boldsymbol{k}$, located in the first Brillouin zone, characterizes the translational symmetry of the basis functions $\varphi^i_{kj}(\boldsymbol{r})$, which are the *linear* combinations of the local atomic functions, associated with the crystalline structure. The index j labels the representations τ_{kj} of G_0, and the index i ($i = 1, \ldots, n$) runs over the distinct basis functions spanning the n-dimensional IR τ_{kj} and the corresponding representation space ε. For a given j, the set of scalar coefficients η^i_{kj} defines the *order-parameter*, which describes the total distortion of the initial structure at the transition. Usually, a single irreducible

OP breaks the symmetry in a phase transition, so we only keep index i in Eq. (4.1).

Naturally, we consider the symmetry identity of η_i and $\phi_i(\boldsymbol{r})$ as a general property, which is also valid for the more general renormalization group approach. Indeed, a linear *projection operator* of a space group induces basis functions for the relevant IR in the form of the linear combinations of local atomic functions. The latter are linked to the phase transition mechanism and were selected as a result of a *regular* renormalization transformation, separating critical and non-critical variables [3–5]. The integral over the non-critical variables gives the equilibrium part of the free energy F_0, and the unintegrated part forms the variational free energy (Landau potential) $F_L(\delta\rho)$ (see Sec. 1.2.3.3).

The substitution in F of the *functions* $\varphi^i_{kj}(\boldsymbol{r})$, by the *scalar* order-parameter components η^i_{kj}, makes the problem of the description of a transition more general and easier to handle, as it does not require specific information about the concrete structural modifications taking place at the transition. One needs only to know the symmetry of Γ_{kj}, or equivalently the corresponding space group change $G_0 \to G$, for a description of the essential features of the transition. The order-parameter components define the order-parameter space ε_n that is irreducible invariant space by the group G_0. The $\delta\rho$ variation of the probability density, associated with a phase transition, can be considered as a vector in the representation ε_n-space, and the components of $\eta = \{\eta_i^{eq}\}$ invariant vector, in the basis of this space, are the values of the order-parameter that minimize the thermodynamic potential.

This is one of the main achievements of the Landau theory of phase transitions, as in such an abstract form, it allows one to describe the whole classes of transitions in a unified way, exclusively on the basis of symmetry considerations, and requires a minimal amount of experimental information about the transition. It leads also to establish universal behaviors depending exclusively on the symmetry of the order-parameter. *However, it is precisely this universal aspect that forbids the applicability of the Landau theory to transitions in which the group–subgroup relationship, between the structures of the phases, is lost.*

Another important limitation of the conventional formulation of the Landau theory of structural phase transitions for its application to group–subgroup unrelated structures is *the space-group description of the structures.* A space-group is defined as a set of *symmetry elements* corresponding to the *spatial transformation rules* exchanging the coordinates of the sites in a given structure. The success and usefulness of the space-group concept is that it allows defining the classes of structures without considering the specific configuration of atoms. By contrast, it forbids accounting for *the transformation of a symmetry element into another*, or for *the onset of new symmetry elements*, which characterize reconstructive transformations, as it will be illustrated in the next section. A more adapted approach to describe the structures in the case of reconstructive transitions is to consider them as a *specific configuration of sites.* Such a representation of a structure leaves the possibility to transform a symmetry element into another (e.g., the six-fold axis of a structure transforms into the four-fold axis of another structure across the transition), by a *continuous* deformation of the initial sites, and thus to expand this deformation in symmetric functions, *without imposing a group–subgroup relationship between the structures of the phases.* Thus, in case of reconstructive phase transitions, we can, first, define an order-parameter in terms of density waves, and then, when the symmetry of corresponding order-parameter (its IR) is identified, use a conventional phenomenological procedure operating scalar order-parameter component. In such form, the advanced phenomenological theory avoids the "group–subgroup" limitation and accounts for *strongly discontinuous* phase transitions with no exception.

In this chapter, we will show how a phenomenological theory can be formulated for strongly discontinuous and even for reconstructive phase transitions by generalizing and modifying a number of fundamental aspects of the conventional Landau scheme [6–10]. In view of constructing this theory, we will use the formalism developed for X-ray analysis in which a structure is expressed in terms of the interference of an infinite number of density waves. This density-wave approach will be first introduced through simple working examples, corresponding to the main classes of reconstructive transitions.

4.2. Introductory Examples to the Density-Wave Approach

4.2.1. *A first example of displacive reconstructive phase transition*

4.2.1.1. *Crystallographic description of the transition mechanism*

Let us consider a crystal of hexagonal $P6_3/mmc$ symmetry in which only the two-fold positions 2(c):(1/3 2/3 1/4; 1/3 2/3 1/4) are occupied. We assume that the degree of freedom associated with the transition mechanism is a set of antiparallel displacements of the atoms, denoted ξ_i with $i = 1, 2$, from their initial 2(c) positions, along the direction Z of the six-fold axis (Fig. 4.1(a)). For small arbitrary displacements of equal magnitudes $\xi_1 = -\xi_2 = \xi$, the displaced atoms do not lie in a single plane, so that the screw axes are lost, and the six-fold symmetry of the crystal is lowered to the rhombohedral symmorphic space group $P\bar{3}m1$, which is a subgroup of $P6_3/mmc$. If ξ increases, the rhombohedral symmetry is preserved until ξ reaches the specific value $\xi_0 = c/4$, where c is the hexagonal lattice parameter. For this value, the z coordinate of the two atoms becomes identical ($z = 0$, or $z = 1/2$), and as they lie in the same plane, the symmetry of the crystal *increases* to the simple hexagonal structure of symmetry $P6/mmm$ (Wyckoff position 2(d)), as illustrated in Fig. 4.1(b). This hexagonal structure possesses two atoms

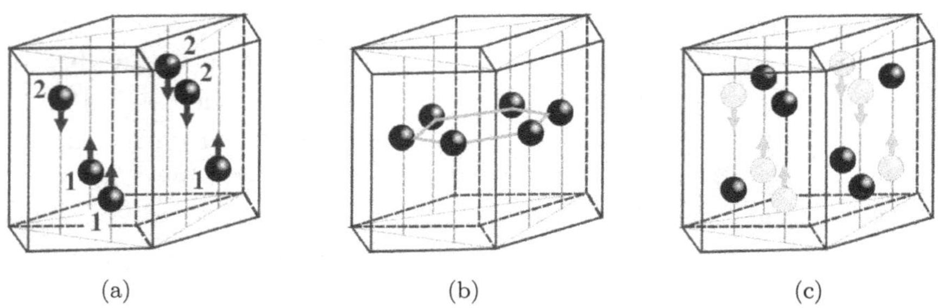

Fig. 4.1. Antiparallel shifting mechanism from the $P6_3/mmc$ structure (a), which leads to $P6/mmm$ (b) and (c) represents the two orientational domains of the $P6_3/mmc$ structure. Black and gray circles pertain to two different domains.

per unit cell, and is equitranslational ($V_0 = V_d$) to the initial hexagonal structure. Accordingly, $P6/mmm$ *is not a subgroup of* $P6_3/mmc$, i.e., *the group–subgroup relationship has been lost in the preceding displacive mechanism for the specific displacement of magnitude* ξ_0.

Let us note that when displacing the atoms in opposite direction (Fig. 4.1(c)), one gets the same sequence of structures: rhombohedral $P\bar{3}m1$ for $|\xi|<c/4$, and hexagonal $P6/mmm$ for $|\xi| = c/4$, but the corresponding crystal lattice is turned by 60° around the six-fold axis, with respect to the lattice obtained for opposite displacements. In other words, one gets *differently oriented domains* of the rhombohedral and simple hexagonal structures. Along the same line, for antiparallel displacements of magnitude $|\xi|= c/2$, the initial symmetry $P6_3/mmc$ is restored, but the corresponding lattice is rotated by $\pi/3$ with respect to the original hexagonal lattice (Fig. 4.1(c)).

4.2.1.2. *Density-wave description of the phase transition*

Let us express the distribution of atoms in the considered structures by a *lattice function* $\psi(\boldsymbol{r})$, defined as the sum of the first harmonics of a Fourier series

$$\psi\left(\boldsymbol{r}\right) = \frac{1}{N}\sum_{j=1}^{N} A_j e^{i\boldsymbol{k}(\boldsymbol{r}-\boldsymbol{R}_j)} + \text{c.c.}, \tag{4.2}$$

where $\boldsymbol{r}$ is the position vector, N is the number of atoms in the unit cell, the A_j are normalized amplitudes, $\boldsymbol{k}$ is the wave vector associated with the first harmonics, and $\boldsymbol{R}_j$ denotes the position of the jth atom. As shown in Fig. 4.2(a), Eq. (4.2) can be interpreted as a sum of *stationary density waves*, which propagate in the direction $\boldsymbol{k}$ perpendicular to the atomic planes forming the structure, i.e., the maxima of the waves coincide with the average atomic positions.

In the initial ($P6_3/mmc$) structure, one has only two atoms per unit cell, located in planes orthogonal to the Z axis at the coordinates z_0^1 and $z_0^2 = -z_0^1$, which are distant by $c/2$. The corresponding lattice function can thus be written in the direction Z of the displacements (neglecting the "noncritical" density waves along X and Y) as

$$\psi_0(z) = \frac{A}{2}\left\{\cos\left[c^*\left(z_0^1 - z\right)\right] + \cos\left[c^*\left(z_0^2 - z\right)\right]\right\}, \tag{4.3}$$

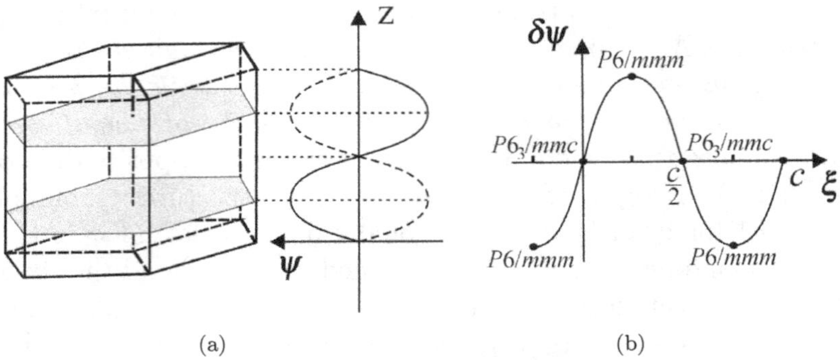

Fig. 4.2. (a) Connection between density waves and atomic planes for the $P6_3/mmc$ structure. (b) Periodic dependence of the lattice function increment $\delta\psi(\xi)$.

where $A = A_1 = A_2$, and $\boldsymbol{k} = \boldsymbol{c}^*$ is the primitive reciprocal lattice translation along Z. When shifting the atoms by the antiparallel displacements of magnitude $\xi_1 = -\xi_2$, the lattice function becomes

$$\psi(z) = \frac{A}{2}\left\{\cos\left[c^*\left(z_0^1 + \xi_1 - z\right)\right] + \cos\left[c^*\left(z_0^2 + \xi_2 - z\right)\right]\right\}. \quad (4.4)$$

When $z_0^1 = c/4 = -z_0^2$ and $\xi_1 = -\xi_2 = \xi$, the increment $\delta\psi(z,\xi) = \psi_0(z,\xi) - \psi(z,\xi)$ takes the form

$$\begin{aligned}\psi(z,\xi) = \frac{A}{2}&\left\{\cos\left[c^*\left(\frac{c}{4} + \xi - z\right)\right] + \cos\left[c^*\left(-\frac{c}{4} - \xi - z\right)\right]\right\} \\ &+ \frac{A}{2}\left\{\cos\left[c^*\left(-\frac{c}{4} - z\right)\right] - \cos\left[c^*\left(\frac{c}{4} - z\right)\right]\right\}. \quad (4.5)\end{aligned}$$

Since $(\boldsymbol{c}^* \cdot \boldsymbol{c})/4 = \pi/2$, one gets

$$\psi(z,\xi) = -2A \cdot \sin(c^*\xi) \cdot \cos(c^* z). \quad (4.6)$$

Whereas for unequal magnitude of the shifts $\xi_1 \neq -\xi_2$, one has

$$\psi(z,\xi_1,\xi_2) = -A \cdot [\sin(c^*\xi_1) - \sin(c^*\xi_2)] \cdot \cos(c^* z). \quad (4.7)$$

Equations (4.6) and (4.7) express *the periodic deformation of the initial hexagonal lattice as a function of the magnitude of the displacements.* One can see in Fig. 4.2(b) following Eq. (4.6) that the $P6_3/mmc$ structure is obtained for $\xi = 0$, $c/2$, c, $3c/2, \ldots$, whereas the $P6/mmm$ structure corresponds to the successive displacements

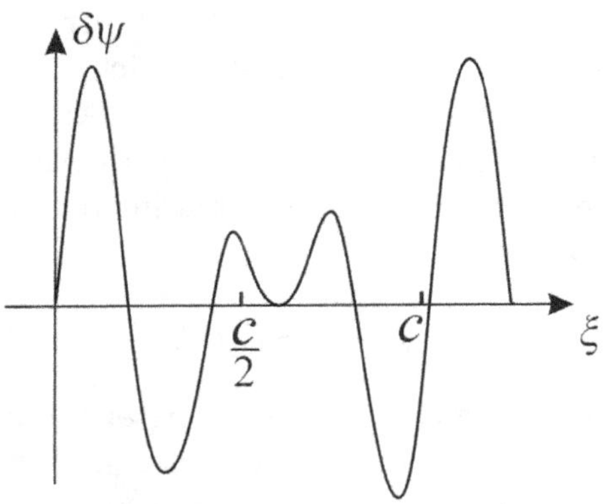

Fig. 4.3. Dependence of $\delta\psi(\xi)$ as given by Eq. (4.7) for two unequal antiparallel shifts ξ_1 and ξ_2.

$\xi = c/4,\ 3c/4,\ 5c/4, \ldots$. For all other displacements, the structure of the crystal is lowered to the rhombohedral $P\bar{3}m1$ structure, which corresponds to the maximal substructure common to the $P6_3/mmc$ and $P6/mmm$ structures. Let us note that for the hexagonal structures, two successive displacements (e.g., $\xi = 0,\ c/2$ for $P6_3/mmc$, and $\xi = (c/4,\ 3c/4)$ for $P6/mmm$) coincide with the *pairs of differently oriented domains* mentioned in Sec. 4.2.1.1. Let us also stress that Eq. (4.6) holds for *infinite* displacements.

For unequal magnitudes of the displacements $\xi_1 \neq \xi_2$, one should use Eq. (4.7), which corresponds to a family of aperiodic curves determined by the respective values of ξ_1 and ξ_2. Figure 4.3 shows the curve corresponding to $\xi_1/\xi_2 = \sqrt{2}$ in which the initial hexagonal symmetry is lowered to $P3m1$ for general displacements. At the extrema of the curve, the crystal structure increases to $P6/mmm$, whereas the initial symmetry is recovered for the points of intersection of the curve with the axis of displacements. At last, let us mention that the situation in which the displacements are not parallel to the Z direction requires a more elaborated approach, but will not lead, in general, to a periodic dependence of the structural deformation as a function of displacements.

4.2.1.3. *Thermodynamic potential and phase diagram*

In the preceding density-wave approach to the $P6_3/mmc$-$P6/mmm$ reconstructive transition, the *order-parameter* corresponding to the transition mechanism (the deformation of the structure) is expressed in function of the atomic displacements, which constitute the *variational parameters*. Equation (4.6) provides the explicit dependence of the structural deformation in function of the displacements of magnitude ξ. Hence, one can define the order-parameter η as the variational part of $\delta\psi(z, \xi)$ namely

$$\eta(\xi) = \eta_0 \sin(c^* \xi), \tag{4.8}$$

where η_0 is a normalizing constant. Parameter η fulfills the standard conditions imposed to the order-parameter in the Landau theory, namely $\eta = 0$ for the initial phase, and $\eta \neq 0$ for the rhombohedral phase as well as for the "limit" hexagonal phase ($P6/mmmc$), which satisfies the additional conditions to coincide with the extrema of $\eta(\xi)$: $\partial\eta/\partial\xi = 0$.

Following the procedure described in Secs 4.2.2 and 4.2.3, one can express the *thermodynamic potential* F associated with the transition from the initial $P6_3/mmc$ phase, by an expansion of the order-parameter $\eta(\xi)$, the form of which is determined by the transformation property of $\eta(\xi)$ under the symmetry operations of $P6_3/mmc$. As $\eta(\xi)$ is an odd function of ξ, the functional $F[\eta(\xi)]$ contains only even powers of the order-parameter. Restricting to the sixth degree, which is the minimal degree necessary to have a direct first-order $P6_3/mmc$–$P6/mmm$ transition, one can write

$$F[\eta(\xi)] = a_1\eta^2(\xi) + a_2\eta^4(\xi) + a_3\eta^6(\xi), \tag{4.9}$$

where the phenomenological coefficients are assumed to obey the standard conditions: $a_1 = a_{10}(T - T_C)$, $a_{10} > 0$, $a_2 = \text{const}$, $a_3 = \text{const} > 0$.

The minimization of F has to be performed with respect to the variational parameter ξ, so that the equation of state can be written under the general form

$$\frac{\partial F}{\partial \eta} \cdot \frac{\partial \eta}{\partial \xi} = 0, \tag{4.10}$$

which gives explicitly

$$2\eta \cdot \frac{\partial \eta}{\partial \xi}\left(a_1 + 2a_2\eta^2 + 3a_3\eta^4\right) = 0. \tag{4.11}$$

The stability condition has the form

$$\frac{\partial^2 F}{\partial \eta^2} \cdot \left(\frac{\partial \eta}{\partial \xi}\right)^2 + \frac{\partial F}{\partial \eta} \cdot \frac{\partial^2 \eta}{\partial \xi^2} \geq 0, \tag{4.12}$$

which is expressed by

$$2\left(a_1 + 2a_2\eta^2 + 3a_3\eta^4\right)\left[\left(\frac{\partial \eta}{\partial \xi}\right)^2 + \eta\frac{\partial^2 \eta}{\partial \xi^2}\right] + 8\eta^2\left(\frac{\partial \eta}{\partial \xi}\right)^2\left(a_2 + 3a_3\eta^2\right) \geq 0. \tag{4.13}$$

Equations (4.11) and (4.13) yield, using the form (4.8) of $\eta(\xi)$, three possible stable states:

(1) the initial $P6_3/mmc$ phase (I) for $\sin(c^*\xi) = 0$, namely for $\xi = 0, c/2, c, \ldots$, and $a_1 \geq 0$;
(2) the hexagonal $P6/mmm$ phase (phase II) for $\cos(c^*\xi) = 0$, i.e., for $\xi = c/4, 3c/4, 5c/4, \ldots$, and $a_1 \leq \left(-2a_2 - 3a_3\eta_0^2\right) \cdot \eta_0^2$;
(3) the rhombohedral $P\bar{3}m1$ phase (phase III) for the equilibrium value of the order-parameter:

$$\eta = \pm\sqrt{\frac{-2a_2 + \sqrt{4a_2^2 - 12a_1a_3}}{6a_3}}, \tag{4.14}$$

which is stabilized for $a_1 \leq a_2^2/3a_3$.

The corresponding phase diagram is shown in Fig. 4.4 in the plane of the phenomenological coefficients (a_1, a_2). Thus, the reconstructive transition I–II is always first-order, whereas the I–III transition can be either first- or second-order, the two regimes being separated by a *tricritical point.* The fact that the II–III transition is always second-order in Fig. 4.4 results from the sixth power assumed for $F\left[\eta\left(\xi\right)\right]$, and also from the condition $a_3 > 0$. As it will be shown in Sec. 4.2.1.5, a higher degree expansion, or a negative value for a_3, would give a possible first-order transition regime between phases II and III.

The phase diagram contains a *three-phase point* at which the three transition lines merge: (i) the first-order transition line between phases I and II. Its equation is $a_1 = -a_2\eta_0^2 - a_3\eta_0^4$, which corresponds *to a straight line*, (ii) the first-order I–III transition line, which is a

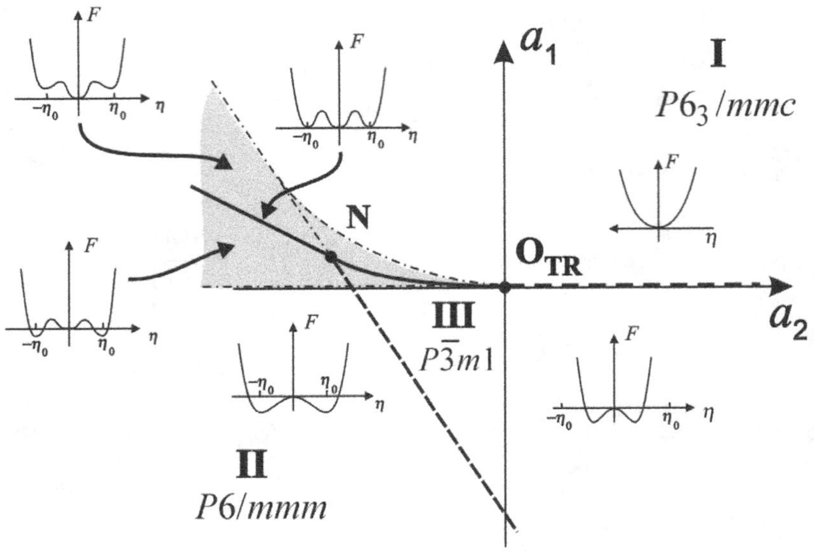

Fig. 4.4. Phase diagram representing the order-parameter expansion (4.9) in the plane (a_1, a_2). Solid and dashed lines are first-order and second-order transition lines, respectively. Dashed-dotted lines are limits of stability lines. I, II and III denote the phases of symmetry $P6_3/mmc$, $P6/mmm$ and $P\bar{3}m1$, respectively. O_{TR} is a tricritical point, and N is a three-phase point.

parabolic branch: $a_1 = a_2^2/4a_3$, and (iii) the line of second-order order transitions II–III, which is also a *straight line*, of equation $a_1 = -2a_2\eta_0^2 - 3a_3\eta_0^4$. Let us note that this line prolongates the limit of stability line between phases I and II. The coordinates of the three-phase point are $a_1 = a_3\eta_0^4$, $a_2 = -2a_3\eta_0^2$. The form of the $F(\xi_e)$ curves, where ξ_e denotes the equilibrium value of the displacements, is represented in each of the relevant region in Fig. 4.4, neglecting the thermal expansion contribution $F_0(T)$ to the thermodynamic potential. The main distinctive feature of the limit $P6/mmm$ phase is that the value of $F(\xi_e)$ corresponding to the minima of $F(\xi)$ remains constant across the phase, as ξ_e corresponds to *fixed* displacements.

4.2.1.4. *Temperature dependence of the order-parameter, and of the critical displacements*

In phases I and II, the order-parameter has temperature-independent values, $\eta = 0$ and $\eta = \eta_0$, respectively, whereas in phase III, η varies with temperature as expressed by Eq. (4.14). The shape of $\eta(T - T_C)$

across different sequences of phases, indicated in Fig. 4.5(a), is given in Fig. 4.5(b). Thus, one has a typical step-like shape for the reconstructive I–II transition (Fig. 4.5(b)-1,2) and the usual second- or first-order transition temperature dependence of $\eta(T - T_C)$ for the I–III transition (Fig. 4.5(b)-3,5). The second-order II–III transition exhibits a change in the slope of the $\eta(T - T_C)$ curve.

Figure 4.5(c) gives the temperature dependence of the critical displacements $\xi(T - T_C)$ corresponding to the preceding thermodynamic paths. One can note that the shape of $\xi(T - T_C)$ is always *symmetric* for the intermediate phase III with respect to the neighboring hexagonal phases.

4.2.1.5. *Other topologies of the phase diagram*

A positive sign was assumed for the a_3 coefficient in $F(\eta(\xi))$ as given by Eq. (4.9). In the framework of the traditional approach to phase transitions, described in Chapter 1, the positivity of the coefficient of the highest degree invariant in the thermodynamic potential insures the global stability of the phases and the convexity of the potential for large values of the order-parameter. However, in the example under consideration, Eq. (4.8) shows that the magnitude of the order-parameter *cannot be arbitrarily large*, and is always smaller than η_0. This restriction, which characterizes the "limit" phases introduced in our approach, opens the possibility of "globally stable" phase diagrams, even for negative values of the highest-degree term coefficient. Figure 4.6 shows the phase diagram associated with $F[\eta(\xi)]$, expanded to the sixth degree, with $a_3 < 0$. It differs from the phase diagram of Fig. 4.4, by the possibility of going from phase II to phase III through a first- *or* a second-order transition, the two regimes being separated by a tricritical point. Correlatively, the tricritical point disappears on the I–III transition line, which is a second-order transition line.

Expanding $F[\eta(\xi)]$ to the *eighth degree* modifies only slightly the phase diagram of Fig. 4.6. Figure 4.7 shows different phase diagrams associated with the thermodynamic potential

$$F\left[\eta\left(\xi\right)\right] = a_1\eta^2\left(\xi\right) + a_2\eta^4\left(\xi\right) + a_3\eta^6\left(\xi\right) + a_4\eta^8(\xi), \qquad (4.15)$$

assuming $a_3 < 0$ and $a_4 > 0$. The diagrams are discriminated by the relative values of η_0 and (η_1, η_2), where η_1 and η_2 are the solutions

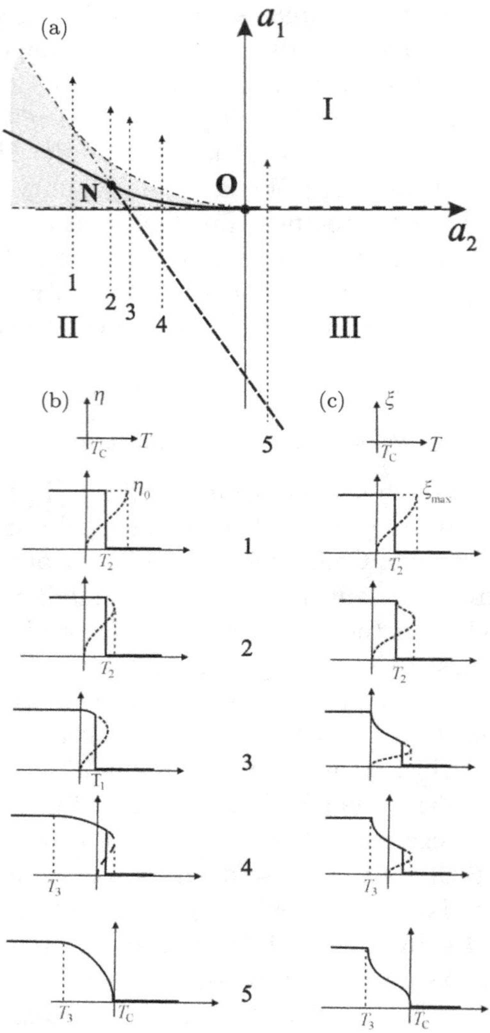

Fig. 4.5. (a) Same phase diagram as in Fig. 4.4 on which are indicated the thermodynamic paths, denoted as 1 to 5, corresponding to different sequences of phases. (b) Temperature dependence of the order-parameter for the various paths shown in (a). (c) Temperature dependence of the critical displacements for the paths indicated in (a).

of the equation

$$a_1 + 2a_2\eta^2 + 3a_3\eta^4 + 4a_4\eta^6 = 0, \tag{4.16}$$

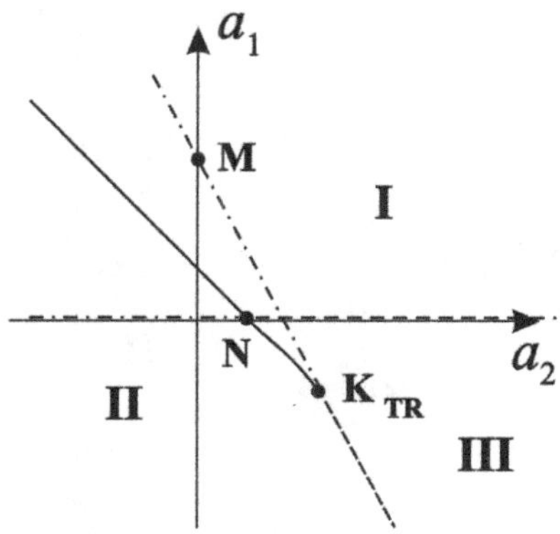

Fig. 4.6. Phase diagram associated with the sixth-degree expansion (4.9), assuming $a_3 < 0$. Same notation as in Fig. 4.4. K_{TR} is a tricritical point.

which are physically acceptable, i.e., which correspond to $\eta_1^2 > 0$, $\eta_2^2 > 0$, $\eta_1 \neq \eta_2$. Thus, Fig. 4.7(a) is obtained when $\eta_0 = \eta_2 > \eta_1$, while the phase diagrams represented in Figs. 4.7(b)–4.7(d) fulfill the conditions: $\eta_0 > \eta_2 > \eta_1$. These diagrams contain a *critical point* C ending an isostructural transition line between two rhombohedral phases, respectively, associated with η_1 and η_2. The three figures are distinguished by the value of η_0 with respect to the equilibrium value of the order-parameter in phase III at point M, which represents the intersection of the a_1 axis with the limit of stability line of phase III with respect to phase II. One has transition line between two rhombohedral phases, respectively, associated with η_1 and η_2. The three figures are distinguished by the value of η_0 with respect to the equilibrium value of the order-parameter in phase III at point M, which represents the intersection of the a_1 axis with the limit of stability line of phase III with respect to phase II. One has transition line between two rhombohedral phases, respectively, associated with η_1 and η_2. The three figures are distinguished by the value of η_0 with respect to the equilibrium value of the order-parameter in phase III at point M, which represents the intersection of the a_1 axis with the limit of stability line of phase III with respect to phase II. One has $\eta_0^2 > \eta_M^2$ (Fig. 4.7(b)), $\eta_0^2 = \eta_M^2 = -a_3/2a_4$ (Fig. 4.7(c)), and $\eta_0^2 > \eta_M^2$ (Fig. 4.7(d)). Let us also note that in Fig. 4.7(a), C_{TR} becomes a tricritical point K_{TR}.

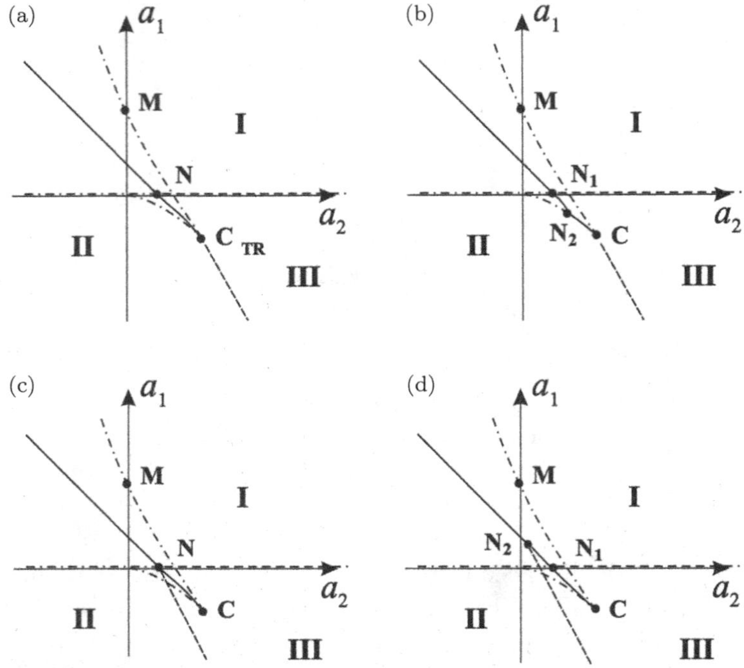

Fig. 4.7. Phase diagrams associated with the eighth-degree expansion (4.15) assuming $a_3 < 0$ and $a_4 > 0$. The figures correspond to different relative magnitudes of the order-parameter indicated in the text. C is a critical point ending a line of isostructural transitions.

4.2.2. *A second example of displacive reconstructive phase transition*

4.2.2.1. *Three possible phase transition mechanisms*

We will describe the transition from a two-dimensional primitive tetragonal lattice of point-group symmetry $4mm$ to a hexagonal lattice of symmetry $6mm$. *Three alternative mechanisms* may possibly account for the transformation of one structure into the other:

(1) A macroscopic deformation that consists in a *shear strain* e_{xy}. Figure 4.8(a) shows the effect of the corresponding *displacement field* on the atoms pertaining to four tetragonal unit cells. Denoting the atoms from 1 to 9, they can be grouped in four pairs, (1, 2), (3, 4), (5, 6), (7, 8), each pair displaying antiparallel displacements with respect to the central atom 9, which is immobile.

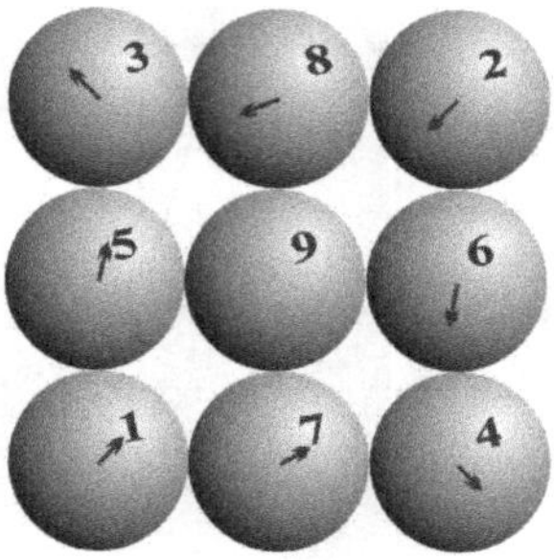

(a)

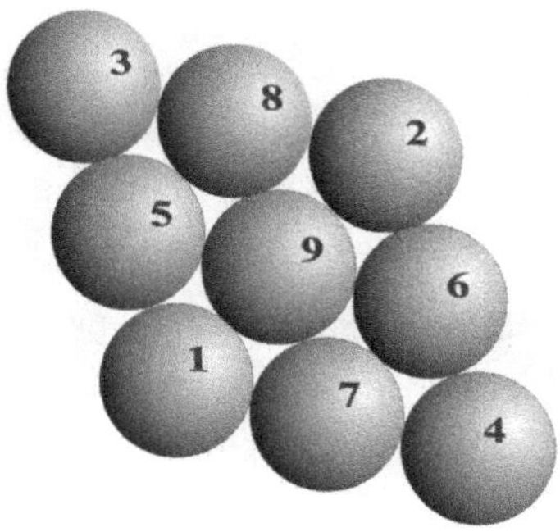

(b)

Fig. 4.8. (a) Displacement field corresponding to the shear strain e_{xy} applied to the atoms of four tetragonal unit cells of the $4mm$ structure. (b) Hexagonal atomic packing corresponding to the $6mm$ structure.

Thus, one has a shifting of the (1, 2) pair of atoms in the [11]-direction *toward* atom 9, whereas the atoms of the (3, 4) pair will be removed from the central atom along $[\bar{1}1]$. On the other hand, the remaining pairs of atoms, (5, 6) and (7, 8), will get closer to the $[\bar{1}1]$ direction, as represented in Figs. 4.8(a) and 4.8(b). If ξ is the magnitude of the displacements of one pair of atoms, let us say (1, 2), the tetragonal lattice will correspond to $\xi = 0$, and for arbitrary small shifts, the four-fold symmetry will decrease to orthorhombic ($2mm$). For a given magnitude ξ^{max} of the displacements, the pair of atoms (1, 2) will get in contact with the central atom, and the lattice will acquire the hexagonal symmetry 6 *mm*. ξ^{max} corresponds to the maximal value of the shear strain $e_{xy}^{\max}$.

(2) In the framework of our approach to the transition mechanism, the preceding description is not satisfactory as the variational parameter e_{xy} has a (maximal) limit value corresponding to a hexagonal close packing of the atoms. Another equivalent description can be given in terms of *parallel shifts of atomic rows* as represented in Fig. 4.9(a). The transition is associated with two correlated (bilinearly coupled) mechanisms: a collective displacement of parallel rows of atoms along one axis (X) and a macroscopic deformation in the perpendicular direction (Y), which keeps the atoms in contact. The advantage of this second

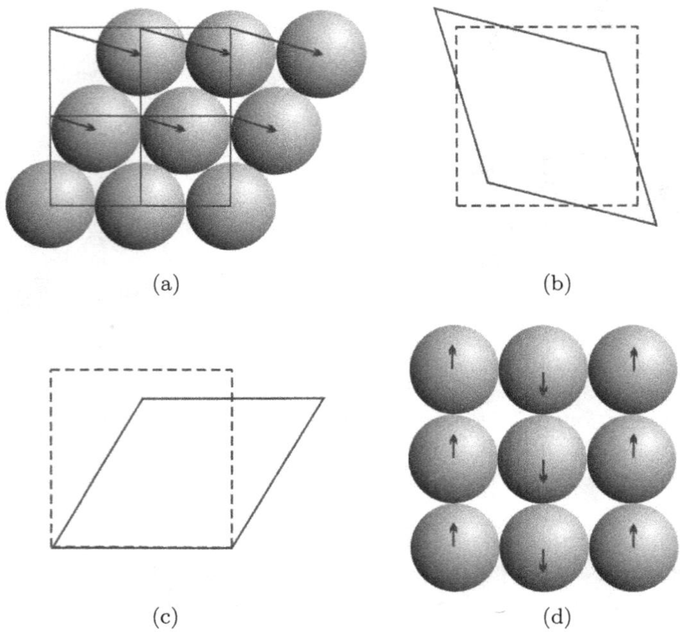

Fig. 4.9. (a) Parallel shifts of atomic rows transforming the $4mm$ into $6mm$. (b) and (c) Modifications of the shape of the crystal (b) subsequent to the displacive field shown in Fig. 4.8(a), and (c) resulting from the parallel shifting of atomic rows. (d) Antiparallel displacement of atomic rows lowering the $4mm$ symmetry to $mm2$.

description of the transformation is that it shows *the periodicity of the mechanism as a function of the displacements.* Thus, for small displacements of parallel rows of atoms, the symmetry decreases to orthorhombic, whereas for displacements corresponding to $a/2$, where a is the tetragonal lattice parameter, the hexagonal lattice is formed. When increasing the displacements, one will get the tetragonal lattice periodically (for displacements of magnitude (a) and again the orthorhombic and hexagonal lattices).

(3) The two preceding mechanisms modify the overall shape of the crystal as illustrated in Figs. 4.9(b) and 4.9(c). A third mechanism can be proposed, which preserves the crystal morphology. It consists in an *antiparallel displacement of rows of atoms* along one of the axes, let us say X, and in a correlated macroscopic deformation (e_{yy}) perpendicular to the atomic displacements,

which preserves the contact between atoms. This third mechanism, which is shown in Fig. 4.9(d), will now be used for a density-wave description of the $4mm$–$6mm$ reconstructive phase transition.

4.2.2.2. *Density-wave approach to the 4mm–6mm phase transition*

Choosing the origin at middle distance, along the Y axis, between two antiparallel atomic sublattices (Fig. 4.10(a)), one can write the lattice function (4.2), in the tetragonal structure, as

$$\psi_0(x,y) = A \cdot \left\{ \exp\left[-i(a^*x + b^*y)\right] \cdot \left[\exp\left(i\frac{b^*}{2} \cdot \frac{y_0}{2} \right) \cdot \exp\left(ia^*x_0\right) + \exp\left(-i\frac{b^*}{2} \cdot \frac{y_0}{2} \right) \cdot \exp\left(ia^*x_0\right) \right] + \text{c.c.} \right\}, \quad (4.17)$$

where a^* and b^* are the tetragonal reciprocal lattice vectors, and $x_0 = 0$, $y_0 = b$, the initial atomic positions. After an antiparallel displacement ξ along the X axis of two consecutive rows of atoms, the lattice function becomes

$$\psi(x,y) = A \cdot \left\{ \exp\left[-i\left(a^*x + b^*y\right)\right] \cdot \left[\exp\left(i\frac{b^*}{2} \cdot \frac{y_0}{2} \right) \cdot \exp\left(ia^*\left(x_0 + \xi\right)\right) + \exp\left(-i\frac{b^*}{2} \cdot \frac{y_0}{2} \right) \cdot \exp\left(ia^*\left(x_0 - \xi\right)\right) \right] + \text{c.c.} \right\}, \quad (4.18)$$

which yields the increment of the lattice function at the transition

$$\delta\psi\left(x,y,\xi\right) = 2A \cdot \cos\left(a^*x + b^*y\right) \cdot \sin a^*\xi. \quad (4.19)$$

It expresses the periodic deformation of the structure in function of the atomic displacements ξ from the initial tetragonal structure. However, as these displacements constitute only part of the transition mechanism, the tetragonal lattice will not be transformed into the hexagonal lattice. This can be viewed on the $\delta\psi(\xi)$ curve, represented in Fig. 4.10(b), which shows that one gets the tetragonal structure periodically, for $\xi = 0, a/2, a, \ldots$, and *two distinct*

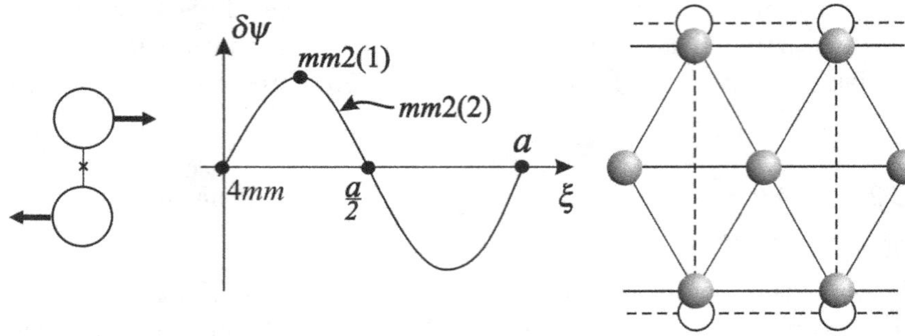

(a) (b) (c)

Fig. 4.10. (a) Two atoms of the antiparallel atomic sublattice with the coordinate of the origin at middle distance between them. (b) The $\delta\psi(\xi)$ curve follows Eq. (4.19). (c) Effect of the strain component e_{yy} on the $4mm$–$6mm$ mechanism. Open and full circles represent, respectively, the initial and final atomic positions.

orthorhombic structures of symmetry $2mm$: a *primitive* orthorhombic structure with *two atoms per unit cell* (i.e., the tetragonal cell is doubled along $\boldsymbol{b}$ by the "optical" antiparallel shifting mechanism), and a *centered* orthorhombic structure with *one atom per unit cell.* This latter structure is a "limit" phase corresponding to the fixed displacements $\xi = a/4, 3a/4, 5a/4, \ldots$, associated with the extrema of $\delta\psi(\xi)$.

In order to obtain the hexagonal structure, one has, in addition, to take into account the macroscopic deformation along Y, which transforms the centered orthorhombic lattice into the hexagonal lattice. Figure 4.10(c) shows the effect of the strain component e_{yy}, which brings about the b/a ratio to its hexagonal value $\sqrt{3}$, or equivalently the angle between the diagonals to $\pi/3$, the transformation occurring continuously without loss of contact between the atoms.

Let us emphasize that in the present example, the simpler situation has been assumed, in which the displacements occur along a single direction, and the deformation is in the orthogonal direction. As a result, the limit hexagonal phase takes place when the atoms get in contact. One may find more complex cases in which either the limit phase is obtained geometrically, i.e., the atoms are in contact without an increase in symmetry, or conversely the symmetry increases without contact between atoms.

Equation (4.19) shows that the effective order-parameter $\eta(\xi)$ of the transition has the same symmetry and form as the order-parameter of the $P6_3/mmc$–$P6/mmm$ transition given by Eq. (4.18). Accordingly, the same form (4.9) of the thermodynamic potential and the same phase diagram topology, as represented in Fig. 4.4, hold for the $4mm$–$mm2$–$6mm$ sequence of phases (in which the phases denoted I, II and III coincide, respectively, with the phases of symmetries $4mm$, $6mm$ and $mm2$). However, it should be stressed that the full symmetry of the order-parameter in the present example may involve *two components*, only one of which has been considered for the $4mm$–$6mm$ transition. Furthermore, in order to obtain the hexagonal structure, one has to include in Eq. (4.9) a *coupling* between the order-parameter $\eta(\xi)$ and the spontaneous strain component e_{yy}. The two first mechanisms, discussed in Sec. 4.2.2.1, require a *bilinear* coupling $\eta \cdot e_{yy}$. In the third mechanism, assumed in our density-wave approach, one has a *nonlinear* coupling of the $\eta^2 \cdot e_{yy}$ type.

4.2.3. *Examples of periodic–aperiodic displacive-type phase transition*

4.2.3.1. *Crystallographic description of the tetragonal–octagonal transformation mechanism*

As an illustrative example of the density-wave approach to a crystal-aperiodic quasicrystal phase transition, we will describe the simple transformation from a planar tetragonal structure of point-group symmetry $4mm$ to an octagonal structure of symmetry $8mm$. The unit cell of the hypothetical tetragonal structure contains eight atoms in general positions $(x,\ y)$, formed by two squares, rotated with respect to each other as represented in Fig. 4.11(a). If we apply a *displacement field* to this structure, corresponding to antiparallel shifts of atoms, which rotate the initial squares with opposite angles $\pm\varphi$, the atoms pertaining to two squares will form, at $\varphi = \pi/8$, a *regular octagon*. The obtained structure remains, however, tetragonal, although it possesses *local* octagonal motifs. In order to break the periodicity of the tetragonal lattice, one has to displace sets of close pairs of atoms, indicated in Fig. 4.11(b), in such a way that they will occupy regular octagonal positions on a given coordination circle. This additional mechanism corresponds to an *inhomogeneous deformation* of the structure, as two different coordination circles involve displacements of different magnitudes.

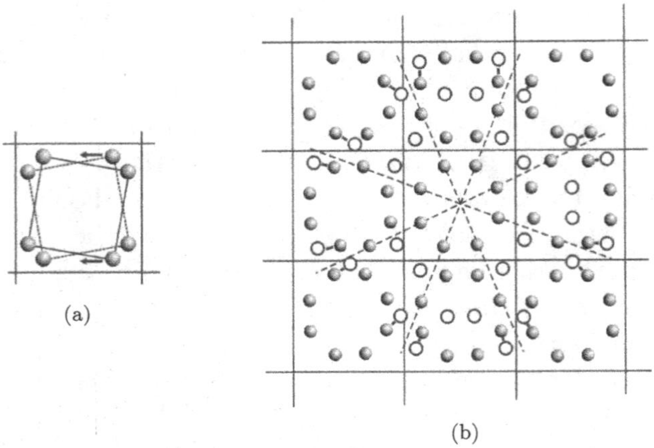

Fig. 4.11. (a) Opposite rotation of the two squares leading to the octagonal basis. (b) Representation of the mechanism leading to the breaking of the translational symmetry. Full and open circles are, respectively, occupied and vacant octagonal positions. The ideal octagonal quasilattice is obtained after the merging at crossed positions of close pairs of full and open circles.

In summary, the $4mm$–$8mm$ transformation can be described as the result of two correlated mechanisms: a homogeneous deformation that increases the local symmetry of the atoms to $8mm$ and an inhomogeneous deformation that achieves the overall quasiperiodic symmetry. Let us stress that the final quasiperiodic structure is *defective*, as shown in Fig. 4.11(b), due to the existence of regularly ordered *vacant* octagonal sites, which are distributed according to the initial tetragonal symmetry. One needs a further *disordering* of the sites through a diffusion process, which will lead to the ideal octagonal quasiperiodic structure.

4.2.3.2. *Density-wave description of the 4mm–8mm mechanism*

Let us show that the preceding combined displacive mechanism can be expressed in terms of *inhomogeneous rotations* of the system of density waves associated with the initial tetragonal structure. The lattice functional corresponding to the tetragonal unit cells is

$$\psi_t(\boldsymbol{r}) = A \sum_{i=1}^{8} \exp(i\boldsymbol{k}_i \boldsymbol{r}) + \text{c.c.}, \qquad (4.20)$$

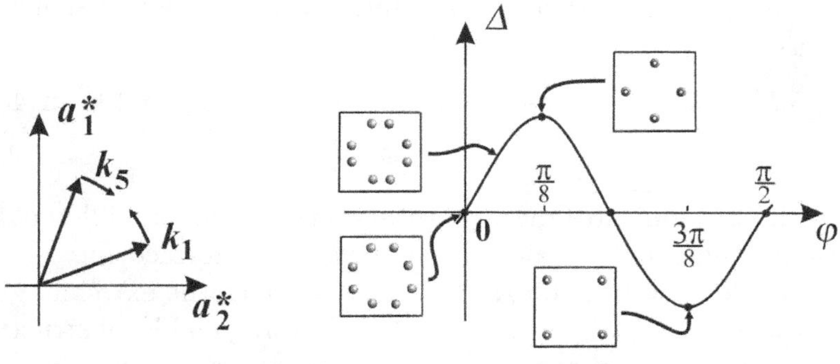

Fig. 4.12. (a) Opposite rotation of a pair of branches of the star $\boldsymbol{k}_1$ associated with the symmetrization process, which transforms the local tetragonal basis into octagonal. (b) $\Delta(\varphi)$ dependence following (4.23). The squares indicate the local basis for relevant values of φ.

where $\boldsymbol{k}_i$ are the eight branches of the star $\boldsymbol{k}'_1$, $\boldsymbol{k}_1$ being a wave vector located in general position in the tetragonal Brillouin zone. Equation (4.20) expresses, in terms of density waves, the family of eight planes forming the tetragonal structure and transforming one into another under the symmetry operations of $4mm$.

The *symmetrization process* that transforms the local tetragonal symmetry into octagonal symmetry can be described in the reciprocal space by opposite rotations of four $\boldsymbol{k}_i$ branches associated with the two systems of planes, which are rotated in opposite directions, as represented in Fig. 4.12(a). For opposite rotation angles $\pm\phi$, one can write

$$\psi(\boldsymbol{r}) = A\left[\sum_{i=1}^{8}\exp\left(i\boldsymbol{k}_i\boldsymbol{r}\right) + \sum_{i=5}^{8}\exp\left(i\boldsymbol{k}'_i\boldsymbol{r}\right) + \text{c.c.}\right], \tag{4.21}$$

where

$$\begin{aligned} \boldsymbol{k}_i\cdot\boldsymbol{k}'_i &= k_i^2\cdot\cos 4\left(\frac{\pi}{8}-\varphi\right) \text{ for } i = 1\text{–}4,\\ \text{and } \boldsymbol{k}_i\cdot\boldsymbol{k}'_i &= k_i^2\cdot\cos 4\left(-\frac{\pi}{8}+\varphi\right) \text{ for } i = 5\text{–}8. \end{aligned} \tag{4.22}$$

The periodicity of the structures as a function of ϕ can be viewed by expressing the variation of the scalar products $(\boldsymbol{k}_i\cdot\boldsymbol{k}'_i)$ in the reciprocal space from their initial values in the tetragonal structure. One has

$$\Delta\left(\boldsymbol{k}_i \cdot \boldsymbol{k}_i'\right) = k_i^2 \cdot \left[\cos 4\left(\frac{\pi}{8} - \varphi\right) + \cos 4\left(-\frac{\pi}{8} + \varphi\right)\right] = 2k_i^2 \sin 4\varphi. \tag{4.23}$$

Figure 4.12(b) indicates the structures corresponding to different values of φ. Thus, an octagonal local structure is formed for the successive angles $\varphi = 0, \pi/4, \pi/2, 3\pi/4, \ldots$, whereas the extrema, given by $\varphi = \pi/8, 3\pi/8, 5\pi/8, \ldots$, coincide with tetragonal local structures formed with four atoms per unit cell and turned alternately by an angle of $\pi/8$. For general values of ϕ, one has the initial local tetragonal basis with eight atoms per unit cell.

The preceding description concerns the homogeneous deformation, which transforms the initial tetragonal structure in a *rational approximant* of the octagonal quasicrystal, i.e., the formation of an octagonal basis *within* the tetragonal unit cell. In order to break the tetragonal translational symmetry and to obtain a long-range octagonal quasilattice, one has to perform an *inhomogeneous deformation* of the structure, which brings a number of atoms to occupy vacant octagonal positions *created* by the homogeneous deformation as shown in Fig. 4.11(b). The description of this symmetry-breaking mechanism in terms of density waves entails to consider *the whole structure of the crystal* in order to determine, numerically, the atomic positions in the quasicrystal state. The corresponding procedure requires a number of experimental data, such as the atomic positions in the crystal structure and the orientation of the wave-vector in the quasicrystal structure, and thus cannot be carried on in the tetragonal–octagonal working example under consideration. However, the necessary steps for finding the atomic positions in the octagonal structure, assuming that the tetragonal atomic positions are known, can be briefly depicted (and will be illustrated in Chapter 7). The lattice function corresponding to the symmetrized tetragonal structure is

$$\psi\left(\boldsymbol{r}, \boldsymbol{r}_k, \boldsymbol{k}_i'\right) = \frac{1}{N}\sum_{k=1}^{N}\sum_{i=1}^{8} A\left\{\exp\left[i\boldsymbol{k}_i'(\boldsymbol{r}_k - \boldsymbol{r})\right] + \text{c.c.}\right\}, \tag{4.24}$$

where the sum runs over the *total number N of atoms contained in the sample.* In order to obtain equal and maximal intensities for the

Bragg reflections in the octagonal quasiperiodic structure, one has to maximize the *amplitude* of ψ

$$A\left(\boldsymbol{r}_k, \boldsymbol{k}'_i\right) = \frac{1}{N}\sum_{k=1}^{N}\sum_{i=1}^{8} A \cdot \cos\left(\boldsymbol{k}'_i \boldsymbol{r}_k\right), \tag{4.25}$$

with respect to the $\boldsymbol{k}'_i$ *and* to the $\boldsymbol{r}_k$. The maximization process, which is performed in the vicinity of the initial tetragonal positions, provides the *critical wave vector* $\boldsymbol{k}_i^{\text{oct}} = \boldsymbol{k}'_i + \Delta\boldsymbol{k}'_i$ and the *critical displacements* $\xi_i^{\text{oct}} = \boldsymbol{r}_k^{\text{oct}} - \boldsymbol{r}_k^{\text{tet}}$ associated with an octagonal quasiperiodic structure. For non-critical wave vectors $\boldsymbol{k}''_i \neq \boldsymbol{k}_i^{\text{oct}}$ or (and) non-critical displacements ξ_k, the resulting structure is incommensurate (non-octagonal) with respect to the initial crystal structure. Accordingly, the octagonal structure appears as a *limit situation* to an incommensurate regime.

4.2.3.3. *Moiré patterns on twisted bilayer lattices*

A phase transition from periodic to aperiodic structure in a system of twisting bilayers could be considered as only another illustrative example, but recent discovery of very special electric and magnetic effects in the bilayer structure built by twisted graphene sheets makes it important in further applications (see Sec. 8.4). In this section, we parameterize the model of the phase transition in two stacked square lattices rotating with respect to each other around the Z axis directed normally to their surface.

Grids. Actually, such a system of rod or line grids has been known for a long time without any regard to phase transitions but as demonstrating so-called *moiré patterns* (Fig. 4.13). The latter is a *visual* effect obtained by superimposing plane motifs, which are obtained from one another through small motions. Figures 4.13(a) and 4.13(b) show superposition of two identical square grids twisted by (a) $2\varphi = \pi/60$ and (b) $2\varphi = \pi/30$ resulting in new patterns with remarkably larger periods consisting of dark and bright bands (or *pail lines*).

The parent structure for the model is a bilayer of stacked square lattices with basic orthogonal translations $\boldsymbol{a}_1$ and $\boldsymbol{a}_2(|\boldsymbol{a}_1| = |\boldsymbol{a}_2|)$ separated by a non-zero gap characterized by the vector $\boldsymbol{Z}$ (Fig. 4.14). Its point symmetry is tetragonal $G_0 = 4mm$. A small

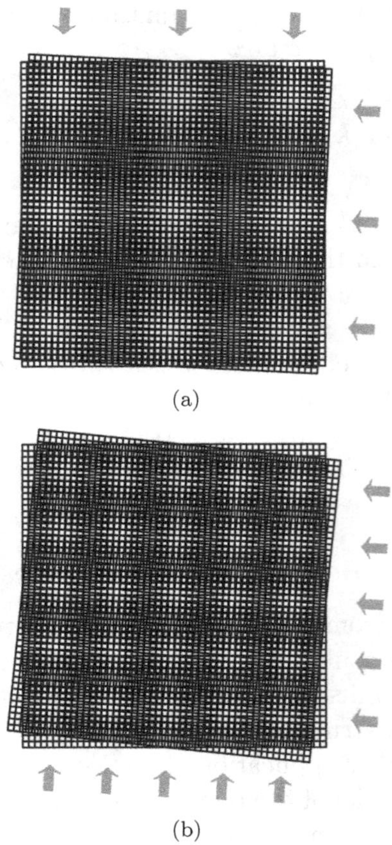

Fig. 4.13. Moiré patterns obtained from two identical square grids overlaid with a relative angle: (a) $2\varphi = \pi/60$; (b) $2\varphi = \pi/30$.

twist of the layers breaks translational symmetry in the bilayer structure, and point group of the twisted bilayer structure lowers to $G_d = 4$. Its new translational symmetry is characterized by orthogonal basis translations $\boldsymbol{L}_1$ and $\boldsymbol{L}_2$ (Fig. 4.14). These latter are spacing-separated pale lines passing through the interceptions of the two lattices (Fig. 4.14).

The transformation mechanism consists, therefore, of the rotation of tetragrids by opposite angles $\pm\varphi$. It results, at $|\varphi| = \pi/8$, in octagonal quasicrystalline bilayer structure breaking group–subgroup relation with respect to the parent phase symmetry (Fig. 4.15).

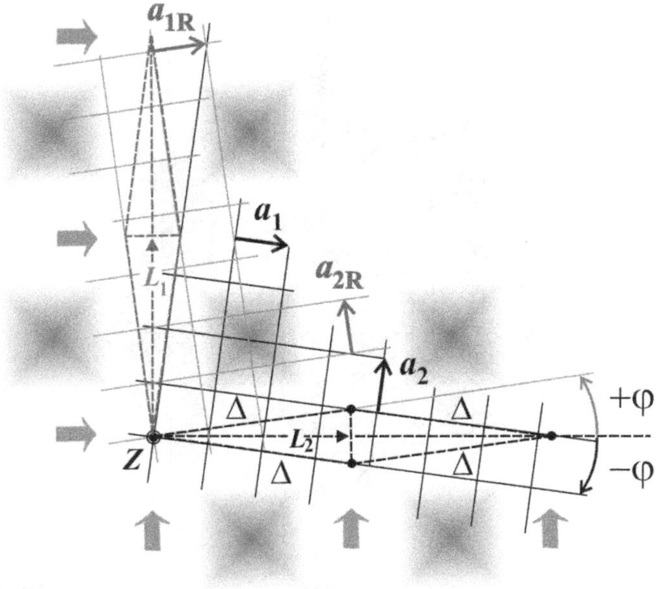

Fig. 4.14. Periodicity of the pale lines superlattice in twisted square bilayer.

We can use the density-wave formalism developed in Sec. 4.2.3.2 for the double sublattice plane structure, keeping in mind, however, that the grids constituting bilayer are absolutely rigid. The latter condition means, in a real layered system, that interlayer coupling is negligible compared to interatomic in-layer interactions. We use a lattice functional identical to Eq. (4.20) and following calculations with eight branches of the vector stars $\pm \boldsymbol{a}_i^*$ and $\pm \boldsymbol{a}_{iR}^*$ (Fig. 4.16(a)) corresponding to the eight planes forming both the parent and distorted tetragonal structures. Periodicity of the structures, as a function of the rotation angle φ, becomes evident if we consider, as an effective order-parameter $\eta(\varphi)$, an axial vector $\eta || \boldsymbol{Z}$, which is a vector product of density-wave vectors: $|\eta| \sim [\boldsymbol{a}_i^* \times \boldsymbol{a}_{iR}^*] \sim |\boldsymbol{a}*|^2 \cdot \sin\varphi$ (see Eq. (4.20)). Figure 4.16(b) indicates the structures corresponding to different values of φ. It is worth noting that the density-wave approach developed for discontinuous and reconstructive phase transitions is well adopted to description of moiré effects in bilayer systems, since the modern moiré theory is largely based on the Fourier approach too.

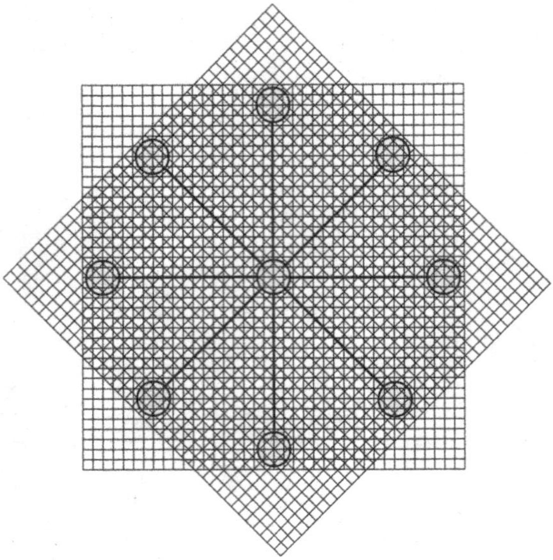

Fig. 4.15. Octagonal quasicrystalline structure obtained in square bilayer twisted by $2\varphi = \pi/4$.

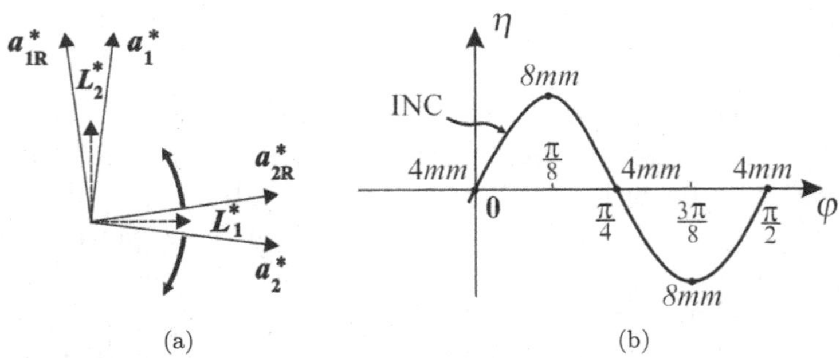

Fig. 4.16. (a) Opposite rotation of reciprocal vectors ($\boldsymbol{a}_i^*$) of the superimposed grids and directions of moiré pattern vectors $\boldsymbol{L}_i^*$. (b) Order-parameter η as a function of the twist angle φ.

Orientation of the orthogonal basis $\boldsymbol{L}_1$–$\boldsymbol{L}_2$ of the distorted Landau phase ($0 < \varphi < \pi/8$) coincides with the directions of pale lines, i.e., $\boldsymbol{L}_1$ and $\boldsymbol{L}_2$ are twisted by angle φ, respectively, to two orthogonal vector stars $\boldsymbol{a}_i$ and $\boldsymbol{a}_{iR}$. The period of the new moiré superlattice is $L = |\boldsymbol{L}| = |\boldsymbol{a}|/2 \sin \varphi$ (Fig. 4.14). Thus, distorted (moiré) structures are either incommensurate or long periodic.

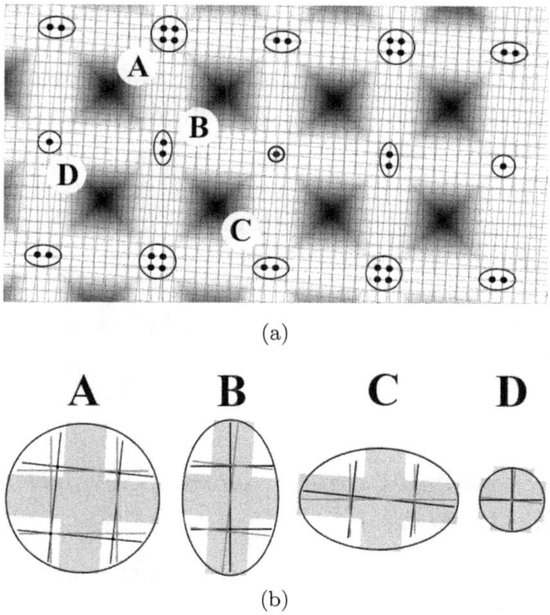

Fig. 4.17. (a) Long-period superstructure of twisted square atomic layers ($2\varphi = \pi/30$). (b) Four types of local nod structures. Gray bands schematically represent pale lines in the corresponding twisted grid bilayer.

Lattices of nods. The example of twisted grids allows us easy understanding and parameterizing of the geometry of long-periodic superlattices; however, it is valid mainly for art and applied optics applications. In physics, a model with crystalline layers containing atoms in the lattice nods is more realistic. Figure 4.17(a) gives more details to the bilayer structure of Fig. 4.13(b) where misalignment of the stacked lattices is equal to $\pi/30$. One sees immediately that the atomic superstructure is more complex than the grid superstructure. Indeed, intersections of pale lines in the crystalline structure are not identical. Figure 4.17(a) identifies four types, A to D, of local nod structures, and Fig. 4.17(b) shows its details.

We see from the figure that periodicity of atomic crystalline superstructure is controlled by minimization of the interatomic distances (its projections on the layer plane in Fig. 4.17(a)), and atomic superstructure basis translations are twice longer than the pale line spacings $\boldsymbol{L}_1$ and $\boldsymbol{L}_2$ (Fig. 4.14).

Thus, the symmetry-based density-wave consideration concludes, on a pure model-free basis, that (i) a twisted bilayer structure is long periodic or incommensurate (especially at small twist angle), and as a consequence, (ii) its energy bands or branches present a result of multiple parent square Brillouin-zone folding; bands and branches with $\partial E(\boldsymbol{k})/\partial \boldsymbol{k} \to 0$ at $\boldsymbol{k} \to 0$ become, therefore, remarkably flat. We are going to discuss the latter result in detail in Sec. 8.4, but here it should be underlined that such a feature of the electron zone, which is considered as a key detail in the mechanism of superconductivity at small twist angles in bilayer graphene, is fully symmetry controlled and model-free, i.e., it does not depend on details of interatomic interaction.

4.2.4. *An example of order–disorder reconstructive phase transition*

4.2.4.1. *Mechanism of the $Im\bar{3}m - P6/mmm$ ordering transition*

Let us assume a transition from a disordered body-centered cubic structure in which one has a random distribution of atoms and vacancies, each BCC position being occupied with a probability of 1/3, to a fully ordered simple hexagonal structure in which the probability of occupancy is 1. As shown in Fig. 4.18(a), the hexagonal unit cell (space group $P6/mmm$, $Z = 1$) corresponds to a *tripling* of the cubic BCC unit cell (space group $Im\bar{3}m$, $Z = 1/3$), with the following relations between the corresponding basic translations

$$\boldsymbol{A}_h = \boldsymbol{a}_c - \boldsymbol{b}_c, \quad \boldsymbol{b}_h = \boldsymbol{b}_c - \boldsymbol{c}_c, \quad \boldsymbol{c}_h = \boldsymbol{a}_c + \boldsymbol{b}_c + \boldsymbol{c}_c. \tag{4.26}$$

Accordingly, the $\boldsymbol{c}_h$ axis coincides with the diagonal of the primitive cubic unit cell $\{111\}_c$, whereas $\boldsymbol{a}_h$ $\{\bar{1}10\}_c$ and $\boldsymbol{b}_h$ $\{10\bar{1}\}_c$ are diagonals of the primitive cubic faces, and one has $|\boldsymbol{c}_h|/|\boldsymbol{a}_h| = \sqrt{3/8}$. The vector star $\{\boldsymbol{a}_h, \boldsymbol{b}_h, \boldsymbol{c}_h\}$ (4.26) defines a primitive hexagonal lattice as $\boldsymbol{a}_h = \boldsymbol{b}_h$, angle between $\boldsymbol{a}_h$ and $\boldsymbol{b}_h \gamma = 2\pi/3$, and $\boldsymbol{c}_h$ is perpendicular to the plane $(\boldsymbol{a}_h, \boldsymbol{b}_h)$.

The preceding ordering mechanism involves exclusively a change in the distribution of atoms and vacancies among the positions, but

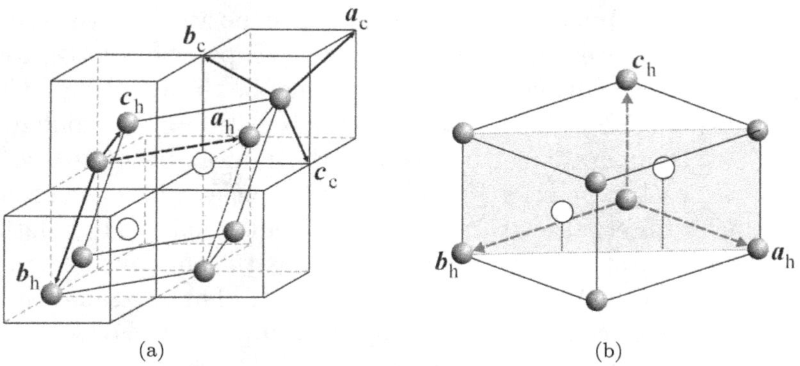

Fig. 4.18. Connection between the cubic (a) and hexagonal (b) unit cells following (4.26). The arrows are the basic lattice vectors of the BCC and hexagonal cells.

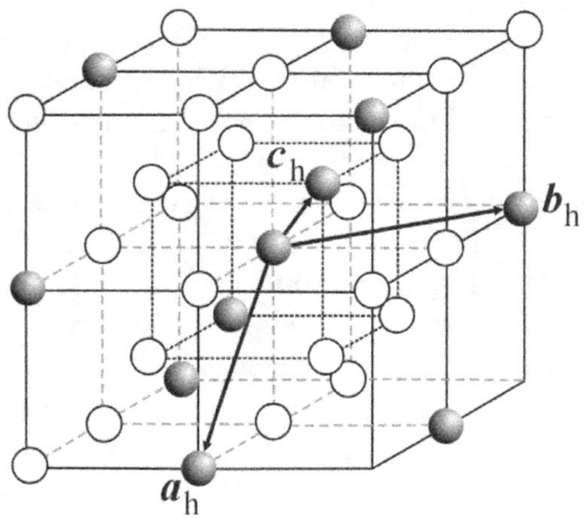

Fig. 4.19. Cubic to hexagonal ordering mechanism. Full and open circles are, respectively, occupied and vacant positions in the hexagonal structure.

does not displace the initial positions, and requires no deformation of the cubic lattice. In the final hexagonal positions, the atoms fill only the full circles shown in Fig. 4.19 instead of occupying both the initial open full circles corresponding to the cubic positions.

At this point, we should address our remark in Sec. 4.1 on definition and identification of symmetry elements and space groups: (i) via the set of three-dimensional matrices describing permutation rules for coordinates or (ii) via the set of characteristic positions of the group. In crystal structures with different space groups, the relative locations of symmetrically equivalent atoms may be the same (e.g., Cl in CsCl and F in CaF_2). Such a set of points is called a point configuration. Point configurations have been grouped into types called *lattice complexes*, each type referring to a Wyckoff position. Following the definition in International Tables for Crystallography [11], a lattice complex is the set of all point configurations that may be generated within one type of Wyckoff sets. Let us stress that the same lattice complexes may occur in different space-group types. For example, the complex "cubic primitive lattice" may be generated on $Pm3m$ – Wyckoff positions 1(a) and 1(b), $Fm\bar{3}m$ – 8(c). All Wyckoff positions, Wyckoff sets, and types of Wyckoff sets are assigned to the 402 lattice complexes listed in Ref. [11]. The existence of a restricted number of occupied positions in the ordered structure makes the characteristic space-group type of the lattice complex important; that is, one stands out because its corresponding Wyckoff position exhibits the highest site symmetry. This latter is called the *characteristic position* of the lattice complex.

The concept of lattice complex is very useful when considering crystal structures with a limited number of sublattices or their *partial filling* by atoms. We should take into account only the characteristic positions being occupied provide experimental possibility to distinguish between space groups. For example, analyzing probable space group $Fm\bar{3}m$ for a crystal structure, one cannot localize single-type atoms in the 8(c) position mentioned above. This is due to the fact that it belongs to the primitive cubic lattice and so will be identified as $Pm3m$ space group in the X-ray diffraction experiment [11].

The above description of atoms and vacancies *exchange* process [12] is formulated in terms of *uncorrelated diffusion* and *jump* of the atoms between *fixed* positions. It *virtually* follows the same scheme described in the preceding examples. The difference is that the incompletely ordered structures (with a probability of occupancy comprised between 1/3 and 1) possess the same symmetry ($P\bar{3}m1$), but the fully ordered structure represents the *limit state* $P6/mmm$. Table 4.1, which is extracted from the Tables for Crystallography [11], supports this description. It represents the symmetry of the $P\bar{3}m1$ rhombohedral structure for different occupancies of the positions. In the table, one can see that for definite occupancies, the

Table 4.1. Possible symmetry of the $P\bar{3}m1$ space group (column (1)) for different occupancies of positions given in column (2).

(1)	(2)
$P6/mmm$	(a) in (a) or (b)
$P6/mmm$	(e) in (c)
$P\bar{3}m1$	(d) in (d)
$P6/mmm$	(f) in (e) or (f)
$P6/mmm$	(j) in (g) or (h)
$P\bar{3}m1$	(i) in (i)

overall structural symmetry increases to the hexagonal space-group $P6/mmm$.

The preceding increase in symmetry can also be understood in terms of probability distribution functions. In the initial cubic structure, one has only one atomic distribution function of symmetry $Im\bar{3}m$. When the initial atomic position splits, the preceding function also splits into three independent functions, the set of which is invariant by the $P\bar{3}m1$ symmetry. When the fully ordered state is reached, the coefficients of two among these three functions vanish, and the distribution function corresponding to the non-vanishing coefficient acquires the symmetry of a special (isolated) atomic position, which is $P6/mmm$.

Let us stress that in diffraction experiments, the "partially ordered–fully ordered" transition coincides with the disappearance of superstructure reflections corresponding to the vacant positions. In the reconstructive phase transition, this is usually interpreted in terms of *stacking faults*, the number of which decreases when the occupancy varies from 1/3 to 1. In other words, the process that brings about a full ordering state, due to its slow diffuse kinetics, may occur continuously and *does not necessarily look as a phase transition.*

4.2.4.2. *Density-wave description of the reconstructive $Im\bar{3}m - P6/mmm$ transition*

Figure 4.19 illustrates the cubic-to-hexagonal change in atomic probability distribution. Using Eq. (4.2), one can express the corresponding reconstructive transition mechanism in terms of density waves, which gives the form of the lattice function, valid for the different structures

$$\psi(z) = e^{-\frac{4\pi i}{a\sqrt{3}}} \cdot \left\{ A_1 e^{\frac{4\pi i}{a\sqrt{3}} z_1} + A_2 e^{\frac{4\pi i}{a\sqrt{3}}\left(z_1 + \frac{a}{2\sqrt{3}}\right)} + A_3 e^{\frac{4\pi i}{a\sqrt{3}}\left(z_1 + \frac{a}{\sqrt{3}}\right)} \right\} + \text{c.c.}, \tag{4.27}$$

where A_1, A_2 and A_3 are the amplitudes associated with the three possibly inequivalent crystallographic positions (0 0 0), (1/3 2/3 1/3) and (2/3 1/3 2/3) written in hexagonal coordinates (Fig. 4.11(b)) and corresponding, respectively, to the z coordinates: $z_1 = 0$, $z_2 = z_1 + \frac{a}{2\sqrt{3}}$ and $z_3 = z_1 + \frac{a}{\sqrt{3}}$, where a is the cubic lattice parameter.

The z-dependent factor ahead of brackets in (4.27) expresses the periodicity of the structures along the Z hexagonal direction. The exponential terms inside the braces account for the geometry of the structure and thus remain unchanged at the $Im\bar{3}m - P6/mmm$ transition. The ordering process modifies exclusively the occupancy of the atomic positions, which determines the values of the density-wave amplitudes $A_i (i = 1 - 3)$. In this respect, one can write

$$A_i^k = A \cdot \delta P_i^k, \tag{4.28}$$

where P_i^k denotes the probability of occupancy of the ith position corresponding either to an atom ($k = \alpha$) or to vacancies ($k = \beta, \gamma$). Thus, in the cubic structure, one has $P_1^\alpha = P_1^\beta = P_1^\gamma = 1/3$ and $P_1^\alpha = P_2^\alpha = P_3^\alpha = 1/3$. The rhombohedral $P\bar{3}m1$ atomic distribution corresponds to $1/3 < P_1^\alpha \leq 1$, $0 \leq P_1^\beta < 1/3$, $0 \leq P_2^\alpha < 1/3$, and $\beta = \gamma$, $A_2^k = A_3^k$ for all k. The limit hexagonal atomic distribution is reached for $A_2^\alpha = A_3^\alpha = 0$, $P_1^\alpha = 1$ and $P_1^\beta = P_1^\gamma = 0$. Let us also note that the more general atomic distribution, which is not stabilized in the present mechanism, involves three inequivalent atomic positions, with an overall symmetry $P3m$, and $A_1^k \neq A_2^k \neq A_3^k$.

Using the preceding numbers one can write the increment of the lattice function (4.27) at the disorder–order transition

$$\psi(z) = e^{-\frac{4\pi i}{a\sqrt{3}}} \cdot \left\{ \eta A^\alpha \left(P_1^\alpha - \frac{1}{3} \right) + (1 - \eta) \times A^\alpha \left[e^{\frac{4\pi i}{a\sqrt{3}} z_2} + e^{\frac{4\pi i}{a\sqrt{3}} z_3} \right] \right\} + \text{c.c.}, \tag{4.29}$$

where η represents the order-parameter of the ordering transition, defined as

$$\eta = \frac{3}{2}\delta P_1^\alpha. \tag{4.30}$$

δP_1^α is the increment of the atomic probability of occupancy P_1^α from the cubic to the ordered phases. As in the initial cubic structure $P_1^\alpha = 1/3$ and in the ordered phase $1/3 < P_1^\alpha \leq 1$, one has thus $0 \leq \delta P_1^\alpha \leq 2/3$ consistently with the standard property of the order-parameter associated with an order–disorder transition, namely to vary from $\eta = 0$ in the disordered cubic phase to $\eta = 1$ in the fully ordered hexagonal phase.

4.2.4.3. *Symmetry of the order-parameter and phase diagram*

The density waves associated with the three atomic sublattices, expressed in Eq. (4.27), transform as the $3m$ point group, which is the group of permutation of three equivalent objects. The phase diagrams associated with the image-group $3m$ have been discussed in detail in Sec. 2.2.1.2. They correspond to a two-component order-parameter (η_1, η_2). Although in the present example the actual dimensionality of the order-parameter is eight (i.e., it transforms as an eight-dimensional IR of the $Im\bar{3}m$ space group associated with the wave vector $\boldsymbol{k} = \frac{1}{3}(\boldsymbol{a}_c^* + \boldsymbol{b}_c^* + \boldsymbol{c}_c^*)$ of the BCC Brillouin zone), one has two non-vanishing *effective* components, which can be written as

$$\eta_1 = \delta P_1^\alpha + \varepsilon^* \cdot \delta P_2^\alpha + \varepsilon \cdot \delta P_3^\alpha; \quad \eta_2 = \delta P_1^\alpha + \varepsilon \cdot \delta P_2^\alpha + \varepsilon^* \cdot \delta P_3^\alpha, \tag{4.31}$$

with $\varepsilon = \exp(2\pi i/3)$.

Among the possible $3m$ phase diagrams shown in Sec. 2.2.1.2, the one that is more adapted to our example is represented in Fig. 4.20(a) in the case of a *fully occupied* $Im\bar{3}m - P6/mmm$ structure, and in Fig. 4.20(b) for the *partially filled* cubic structure assumed in our description. One can see that the two anti-isostructural low-symmetry structures of symmetry $P\bar{3}m1$ in Fig. 4.20(a), which correspond to distinct binary alloys AB_2 and A_2B transforming one into the other by chemical decomposition across a line of first-order transition, become respectively in Fig. 4.20(b) hexagonal ($P6/mmm$)

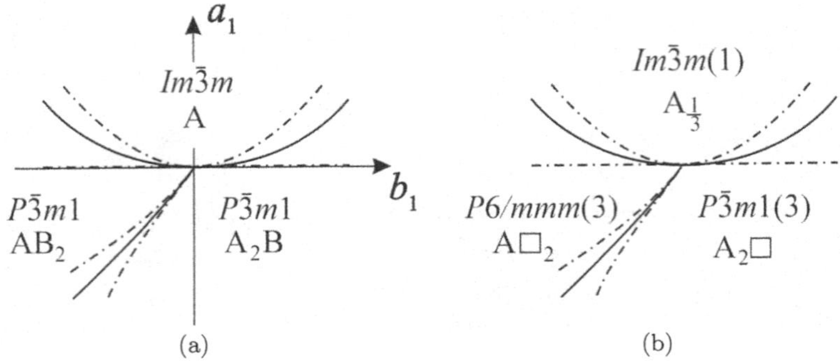

Fig. 4.20. Phase diagram corresponding to the image-group $3m$ associated with (a) the $Im\bar{3}m - P\bar{3}m1$ transition (full occupancy of the lattice) and (b) the $Im\bar{3}m - P6/mmm$ transition (fractional occupancy of the lattice). Full lines and dashed-dotted lines are, respectively, first-order and limit-of-stability lines.

and rhombohedral ($P\bar{3}m1$) *binary mixtures* of atoms (A) and vacancies ($\square$). The occupancy of atoms and vacancies in the structures of the $A\square_2$ and $A_2\square$ mixture can be easily deduced from Fig. 4.19.

4.2.4.4. *Order-parameter for ordering-type transformation: Statistical approach*

In Secs. 4.2.4.1–4.2.4.3, we geometrically considered the possibility of reconstructive transformations (onset of new symmetry elements breaking the group–subgroup relation) of the ordering type, which occur to the non-Landau (limit $\eta = 1$) phases, intuitively assuming existence of the corresponding minimum in the free-energy. Let us do it now in a regular way showing a functional dependence of the ordering-type phenomenological order-parameter on essential parameters — site occupancy numbers. We follow a standard procedure and consider ordering system as an ensemble of discrete states or sites homogeneously populated by objects of different nature (for instance, metal atoms and vacancies). Two competing states at any point are (i) presence $|1\rangle$ or (ii) absence $|0\rangle$ of the atom on the site. Thus, $\langle\alpha|1\rangle$ is a function describing the state corresponding to filling identical points by objects (let it be A-type atoms) in concentration α. The probability to reach the state is $P_\alpha^A = |\langle\alpha \mid 1\rangle|^2$, the probability to avoid it is $P_\beta^A = |\langle\beta \mid 1\rangle|^2$. Two evident equations are

$$P_\alpha^A + P_\beta^A = |\langle\alpha \mid 1\rangle|^2 + |\langle\beta \mid 0\rangle|^2 = 1, \tag{4.32}$$

and

$$P_\alpha^A + P_\alpha^B = |\langle \alpha_A \mid 1\rangle|^2 + |\langle \alpha_B \mid 1\rangle|^2 = 1. \tag{4.33}$$

Both Eqs. (4.32) and (4.33) describe the circles, spheres or hyperspheres, depending on the number of segregating components in the mixture, and we can express the probabilities of the events as follows:

$$P_\alpha = \sin^2(\Phi + \Omega\alpha), P_\beta = \cos^2(\Phi + \Omega\alpha). \tag{4.34}$$

A general form of the argument is used anticipating further normalization, and factor π is included. Population of a certain site is

$$\alpha = \frac{N_\alpha}{N_\alpha + N_\beta}, \quad N_\alpha + N_\beta = N_0 = \text{const.} \tag{4.35}$$

Thus, the microscopic (essential) parameter that characterizes the ordering system is the occupation ratio of a site. The population of an individual site, via the amplitude of probability $\langle\alpha|1\rangle$, should be averaged through the ensemble. This results in macroscopic (measurable) probability density serving as an order-parameter of ordering systems.

Equation (4.34) allows reducing the phenomenological order-parameter to a periodic function

$$\eta = \frac{|P_\alpha - P_\beta|}{|P_\alpha + P_\beta|} = \cos\{2(\Phi + \Omega\alpha)\}. \tag{4.36}$$

Boundary conditions for *disordered* state

$$\eta = \cos\{2(\Phi + \Omega\alpha)\} = 0, \tag{4.37}$$

and for *fully ordered* state ($\alpha = 1$)

$$\eta = \cos\{2(\Phi + \Omega)\} = 1, \tag{4.38}$$

allow one to calculate $\Phi = -\Omega = \frac{\pi}{4} \cdot \frac{1}{1-\alpha_0}$. Then, recalculating Eq. (4.36), one obtains

$$\eta = \cos\frac{\pi}{2}\left(1 - \frac{\alpha - \alpha_0}{1 - \alpha_0}\right) = \sin\frac{\pi}{2}\xi_\alpha, \tag{4.39}$$

where $\xi_\alpha = \frac{\alpha - \alpha_0}{1-\alpha_0}$ is the microscopic local order-parameter ($0 \leq \xi_\alpha \leq 1$). Thus, for ordering-type phase transitions, a transcendental form of the order-parameter function $\eta(\xi)$ becomes evident. Two points should be noticed in conclusion. Firstly, unlikely a displacive mechanism, $\eta(\xi)$ in the latter case is not periodic since its argument, which is the occupancy number, is limited ($0 \leq \xi_\alpha \leq 1$). Secondly, one concludes that a full ordering in an alloy can occur at a finite temperature, i.e., the corresponding order-parameter can reach saturation at certain temperature $T \neq 0$ K but not to be asymptotically approaching $\eta = 1$ at $T = 0$ K, as it appears in a classic statistical approach.

4.3. Generalization of the Preceding Examples: Density-Wave Theory of Reconstructive Phase Transitions

4.3.1. *Differences in the phenomenological descriptions of displacive and order–disorder reconstructive phase transitions*

Let us summarize the main features characterizing the phenomenological approach to the preceding working examples of reconstructive phase transitions. In this approach, the structures of the phases surrounding a transition are described in terms of *lattice functions*, which are the sums of stationary plane waves, associated with atomic planes and corresponding to the first harmonics of the Fourier expansions of the respective lattice functions. The form of the lattice function *increment* $\delta\psi$ expresses the transition mechanism. In this respect, it has been shown through the considered examples that *the transition order-parameter can be identified to the* ξ*-dependent part (phase) of* $\delta\psi$ *for displacive mechanisms and to the amplitude of* $\delta\psi$ *for order–disorder mechanisms.* The fact that the two classes of mechanisms are related to *distinct physical quantities defining* $\delta\psi$ has the following consequences on their respective phenomenological approach:

(1) *Displacive mechanisms* give rise to a periodic dependence of the order-parameter η as a function of the critical atomic displacements ξ and consequently to a *periodic* thermodynamic potential $F(T, \eta(\xi))$. This potential has the meaning of a *truncated Fourier series,* since it is formed by powers of the increment of the first harmonics of a Fourier expansion. Even the loss

of group–subgroup relationship between the symmetries of the structures is possible for certain definite displacements at which the system acquires new symmetry elements. The minimization of $F(T,\eta(\xi))$ with respect to ξ in the advanced approach yields two types of stable states: "Landau" states that are strictly group–subgroup related to the parent state symmetry and "limit" states that, in some structures, have no group–subgroup relation with the parent structure and coincide with the extrema of $\eta(\xi)$. For these limit states, the temperature dependence of the corresponding order-parameters displays a strongly discontinuous step-like behavior at the transition point, and then remains constant in full stability range.

For displacive reconstructive transitions to aperiodic structures, one must take into account the additional variation (in length and orientation) of the transition wave vector $\boldsymbol{k}$. In this case, the stable states are obtained by a maximization of the lattice functional increment $\delta\psi(\boldsymbol{k}, \xi)$ of the whole structure, both with respect to $\boldsymbol{k}$ and to the displacements ξ (see Sec. 6.4).

(2) For *order–disorder* mechanisms, the order-parameter η is a function of the atomic occupancy of the structure and is proportional to the increment of probability of occupancy of definite atomic sites. The loss of group–subgroup relationship with respect to the disordered structure is the result of a selective occupancy of the preceding sites, which corresponds to the fully ordered state. This latter state coincides with a limit situation corresponding to a maximal probability of occupancy of the sites equal to unity. In this framework, "Landau" partially ordered states, which are group–subgroup related to the disordered structure, can be interpreted as *defective limit states.*

The thermodynamic potential $F(\xi)$ associated with order–disorder reconstructive transitions can be a standard Taylor expansion of the order-parameter, and the stable (Landau and limit) states are obtained by minimization of F with respect to η and ξ. As fully ordered states correspond to fixed (temperature-independent) probabilities of occupancy of given sites, they will be distinguished from partially ordered state by a step-like dependence of the order-parameter in function of temperature.

In contrast to the content of this section, which underlines the differences in the approaches to the various categories of discontinuous transitions, the next section stresses the unifying character of the density wave theory of phase transitions by providing its general formulation.

4.3.2. *Formulation of the density-wave theory, and its connection with the Landau theory*

4.3.2.1. *Lattice function, probability density, order-parameter*

The electron density in a crystal $\rho(x,\ y,\ z)$ varies periodically along any direction through the lattice, going across a complicated cycle of peaks and minima, and is thus described in structure analysis, by Fourier series as can be done for any periodic function. The electron density at the point of coordinates $(x,\ y,\ z)$ can be written for a unit cell of volume V [13] by

$$\rho\,(x,y,z) = \frac{1}{V}\sum_{h,k,l=-\infty}^{+\infty} F(h,k,l)\cdot e^{2\pi i(hx+ky+lz)}, \tag{4.40}$$

where each Fourier term in the series, associated with the coefficient $F(h,\ k,\ l)$, represents a stationary system of density waves parallel to the $(h,\ k,\ l)$ family of atomic planes. Each set of waves is capable of diffracting incident rays with a certain intensity $|F(h,\ k,\ l)|^2$, for the successive parallel sheets of electron density act as reflecting planes.

Although the sum (4.40) is infinite for a full determination of the atomic positions, a number of simplifications [13, 14] using the symmetry of the crystal allow to restrict to a finite number of harmonics for each sublattice. Furthermore, when dealing with the sole description of the structural modifications occurring at a phase transition, one can restrict to the critical sublattice primarily involved in the transition mechanism and to the first harmonics that contain the relevant information on the periodicity of the lattice and on the average positions of the atoms. Summing up over the equivalent atoms contained in the unit cell in real space instead of the atomic planes in reciprocal space, the probability density given by Eq. (4.40) reduces to the lattice function defined by Eq. (4.2).

At a structural transition, which may involve in the more general case a change in the wave vector *and* in the atomic positions, the increment $\delta\psi$ of the lattice function from the initial structure (with atomic positions $\boldsymbol{R}_j^0$ and k-vector $\boldsymbol{k}_0$) to the final structure ($\boldsymbol{R}_j = \boldsymbol{R}_j^0 + \Delta\boldsymbol{R}_j$, $\boldsymbol{k} = \boldsymbol{k}_0 + \Delta\boldsymbol{k}$), at a point defined by the position vector $\boldsymbol{r}$, is

$$\delta\psi\left(\boldsymbol{r}\right) = \frac{1}{N}\sum_{j=1}^{N}\left[A_j' e^{i\boldsymbol{k}(\boldsymbol{R}_j-\boldsymbol{r})} - A_j e^{i\boldsymbol{k}_0\left(\boldsymbol{R}_j^0-\boldsymbol{r}\right)} + \text{c.c.}\right], \qquad (4.41)$$

where the sum runs over the N atoms of the unit cell of the *final structure* in which the normalizing density wave coefficients are A_j'. Two main specific forms can be used for Eq. (4.41) depending on the nature of the transition mechanism:

(1) For order–disorder transitions between periodic structures, one has fixed positions $\boldsymbol{R}_j = \boldsymbol{R}_j^0$ and $\Delta\boldsymbol{k} = 0$. Equation (4.41) simplifies to

$$\delta\psi\left(\boldsymbol{r}\right) = \frac{1}{N}\sum_{j=1}^{N}\left[\left(A_j' - A_j\right) e^{i\boldsymbol{k}_0\left(\boldsymbol{R}_j^0-\boldsymbol{r}\right)} + \text{c.c.}\right]. \qquad (4.42)$$

Besides, one can write $A_j' - A_j = A^{\alpha}\cdot\Delta P_j^{\alpha}$, where ΔP_j^{α} represents the variation at the transition of the probabilities of occupancy of an atom of type α in position j. Using the property that the ΔP_j^{α}, for different j, differ only by their sign, Eq. (4.42) becomes

$$\delta\psi\left(\boldsymbol{r}\right) = A^{\alpha}\left|\Delta P^{\alpha}\right|\left[\frac{1}{N}\sum_{j=1}^{N} e^{i\boldsymbol{k}_0\left(\boldsymbol{R}_j^0-\boldsymbol{r}\right)}\right]. \qquad (4.43)$$

In (4.43), $\eta = A^{\alpha}\left|\Delta P^{\alpha}\right|$ defines the (scalar) transition order-parameter, and the multiplying factors $\frac{1}{N}\sum_{j=1}^{N} e^{i\boldsymbol{k}_0\left(\boldsymbol{R}_j^0-\boldsymbol{r}\right)}$ represent the normalized basic functions spanning the order-parameter space ε. These functions possess the same symmetry as the δ-atomic functions and transform as IRs of the initial crystal structure.

(2) For displacive reconstructive transitions between periodic crystal structures $A_j' = A_j$ for all j and $\Delta\boldsymbol{k} = 0$, Eq. (4.41) can be

written as

$$\delta\psi\left(\boldsymbol{r}\right)=\frac{1}{N}\sum_{j=1}^{N}A_j\cdot\left\{e^{-i\boldsymbol{k}_0\boldsymbol{r}}\cdot\left[e^{i\boldsymbol{k}_0\left(\boldsymbol{R}_j^0+\Delta\boldsymbol{R}_j\right)}-e^{i\boldsymbol{k}_0\boldsymbol{R}_j^0}\right]+\text{c.c.}\right\}. \tag{4.44}$$

Assuming only one type of atom ($A_j = A$ for all j), the collective displacements of the atoms at the transition can be decomposed into parallel displacements, which do not modify the relative structural arrangement of atoms and *antiparallel displacements of pairs of atoms.* The critical sublattice can thus be subdivided into two antiparallel sublattices, indexed j_1 and j_2, with $\boldsymbol{R}_{j1}^0 = -\boldsymbol{R}_{j2}^0$, and $\Delta\boldsymbol{R}_{j1}^0 = -\Delta\boldsymbol{R}_{j2}^0$. Equation (4.44) is transformed into

$$\delta\psi\left(\boldsymbol{r}\right)=\frac{A}{N}\cos\left(\boldsymbol{k}_0\boldsymbol{r}\right)\cdot\sum_{j1}^{N}\left\{\sin\left[\boldsymbol{k}_0\left(\boldsymbol{R}_{j1}^0+\Delta\boldsymbol{R}_{j1}\right)\right]-\sin\left(\boldsymbol{k}_0\boldsymbol{R}_{j1}^0\right)\right\}. \tag{4.45}$$

In Eq. (4.45), the position of the atoms coincides with the maxima of the density waves, so for different indices j_1, the arguments of the corresponding functions differ only by periods of these functions. Accordingly, one has the final expression of $\delta\psi$

$$\delta\psi\left(\boldsymbol{r}\right)=\frac{A}{N}\cos\left(\boldsymbol{k}_0\boldsymbol{r}\right)\cdot\left\{\sin\left[\boldsymbol{k}_0\left(\boldsymbol{R}_1^0+\Delta\boldsymbol{R}_1\right)\right]-\sin\left(\boldsymbol{k}_0\boldsymbol{R}_1^0\right)\right\}, \tag{4.46}$$

where $\boldsymbol{R}_1^0$ is an arbitrary atomic position in the critical sublattice. In Eq. (4.46), the set of $\cos(\boldsymbol{k}_0\boldsymbol{r})$ functions identifies to the basic atomic functions spanning the representation space ε of the initial structure. Thus, each of the transition order-parameter components has the general form

$$\eta=A\cdot\left\{\sin\left[\boldsymbol{k}_0\left(\boldsymbol{R}_1^0+\Delta\boldsymbol{R}_1\right)\right]-\sin\left(\boldsymbol{k}_0\boldsymbol{R}_1^0\right)\right\}, \tag{4.47}$$

i.e., it is a scalar periodic function of the magnitude of the atomic antiparallel displacements that constitute the transition mechanism.

In the case of a purely displacive periodic–aperiodic transition, one has again $A'_j = A_j =$ for all j but $\Delta \boldsymbol{k} \neq 0$. Analogous considerations lead to the increment of the lattice functional for the whole structure

$$\delta\psi(\boldsymbol{r}) = \frac{1}{N}\sum_{j1}^{N}\sum_{k} A_{j1} \cdot \cos(\boldsymbol{k}_0 + \Delta\boldsymbol{k})\boldsymbol{r} \cdot \left\{\sin\left[(\boldsymbol{k}_0 + \Delta\boldsymbol{k})\left(\boldsymbol{R}^0_{j1} + \Delta\boldsymbol{R}_{j1}\right)\right] - \sin\left(\boldsymbol{k}_0\boldsymbol{R}^0_{j1}\right)\right\}, \quad (4.48)$$

which generalizes Eq. (4.46) to an inhomogeneous (aperiodic) structure. The sinusoidal factor in Eq. (4.47) formally represents the order-parameter of the crystal-aperiodic quasicrystal reconstructive phase transition. In real space, the order-parameter is a quasi-periodic function with respect to the initial crystal lattice. *It becomes a periodic function in the multidimensional space spanned by the* $(\boldsymbol{k}_0 + \Delta\boldsymbol{k})$ *wave vectors.*

The definition of the order-parameter given before for order–disorder-type reconstructive phase transitions, namely as a linear combination of increments of probability of occupancy for given atomic sites, is the standard definition assumed in the approaches to order–disorder transitions [15]. It coincides with the "degree of order" introduced in the original papers of Landau [1] for positional ordering in binary mixtures and for the orientational ordering in NH_4Cl. By contrast, for displacive-type reconstructive phase transitions, our definition of the order-parameter, given by Eq. (4.47), as a periodic function of the critical displacements, represents obviously a *generalization for large displacements* of the *linear relationship* between order-parameter and small displacements [16], usually assumed for second- or slightly first-order displacive transitions. For a small magnitude of $\boldsymbol{\Delta R}^0_1$ in Eq. (4.47), one has as a first approximation

$$|\eta| \approx Ak_0|\Delta\boldsymbol{R}^0_1|. \quad (4.49)$$

The expressions (4.43) and (4.46) found for the increment of lattice functional $\delta\psi(\boldsymbol{k})$ appear as formally similar to the increment of probability density $\delta\rho(\boldsymbol{r})$, given by Eq. (4.1) in the Landau theory, as the two quantities are bilinear product of the order-parameter and of atomic functions, which span the representation space of the initial crystal structure. However, they are obtained from different premises. In the Landau theory [1], $\delta\rho(\boldsymbol{r})$ is assumed to be small. This

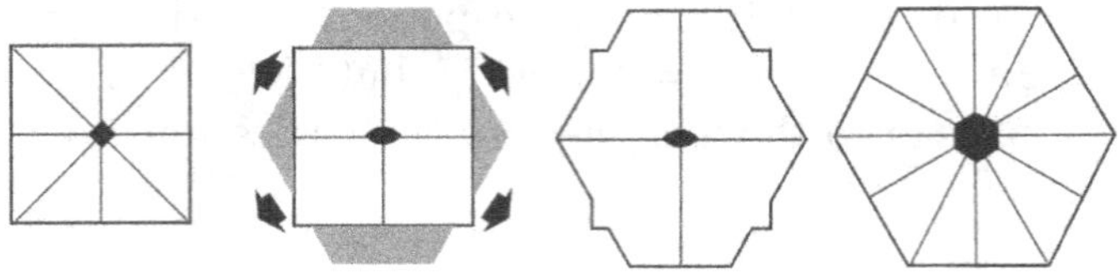

Fig. 4.21. Illustrative example showing that for a *large and specific* increment $\delta\rho$, symmetries of the densities ρ and ρ_0 in Eq. (4.1) *may not be group–subgroup related.* Thus, for the large increment shown in (b), which has an orthorhombic ($2mm$) symmetry, the initial figure (a) of tetragonal symmetry ($4mm$) may either decrease in symmetry (c) or change to the hexagonal symmetry ($6mm$) for a specific size of the increment (d).

has the correlated consequences that a group–subgroup relationship necessarily exists between the initial and final structures, and that the order-parameter components (the η^i_{kj} in Eq. (4.1)) are necessarily small quantities. In our definition, no restriction is made on the magnitude of $\delta\psi(\boldsymbol{r})$, and as a consequence, the symmetry groups that leave invariant $\psi(\boldsymbol{r})$ and $\psi_0(\boldsymbol{r})$ are not necessarily related by a group–subgroup relationship. This is illustrated in Fig. 4.21, which shows that *for a large increment, a system may acquire new symmetry elements.*

Another consequence of the group–subgroup relationship assumed in the Landau theory between $\delta\rho(\boldsymbol{r})$ and $\rho_0(\boldsymbol{r})$ is that the bilinear form found for $\delta\rho(\boldsymbol{r})$ can be obtained directly from general group theoretical considerations, namely from the property that $\delta\rho(\boldsymbol{r})$ can be expanded into a sum of functions, the number of which is equal to the number of symmetry elements of G_0.

Accordingly, the Landau theory is formulated without considering any specific transition mechanism and *without expressing* $\rho(\boldsymbol{r})$ and $\rho_0(\boldsymbol{r})$. In our formulation, the form of $\delta\psi(\boldsymbol{r})$ is *deduced* from the calculated forms of $\psi_0(\boldsymbol{r})$ and $\psi(\boldsymbol{r})$ *assuming a given type of mechanism.* If such formulation loses in generality, it has the advantage to *relate explicitly the order-parameter to the physical mechanism.*

In the following two sections, we will show that essential distinctive features of our approach, with respect to the Landau approach, are reflected in the structure of the order-parameter space and in the mathematical nature of the thermodynamic potential associated with a reconstructive phase transition.

4.3.2.2. *Order-parameter space for strongly discontinuous and reconstructive phase transitions*

Equations (4.43) and (4.46) reveal that the order-parameter space of reconstructive phase transitions, denoted σ_n hereafter, differs in an essential manner from the order-parameter space ε_n used in the description of second- and weakly first-order transitions. In the standard Landau theory, the order-parameter is represented by a *fixed vector* $\boldsymbol{\eta}$ in the n-dimensional *vector space* ε_n. Every vector direction η in ε is left invariant by the isotropy group I_η of $\boldsymbol{\eta}$, and the symmetry group G of the phase stable below the transition is related to the isotropy group I_η^{eq} of the vector $\boldsymbol{\eta}^{\mathrm{eq}} = (\eta_1^{\mathrm{eq}}, \ldots, \eta_n^{\mathrm{eq}})$, which *fixes the corresponding symmetry direction in* ε. For strongly discontinuous and reconstructive transitions, the magnitude of $\boldsymbol{\eta}^{eq}$ (and of the components η_i^{eq}) *is not determined* but *is a function of a variational parameter associated with the transition mechanism.* The order-parameter $|\boldsymbol{\eta}^{eq}(\xi_j)|$ is a function of the local variables ξ_j. It is a *transcendental, bounded, periodic* function, and depending on the magnitude of $|\boldsymbol{\eta}^{eq}(\xi_j)|$, the order-parameter exhibits different symmetries, which may correspond to a subgroup G of G_0 but also to a supergroup G' of G, G being, therefore, a common subgroup for G_0 and G'. The increase in symmetry from G to G' occurs for specific points of the distribution determined by $|\boldsymbol{\eta}^{eq}(\xi_j)|$, *which correspond to the holohedry* (maximal symmetry of the Bravais lattice) *of G.* These special points of symmetry degeneracy of the structure are located on the *edges* of the σ-space, which coincide with the *maximal values* of the $|\boldsymbol{\eta}^{eq}(\xi_j)|$ function.

Thus, the essential difference in the advanced description of strongly first-order, in general, and reconstructive, in particular, transition [17] as compared to the standard Landau description of structural transitions [1, 16] is due to the fact that *the magnitude* of the variational parameters associated with the transition mechanism *influences the symmetry of the order-parameter*, and thus it has to be taken into account in the symmetry approach to the transition. By contrast, in the standard Landau theory, the dependence of $|\eta|$ as a function of the variational parameters is a trivial (linear) function of the variational parameter, which has no effect on its symmetry, and accordingly, *the description of the transition mechanism* can be separated from the considerations that lead to the order-parameter symmetry.

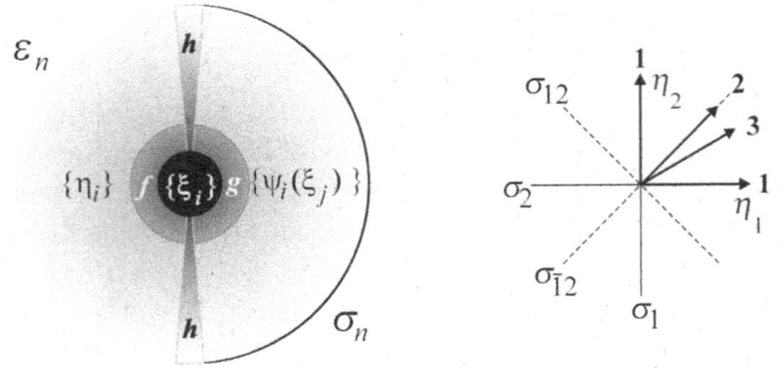

(a) (b)

Fig. 4.22. (a) Theoretical scheme showing the connection between ε and σ, i.e., between the set of m displacements ξ_i, the n scalar order-parameter components η_i, and the lattice functional ψ, which depends on the linear combinations of basic density waves $\exp\{i\boldsymbol{k}(\boldsymbol{R}_0 + \xi_i)\}$. (b) Orientations of the "vector" order-parameter $\boldsymbol{\eta}$ in the ε_2 space associated with the $4mm$–$6mm$ transition. 1, 2 and 3 are the orientations corresponding, respectively, to the symmetries $6mm$, $4mm$ and $mm2$.

Figure 4.22(a) illustrates the mathematical difference existing between the ε and σ order-parameter spaces. One can see that while ε is an n-dimensional vector space, σ is an n-dimensional *functional space*, the structure of which *depends on the periodicity of the function of the variational parameter describing the transition mechanism.* The operator h transforming ε into σ can be written

$$h = g \circ f^{-1}, \tag{4.50}$$

where f is a linear projection operator transforming a set of displacements into the basis of an IR τ_{kj}, and g is a transcendental operator, which associates the set of displacements to the *variation* of the basic density waves (or simply to density waves for ordering transitions). Both g and h operators act as Fourier transformers: g transforms atomic planes of the parent structure to their Fourier harmonic images, while h calculates difference between parent and distorted images. In our above operations with free energy and order-parameter, a real-space renormalization h transforms a set of variables $\{\eta_i\}$ to $\{\xi_j\}$ one.

As an illustration of the preceding considerations, let us come back to the examples discussed in Sec. 4.2. In the first example of displacive ($P6_3/mmc$–$P6/mmm$) transition, the subgroup $G = P\bar{3}m1$

is induced by a one-dimensional IR (B_{1g}) associated with the center of the hexagonal Brillouin zone (denoted $\tau_5(\boldsymbol{k}_{16})$ in Kovalev's notations [18]). The σ-space is formed by the function (4.8) of critical displacements along Z, which has two symmetrical edges $\pm\eta_0$. The limit states have the symmetry $G' = P6/mmm$ corresponding to the holohedry of the $P\bar{3}m1$ lattice. They occur at the special points of the $P6/mmm$ unit cell shown in Fig. 4.2.

The second case of displacive reconstructive transition described in Sec. 1.2 shows explicitly the difference and connection existing *between the σ-space and the critical displacement space.* Here, the ε-space is two-dimensional, as the group $G = mm2$ is associated with a bidimensional IR ($\tau_3(\boldsymbol{k}_{15})$, with $\boldsymbol{k}_{15} = \boldsymbol{a}^*/2$) of the initial $P4mm$ planar space-group. In the ε-space, three distinct *orientations* of the order-parameter $\boldsymbol{\eta}$ are found as shown in Fig. 4.22(b).

(1) For $\eta_1 \neq 0$, $\eta_2 = 0$ (or for $\eta_1 = 0$, $\eta_2 \neq 0$), the direction of $\boldsymbol{\eta}$ coincides with critical displacements (ξ_a *or* ξ_b) along X *or* Y, and its *magnitude* is $|\boldsymbol{\eta}(\underline{\xi})| = \eta_0 \cdot \sin(a^*\xi)$. The holohedry of the $mm2$ lattice increases to $6mm$, when taking into account the tensile deformation (e_{xx} *or* e_{yy}), for the specific displacements of magnitude $\xi = a/4$, $3a/4$, $5a/4$,... along X *or* Y.

(2) For $\eta_1 = \eta_2 = \eta \neq 0$, $\boldsymbol{\eta}$ is oriented along the diagonals of the (η_1,η_2) plane in the ε-space, and its magnitude is $|\boldsymbol{\eta}(\xi)| = \sqrt{2} \cdot \eta_0 \cdot \sin(a^*\xi)$, i.e., it corresponds to *equal* critical displacements $\xi_a = \xi_b = \xi$, along X *and* Y. For arbitrary values of ξ, one has $G = mm2$ with *four* atoms per unit cell, whereas for the special displacements $\xi_a = \xi_b = a/2$, the preceding symmetry increases to $G' = 4mm$, *with one atom per unit cell*, as shown in Fig. 4.10(b). The G'-lattice is turned by $\pi/4$ with respect to the initial tetragonal lattice.

(3) For $\eta_1 \neq \eta_2 \neq 0$, $\boldsymbol{\eta}$ has a general direction in the (η_1,η_2) plane and corresponds to *unequal* critical displacements $\xi_a \neq \xi_b$ along the X and Y axes. The magnitude of the $|\boldsymbol{\eta}(\underline{\xi})|$ function is $\eta_0 \cdot [\sin^2(a^*\xi) + \sin^2(b^*\xi)]^{1/2}$, and there are no special points at which the lowest $2mm$ symmetry may increase.

In the example of order–disorder $Im\bar{3}m$–$P6/mmm$ transition discussed in Sec. 4.2.4, the ε-space is eight-dimensional, although the orientation of $\boldsymbol{\eta}$ is invariant by a two-dimensional subspace ε^G, where

$G = P\bar{3}m1$. Besides, G represents a distribution of positions never realized as a stable state due to the existence of vacancies determined by the initial concentration 1/3. This specific value increases the symmetry of G to $P6/mmm$, which *is* the holohedry of the structure. The magnitude of $\boldsymbol{\eta}$ varies from 0 to 1 when $1/3 \leq P^{\alpha} \leq 1$. All other fractional concentrations of the initial lattice (1/2, 2/3,...) will *preserve* the rhombohedral symmetry $P\bar{3}m1$, and the $P6/mmm$ symmetry will not take place in this case.

In the tetragonal–octagonal transition described in Sec. 4.2.4 (Fig. 4.12), the ε-space is eight-dimensional, as the $4mm$–$8mm$ symmetry change requires a star $\boldsymbol{k} = \boldsymbol{a}^*$ of the wave vector with eight branches. However, while starting from the crystal structure in the description of the crystal-quasicrystal transition is physically justified (we usually *know* the structure and properties of the crystal, and try to work out the structure and properties of the quasicrystal), it corresponds to an unusual scheme with respect to the properties of the σ-space. As shown in Fig. 4.12(b), the lowest symmetry G *coincides* with the crystal structure ($4mm$ symmetry with eight atoms per unit cell). The symmetry *increases* to $8mm$ or $4mm$ (with four atoms per unit cell) for specific displacements $\xi_{\max}$ *and specific orientations of the transition wave vector* $\Delta\boldsymbol{k}_{\max}$, which are obtained by maximizing both the magnitude and orientation of the order-parameter with respect to ξ and $\boldsymbol{k}$. The latter example relates to the class of "commensurate–incommensurate" transformations in classic phenomenological theory where not only the magnitude of the order-parameter is variable but also the corresponding k-vector of reciprocal space varies.

4.3.2.3. *Thermodynamic potential, equations of state, stable states*

The non-equilibrium thermodynamic potential associated with a strongly first-order transition has the general form

$$F\left[T, P, \delta\psi\left(x_j\right)\right] = F_0\left(T, P\right) + F[\eta_{ki}\left(x_j\right)], \qquad (4.51)$$

where the n-components η_{ki} transform as an IR τ_{kj} of the space group G_0 of the initial structure and determine the possible stable directions of the order-parameter $\boldsymbol{\eta}$ in the ε-space. The magnitude of $\boldsymbol{\eta}$

$$|\boldsymbol{\eta}(x_j)| = \left[\sum_{i=1}^{n} \eta_{ki}^2(x_j)\right]^{1/2}, \tag{4.52}$$

depends on the m variational parameters.

The *form* of the order-parameter expansion $F[\eta_{ki}(x_j)]$ is determined, as usually, by the symmetry of τ_{kj} (see Sec. 1.2.3). By contrast, *the minimization of the expansion has to be performed with respect to the variational parameters x_j, and not with respect to the η_{ki}*, as in the standard procedure. This leads to m equations of state

$$\frac{\partial F}{\partial x_j} = \sum_i \frac{\partial F}{\partial \eta_{ki}} \cdot \frac{\partial \eta_{ki}}{\partial x_j} = 0 \quad (j = 1 - m), \tag{4.53}$$

and to stability conditions, which consist in finding the range of positiveness of the eigenvalues of the $(m \times m)$ matrix built from the second derivatives of $F[\eta_{ki}(x_j)]$ with respect to x_j

$$\left\| \frac{\partial^2 F}{\partial x_j \partial x_k} \right\| \geq 0. \tag{4.54}$$

Equation (4.53) clearly shows the existence of *two different sorts of stable states*: the "Landau" states, which correspond to the "classic" minima determined by the conditions $\partial F/\partial \eta_{ki} = 0$ $(i = 1{-}n)$, and the limit (saturated) states associated with the conditions $\partial \eta_{ki}/\partial x_j = 0$ $(j = 1{-}m)$.

The minimization procedure corresponding to Eqs. (4.53) and (4.54) is the more general, and it allows to work out all the Landau *and* limit states, which may take place as the result of a transition (continuous, discontinuous or even reconstructive) from a given structure, of symmetry group G_0, for a given IR τ_{kj}. However, the problem posed by the interpretation of reconstructive phase transitions consists of not only operating with the free-energy depending on non-small variational parameters. The problem usually appears under a more restricted, but more complex, form: *Given a reconstructive phase transition between two known structures (associated with the unrelated group–subgroup symmetries G_0 and G'), (1) what is the* *microscopic (atomistic) mechanism of the transition? (2) Which theoretical framework allows describing the specific macroscopic features (phase diagram, physical properties) of the transition?* The practical procedure, which allows answering these questions, can be decomposed as follows:

(1) Determination of the maximal substructure, of space group $G = G_0 \cap G'$, common to the structures surrounding the transition, from which one can deduce, following the standard procedure indicated in Chapter 1, the n-dimensional IR τ_{kj} of the G_0 space-group associated with the virtual $G_0 \to G$ "Landau" transition. G corresponds to given relationships between the n order-parameter components, *which fix the orientation of* η *in the* ε*-space.*
(2) Taking into account the symmetry of the degree of freedom τ_{kj}, one can find the *possible* atomic mechanisms transforming the intermediate virtual structure (G) into the final one (G'). The *nature* of the mechanism (displacive, order–disorder, or a combination of the two types of processes) and the selection of the *more probable* steps of reorganization from one structure to another have to be chosen taking into account their consistency with the experimentally known properties of the transition. For example, in the case described in Sec. 4.2.2.1, three possible mechanisms correspond, *a priori*, with the $4mm$–$6mm$ structural change, but only one preserves the crystal shape.
(3) Once the effective transition mechanism is singled out, one expresses the structures in terms of the lattice functions ψ_0 and ψ and deduces the form of the increment $\delta\psi(x_j)$, which provides the magnitude of the order-parameter $|\boldsymbol{\eta}(x_j)|$ in function of the relevant variational parameters x_j. *This determines the structure and symmetry of the corresponding functional space* σ, and the critical variational parameters x_j^c at which the $G_0 \to G$ group–subgroup relationship is lost.
(4) For a thermodynamic description of the transition, which gives the phase diagram in which the various structures are inserted, as well as the macroscopic properties of the phases, one can then restrict to the use of the effective order-parameter expansion

$$F\left[\left|\boldsymbol{\eta}\left(x_j\right)\right|\right] = a_1\left|\boldsymbol{\eta}\left(x_j\right)\right|^2 + b_1\left|\boldsymbol{\eta}\left(x_j\right)\right|^3 + a_2\left|\boldsymbol{\eta}\left(x_j\right)\right|^4 + \cdots, \quad (4.55)$$

in which the allowed non-vanishing successive powers of $|\boldsymbol{\eta}(x_j)|$ are determined by the symmetry of τ_{kj}.

The various stages of the phenomenological approach to discontinuous phase transitions may be simplified, or on the contrary complicated, depending on the nature of the mechanism.

(i) For displacive transitions between strictly periodic structures, the wave vector is fixed ($k = k_0$), and for a given direction $\boldsymbol{\eta}$ in the ε_n-space, the order-parameter is a sinusoidal function of the critical displacements given by Eq. (4.47). The extrema of this function, given by $\partial\eta_{ki}/\partial x_j = 0$, provide directly the special points of the lattice at which the structure of space-group G' may arise, whereas the intersections of the sinusoid with the displacement axis coincide with different orientational domains of the initial structure. The effective order-parameter expansion (4.55) is a sum of powers of sinusoidal functions, i.e., it represents a Fourier series corresponding to the successive higher harmonics of the series. One can generally restrict to the lowest degrees (fourth or sixth) of $F[\eta_{ki}(x_j)]$, as the sinusoidal arc separating the initial structure from the first maximum (one-fourth of a period) expresses a fifth-degree approximation of the sinus with respect to the x_j as illustrated in Fig. 4.23.

(ii) For displacive reconstructive transitions to aperiodic structures, $\boldsymbol{k}$ is also a variable, which determines the orientation of $\boldsymbol{\eta}$ in the ε-space and the magnitude of $|\boldsymbol{\eta}(x_j)|$ in the σ-space. Thus, the determination of the equilibrium, aperiodic, state requires a minimization procedure with respect to $\boldsymbol{k}$ and not only with respect to the displacements. This can be performed either by minimizing the lattice functional increment, given by Eq. (4.48), or by a minimization of the non-equilibrium thermodynamic potential $F[\eta_{ki}(x_j)]$ with respect to $\boldsymbol{k}$ and to the x_j. In this case, the dependence on $\boldsymbol{k}$ is taken into account by a renormalization in k-space (see Sec. 2.3) of the coefficient $\alpha(k)$ of the second-degree invariant $\sum_i \eta_{ki}^2$ as in the current approach to incommensurate structures [19].

(iii) For order–disorder transitions, the x_j variables identify to the increments δP_j^α of probability of occupancy of the j position by the α-atom. However, in our description, we have not considered the number of atoms in the unit cell of the initial structure (the

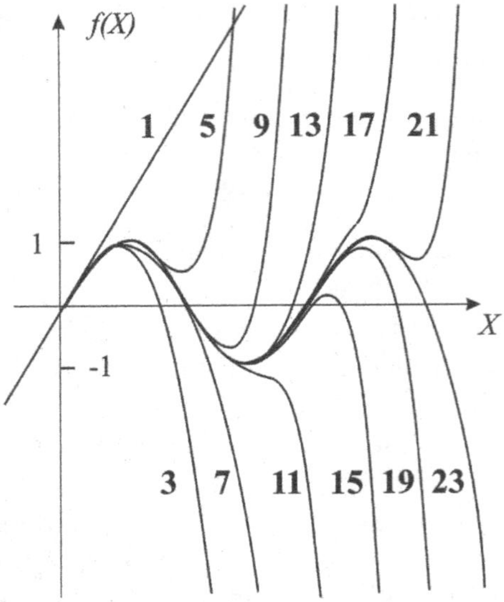

Fig. 4.23. Approximation of the $\sin(x)$ function by successive functions Σx^n of increasing power.

concentration) as a variable but as a fixed number, which determines the lower limit of the δP_j^α in the ordered states. The loss of group–subgroup relationship for specific fractional concentrations, which is typical of morphotropic transitions [20], reveals that the temperature–concentration phase diagram of order–disorder transitions *would correspond to a nonlinear dependence of the order-parameter* with a more complex structure than the linear function $|\eta(\delta P_j^\alpha)|$ found for *temperature-dependent* reconstructive order–disorder mechanisms.

4.4. Specific Properties of Direct Transitions to the Non-Landau States: Reconstructive Transformations

4.4.1. *Temperature dependence of the order-parameter and related physical quantities*

Since a displacive reconstructive transition to a limit state occurs for fixed critical displacements (given by the extrema $\partial\eta/\partial\xi = 0$), the corresponding temperature dependence of the order-parameter

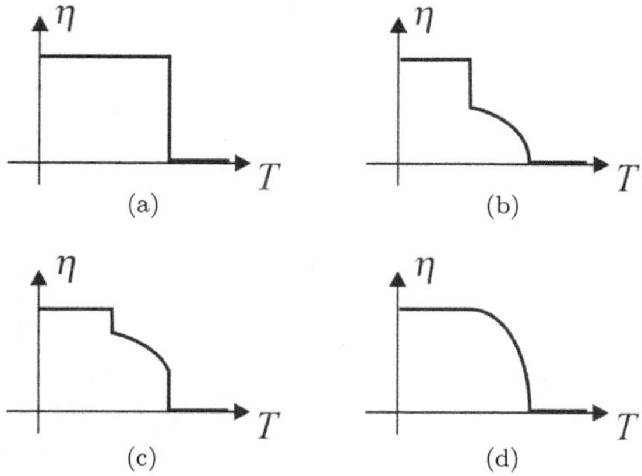

Fig. 4.24. (a) Step-like dependence of $\eta(T)$ at a direct transition to a limit state. (b) Temperature dependence of $\eta(T)$ for a second-order transition to a Landau state followed by a transition to a limit state, (c) for a first-order transition to a Landau state, followed by a transition to a limit state, (d) for an ordering transition to a fully ordered state going through the partially ordered state.

displays a typical step-like behavior shown in Fig. 4.24(a). Reciprocally, the constancy of $\eta(T)$ below the transition reflects only the existence of a limit state, but does not imply necessarily the absence of a group–subgroup relationship between the phases. For example, a step-like behavior is found for the liquid–solid transition [21] and also for a number of first-order transitions in ferroic materials (e.g., in langbeinites [22]) in which the complexity of the structure preserves the group–subgroup connection of the phases (see Sec. 4.6). When a displacive reconstructive transition occurs via an intermediate Landau state, one may have different behaviors for $\eta(T)$ as shown in Figs. 4.24(b) and 4.24(c).

A constant limit $\eta = 1$ is also reached by the order-parameter in the fully ordered state resulting from a disorder–order transition. Thus, the same step-like shape represented in Fig. 4.24(a) should be observed for reconstructive order–disorder transitions. However, due to the fact that partially ordered states, with $0 < \eta < 1$, do not differ in their symmetry from the fully ordered state, the $\eta(T)$ curve is more liable to display the dependence shown in Fig. 4.24(d) in which the partially ordered–fully ordered limit corresponds to a *break* between the "Landau" and limit regimes. A comparison with experiment here is uneasy (see Chapter 6), and to our knowledge,

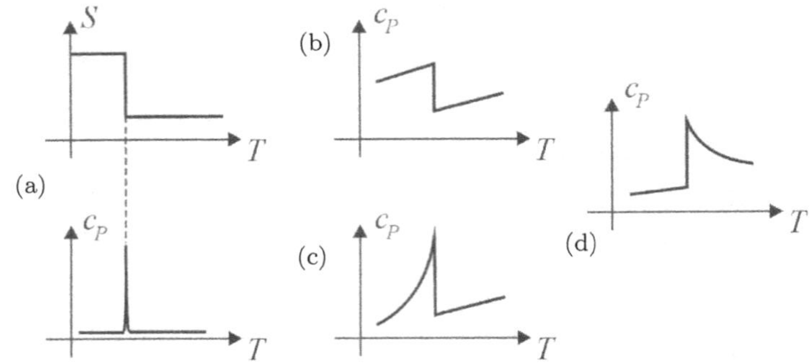

Fig. 4.25. (a) Dependence on temperature of the entropy $S(T)$ and specific heat $C_P(T)$ at a reconstructive transition. (b) $C_P(T)$ for a second-order and (c) a first-order transition to a Landau state. (d) $C_P(T)$ for a second-order transition between a Landau state and a limit state.

the more generally observed behavior coincides with the curve represented in Fig. 4.24(d) in which the limit regime is reached asymptotically without break.

Another distinctive property of reconstructive transitions is the temperature dependence of the specific heat $c_P = -T \cdot (\partial^2 F/\partial T^2)$. From the constant values of the order-parameter in the two phases surrounding a reconstructive transition, one can easily deduce that the corresponding values of their specific heats are *equal*, and thus *no jump* Δc_P should take place at the transition. This is in contrast with the *upward* jump observed on cooling for first-order (or second-order) transition to Landau states (Fig. 1.11(b)). However, from the finite discontinuity in entropy $S = -\partial F/\partial T$ at the transition, one can infer that $c_P(T)$ will display a *narrow peak* at the transition temperature as shown in Fig. 4.25(a).

Table 4.2 summarizes the expressions, in function of the phenomenological coefficients of the thermodynamic potential (4.9), of the latent heats $L = \frac{\Delta S}{T_{tr}} = -\frac{1}{T_{tr}} \cdot \frac{\partial \Delta F_{eq}}{\partial T}$ and specific heat jumps Δc_P, for the various transitions occurring in the phase diagram of Fig. 4.4. Figures 4.25(b) and 4.25(c) illustrate the corresponding behaviors of $c_P(T)$, respectively, for a second- and first-order transition to a Landau state (e.g., between states I and III in Fig. 4.4). Figure 4.25(d) shows the variation of $C_P(T)$ for the

Table 4.2. Latent and specific heats for the phase transitions represented in Fig. 4.4.

	Transition temperature	L	Δc_P
I–III (second order)	T_C	0	$\frac{T_C a_{10}^2}{2a_2}$
I–III (first order)	$T_1 = T_C + \frac{a_2^2}{4a_{10}a_3}$	$-\frac{a_{10}}{T_1} \cdot \frac{a_2}{6a_3}$	$-\frac{T_1 a_{10}^2}{a_2}$
I–II	$T_2 = T_C - \frac{a_2 + a_3}{a_{10}}$	$-\frac{a_{10}}{T_2}$	0
II–III	$T_3 = T_C - \frac{2a_2 + 3a_3}{a_{10}}$	0	$\frac{T_3 a_{10}^2}{2(2a_2 + 3a_3)}$

transition (II–III in Fig. 4.4) between a Landau state and a limit state.

Using the forms (4.8) and (4.9) of the order-parameter η and thermodynamic potential $F(\eta)$, respectively, one can calculate the temperature dependence of the susceptibility associated with the order-parameter $\chi = \lim_{\bar{\eta}\to 0} (\partial\eta/\partial\bar{\eta})$, where $\bar{\eta}$ is the field conjugated to η. One finds for the phases I, II and III,

$$\chi_I = (2a_1)^{-1}, \chi_{II} = (2a_1 + 12a_2\eta_0^2 + 30a_3\eta_0^4)^{-1},$$
$$\chi_{III} = \left[-8\left(a_1 + a_2\eta_0^2\right)\right]^{-1}.$$

The corresponding temperature dependencies $\chi(T)$ across the I–II, I–III and II–III transitions and the inverse susceptibilities $\chi^{-1}(T)$ are shown, respectively, in Figs. 4.26(a)–4.26(c). One can see that the reconstructive I–II transition is characterized by a *downward* jump of $\chi(T)$, while an upward jump is usually found for the first-order transition I–II to a Landau state as shown in Sec. 1.3.3.2.

In summary, we have shown that *the anomalies found at reconstructive transitions for the order-parameter $\eta(T)$, the specific heat $C_P(T)$, and susceptibility $\chi(T)$ differ qualitatively from the anomalies exhibited by these quantities for standard first-order transitions to Landau states.* Reciprocally from the typical shapes of the anomalies displayed by $\eta(T)$, $C_P(T)$ and $\chi(T)$, one may infer its reconstructive character.

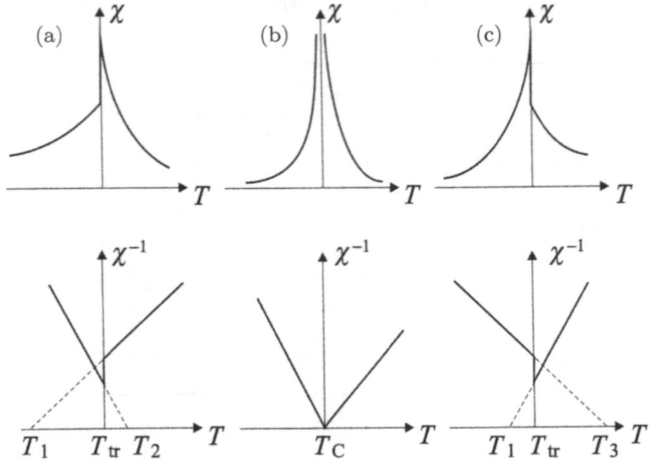

Fig. 4.26. Temperature dependences of the susceptibility $\chi(T)$ and its inverse $\chi^{-1}(T)$ for (a) a reconstructive transition, (b) a second-order transition to a Landau state, (c) a transition between a Landau state and a limit state.

4.4.2. *Specificity of the phase diagrams involving non-Landau phases and reconstructive transitions*

With respect to the phase diagrams containing only group–subgroup-related phases, the phase diagrams involving reconstructive transitions contain *additional limit states*, which may be adjacent to *Landau states*, representing an intermediate structure with respect to the parent structure. For example, in the pressure–temperature phase diagrams of pure elements, which are discussed comprehensively in the following chapters, Landau states are found in a number of elements, but the more common situation encountered in the periodic table corresponds to the stabilization of one or two contiguous limit phases below the initial structure.

The fact that the order-parameter has a constant value in a limit phase results in the general properties of the phase boundaries, which in a two-dimensional (pressure–temperature) phase diagram can be expressed as follows:

Property 1. The lines of first-order transitions between limit phases as well as the lines limiting their stability regions are *straight lines.*

Proof: The effective order-parameter expansion (4.55) can be written under the general form

$$F(x_j) = \sum_i a_i(P,T,\ldots) \cdot f^{n_i}(x_j), \tag{4.56}$$

where the $f^{n_i}(x_j)$ are n_ith-degree monomials of the x_j, and the a_j are phenomenological coefficients, which depend linearly on the external variables P, T, ... The equations of the limit of stability lines, which are $\left\| \frac{\partial^2 F}{\partial x_j \partial x_k} \right\| = 0$ (j,$k = 1, \ldots, m$), take the explicit form

$$\sum_i n_i \cdot (n_i - 1) \cdot a_i(P,T,\ldots) \cdot f^{n_i-2}(x_j) = 0. \tag{4.57}$$

As the $(n_i - 2)$th degree monomials $f^{n_i-2}(x_j)$ are constant in the limit phases, the *linear* equations of the a_j (4.57) define *hyperplanes* in the n-dimensional space A_n of the coefficients a_j. The hyperplanes are straight lines in two-dimensional *cross-sections* of the A_n-space. As the pressure–temperature plane can be obtained by a linear transformation of any of the preceding cross-sections, the same property holds in the (P, T) plane in which the limit of stability lines of the limit phases appears as straight lines.

First-order transition lines are lines of equal energies. If $F(\xi_j^I)$ and $F(\xi_j^{II})$ are the equilibrium values of the order-parameter expansions corresponding to the limit phases denoted I and II, the equality of the expansions on the first-order transition line I–II gives

$$\sum_i a_i(P,T,\ldots) \cdot \left[f^{n_i}(x_j^I) - f^{n_i}(x_j^{II})\right] = 0.$$

As the term between brackets is constant, one has again a linear equation relating the a_j. The same arguments given above lead to the conclusion that the first-order transition line I–II is a straight line.

It is worth noting that the prediction of the advanced phenomenological theory matches the experimental data obtained for the elemental metals (see Sec. 7.3.4.1).

Property 2. The limit of stability and the transition lines of a limit phase, with respect to the initial phase, are *tangent*, respectively, to the limit of stability and the transition lines of the adjacent Landau state with respect to the initial phase. The point of tangency of the transition lines coincides with the three-phase point, at which the initial, limit and Landau phases merge.

Proof: In the A_n-space of the phenomenological coefficients a_j, the equation of the hyperspace $\Omega(a_i) = 0$ limiting a Landau phase can be approximated by a polynomial

$$\Omega\left(x_j\right) = \sum_i a_i \cdot f^{n_i}\left(x_j\right) = 0, \tag{4.58}$$

where f is a nonlinear function of the variational parameters x_j, which are functions of the a_i. The three-phase point N at which the three phases merge corresponds to constant values of the x_j and thus to constant $f(x_j)$. Thus, at the N-point, Eq. (4.58) becomes the equation of a hyperplane $\sum_i a_i = 0$, which is necessarily tangent to the hypersurface $\Omega(a_i) = 0$. Besides, as the preceding hyperplane contains the N-point, it identifies to the hyperplane separating, in the A_n-space, the limit and initial phases. In projection in a two-dimensional section of A_n, the hyperplane and hypersurface become transition *lines*. The proof for the limit of stability lines follows an identical scheme.

Properties 1 and 2 are clearly illustrated in the phase diagram of Fig. 4.4, in the (a_1, a_2) plane of the phenomenological coefficients pertaining to the thermodynamic potential (4.9). Thus, the first-order transition line separating the initial (I) and limit (II) phases has the straight-line equation: $a_1 = -a_2\eta_0^2 - a_3\eta_0^4$. It prolongates the (parabolic) line of transitions between phases I and III at the three-phase point of coordinates: $a_1 = a_3\eta_0^4$, $a_2 = -2a_3\eta_0^2$. On the other hand, the limit of stability line of phase II with respect to phase I also obeys the equation of a straight line ($a_1 = -2a_2\eta_0^2 - 3a_3\eta_0^4$). It is tangent to the line of stability limit of phase III with respect to phase I at the point of coordinates ($a_1 = 3a_3\eta_0^4$, $a_2 = -3a_3\eta_0^2$).

4.4.3. *Phonon spectra and reconstructive phase transitions*

A temperature-dependent softening of phonon branches corresponding to the critical modes is expected, and generally observed, for second-order or weakly first-order displacive structural transitions. For more pronounced first-order displacive transitions, even between group–subgroup-related structures, the softening decreases with the decrease in amplitude of the critical oscillations. At reconstructive transitions, for which the critical displacements between the initial and limit phases are *fixed*, there is *a priori no critical fluctuations*, and thus *no softening should be observed.* However, this property should be verified only far from the domain of stability of the intermediate Landau state, and provided a secondary, temperature-dependent and displacive mechanism should not be involved in the transition.

No softening is observed in the phonon spectra of second-order order–disorder transitions, as the average atomic positions remain fixed across such transitions, and the jumps of atoms between the sites are supposed to be uncorrelated. This picture is still reinforced for discontinuous order–disorder transitions between group–subgroup-related and unrelated structures. The absence of mode softening holds, however, for purely ordering mechanisms. In Chapter 6, we show that ordering processes often take place with correlated displacive mechanisms, which may result in slight softening of some phonon branches.

The absence of temperature dependence of the phonon spectrum for reconstructive transitions constitutes, however, only a negative test. A more positive confirmation of the mechanisms proposed in our theoretical description of these transitions can be found in the frequency dependence on the reduced $\boldsymbol{k}$-vector of the phonon spectra, as it can be used for verifying the coincidence of the structures between the initial and limit phases in the directions of space *preserved* by the reconstruction of the lattice, i.e., one expects to find a coincidence of the phonon branches for the two structures in these directions and a softening of $\omega(k)$ at special points corresponding to the translational connection of the structures. The phonon spectrum may also serve as a test to check the influence of macroscopic deformations, which are often involved in reconstructive transition mechanisms, as such deformations modify the distance between equivalent sites in the initial and final structures (the *metric* of the structures).

4.5. Fluctuations and Temperature Scaling in the Advanced Phenomenological Theory of Phase Transitions

4.5.1. *Phenomenological theory versus fluctuation theory of phase transitions*

It is evident from the preceding considerations (see Chapters 1 and 2) that the Landau theory of phase transitions does not account for long-wave fluctuations of the order-parameter. Despite the fact that the theory deals with various types of variational free-energy, it predicts, in fact, a single set of critical point exponents ($\alpha = 0$, $\beta = 1/2$, $\gamma = 1$ and $\delta = 3$) and thus is assigned to the single "mean field" universal class [23]. The phenomenological theory characterizes a system by the *same critical exponents* both outside and within critical region, where, following the fluctuation theory, long-wave fluctuations of the order-parameter become significant affecting the temperature behavior of measurable thermodynamic parameters and resulting in its deviation from the predicted laws.

The *phenomenological* theory approximates the temperature behavior of a general function $f(t)$ by simple power function $f(t) = A \cdot t^{\lambda}$, with λ as critical point exponent, and $t = |T - T_C|/T_C$ as a dimensionless variable to measure the temperature difference with the critical temperature T_C (see Sec. 1.2). The *fluctuational* theory introduces a set of *critical indices* to describe the quantities that have singularities at T_C and to assign systems to *universality classes*: for the specific heat $c \sim t^{-\alpha}$, order-parameter $\eta \sim |t|^{\beta}$, isothermal susceptibility $\chi \sim |t|^{-\gamma}$, correlation length $r \sim t^{-\nu}$ [23, 24]. We focus here on the order-parameter critical exponent β as it is directly linked to a transformation mechanism.

A considerable body of experimental data indicates, however, that the real systems show regular deviation from the behavior predicted in the framework of the Landau phenomenological theory, and different universality classes were found experimentally in such systems. It is convenient to consider such a discrepancy as caused by two main reasons: (i) the phenomenological theory neglects, as it was said above, critical fluctuations, and (ii) the above theory uses the dimensionality for the fluctuation space lower than the marginal dimensionality ($d < d^*$) (see, e.g., Refs. [3, 25, 26]). It is worth noting that both these reasons relate to the critical fluctuations and normally should be valid only in the critical region.

Another group of authors, analyzing the applicability of the Landau theory, argues that one can find incorrect statements concerning its applicability even in the case where the theory is actually valid. This is especially typical of studies of *structural phase transitions*, i.e., of the main subject of our book. The authors, among them even inventors of the non-applicability criterion [27, 28], remind that for many quantities the results of the fluctuation theory are not so different from the predictions of the Landau theory [29]. Considering, for instance, the temperature dependence $\eta^{\text{eq}}(t)$ for $t < 0$, one usually approximates it by the power law $\eta^{\text{eq}} = \eta_0 \cdot |t|^{\beta}$, and if exponent β proves to be close to 1/3 while in the Landau theory $\beta = 1/2$, one takes this as an evidence in favor of non-applicability of the Landau theory. However, the dependence $\eta^{\text{eq}} = \eta_0 \cdot |t|^{1/2}$ *is not a general result* of the phenomenological theory. It holds virtually only for one limiting case: for second-order phase transition *far from the tricritical point* (see Figs. 1.12 and 4.5, and Eqs. (1.92) and (1.93)). In the opposite limiting case, where the tricritical point is close, $\eta^{\text{eq}} \sim \eta_0 \cdot |t|^{1/4}$ (see, e.g., Ref. [29]). It seems that we still do not know many examples of reliably revealing critical fluctuation regions for structural phase transitions. As a likely case for the large critical region, the structural phase transition in strontium titanate at $T_C \cong 100$ K was intensively discussed. However, the authors of Ref. [30] obtain a negligible value $\Delta t_{fl} \leq 10^{-5}$ for the width of the critical region.

Summarizing for structural phase transitions, the critical region is very narrow (<1 K), and experimentally revealing this region is not a trivial task. Thus, when one observes some deviation from the Landau theory for structural second-order phase transitions, one has first to try finding different reasons than the critical heterophase fluctuations developing in the system.

We should remember that classical phenomenological approach does not provide a way of calculating the critical indices accounting for order-parameter fluctuations. The calculation of the indices is a problem of microscopic theory. Nevertheless, it is possible to give an intimation of the dependence of the indices remaining in the framework of the advanced phenomenological approach.

The existence of background hetero- and homophase fluctuations [31, 32] was beyond both the phenomenological and fluctuational theories of phase transitions. In the following sections, we consider yet another reason for the divergence of the phenomenological calculations and corresponding experimental data. As revealed above, the transcendental relation between a phenomenological order-parameter and the corresponding local atomic variables along with the functional, bounded form of the order-parameter space σ provide us with another way to correct the results of classical phenomenological theory.

4.5.2. *Phenomenological order-parameter and essential variational parameters*

We derive the general form of the function $\eta_i(\xi_j)$, where the set of η_i is a long-range phenomenological order-parameter and ξ_j represent variational local atomic parameters, i.e., short-range order-parameters (variation of probability for the segregation or disorder-order transformation, magnitude of atomic displacements for displacive-type transitions, etc.), by considering the problem in the functional order-parameter space σ_n.

We make use of the usual scheme for calculating $\eta(\xi)$ by finding the solution of the Euler's variational equation $\{\delta\Phi/\delta\eta(\xi)\}= 0$ that minimizes the free energy functional. The appropriate choice for the latter in the case of continuous phase transition is the classical Landau–Ginzburg functional

$$\Phi(\eta) = \int_{V_\xi} d\xi \left[a_1\eta^2(\xi) + a_2\eta^4(\xi) + g\left(\frac{d\eta}{d\xi}\right)^2\right], \tag{4.59}$$

where the integral is over a volume in the *order-parameter space* but not in the real coordinate space. For the sake of simplicity, we treat a single-component or effective OP, while the conjugate external field is neglected. The coefficient a_1 is conveniently assumed to be a regular function of thermodynamic variables (temperature, pressure, etc.), and the remaining coefficients are regarded as temperature- and pressure-independent parameters. The corresponding Euler's equation takes the form

$$\frac{\delta\Phi}{\delta\eta(\xi)} = a_1\eta(\xi) + 2a_2\eta^3(\xi) + g\frac{d^2\eta}{d\xi^2} = 0. \tag{4.60}$$

The boundary conditions are $\eta(0) = 0$ and $\eta'(0) = 1$. The first condition indicates a coincidence of the origin points for the variables η and ξ. The second one ensures their identical behavior close to T_C, i.e., it justifies the change of variables $\xi \to \eta$ in the Landau theory and making η a phenomenological order-parameter. The differential equation (4.60) has exact general solution expressed as

$$\eta = \eta_0 \cdot sn[\mu(\xi - \xi_0), \kappa]. \tag{4.61}$$

In the above equation, $\mathrm{sn}[\mu(\xi - \xi_0),\ \kappa] = \mathrm{sn}\ u$ is the elliptic sine of Jacobi, and μ and ξ_0 are the arbitrary constants [33]. With the applied boundary conditions, parameters $\mu = 1$ and $\xi_0 = 0$. The parameter $\kappa = (a_2/|g|)^{1/2}$ is the *modulus* of the elliptic integral of the first kind

$$u = F(\xi, \kappa) = \int_0^{\xi} \frac{d\xi}{\sqrt{1 - \kappa^2 \sin^2\xi}}. \tag{4.62}$$

The phase diagram of this system can be obtained by minimizing the Landau potential

$$\Phi_L(\eta) = a_1\eta^2 + a_2\eta^4, \tag{4.63}$$

where the OP η has the form of Eq. (4.61). It reminds Eq. (4.9) but with coefficient $a_3 = 0$. The minimization of Φ_L with respect to the actual variational parameter ξ is expressed by $\frac{\partial\Phi_L}{\partial\xi} = \frac{\partial\Phi_L}{\partial\eta} \cdot \frac{\partial\eta}{\partial\xi}$, giving the equation of state

$$\eta'\{a_1 + 2a_2\eta^2\} = \mathrm{sn}\,u \cdot \mathrm{cn}\,u \cdot \mathrm{dn}\,u\{a_1 + 2a_2\mathrm{sn}^2\,u\} = 0, \tag{4.64}$$

where $\mathrm{cn}u = \sqrt{(1 - \mathrm{sn}^2 u)}$ and $\mathrm{dn}u = \sqrt{(1 - \kappa^2 \sin^2\xi)}$. No surprise that, like Eq. (4.9), Eq. (4.64) yields three possible stable states: (i) the parent phase I for $\mathrm{sn}u = 0$ (origin of the space σ); (ii) the limit, "non-Landau", phase II, given by cn $u = 0$ (sn $u = 1$) (boundary of the space σ), corresponding to the fixed values η_0 of the OP; (iii) "Landau" phase corresponding to the standard minimization of Φ_L with respect to the OP η, whose value $\eta_2 = -a_1/2a_2$ varies between 0 and η_0 (interior of the space σ). The function dn u has zeros only if $\kappa = 1$, however, even in this case, dn$u$ vanishes simultaneously with cn$u$, and no different solution of Eq. (4.64) exists.

The stability condition has the form

$$\frac{\partial^2 \Phi_L}{\partial \eta^2}\left(\frac{\partial \eta}{\partial \xi}\right)^2 + \frac{\partial \Phi_L}{\partial \eta} \cdot \frac{\partial^2 \eta}{\partial \xi^2} \geq 0. \tag{4.65}$$

The resulting phase diagram in the plane of thermodynamic parameters (a_1, a_2) is shown in Fig. 4.27(a). As shown in Sec. 4.2.1.3, the second-order phase transition line $a_1 = 0$ separates parent I and Landau III phases. The stability regions of Landau III and limit II phases adjoin along the line $a_1 = -2a_2$. The first-order transition lines $a_1 = -a_2$ between parent I and limit II phases meet at the three-phase point O, two latter lines. Notably, in the classical Landau theory, the mandatory positivity of the coefficient a_2 in the potential (4.63) ensures the global stability of the phases and the convexity of the potential for large values of the OP. However, Eq. (4.61) shows that magnitude of the OP cannot be arbitrarily large. This restriction raises the possibility of the global stable phase diagrams, even for negative values of a_2 (Fig. 4.27(a)).

Importantly, one can see in Fig. 4.27(a) that the temperature range of the OP variation, that is, the stability region of Landau phase III, is limited. Denoting the low-temperature stability limit for the Landau phase by T_L (Fig. 4.27(a)), the order-parameter has the value 0 above T_C and varies between 0 and η_0 for $T_C > T > T_L$, finally reaching the saturation value η_0 (conventionally, $\eta_0 = 1$) at T_L, and then remains invariable for $T < T_L$ (Fig. 4.27(b)). As no conventional reasons give $T_L = 0$ K, one makes use of a generalized form for the dimensionless variable $t_n = (T - T_C)/(T_C - T_L)$. The latter distinguishes the Landau approach in which any phase has its thermodynamic limit of stability at a finite temperature (or pressure) $T_L \neq 0$ K, defined by Eq. (4.65) $\partial^2\Phi_L/\partial\xi^2 = 0$, where the corresponding energy minima disappear, and the Gibbs-type consideration where the set of minima exists at any temperature up to $T =$ 0 K, and the system finds a lowest one for any given T.

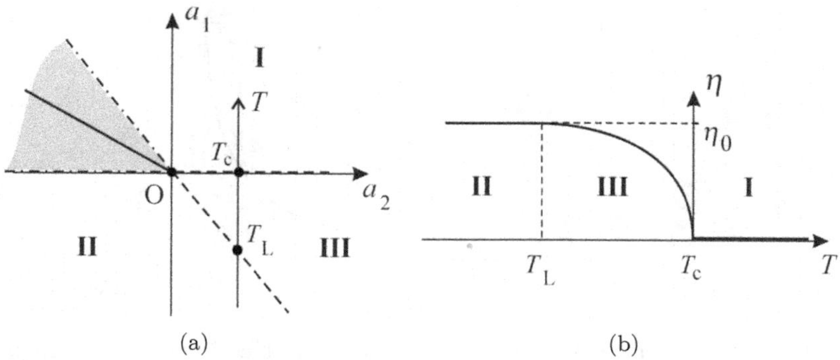

Fig. 4.27. (a) Phase diagram of the model Eq. (4.63). Solid, dashed and dashed-dotted lines are, respectively, discontinuous first-order transition, continuous transition, and limit of stability lines. O is the three-phase point. (b) Temperature dependence of the phenomenological order-parameter for the path indicated in (a).

4.5.3. *Temperature scaling transformation*

4.5.3.1. *OP temperature behavior*

As discussed above, the Landau theory uses identical temperature dependencies for both a phenomenological OP η and atomic variable parameters ξ. This means that the Landau potential (Eq. (4.63)) can be introduced for the latter as well: $\tilde{\Phi}_{\mathrm{L}}(\xi) = \alpha_1\xi^2 + \alpha_2\xi^4$. By minimizing variational free energy $\tilde{\Phi}_{\mathrm{L}}$ and assuming $\alpha_1 = \alpha_{10} \cdot t(\alpha_{10} = \text{const})$, one finds $\xi = (-\alpha_1/2\alpha_2)^{1/2} \sim t^{1/2}$. In Sec. 4.5.2, we stressed the existence of the complex relation between essential variational parameters ξ_i and measurable phenomenological OP η_j. Thus, if one approximates $\xi = \tilde{\alpha}_{10}\sqrt{t_n}$, then a phenomenological OP η from Eq. (4.61) takes form

$$\eta(t_n) = \eta_0 \cdot \mathrm{sn}(\tilde{\alpha}_{10}\sqrt{t_n}; \kappa). \tag{4.66}$$

We consider the elliptical sine as a circular sine with the argument scaled with the integral transformation (4.62): $\mathrm{sn}(x, \kappa) = \sin\tilde{x}$ (factor $\pi/2$ is included in the definition of the argument) [33]. Making use of temperature as a variational thermodynamic parameter, one can convert Eq. (4.62) into nonlinear temperature scaling transformation

$$\sqrt{\tilde{t}} = \int_0^{t_n} \frac{d\left(\sqrt{t_n}\right)}{\sqrt{1 - \kappa^2 \sin^2\left(\frac{\pi}{2}\sqrt{t_n}\right)}}. \tag{4.67}$$

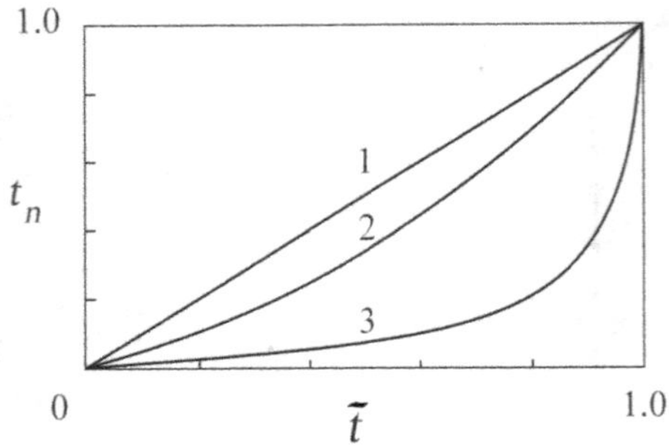

Fig. 4.28. Scaling ratio between temperature variables t_n and $\tilde{t}$ for different values of the modulus κ: 1 – 0, 2 – 0.8, 3 – 0.999.

Nonlinearity of such a temperature scaling transformation is fully defined by the modulus κ varying from 0 to 1. Figure 4.28 shows the scaling ratios between variables t_n and $\tilde{t}$ for different values of the modulus κ.

To confirm the applicability of the presented approach to real physical systems, we choose three crystals undergoing continuous phase transitions of different nature.

(i) In metallic Sn, the transition to superconducting state takes place at $T_C = 3.782$ K [34]. The microscopic OP for such a transition is an effective wave function $\Psi(\boldsymbol{r})$, which is proportional to the local value of the energy-gap parameter Δ [35,36]. The measurable phenomenological OP is then the average normalized energy gap in the elementary excitation spectrum of a superconductor [34].
(ii) The cooperative Jahn–Teller transition with the ordering in the electronic degrees of freedom reduces the symmetry of the perovskite ($PrAlO_3$) crystal from tetragonal to monoclinic at $T_C = 151$ K. The electronic functions of the lowest crystal-field doubly degenerated level E_g of the Pr^{3+} (in the parent cubic field) form the basis functions of the relevant IR. The observable phenomenological OP is then either the splitting between the doublet levels in the ordered phase or the linearly coupled acoustic- and optical-phonon modes of the same symmetry [37].
(iii) Disorder-to-order transformation in β-brass at $T_C = 740$ K [38–40]. The deviation of average probability of atoms to occupy

> positions in a crystal lattice can be considered as the phenomenological OP for ordering-type transitions. The local variation of the probability of occupation, or equivalently, population of the corresponding sites, is then the essential variable parameter for the transition.

The temperature T_L, limiting the stability region of the Landau phase, and the modulus κ, characterizing nonlinearity of the temperature scaling transformation, are used as the fitting parameters. Figure 4.29 shows the best fit for different OPs plotted versus *scaled normalized* temperature $\tilde{t}$ and compared with the worked-out function $\eta(\tilde{t}) = \sin\frac{\pi}{2}\sqrt{\tilde{t}}$. Only the temperature range where $\eta \neq$ const is displayed. The corresponding fitting parameters were fixed as follows: Sn ($T_L = 1.236$ K; $\kappa = 0.05$), $PrAlO_3$ ($T_L = 61$ K; $\kappa = 0$) and β-brass ($T_L = 490$ K; $\kappa = 0.77$).

Clearly, the predicted uniform function reproduces the experimentally observed behavior of order-parameters well in the entire range of its variation. To emphasize the role of the temperature scaling transformation, the experimental spots for the β-brass order-parameter [39, 40] are plotted (Fig. 4.29(b)) versus the *non-scaled* normalized temperature t_n. Two curves in Fig. 4.29(b) display functions $\eta_1 = \sin(\frac{\pi}{2}t_n^{1/2})$ and $\eta_2 = \sin(\frac{\pi}{2}t_n^{1/3})$. It is clear that close to T_C the approximate function $\eta_2 \propto t_n^{1/3}$ better reproduces experimentally observed OP behavior, and one deduces the critical exponent $\beta \approx 1/3$ for β-brass.

4.5.3.2. *Modulus of the temperature scaling transformation*

At this point, we see that the phenomenological approach deals with two temperature scales. One can be termed *real-temperature.* This temperature t (or, equivalently, t_n) is the *measurable* thermodynamic variable for the observable functions describing the behavior of a system. Another, termed *rescaled* temperature $\tilde{t}$, appears in the phenomenological approach after the *renormalization procedure that eliminates the local microscopic variables and replaces them with the macroscopic averaged variables*, the components of the phenomenological order-parameter. More precisely, they are the rescaled variables in the Landau theory for the temperature behavior of

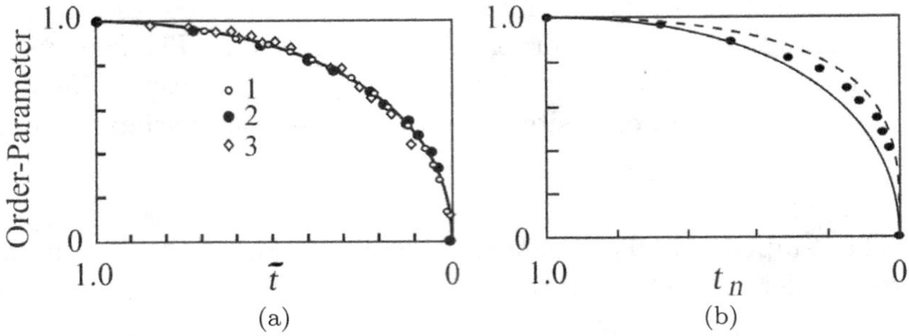

Fig. 4.29. (a) Reduced values of the order-parameters for (1) Sn [34], (2) β-brass [40] and (3) $PrAlO_3$ [37] as a function of rescaled temperature, and compared to the theoretical curve (solid line). (b) Experimental order-parameter data points in β-brass compared to curves with critical indices 1/2 (solid line) and 1/3 (dashed line).

different physical quantities or the corresponding critical exponents. The example of β-brass (Fig. 4.29(b)) clearly shows that the values for the order-parameter critical exponent depend on the temperature scale. To compare experimental results and phenomenological predictions, it is important to take into account the temperature scaling transformation (Eq. (4.67)). Figure 4.30 shows the real-temperature order-parameter critical exponent β as a function of the modulus κ of such transformation.

Despite the fact that $\beta = 1/2$ is unambiguously predicted in the Landau theory, we expect to obtain the *different* values for the critical exponent measured in the real-temperature scale in the real fluctuating materials. Figure 4.31 summarizes the modification steps passed by the experimentally measured curve $\eta(t)$ to be compared with the predictions of a phenomenological theory. Customary normalization of the temperature scale by the corresponding critical temperature T_C ($T \to t$) brings different curves to the single origin (Fig. 4.31(a)). Then, modified normalization introduced above merges the curves at the origin $t_n = 0$ ($T = T_C$) and at the saturation point $t_n = 1$ ($T = T_L$) of the order-parameter variation range (Fig. 4.31(b)). Finally, nonlinear temperature scaling transformation $t_n \to \tilde{t}$ fits experimental points for different systems in the uniform curve $\eta\left(\tilde{t}\right) = \sin\frac{\pi}{2}\sqrt{\tilde{t}}$ (Fig. 4.31(c)) predicted in the extended phenomenological approach.

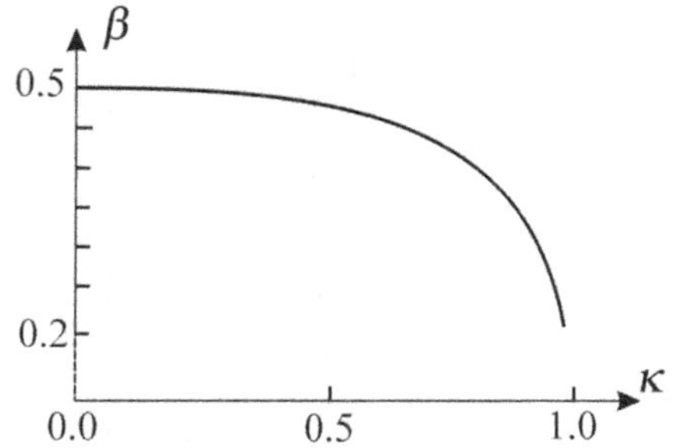

Fig. 4.30. Measurable real temperature order-parameter critical exponent as a function of the transformation modulus κ.

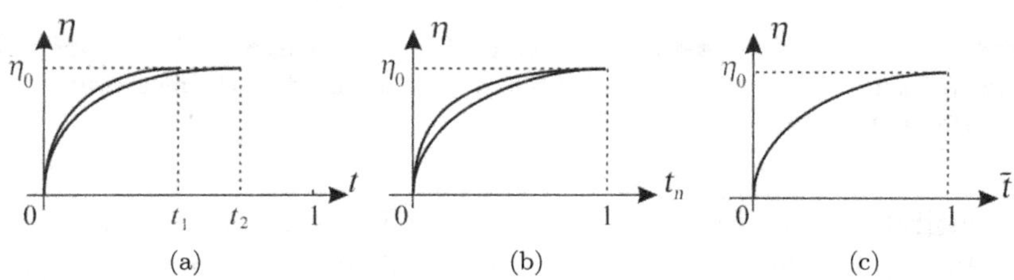

Fig. 4.31. Order-parameter variation curves presented in different temperature scales. (a) Order-parameter as a function of a dimensionless variable $t = |T - T_C|/T_C$, (b) as a function of $t_n = (T - T_C)/(T_C - T_L)$, and (c) in function of the nonlinearly rescaled variable $\tilde{t}$.

A considerable amount of experimental data unambiguously indicates that such limit states do exist in various physical systems. The examples of Sec. 4.5.3.1 provide typical cases of the temperature behavior predicted by the advanced phenomenological approach presented in this chapter. It should be stressed that for a given system in the possible but not necessarily perfect limit state, the degree of order $\eta_0 \leq 1$ can be realized. This means that the limit phase is the state with the *temperature-independent* stable equilibrium between perfect order and partial disorder. One deals with a *saturated* state perfectly ordered only in a particular case.

4.5.3.3. *Long-wave background fluctuations*

To elucidate the physical meaning of the modulus κ of the nonlinear transformation (4.67), let us consider the response function $\tilde{G}(\xi) = (4\pi g\xi)^{-1} e^{-q\xi}$, where $q^{-2} = g\chi$ and χ is the generalized

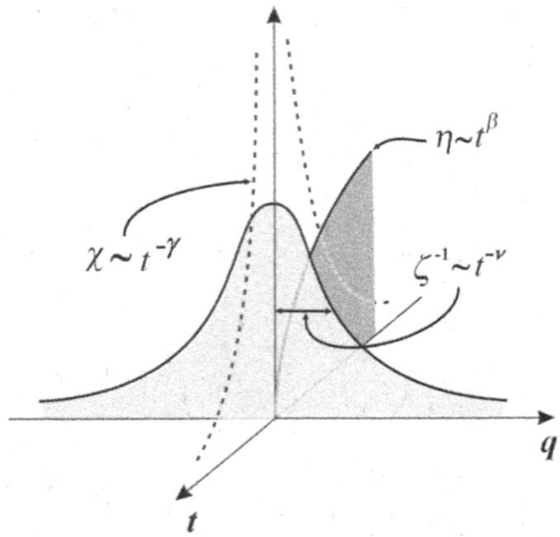

Fig. 4.32. Order-parameter η, wave vector-dependent susceptibility χ, and scattering peak profile (ζ is half-width at half maximum) directly measurable in a scattering experiment.

susceptibility [24]. The quantity $r_c = q^{-1}$ is the correlation length of the short-range order. The latter can be directly measured in a scattering experiment when the radiation couples to the order-parameter (Fig. 4.32, Ref. [26]). By incorporating a generalized external field in Eq. (4.63), it is easy to calculate the function $\chi = \left(a_1 + 3a_2\eta^2\right)^{-1}$ and then derive the modulus $\kappa^2 = a_2/|g| = \varepsilon \cdot r_{c0}^{-2}$, where r_{c0} is the correlation length far from the phase transition point ($t \sim 1$) and ε is the normalizing factor. The modulus κ thus characterizes the *background* fluctuational property of a physical system (i.e., far beyond the critical region).

4.5.4. *Temperature scaling as another method of fluctuation correction*

Two assumptions should be considered to ensure the generality of the presented approach in the analysis of non-critical fluctuation impact on temperature behavior of thermodynamic parameters.

Firstly, only the case of the single-dimensional order-parameter was treated in Sec. 4.5. However, the temperature behavior of the order-parameter in the Landau-type phenomenological approach is controlled by the coefficient a_1 (see Eqs (4.59) and (4.63)). It is associated with the invariant quadratic in order-parameter components $a_1 I_1 = a_1\left(\eta_1^2 + \cdots + \eta_n^2\right)$ of the free energy expansion, and a_1

specifies the uniform functional form of the temperature dependence $\eta_i(t)$ for each of the multiple order-parameter components as well as for the corresponding single-component effective order-parameter. Figure 4.29 further confirms this, as the uniform function there reproduces the temperature behavior of the single- (β-brass), two- (Sn) and three-component ($PrAlO_3$) phenomenological order-parameters.

Secondly, we deal here with the free energy expansion restricted to the fourth-degree term in order-parameter components, while the above argument on the uniqueness of the coefficient a_1 is still valid. Including the higher-order terms modifies the stability conditions and the equations of stability boundaries and transition lines in the phase diagram, but does not affect the temperature-dependent part of the free energy.

Summarizing, we have to conclude that the *real-space renormalization* procedure, eliminating the local atomic variables and replacing them by the phenomenological order-parameter components, is accompanied by the nonlinear *temperature scaling* transformation.

The classical Landau theory with small order-parameter is a linear theory, and this is why both, local and integral phenomenological, order-parameters renormalized by linear operators depend on temperature as $t^{1/2}$. The large order-parameter makes renormalization (and the theory) nonlinear.

Rescaled temperature is an essential thermodynamic variational parameter of the advanced phenomenological theory with respect to which the critical exponents are predicted unambiguously, while the corresponding experimental values are obtained in the *real-temperature* scale. The general form of such transformation is the elliptic integral of the first kind.

4.6. Applicability of the Phenomenological Approach to Strongly Discontinuous and Reconstructive Phase Transitions

4.6.1. *Summary: Specificity of reconstructive phase transitions*

The phenomenological description of the second- and first-order phase transitions with large order-parameter, including reconstructive phase transitions, presented in this chapter, has disclosed the distinctive properties of these transitions on a theoretical ground. Thus, when starting from a parent structure, a phase transition to a limit phase occurs for definite *fixed* values of the variational parameters, for which a simple (group–subgroup) symmetry relationship with the initial structure can be lost. The physical anomalies that characterize the transition display-typical features, such as the step-like temperature (or pressure) dependence of the order-parameter, the absence of jump at the transition for the specific heat, and the corresponding drop of the susceptibility. The particularities of the phase diagrams, and of the dynamical properties (phonon spectra) of reconstructive phase transitions, have also been stressed.

The specificity of transitions to the limit states is also reflected in the mathematical structure of the quantities, which are used in the approach. Thus, the order-parameter space *includes* the periodicity of the transition mechanism, as this mechanism also determines the symmetry of the order-parameter. For a given *direction* of the order-parameter in the representation space, the transition mechanism is expressed by the magnitude of the order-parameter, which is either a periodic function of the critical displacements or a saturating scalar function of the increment of occupancy of probability of definite atomic sites. Consequently, the *effective* order-parameter expansion depends exclusively on the magnitude of the order-parameter, and is thus a Fourier series for displacive transitions. The various stable states arising from the initial state are obtained by minimizing the order-parameter expansion with respect to the variational parameters. This is made possible by the fact that the *explicit form* of the magnitude of the order-parameter in function of these parameters has been *deduced* from the increment of the lattice functional between the initial and limit structures, which are calculated using a density-wave formalism.

For the interpretation of reconstructive transitions in concrete systems, a number of points in our approach still require to be clarified as, e.g., *the choice of the initial structure.*

4.6.2. *Parent phase and initial structure*

In the usual Landau theory of structural phase transitions, the parent phase is generally chosen as the crystal phase of *maximal symmetry*, which is also, with a few well-understood exceptions, the phase that is more stable at high temperature below the melt. The lower symmetry phases, appearing at lower temperatures, correspond to subgroups (G_i) of the parent phase space-group G_0, and the G_i transform as irreducible (or reducible) representations of G_0. The consistency of this simple scheme is attested by the observed properties of the low symmetry phases (domain structure, spontaneous macroscopic tensors), which can be *deduced* from the $G_0 \to G_i$-broken symmetry.

For reconstructive transitions, the situation is less comfortable, and the choice of the parent phase is not obvious and *may be arbitrary*. In most cases, the experimental observations *cannot* discriminate between the parent and "daughter" phases: absence of group–subgroup relation between the phases, existence of orientational domains in both phases, each of which may be considered as the *non-deformed* state, for the macroscopic strains, which are generally the sole spontaneous tensor involved in reconstructive transitions.

Arbitrariness in the choice of the *initial* structure mainly occurs for displacive reconstructive transitions. For instance, in the two examples of transitions discussed in Secs. 4.2.1 and 4.2.2, one may *choose*, as initial structures indifferently, the $P6/mmm$ and $6mm$ phases *instead* of the $P6_3/mmc$ and $4mm$ phases, respectively, as taken in our description. In this case, there is no "natural" parent phase, and one may better speak of *initial structure*. The choice of the "best" initial structure in concrete cases must be then determined by physical considerations, such as the location of the phases in the phase diagram (the highest temperature phase) or by specific geometrical considerations (as the closest packing of atoms in one phase).

In the framework of our approach, the indetermination in the existence of an *actual* parent phase can be foreseen in the *dependence of the order-parameter in function of the critical displacements* and in the *form* of the effective order-parameter expansion. Thus, in the case of the $P6_3/mmc \to P6/mmm$ transition, Eq. (4.8) and Fig. 4.2 show that the sinusoidal dependence of $\eta(\xi)$ is *symmetric* with respect to the displacement axis, namely that the maxima *and* minima $\pm\eta_{\max}(\xi)$ both correspond to *the same* $P6/mmm$ structure. In other words, one gets periodically the same sequences of phases for opposite critical shifts. As a result, the effective thermodynamic potential (4.9) contains only *even* powers of $|\eta(\xi)|$. A similar situation is found for the $4mm \to 6mm$ order-parameter dependence on ξ (Eq. (4.19) and Fig. 4.10b) and order-parameter expansion.

By contrast, when a displacive reconstructive transition is associated with an *asymmetric* periodic dependence of the order-parameters in function of the critical displacements, or in other words, when the maxima and minima of $\eta(\xi)$ correspond to *different* structures, *the choice of the initial structure is not free, and only one of the two structures surrounding the transition may play this role.* In this case, one finds that $\eta(\xi)$ has a more complex form, and the effective order-parameter expansion necessarily contains a *cubic invariant.* In such a situation, which is found for the β–ω transition discussed in Chapter 5, the initial structure identifies unambiguously with the parent phase.

By contrast, no ambiguity exists in the choice of the parent phase for order–disorder reconstructive transitions, as the diffraction patterns clearly distinguish the disordered phase from the ordered phase. Even when the transition takes place between two differently ordered structures, our description (see Chapter 6) assumes a *latent disordered parent phase*, which is the *maximal substructure* common to the two ordered structures.

4.6.3. *Limit non-Landau states and limit-Landau states*

In phase diagrams involving reconstructive transitions, two types of phases are distinguished:

(1) *Landau phases*, which are group–subgroup related to the initial structure. They display a usual temperature dependence of the order-parameter below the transition and are obtained by a minimization of the thermodynamic potential with respect to the phenomenological order-parameter.
(2) *Limit (non-Landau) phases*, which possess no group–subgroup relationship with respect to the initial structure. They are characterized by a *constant* magnitude of the order-parameter (temperature-independent) and correspond to the *extrema* of $|\boldsymbol{\eta}(\xi)|$with respect to the variational parameter ξ. Note that these phases fulfill the condition $\frac{\partial^2 F}{\partial \xi^2} \geq 0$ and thus *coincide with the boundaries of the stability regions of Landau phases.*

While the absence of group–subgroup relationship between the two structures on either side of a transition is a clear-cut signature of its reconstructive character, the constancy of the order-parameter below a transition does not imply *a contrario* that the transition is of the reconstructive type. For example, in the first-order transitions to Landau phases, the transition mechanism may reach its limit below the transition discontinuity. This is the case for the liquid–solid transition [21] and for some ferroelectric, or ferroelastic, transitions [22] where the spontaneous polarization, or strain, appears as saturated below the transition. This class of "limit-Landau phases", which fulfill the simultaneous conditions:

$$\frac{\partial \eta}{\partial \xi} = 0 \quad \text{and} \quad \frac{\partial^2 F}{\partial \xi^2} = 0, \tag{4.68}$$

below the transition point, also contains a number of "unachieved" reconstructive transitions for which the preservation of the group–subgroup relationship, with respect to the initial structure, is related to specific crystal-geometrical restrictions. This may occur, e.g., when a structure reaches a geometrical limit imposed by a close packing of the atoms. Another illustration is found in some complex structures where "non-critical" sublattices preserve the overall symmetry of the structure, while the displacements of "critical" sublattices reach a limit giving rise to new symmetry elements for the corresponding substructure. It underlines that the saturation of the order-parameter, but not breaking group–subgroup relationship, identifies the non-Landau state. In Refs. [41,42], we describe concrete examples of "limit-Landau phases" among actinide elements in which it is the non-spherical shape of the atoms that preserves the group–subgroup relationship with the parent (BCC) structure. Let us stress that one can find situations in which a limit-Landau phase transforms into a non-Landau phase, as it will be shown in the description of the BCC–HCP transition (Sec. 5.3.3).

4.6.4. *Intermediate nature of the mechanism in reconstructive transitions*

In this chapter, in order to disclose the theoretical behaviors of reconstructive transitions, we have insisted upon the differences existing between displacive and order–disorder mechanisms. However, in the interpretation of concrete examples of reconstructive transitions, the two mechanisms often appear as correlated. This is mainly true for the ordering processes, which are almost always accompanied by collective displacements (see Chapter 6). The connection between ordering and displacive mechanisms in reconstructive transition stems from the fact that these transitions realize, in some respect, an intermediate situation between ordering and displacive processes. In order to make this point clearer, we will first briefly recall the current description of ordering and displacive mechanisms.

A system undergoing a purely ordering type of transition can be considered as a set of positions *fixed in space* and filled by heterogeneous objects: atoms and vacancies, atoms and ions of different elements, etc. The ordering process causes a change in the distribution of objects among the positions, but in zero approximation it does not displace the positions. Strictly speaking, the symmetry of the probability distribution function of the objects and the corresponding symmetry group obtained from the experiment is the result of averaging of the real atomic distribution. The coincidence between the distribution functions of the different types of particles and the actual distribution is obtained for the fully ordered state. In systems undergoing a purely displacive transformation, the crystallographic positions are fully occupied by identical particles, which are shifted in the transition process. This process induces a *crystallographic nonequivalence* in the atomic sites (Wyckoff positions), which were initially equivalent and realized even among chemically identical atoms or ions. In all cases, the atoms do not exchange diffusionally. There is also an important difference in the dynamics of the two types of transitions. The diffusional motion of atoms in ordering processes is not correlated, whilst the atomic shifts in a displacive mechanism are a collective process.

Within the preceding classical scheme, discontinuous parent-limit phase transitions correspond to an *intermediate situation* between the two types of mechanisms. On the one hand, a displacive reconstructive transition, directly from the initial structure to a non-Landau phase, transfers atoms from one fixed site to another fixed site without any intermediate position, but the corresponding atomic displacements occur collectively. On the other hand, in ordering reconstructive transitions, the uncorrelated character of the atomic diffusion is only partially verified.

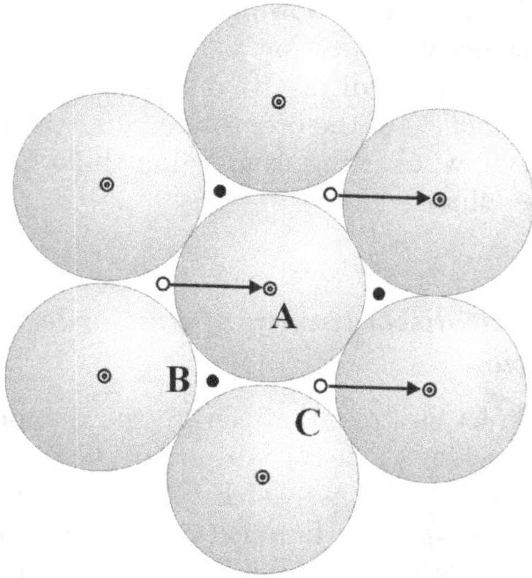

Fig. 4.33. Close atomic packing. A, B and C represent three atomic positions within the atomic layer (A) and in the nearest adjacent layers (B and C).

This appears, e.g., in the reordering mechanism between close-packed structures. In this mechanism, the change in position for any individual atom from position A to position B requires creation of high-energy defects connected with the absence of vacant interstices. It is thus energetically more advantageous to shift simultaneously (collectively) all the atoms of the layer A. Besides, a relative displacement of layers appears as uncorrelated but actually has correlations: if the original stacking of layers is random, e.g., ABCBA..., and the second layer has to be shifted from position B to final position C, it causes a displacement of the successive layers, which cannot occupy position C in the new structure due to geometrical constraints.

For example, in the reordering FCC–HCP transition mechanism described in Chapter 6, the reordering of the layers between the two structures appears as clearly correlated. The three-layer FCC periodic stacking ABCABC... must be replaced by the two-layer HCP ABABAB... stacking. This can be realized by periodically repeated shifts of each AB unshifted pair: CA moves in the direction C $\rightarrow$A (Fig. 4.33) and becomes AB, while BC displaces from B to A and also becomes AB. The next AB pair is unshifted, etc. Thus, *the reordering process becomes periodically correlated.* Other concrete examples, described in the following chapters, illustrate the property that reconstructive transition mechanisms allow a continuous *crossover* from ordering to displacive mechanisms. This is made possible by the *invariability* of the order-parameter value, which is *constant and maximal*, when going from a limit fully ordered phase, to a limit displacive phase.

4.6.5. *Degree of discontinuity of first-order phase transition*

As a final remark, let us stress that the advanced approach described in this chapter provides a general *measure* of the discontinuous character of first-order transitions, which consists in the *magnitude of the order-parameter* $|\boldsymbol{\eta}(\xi)|$ as function of the variational parameter. This degree of discontinuity of first-order transitions can be calculated explicitly from the initial and final structures of the phases. It varies nonlinearly between zero (for second-order transitions) and one (for transitions to non-Landau state) and usefully completes the numbers, which are usually quoted as a measure of the discontinuity of a phase transition (latent heat, discontinuity of the lattice parameters, magnitude of the shifts, jump of entropy, etc.). In the following chapters, some quantitative applications of this criterion are given.

References

[1] L. D. Landau, *Zh. Eksp. Teor. Fiziki* **7** (1937) 19; and **7** (1937) 627. Transl. in *Collected Papers of L. D. Landau*, ed. D. Ter Haar (Pergamon Press, Oxford, 1965), p. 193, 216.
[2] L. D. Landau and E. M. Lifshitz, *Statistical Physics*, 3rd ed. (Pergamon, Oxford, 1980), Vol. 1.
[3] J. J. Binney, N. J. Dowrick, A. J. Fisher and M. E. J. Newman, *The Theory of Critical Phenomena* (Clarendon Press, Oxford, 1992).
[4] K. G. Wilson and J. Kogut, *Phys. Rep.* **12C** (1974) 75.
[5] Th. Niemeijer and J. M. J. van Leeuwen, in *Phase Transitions and Critical Phenomena*, ed. C. Domb and M. S. Green (Academic Press, London, 1976), Vol. 6, p. 425.
[6] V. P. Dmitriev, S. B. Rochal, Yu. M. Gufan and P. Toledano, *Phys. Rev. Lett.* **60** (1988) 1958.

[7] V. P. Dmitriev, S. B. Rochal, Yu. M. Gufan and P. Toledano, *Phys. Rev. Lett.* **62** (1989) 844.
[8] V. P. Dmitriev, S. B. Rochal, Yu. M. Gufan and P. Toledano, *Phys. Rev. Lett.* **62** (1989) 2495.
[9] Yu. M. Gufan, V. P. Dmitriev and P. Toledano, *Sov. Phys.-Solid State* **30** (1988) 613.
[10] V. P. Dmitriev, Yu. M. Gufan and P. Toledano, *Phys. Rev. B* **44** (1991) 7248.
[11] W. Fisher and E. Koch, in *International Tables for Crystallography*, ed. T. Hahn (Kluwer Academic Press, Dordrecht, 1996), Vol. A, p. 825.
[12] N. G. Parsonage and L. A. K. Staveley, *Disorder in Crystals* (Clarendon Press, Oxford, 1978).
[13] C. S. Barrett, *Structure of Metals* (McGraw-Hill, New York, 1952), p. 622.
[14] W. L. Bragg, *The Crystalline State* (MacMillan, New York, 1934).
[15] H. S. Green and C. A. Hurst, *Order-Disorder Phenomena* (Interscience, New York, 1964).
[16] J. C. Toledano and P. Toledano, *The Landau Theory of Phase Transitions* (World Scientific, Singapore, 1987), Ch. III.
[17] P. Toledano and V. Dmitriev, *Reconstructive Phase Transitions in Crystals and Quasicrystals* (World Scientific, Singapore, 1996).
[18] O. V. Kovalev, *Irreducible Representations of the Space-Groups* (Gordon and Breach, New York, 1965).
[19] S. T. Bowden, *The Phase Rule and Phase Relations* (MacMillan, London, 1945).
[20] A. P. Levanyuk and D. G. Sannikov, *Sov. Phys.-Solid State* **18** (1976) 1122.
[21] H. N. V. Temperley, *Changes of State* (Interscience, New York, 1956).
[22] *Landolt-Bornstein Tables for Ferro- and Antiferroelectric Substances* (Springer-Verlag, Berlin, 1969), Vol. 3, group 3.
[23] H. E. Stanley, *Introduction to Phase Transitions and Critical Phenomena* (Oxford University Press, Oxford, 1971).
[24] A. Z. Patashinskii and V. L. Pokrovskii, *Fluctuation Theory of Phase Transitions* (Pergamon Press, Oxford, 1979).
[25] A. Aharony and B. I. Halperin, *Phys. Rev. Lett.* **35** (1975) 1308.
[26] J. Als-Nielsen and R. J. Birgeneau, *Amer. J. Phys.* **45** (1977) 554.
[27] A. P. Levanyuk, *Sov. Phys.: JETP* **9** (1959) 571.
[28] V. L. Ginzburg, *Sov. Phys.-Solid State* **2** (1961) 1824.
[29] V. L. Ginzburg, A. P. Levanyuk and A. A. Sobyanin, in *Modern Problems in Condensed Matter Sciences*, ed. V. M. Agranovich and A. A. Maradudin (North-Holland, Amsterdam, 1983), Vol. 5, Ch. 1.

[30] V. L. Ginzburg, A. P. Levanyuk and A. A. Sobyanin, *Ferroelectrics* **73** (1987) 171.
[31] V. I. Yukalov, *Phys. Rep.* **208** (1991) 395.
[32] V. I. Yukalov, *Intern. J. Mod. Phys.* **17** (2003) 2333.
[33] H. T. Davis, *Introduction to Nonlinear Differential and Integral Equations* (Dover, New York, 1962).
[34] P. Townsend and J. Sutton, *Phys. Rev.* **128** (1962) 591.
[35] J. Bardeen, L. N. Cooper and J. R. Shriffer, *Phys. Rev.* **108** (1957) 1175.
[36] L. P. Gor'kov, *Sov. Phys.: JETP* **9** (1959) 1364.
[37] R. J. Birgeneau, J. K. Kjems, G. Shirane and L. G. Van Uitert, *Phys. Rev. B* **10** (1974) 2512.
[38] J. C. Norvell and J. Als-Nielsen, *Phys. Rev. B* **2** (1970) 277.
[39] D. R. Chipmann and C. B. Walker, *Phys. Rev. B* **5** (1972) 3823.
[40] O. Rathmann and J. Als-Nielsen, *Phys. Rev. B* **9** (1974) 3921.
[41] L. Chachkhiani, V. Dmitriev and P. Toledano, *Sov. Phys.-Solid State* **32** (1990) 1334.
[42] B. Mettout, V. Dmitriev, M. Ben-Jaber and P. Toledano, *Phys. Rev. B* **48** (1993) 6908.

Chapter 5

Displacive Reconstructive Phase Transitions

5.1. Introduction

In our theoretical description, displacive transitions reflect most strikingly the specificity of the reconstructive transition mechanism. In the same way, it is from the experimental results reported on typical examples of displacive mechanisms that our understanding of the particular features of reconstructive transitions was made possible: stabilization of "Landau" and limit phases in the phase diagrams of uranium, existence of two orientational domains of the BCC parent structure in iron, X-ray and electron microscopy diffraction data showing the direction of the atomic shifting mechanism and the special positions corresponding to an increase of the rhombohedral to hexagonal ω symmetry in titanium, absence of phonon softening for the $\beta{-}\omega$ transition or incomplete softening in the vicinity of specific points of the BCC Brillouin zone for the martensitic transitions in lithium and sodium.

Displacive reconstructive transitions display a large variety of complex behaviors, and although our approach offers a unifying framework for their properties, we do not pretend in this chapter to discuss in detail the set of different mechanisms evidenced in the field. However, in the study of particular examples, we try to show how the main current problems are inserted in our phenomenological description: important role of elastic strains, defects, some lattice instabilities, thermodynamics of martensitic transformations, premartensitic behaviors, etc.

In the first place, we have chosen to illustrate the applicability of our approach to allotropic transitions of elemental crystals, because the simplicity of their structures allows more clear considerations on the transition mechanisms. Some examples of alloys and dielectrics are also discussed. The succession of the treated examples goes from the simpler situation (the β–ω transitions in Ti and Zr) to the more complicated one (the martensitic transformations in Li and Na).

5.2. The β–ω Transition

5.2.1. *The transition mechanism*

The β–ω transition takes place in titanium, zirconium, hafnium, and in some of their alloys (TiCr, TiV, TiMo, TiNb, ZrNb,...) [1]. Ti and Zr display a very similar phase diagram, shown schematically in Fig. 5.1(a) [2]. Thus, the β–ω transition occurs directly at high pressures and high temperatures (above 9 GPa and 940 K in Ti, and above 6 GPa and 970 K in Zr), whereas the α (HCP) phase appears as an intermediate structure at room temperature and atmospheric pressure. The transition is found to be completely reversible, with a small latent heat and a slight hysteresis [3].

The β-phase is body-centered cubic, and the ω-phase can be found in two slightly different structures: a hexagonal form (space group $P6/mmm, Z = 3$) [4], and a rhombohedral modification (space group $C\bar{3}m, Z = 3$) [5], which is obtained from the "ideal" hexagonal ω-phase with a low concentration of impurities (1.2% of oxygen in Ti_3V) or in more concentrated alloys (TiCr, ZrNb) [1].

In Ti, the average shifting of the Ti atoms from the β to the ω positions can be deduced from the superstructures observed in the electron diffraction and X-ray patterns [6,7]. It takes place along the [111] cubic direction, which is parallel to the c axis of the hexagonal unit cell (Fig. 5.1(b)). The antiparallel displacements of the atoms along the preceding direction and the formation of the hexagonal ω-layers are shown in Fig. 5.1(c). Figure 5.1(d) represents the $(0\bar{1}1)$ cubic plane, in which one can see that the corner atoms of the unit cell do not move at the transition, but only the two atoms located

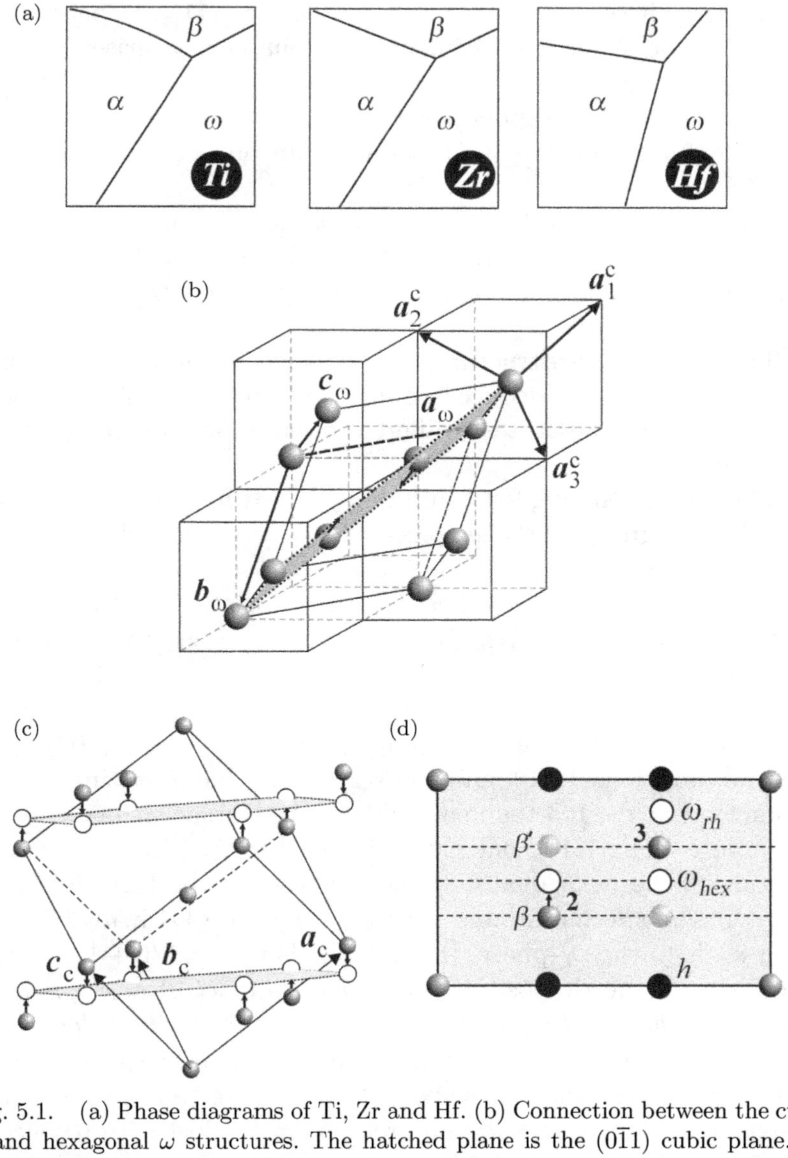

Fig. 5.1. (a) Phase diagrams of Ti, Zr and Hf. (b) Connection between the cubic β and hexagonal ω structures. The hatched plane is the $(0\bar{1}1)$ cubic plane. (c) Antiparallel displacements from the β structure leading to the formation of the hexagonal ω layers. (d) Critical shifts along the [111] cubic direction, leading to the formation of the ω phase. Full, double, hatched and open circles represent the atomic positions in the β, ω_h, β' and h phases, respectively. The crossed circle indicates the general position in the $(0\bar{1}1)$ plane, corresponding to the atomic positions in the ω_{rh} phase.

within the unit cell, which are displaced along $[111]_c$. Depending on the magnitude of the displacements, five phases may arise:

(1) The β-phase, corresponding to unshifted atoms. It has the parent $Im\overline{3}m$ symmetry, with one atom in the elementary unit cell ($Z = 1$).
(2) The hexagonal ω-phase (denoted ω_h hereafter), which appears as the result of a shift of $a\sqrt{3}/12$, where a is the cubic (β) lattice parameter. It possesses the symmetry $P6/mmm$ with $Z = 3$.
(3) The β'-phase, isostructural to β, which is induced by a shift of $a\sqrt{3}/6$, can be deduced from β by a rotation of $\pi/3$ around [111], i.e., it constitutes a different orientational domain of the β structure.
(4) The h-phase arising for a shift of $a\sqrt{3}/3$. It has the same hexagonal space group as the ω_h-phase ($P6/mmm$), but with one atom per unit cell.
(5) The rhombohedral ω-phase (ω_{rh}) associated with any general shift along [111]. It has the space group $C\overline{3}m$ with $Z = 3$.

In the ω_{rh}-phase, the initial one-fold cubic position (000) splits into two positions (000) and (1/3 2/3 z). Characterizing by the parameter $\xi = z - 1/3$ the magnitude of the displacements from the β structure along [111], one can see from Fig. 5.1(d), which is also a section of the hexagonal $C\overline{3}m$ cell parallel to the three-fold axis, that it is the specific atomic shift $\xi = a\sqrt{3}/12\,(1/6 \text{ in } a\sqrt{3}/2 \text{ unit})$ which leads to the ω_h-phase. It corresponds to $z = 1/6 + 1/3 = 1/2$, which is a symmetric position *giving rise to a set of new symmetry elements of the resulting structure, which pertain to the holohedry of the lattice.* In particular, we obtain a mirror plane, perpendicular to the three-fold axis, and also a six-fold axis oriented along $[111]_c$. As a consequence, the group–subgroup relationship is lost with respect to the initial cubic structure, although this relationship is preserved for the general displacements leading to the ω_{rh}-phase. Besides, the displacements which give the succession of high-symmetry phases can be repeated periodically in the neighboring cells as shown in Fig. 5.1(d). Thus, for shifts $a\sqrt{3}/2, 7a\sqrt{3}/12, 4a\sqrt{3}/6$ and $5a\sqrt{3}/6$, one gets the β, ω, β' and h-phases.

5.2.2. *Direction of the order-parameter in the* ε*-space*

In order to formalize the preceding crystallographic description, let us express the basic vectors of the $\omega_{\rm h}$ unit cell, in function of the basic vectors of the BCC (primitive) β-unit cell, which are shown in Fig. 5.1(b). One has [8]

$$\boldsymbol{a}_\omega = \boldsymbol{a}_\beta - \boldsymbol{b}_\beta, \boldsymbol{b}_\omega = \boldsymbol{b}_\beta - \boldsymbol{c}_\beta, \boldsymbol{c}_\omega = \boldsymbol{a}_\beta + \boldsymbol{b}_\beta + \boldsymbol{c}_\beta. \tag{5.1}$$

The wave-vector of the BCC Brillouin zone, which expresses the breaking of the translational symmetry (5.1), is in Kovalev's notation [9]

$$\boldsymbol{k}_7 = \frac{1}{3}(\boldsymbol{a}_\beta^* + \boldsymbol{b}_\beta^* + \boldsymbol{c}_\beta^*), \tag{5.2}$$

where $\boldsymbol{a}_\beta^*$, $\boldsymbol{b}_\beta^*$ and $\boldsymbol{c}_\beta^*$ are the reciprocal lattice vectors of the β structure. $\boldsymbol{k}_7$ is located on the Λ-line, the BCC Brillouin zone, and its *star* $\{\boldsymbol{k}_7\}$, which is invariant by the $3mm$ point group [8], has *eight branches*:

$$\boldsymbol{k}_7^1 = \frac{1}{3}(\boldsymbol{a}_\beta^* + \boldsymbol{b}_\beta^* + \boldsymbol{c}_\beta^*), \boldsymbol{k}_7^2 = \frac{1}{3}(-3\boldsymbol{a}_\beta^* + \boldsymbol{b}_\beta^* + \boldsymbol{c}_\beta^*),$$

$$\boldsymbol{k}_7^3 = \frac{1}{3}(\boldsymbol{a}_\beta^* - 3\boldsymbol{b}_\beta^* + \boldsymbol{c}_\beta^*), \boldsymbol{k}_7^4 = \frac{1}{3}(\boldsymbol{a}_\beta^* + \boldsymbol{b}_\beta^* - 3\boldsymbol{c}_\beta^*), \boldsymbol{k}_7^{i+4} = -\boldsymbol{k}_7^i.$$

The irreducible representations corresponding to the shifts of the atoms of the β structure, which involve all the branches of $\{\boldsymbol{k}_7\}$, are constructed from the two "little" representations $\tau_1(\mathrm{A}_{1z})$ and $\tau_3(\mathrm{E}_{\rm xy})$ of the vector representation of the $3mm$ point group, the matrices of which are given in Table 5.1. The corresponding 8- and 16-dimensional irreducible representations, denoted $\tau_1(\boldsymbol{k}_7)$ and $\tau_3(\boldsymbol{k}_7)$,

Table 5.1. Matrices of the irreducible representations τ_1 and τ_3 (in Kovalev's notation [8]) of the $3mm$ point group, which form the vector representation. The notation of the symmetry elements corresponds to the cubic $(\beta)m3m$ point group. $\varepsilon = \exp(2\pi i/3)$.

$3mm$	1	3_2^{xyz}	3^{xyz}	$m_{\bar{x}y}$	$m_{\bar{y}z}$	$m_{\bar{x}z}$
τ_1	1	1	1	1	1	1
τ_3	$\begin{bmatrix}1 & 0\\0 & 1\end{bmatrix}$	$\begin{bmatrix}\varepsilon & 0\\0 & \varepsilon^2\end{bmatrix}$	$\begin{bmatrix}\varepsilon^2 & 0\\0 & \varepsilon\end{bmatrix}$	$\begin{bmatrix}0 & 1\\1 & 0\end{bmatrix}$	$\begin{bmatrix}0 & \varepsilon\\\varepsilon^2 & 0\end{bmatrix}$	$\begin{bmatrix}0 & \varepsilon^2\\\varepsilon & 0\end{bmatrix}$

Table 5.2. Matrices generating the irreducible representation $\tau_1(\boldsymbol{k}_7)$. Columns represent diagonal matrices. The zero elements are not indicated in non-diagonal matrices. $\varepsilon = \exp(2\pi i/3)$.

$\boldsymbol{a}_\beta$	$\boldsymbol{b}_\beta$	$\boldsymbol{c}_\beta$	$\{2_x\|0\}$	$\{2_y\|0\}$
$\begin{bmatrix} \varepsilon \\ 1 \\ \varepsilon \\ \varepsilon \\ \varepsilon* \\ 1 \\ \varepsilon* \\ \varepsilon* \end{bmatrix}$	$\begin{bmatrix} \varepsilon \\ \varepsilon \\ 1 \\ \varepsilon \\ \varepsilon* \\ \varepsilon* \\ 1 \\ \varepsilon* \end{bmatrix}$	$\begin{bmatrix} \varepsilon \\ \varepsilon \\ \varepsilon \\ 1 \\ \varepsilon* \\ \varepsilon* \\ \varepsilon* \\ 1 \end{bmatrix}$	$\begin{bmatrix} & 1 & & & & & & \\ 1 & & & & & & & \\ & & & 1 & & & & \\ & & 1 & & & & & \\ & & & & & 1 & & \\ & & & & 1 & & & \\ & & & & & & & 1 \\ & & & & & & 1 & \end{bmatrix}$	$\begin{bmatrix} & & 1 & & & & & \\ & & & 1 & & & & \\ 1 & & & & & & & \\ & 1 & & & & & & \\ & & & & & & 1 & \\ & & & & & & & 1 \\ & & & & 1 & & & \\ & & & & & 1 & & \end{bmatrix}$

$\{3_2^{xyz}\|0\}$	$\{2_{\overline{x}y}\|0\}$	$\{\overline{1}\|0\}$
$\begin{bmatrix} 1 & & & & & & & \\ & & & 1 & & & & \\ & 1 & & & & & & \\ & & 1 & & & & & \\ & & & & 1 & & & \\ & & & & & & & 1 \\ & & & & & 1 & & \\ & & & & & & 1 & \end{bmatrix}$	$\begin{bmatrix} & & & & 1 & & & \\ & & & & & & 1 & \\ & & & & & 1 & & \\ & & & & & & & 1 \\ 1 & & & & & & & \\ & & 1 & & & & & \\ & 1 & & & & & & \\ & & & 1 & & & & \end{bmatrix}$	$\begin{bmatrix} & & & & 1 & & & \\ & & & & & 1 & & \\ & & & & & & 1 & \\ & & & & & & & 1 \\ 1 & & & & & & & \\ & 1 & & & & & & \\ & & 1 & & & & & \\ & & & 1 & & & & \end{bmatrix}$

account, respectively, for the atomic shifts along [111] (longitudinal phonons) and within the cubic plane (111) (transverse phonons). $\tau_3(\boldsymbol{k}_7)$ *is not* associated with the β–ω transition mechanism, and one has to consider only $\tau_1(\boldsymbol{k}_7)$. From the matrices generating this representation, which are listed in Table 5.2, one can determine (using the geometrical method of Chapter 1, Sec. 1.3.3.2) the possible low-symmetry groups, subgroups of $Im\overline{3}m$, which may arise below a transition from the β structure, for any shifts of the atoms parallel to [111] and to the other corresponding diagonals. Among the 21 possible equilibrium states, given in Ref. [9] (p. 199), one can see that only two states possess a symmetry involving the three-fold multiplication of the BCC unit cell, following Eq. (5.1). These states possess the symmetry groups $P3m1$ and $C\overline{3}m$, and the corresponding basic vectors φ_{ki} $(i = 1 - 8)$ have zero coefficients η_i except φ_{k1} and φ_{k5}.

The basis functions φ_{k1} and φ_{k5} can be written in terms of the shifts ξ_i $(i = 1 - 27)$ of the 27 atoms, numbered from 1 to 27 in

Fig. 5.1(b), which pertain to the 27-fold unit cell involving all the $\boldsymbol{k}_7^{(i)}$ components. One has

$$\begin{aligned}\boldsymbol{\varphi}_{k1} &= \boldsymbol{c}_\omega \cdot \{(\xi_1 + \xi_8 + \xi_{12} + \xi_{13} + \xi_{14} + \xi_{15} + \xi_{17} + \xi_{20} + \xi_{21}) \\ &\quad + \varepsilon(\xi_2 + \xi_3 + \xi_4 + \xi_{16} + \xi_{18} + \xi_{19} + \xi_{22} + \xi_{23} + \xi_{25}) \\ &\quad + \varepsilon^* \, (\xi_5 + \xi_6 + \xi_7 + \xi_9 + \xi_{10} + \xi_{11} + \xi_{24} + \xi_{26} + \xi_{27})\}\,; \\ \boldsymbol{\varphi}_{k5} &= \boldsymbol{c}_\omega \cdot \{-(\xi_1 + \xi_8 + \xi_{12} + \xi_{13} + \xi_{14} + \xi_{15} + \xi_{17} + \xi_{20} + \xi_{21}) \\ &\quad - \varepsilon(\xi_2 + \xi_3 + \xi_4 + \xi_{16} + \xi_{18} + \xi_{19} + \xi_{22} + \xi_{23} + \xi_{25}) \\ &\quad - \varepsilon^*(\xi_5 + \xi_6 + \xi_7 + \xi_9 + \xi_{10} + \xi_{11} + \xi_{24} + \xi_{26} + \xi_{27})\}. \quad (5.3)\end{aligned}$$

From Ref. [9] (p. 199), one can see that the equilibrium state of symmetry $P3m1$ corresponds to the relationship: $\eta_1 \neq \eta_5 \neq 0, \eta_{\iota\neq 1,5} = 0$. Thus, it is associated with *non-symmetric* antiparallel shifts of atoms denoted 2 and 3 in Fig. 5.1(d). The state of symmetry $R\bar{3}m$, *which identifies to the* $\omega_{\rm rh}$*-phase*, is associated with *symmetric shifts* determined by the equilibrium conditions

$$\eta_1 = \eta_5 \neq 0, \quad \eta_{\iota\neq 1,5} = 0, \tag{5.4}$$

which can also be written, considering the translationally equivalent atoms in Eq. (5.3), as

$$\boldsymbol{\varphi}_{k1} = \boldsymbol{\varphi}_{k5} = (\xi_{19} - \xi_{27}) \cdot \boldsymbol{c}_\omega. \tag{5.5}$$

In other words, *the direction of the eight-dimensional order-parameter* $\boldsymbol{\eta}$*, associated with the* $\beta \to \omega_{\rm rh}$ *transition, in the* ε_8 *representation space of* $\tau_1(\boldsymbol{k}_7)$*, is fixed by the equilibrium conditions* (5.4). In the next section, we will use this result to find the dependence of the order-parameter modulus $|\eta|$ in function of the magnitude ξ of the antiparallel shifts.

5.2.3. *Periodicity of the order-parameter magnitude*

The lattice function describing the extended unit cell involved by the variable displacements $\boldsymbol{r}_j$ along [111], from the β structure, can be written as

$$\psi(\boldsymbol{r}_j) = \frac{1}{27}\sum_{j=1}^{27}\sum_{m=1}^{8}(A_j \cdot \exp\{i\boldsymbol{k}_7^{(m)}(\boldsymbol{R}_j^0 - \boldsymbol{r}_j\} + \text{c.c.}, \tag{5.6}$$

where the sum runs over the 27 atoms contained in the extended cell, the positions of which are characterized by the vectors $\boldsymbol{R}_j^0$, and over the eight branches of $\boldsymbol{k}_7^{(m)}$. Restricting to the substructure determined by the two "active" branches $\boldsymbol{k}_7^{(1)} = -\boldsymbol{k}_7^{(5)}$, corresponding to the antiparallel shifts ξ_1 and ξ_2 associated with the $\beta \to \omega_{\text{rh}}$ transition, one needs to consider only the three atoms contained in the ω_{rh} unit cell. Equation (5.6) takes the following simplified form:

$$\psi(\xi_1, \xi_2) = \frac{A}{3}\left[1 + \exp\frac{4\pi i}{a\sqrt{3}}\left(\frac{a}{2\sqrt{3}} + \xi_1\right)\right.$$
$$\left. + \exp\frac{4\pi i}{a\sqrt{3}}\left(\frac{a}{\sqrt{3}} - \xi_2\right)\right] + \text{c.c.}, \tag{5.7}$$

where the first term in the bracket corresponds to the atoms denoted 1 in Fig. 5.1(d), which are unshifted after the transition, while the second and third terms in Eq. (5.7) relate to atoms of types 2 and 3, which are shifted in opposite directions. In the ω_{rh}-phase, one has $\xi_1 = \xi_2 = \xi$, whereas in the β-phase, $\xi_1 = \xi_2 = 0$. Accordingly, the corresponding lattice functions are

$$\psi_{\omega_{\text{rh}}}(\xi) = A\left[\sin\left(\frac{4\pi\xi}{a\sqrt{3}} + \frac{\pi}{6}\right) - \frac{1}{2}\right], \tag{5.8}$$

and $\psi_\beta = 0$. Thus, the magnitude of the order-parameter $|\boldsymbol{\eta}(\xi)|$ in function of the critical displacements ξ, which is determined by the increment $\delta\psi(\xi) = \psi_{\omega\text{rh}} - \psi_\beta = \psi_{\omega\text{rh}}$, has, up to a normalizing constant η_0, the same form (5.8):

$$|\boldsymbol{\eta}(\xi)| = \eta_0\left\{\sin\left[f(\xi)\right] - \frac{1}{2}\right\}, \tag{5.9}$$

with $f(\xi) = \{(4\pi\xi/a\sqrt{3}) + \pi/6\}$.

Figure 5.2 represents the dependence of $|\boldsymbol{\eta}(\xi)|/\eta_0$. It shows that for successive specific critical shifts ξ, one gets periodically the sequence of β, $\omega_{\rm h}$, β' and h-phases. More precisely, for $\xi = 0, 1, 2, 3, \ldots$ (in $a\sqrt{3}/2$ units), one has $\eta = 0$ for the parent β-phase; for $\xi = 1/6, 7/6, 13/6, \ldots, \eta/\eta_0 = 1/3$, and $\omega_{\rm h}$-phase is obtained; for $\xi = 1/3, 4/3, 7/3, \ldots$, one gets again $\eta = 0$ (β'-phase); and for $\xi = 2/3, 5/3, 8/3, \ldots, \eta/\eta_0 = -1$, which corresponds to the h-phase. For all other arbitrary displacements ξ, the $\omega_{\rm rh}$-phase is stabilized.

5.2.4. *Theoretical phase diagram for the β–ω transition*

The dependence of $|\eta(\xi)|$, as expressed by Eq. (5.9) and shown in Fig. 5.2, *is not symmetric* with respect to the β and $\omega_{\rm h}$-phases. As underlined in Chapter 4 (Sec. 4.6.2), this asymmetry is indicative of the fact that β represents the actual parent structure, and that $\omega_{\rm h}$ cannot be used as the initial structure for describing the β–ω transition. This must also be reflected in the form of the effective order-parameter expansion, which will be shown to contain a *cubic* term, and in the asymmetry of the corresponding phase diagram.

When restricting to *fourth-degree terms*, one can construct, using the transformation properties of the matrices given in Table 5.2, four independent invariants of the order-parameter components, as follows:

$$I_0 = \sum_{i=1}^{8} \eta_i^2, \quad I_1 = \sum_{i=1}^{8} \eta_i^3, \quad I_2 = \eta_1\eta_2\eta_3\eta_4 + \eta_5\eta_6\eta_7\eta_8,$$
$$I_3 = \sum_{i \neq j \neq k \neq l} \eta_i\eta_j\eta_k\eta_l - I_2. \tag{5.10}$$

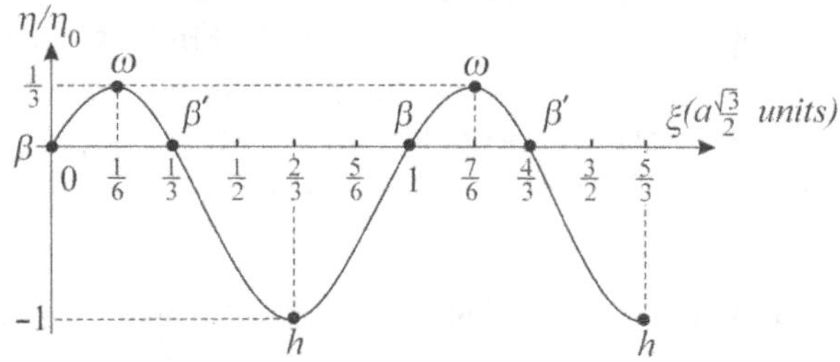

Fig. 5.2. $|\boldsymbol{\eta}(\xi)|$ dependence at the β–ω transition following Eq. (5.9).

Considering the equilibrium conditions (5.4), one has $I_2 = I_3 = 0$, $I_0 = \eta^2, I_1 = \eta^3$. The effective thermodynamic potential can be written as

$$F = F_0(P,T) + a_1|\boldsymbol{\eta}(\xi)|^2 + b_1|\boldsymbol{\eta}(\xi)|^3 + a_2|\boldsymbol{\eta}(\xi)|^4. \tag{5.11}$$

Minimization with respect to ξ, using Eq. (5.9), yields the equation of states

$$\left\{\sin[f(\xi)] - \frac{1}{2}\right\} \cdot \cos[f(\xi)] \cdot \left\{2a_1 + 3b_1 \cdot \left[\sin\left(f(\xi)\right) - \frac{1}{2}\right]\right\}$$
$$+ 4a_2\left[\sin[f(\xi)] - \frac{1}{2}\right]^2 = 0. \tag{5.12}$$

When using also the stability condition $\partial^2 F/\partial\xi^2 \geq 0$, one has the following equilibrium states:

(1) The parent β (or β') phase corresponding to $\eta = 0$, $\sin[f(\xi)] = 1/2$ and $a_1 \geq 0$;
(2) The ω_{h}-phase given by $\partial\eta/\partial\xi = 0$ and $\eta = 1/3$, namely for $\cos[f(\xi)] = 0$ and $\sin[f(\xi)] = +1$. Its domain of stability is determined by $a_1 \leq (-3b_1 - 2a_2)/4$;
(3) The h-phase which obeys the conditions $\partial\eta/\partial\xi = 0$, $\eta = -1$, expressed by $\sin[f(\xi)] = 0$ and $\sin[f(\xi)] = -1$. It is stabilized for $a_1 \leq 9(b_1 - 2a_2)/4$;
(4) The ω_{rh}-phase coinciding with the Landau solution $\partial F/\partial\eta = 0$, and $-1 < \eta < 1/3$, $\eta \neq 0$. It gives two (temperature-dependent) solutions:

$$\sin(f(\xi)) = \frac{1}{2} + \{-3b_1 \pm \sqrt{9b_1^2 - 32a_1a_2}\}/8a_2,$$

associated with two (ω_{rh}^+ and ω_{rh}^-) *anti-isostructural* phases determined by opposite shifts in Fig. 5.1(d). The stability region, in which ω_{rh}^+ and ω_{rh}^- are located, is bounded by three limit of stability lines: with respect to the β-phase ($a_1 \leq 9b_1^2/32a_2$), the ω_{h}-phase ($a_1 \geq (-3b_1 - 2a_2)/4$)), and the h-phase ($a_1 \geq (6b_1 - a_2)/8$).

The corresponding phase diagram (Fig. 5.3) is represented in the plane (a_1, b_1). One can verify that the β-phase is separated from the *limit states* ω_{h} and h by the first-order transition *straight lines*, $a_1 = (-2b_1 - a_2)/4$ and $a_1 = 3(2b_1 - 3a_2)/4$, respectively, whereas the

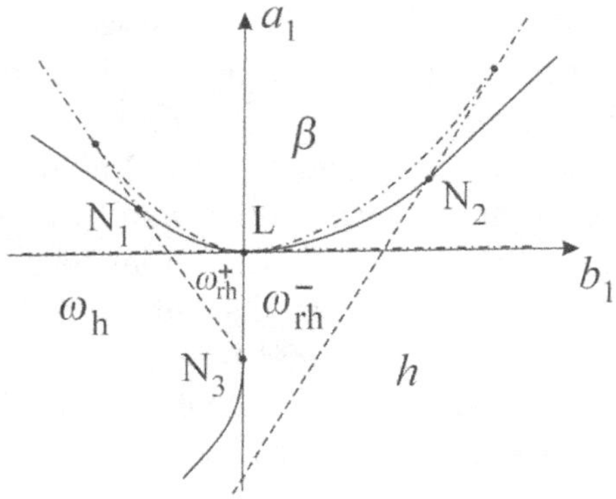

Fig. 5.3. Phase diagram associated with the β–ω transition, corresponding to the thermodynamic potential (5.11). Full, dashed and dashed-dotted lines represent first-order, second-order and limit of stability lines, respectively. N_1, N_2 and N_3 are three-phase points, whereas L is a triple point. ω^+_{rh} and ω^-_{rh} are anti-isostructural phases.

β–ω_{rh} first-order transition line is the parabola $a_1 = b_1^2/4a_2$. Between the group–subgroup-unrelated phases ω^+_{rh} and ω^-_{rh}, one has also a first-order transition line ($b_1 = 0$), whereas the lines $\omega_h - \omega^+_{rh}$ and $\omega_h - \omega^-_{rh}$ correspond to the second-order transitions. Their equations, $a_1 = (-3b_1 - 2a_2)/4$ and $a_1 = (6b_1 - a_2)/8$, are straight-line equations, as the result of the fourth-degree power assumed for $F[\boldsymbol{\eta}(\xi)]$. Finally, there are four singular points in the phase diagrams: (i) a triple point, located on the $\beta - \omega_{rh}$ first-order transition line, and ending the $\omega^+_{rh} - \omega^-_{rh}$ transition line; (ii) three *three-phase* points N_1, N_2 and N_3 at which merge three lines of first- *and* second-order transitions.

5.2.5. *Discussion*

The displacive mechanism which has been used in our density-wave approach to the β–ω transition [9, 10] is based on clear-cut electron and X-ray diffraction data [6] and confirms previous theoretical descriptions of this transition [11, 12]. The mechanism is also attested by the $\omega(\boldsymbol{k})$ phonon dispersion curves reported for Ti, Zr and Hf [13–15], which reveals a strong dip close to $\boldsymbol{k} = \frac{2}{3}(111)$ (the longitudinal phonon mode), in agreement with the lattice instability expressed by Eq. (5.1). A temperature dependence of this mode, when approaching the transition from the β-phase, has been vainly searched [13, 15], confirming the absence of softening with temperature (or pressure) as predicted in our model for direct transitions to limit states. In this respect, it can be noted that the reconstructive character of the β–ω transition is more reliably reflected in the magnitude of the critical displacements (and of the order-parameter) reported experimentally [6] than by the latent heat measurement [3] which suggests a weakly first-order discontinuity for the transition. The observation of four orientational domains of the ω_{h}-phase, corresponding to the four (111) cubic planes in titanium alloys [4, 5], brings another confirmation of the assumed connection between the β and ω-phases.

The fact that the ω_{rh}-phase replaces the ω_{h}-phase when adding a small concentration of oxygen, or when doping with a small amount of metal (Cr, Nb) [5,16], suggests that the corresponding experimental thermodynamic paths are close to the region of the three-phase point N_1, in the theoretical phase diagram of Fig. 5.3. This may also be confirmed by the fact that the h-phase, which is located in a different region of the phase diagram, was not found experimentally. Besides, when considering a model of hard-sphere atoms, instead of points as in Fig. 5.1(d), the atoms would realize in the h-phase a more unstable configuration than the close-packing structure of the α-phase. Thus, the presence of this latter phase in the phase diagram of Ti, Zr and Hf may have the result of destroying the stability of the h-phase.

5.3. The BCC–HCP Transition: The Role of Macroscopic Deformation

5.3.1. *Burgers mechanism*

The BCC–HCP transition is, together with the BCC–FCC transition, the most commonly encountered type of reconstructive transition in crystals of elements. It is found in about 20 elements (see Table 5.3). The BCC phase is, with some exceptions, more stable at high temperatures and high pressures. Figures 5.4(a)–5.4(f) represent schematically some typical phase diagrams containing a direct BCC–HCP transition [17]. The transition is always strongly first-order and corresponds to large atomic displacements of the order of 1 Å [18].

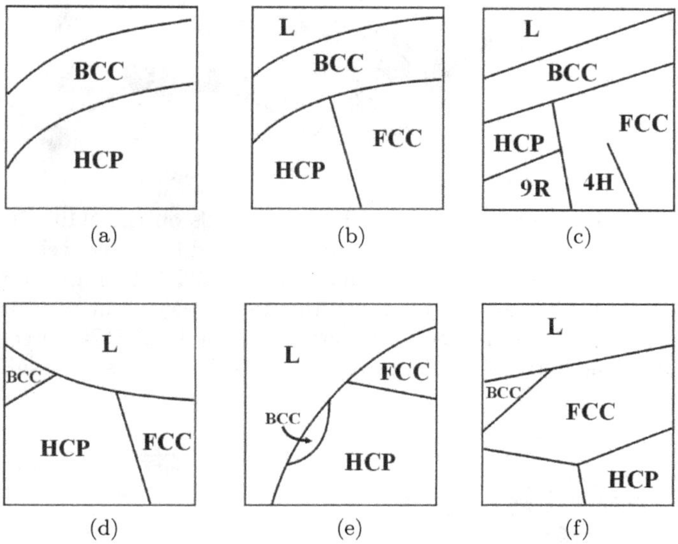

Fig. 5.4. Topology of the BCC–HCP transition line in elemental crystals. The represented phase diagrams are found, e.g., in Be (a), heavy lanthanides (b), Sm (c), Ba (d), He (e) and Fe (f).

The mechanism generally assumed for the BCC–HCP transition is the so-called Burgers mechanism [19], proposed for the $\beta-\alpha$ transition in Zr. It can be divided into three stages, which are represented in Figs. 5.5(a)–5.5(c). In the first stage, the (110) BCC planes transform into the (001) HCP planes via a shearing mechanism (Fig. 5.5(a)). Second, a homogeneous shifting of the atoms in the (110) planes leads to an FCC structure (Fig. 5.5(b)). Third, the displacement of one of two planes in the $[110]_{BCC}$ direction leads to the formation of the HCP structure (Fig. 5.5(c)). In other words, Burgers mechanism involves three different and simultaneous instabilities associated with distinct primary (symmetry-breaking) order-parameters. We will first reformulate this mechanism in a phenomenologically simpler manner [20].

5.3.2. *Burgers mechanism reformulated*

In contrast with the preceding scheme, we propose a BCC–HCP transition mechanism based on a single *primary* order-parameter. It consists in an *antiparallel shifting* of the atoms lying in the $(110)_{BCC}$

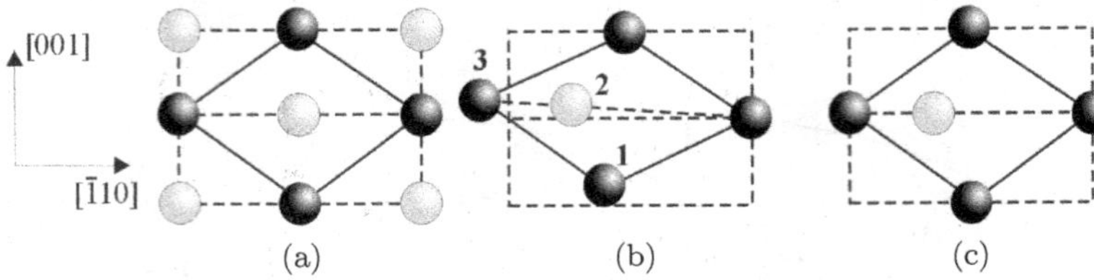

Fig. 5.5. Projection of the BCC and HCP unit cells on the (110) BCC plane. Solid and open circles are atoms in positions 1/2 and 0, respectively. (a) Initial BCC structure. (b) HCP structure: the atoms denoted 1, 2, 3 are shifted along the same [1$\bar{1}$2] BCC direction. (c) HCP structure, following our reformulation of Burgers mechanism: the displacements are along the [1$\bar{1}$0] BCC direction. The deformation along [001] BCC is induced by the shifting of the open sublattice.

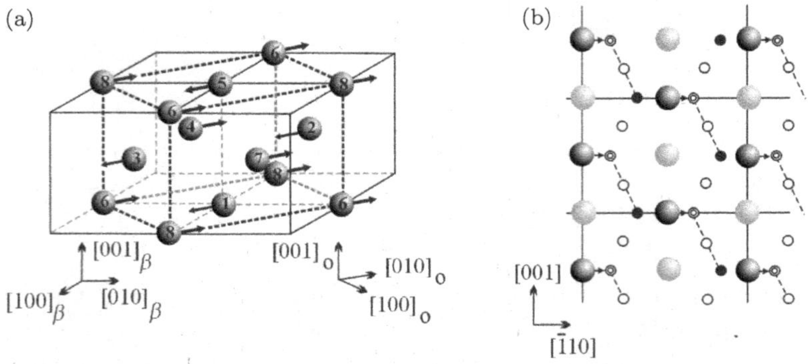

Fig. 5.6. Average shifting of the atoms from the BCC to the HCP structures. Solid and open circles are atoms located in adjacent (0 and 1/2) layers. (a) The small cube (thick lines) is the BCC unit cell. The dotted parallelepiped is the unit cell of the orthorhombic (*Cmcm*) structure. (b) Projection on the (110) BCC plane. The shifting from the BCC (large circles) to the HCP structure (small full dots) is $a\sqrt{2}/12$. The open and double dots correspond to virtual positions at which the symmetry would increase to $Im\bar{3}m$ and $P6_3/mmc$, respectively.

plane, along the $\pm[1\bar{1}0]$ directions, which *induces* as a *secondary* effect, a *compression* of the cubic unit cell along the four-fold axis [001], and a simultaneous decompression along the two other ([100] and [010]) four-fold axes. Figure 5.6(a) shows the average shifting of the atoms from the BCC to the HCP structure, whereas Fig. 5.6(b) represents the projection of the HCP unit cell on the (110) BCC plane.

A more concrete view of the assumed correlated mechanisms at the atomistic level is shown in Figs. 5.7(a) and 5.7(b). Each (110)

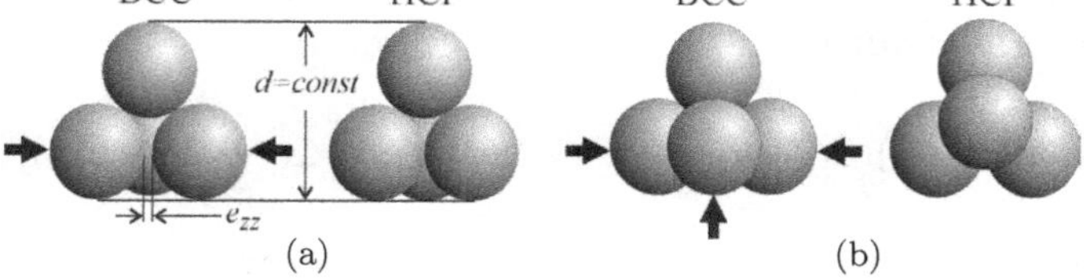

Fig. 5.7. Atomic model illustrating the improper character of the macroscopic deformation induced by the atomic displacements represented by arrows. (a) Projections of the BCC and HCP structures, respectively, on the $(001)_{\mathrm{BCC}}$ and $(\bar{1}10)_{\mathrm{HCP}}$ planes. d is the interplane distance. (b) Projections of the BCC and HCP structures, respectively, on the $(110)_{\mathrm{BCC}}$ and $(001)_{\mathrm{HCP}}$ planes.

BCC layer is organized in such a manner that, in any triangle formed by neighboring atoms, two sides are formed by atoms which touch each other, and the third side is formed by separated atoms. It is on the third side that the atoms of the next layer are located. The displacement of one layer with respect to the following one in the $[1\bar{1}0]$ BCC direction (i.e., the shifting of each atom of the upper layer toward the center of the corresponding triangle, which is formed by the atoms of the lower layer) allows the separated atoms to come in contact, and thus to form a regular double-layered hexagonal close-packed structure. The *reverse* sliding of one close-packed layer with respect to the next layer, at *fixed interlayer distance*, breaks one of each of the three pairs of atoms touching each other, again forming a BCC lattice.

In order to formalize the preceding crystallographic description, let us first write the relationships between the basic vectors of the HCP (space group $P6_3/mmc, Z = 2$) and BCC ($Im\bar{3}m, Z = 1$) unit cells, which are shown in Fig. 5.6(a):

$$\boldsymbol{a}_h = \boldsymbol{a}_C + \boldsymbol{b}_C + \boldsymbol{c}_C, \boldsymbol{b}_h = -\boldsymbol{c}_C, \boldsymbol{c}_h = \boldsymbol{a}_C - \boldsymbol{b}_C. \tag{5.13}$$

The *doubling* of the BCC unit cell involved by Eq. (5.13) corresponds to *one branch*, $\boldsymbol{k}_9^{(2)}$, of the star of the wave-vector $\boldsymbol{k}_9$ [8], of the BCC Brillouin zone, ending at the zone-boundary N-point. The star $\{\boldsymbol{k}_9\}$ has *six* branches, the coordinates of which are

$$\boldsymbol{k}_9^{(1)} = \left(\frac{\pi}{a}\frac{\pi}{a}0\right), \boldsymbol{k}_9^{(2)} = \left(\frac{\pi}{a}\frac{-\pi}{a}0\right), \boldsymbol{k}_9^{(3)} = \left(\frac{\pi}{a}0\frac{\pi}{a}\right),$$

$$\boldsymbol{k}_9^{(4)} = \left(\frac{\pi}{a}0\frac{-\pi}{a}\right), \boldsymbol{k}_9^{(5)} = \left(0\frac{\pi}{a}\frac{\pi}{a}\right), \boldsymbol{k}_9^{(5)} = \left(0\frac{\pi}{a}\frac{-\pi}{a}\right). \tag{5.14}$$

On the other hand, the antiparallel displacement field, represented in Fig. 5.6(a), transforms as the basic function

$$\varphi(x, y) = (x+y)(\Delta_1 - \Delta_2 - \Delta_3 + \Delta_4 + \Delta_5 - \Delta_6 + \Delta_7 - \Delta_8), \quad (5.15)$$

where the Δ_i are the shifts associated with the atoms numbered 1 to 8 in Fig. 5.6(a). $\varphi(x, y)$ transforms as the one-dimensional little irreducible representation, labeled $\hat{\tau}_4$ in Kovalev's tables [9], the matrices of which are, for the invariance group mmm ($m_{xy}m_{\bar{x}y}m_z$) of $\boldsymbol{k}_9^{(1)}$:

	1	2_x	$2_{\bar{x}y}$	2_{xy}	$\bar{1}$	m_z	$m_{\bar{x}y}$	m_{xy}
$\hat{\tau}_4$	1	-1	1	-1	-1	1	1	-1

(5.16)

Using Eqs. (5.14) and (5.16), one can construct the generating matrices of the *six-dimensional* irreducible representation $\hat{\tau}_4(\boldsymbol{k}_9^*)$ of the (symmorphic) space group $Im\bar{3}m$:

$$\{2_x\,|\,0\}: \begin{bmatrix} & 1 & & & & \\ 1 & & & & & \\ & & & 1 & & \\ & & 1 & & & \\ & & & & 1 & \\ & & & & & 1 \end{bmatrix} \quad \{2_y\,|\,0\}: \begin{bmatrix} & 1 & & & & \\ 1 & & & & & \\ & & 1 & & & \\ & & & 1 & & \\ & & & & & 1 \\ & & & & 1 & \end{bmatrix} \quad \{3_2^{xyz}\,|\,0\}: \begin{bmatrix} & & & & 1 & \\ & & & & & 1 \\ 1 & & & & & \\ & 1 & & & & \\ & & 1 & & & \\ & & & 1 & & \end{bmatrix}$$

$$\{2_{\bar{x}y}\,|\,0\}: \begin{bmatrix} 1 & & & & & \\ & 1 & & & & \\ & & & & 1 & \\ & & & & & 1 \\ & & 1 & & & \\ & & & 1 & & \end{bmatrix} \quad \{\bar{1}\,|\,0\}: \begin{bmatrix} 1 & & & & & \\ & 1 & & & & \\ & & 1 & & & \\ & & & 1 & & \\ & & & & 1 & \\ & & & & & 1 \end{bmatrix}. \quad (5.17)$$

The geometrical method described in Chapter 1 (Sec. 1.3.3.2) provides, using Eq. (5.17), the set of 33 low-symmetry groups, subgroups of $Im\bar{3}m$, which are induced by $\hat{\tau}_4(\boldsymbol{k}_9^*)$. They are listed in tables of Refs. [9] (p. 208) and [21] (p. 10415). One can verify in these tables that only one low-symmetry state, of space-group $Cmcm$, involves a *doubling* of the $Im\bar{3}m$ unit cell. It coincides with the branch $\boldsymbol{k}_9^{(2)}$ and to the equilibrium solution $(0\eta 0000)$ of the *six-dimensional order-parameter*, transforming as $\hat{\tau}_4(\boldsymbol{k}_9^*)$.

In summary, the direction in the ε_6-space of the six-dimensional order-parameter $\boldsymbol{\eta}$, which corresponds to the primary mechanism proposed for the BCC–HCP transition (the antiparallel atomic shifts along [110]), is given by

$$\eta_2 = \eta \neq 0, \eta_1 = \eta_3 = \eta_4 = \eta_5 = \eta_6 = 0. \tag{5.18}$$

However, this primary order-parameter leads to a phase of orthorhombic symmetry *Cmcm*.

The additional transformation which brings about an increase of the *Cmcm* symmetry to $P6_3/mmc$ (the HCP structure) corresponds to the following requirements:

(1) A compression of the cubic unit cell along one of the four-fold axes. This *secondary deformation mechanism* has the symmetry of the two-dimensional irreducible representation of the $Im\bar{3}m$ space-group at the center (Γ-point) of the BCC Brillouin zone ($\hat{\tau}_3(\boldsymbol{k}_{11}^*) \equiv \mathrm{E_g}$) in Kovalev's notation [8], which is spanned by the two following combinations of the strain-tensor components:

$$\zeta_1 = (e_{xx} + e_{yy} - 2e_{zz})/\sqrt{6}, \zeta_2 = (e_{xx} - e_{yy})/\sqrt{2}. \tag{5.19}$$

(2) A *specific magnitude* for the antiparallel shifts along [110]:

$$|\xi_{HCP}| = a\sqrt{2}/12, \tag{5.20}$$

where a is the cubic lattice parameter.

When the two preceding requirements are fulfilled, the *Cmcm* symmetry *enlarges* to $P6_3/mmc$. This enlargement of symmetry mechanism is illustrated in Fig. 5.8. This figure, which is a projection of the BCC structure in the $(1\bar{1}0)_{\mathrm{BCC}}$ plane, shows that the angle

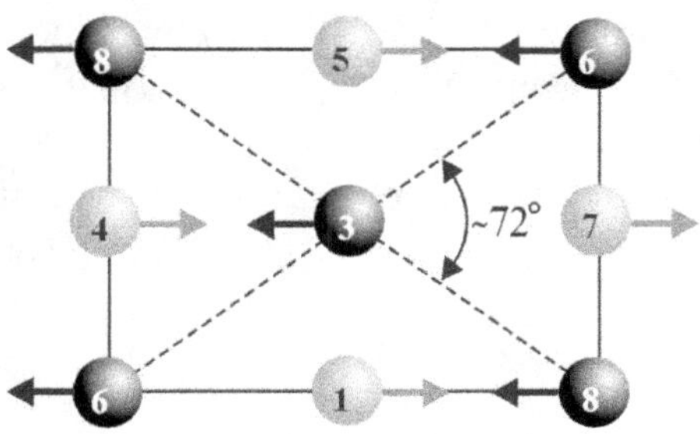

Fig. 5.8. Shifting mechanism at the BCC–HCP transition in projection on the plane $(1\bar{1}0)$. After the shifting, the angle between the diagonals becomes $\pi/3$. Solid and open circles are atoms located in two adjacent layers.

between the three-fold axes, corresponding to arccos $1/3 = 70°32' \approx 0.392\pi$ in the cubic cell, *becomes* $\pi/3$ in the hexagonal cell.

A striking property, which can be foreseen from Fig. 5.6(b), is the *periodic connection* that can be found between the overall symmetry undertaken by the structure and the virtual displacements of the atoms along the [110] direction. Thus, for general arbitrary displacements $|\xi|$ along [110], the system exhibits the symmetry $Cmcm$, while for the specific displacements $|\xi_{HCP}| = a\sqrt{2}/12$, $7a\sqrt{2}/12$, $13a\sqrt{2}/12, \ldots$, the crystal acquires the $P6_3/mmc$ symmetry. For other specific displacements, namely $a\sqrt{2}/4, 3a\sqrt{2}/4, 5a\sqrt{2}/4, \ldots$, the symmetry increases to $P6/mmm$ ($Z = 1$). We will now express this periodicity of the magnitude of $|\boldsymbol{\eta}|$ as a function of ξ within a density-wave description of the BCC–HCP transition.

5.3.3. *Periodic dependence of the order-parameter magnitude*

The lattice function ψ of a unit cell, associated with a general shift of magnitude $\pm\xi$ in the direction [110], can be written in the (x, y) plane as

$$\psi(\xi) = A\Bigg[\exp\left(\frac{2\pi i}{a}\cdot\frac{a}{4}\right)\cdot\exp\frac{4\pi i}{a\sqrt{2}}\xi + \exp\left(-\frac{2\pi i}{a}\cdot\frac{a}{4}\right)\cdot\exp\left(-\frac{4\pi i}{a\sqrt{2}}\xi\right) + \text{c.c.}\Bigg], \quad (5.21)$$

where the origin of the coordinates has been chosen at middle distance between the two atoms shifted in antiparallel directions. In the BCC structure, $\xi = 0$ and $\psi_{\text{BCC}}(\xi) = 0$. Thus, the increment $\delta\psi(\xi)$, corresponding to the variation of $\psi(\xi)$ from BCC to the HCP structure, has the same form (5.21), which can be simplified, giving the following dependence of the order-parameter magnitude, in function of the critical shifts:

$$|\boldsymbol{\eta}(\xi)| = \eta_0 \cdot \sin\frac{4\pi}{a\sqrt{2}}\xi. \quad (5.22)$$

The periodic dependence on ξ of $|\boldsymbol{\eta}|$ is shown in Fig. 5.9(a). Starting from the initial $Im\bar{3}m$ structure, which is realized for $\xi = 0, 1/2, 1, \ldots$, (in $a\sqrt{2}$ unit), one gets a two-layer orthorhombic

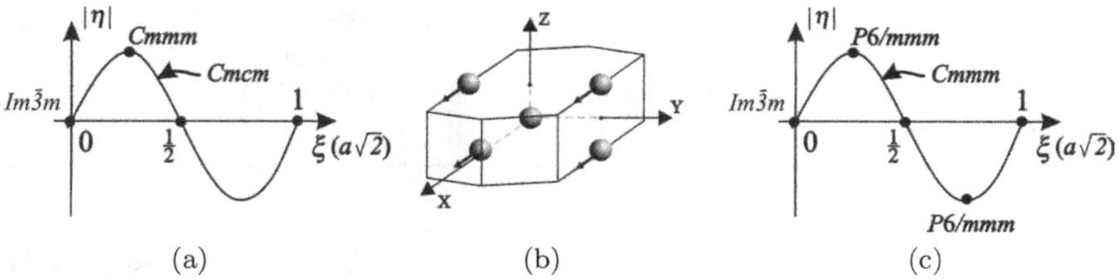

Fig. 5.9. (a) Periodic dependence of the order-parameter following Eq. (5.22). (b) Holohedry of the orthorhombic lattice in the direction of the shifts. (c) Periodic dependence of the order-parameter at the BCC–HCP transition, when the preservation of the interlayer distance is taken into account.

structure of space group $Cmcm$ ($Z = 2$) for general shifts, whereas the extrema of the curve correspond to a *monolayer* orthorhombic structure of *symmorphic* space group $Cmmm$ ($Z = 1$). This latter structure has the maximal symmetry compatible with the holohedry of the orthorhombic lattice, in the direction of the shifts, as shown in Fig. 5.9(b), which also reveals that the HCP symmetry ($P6_3/mmc$) *does not coincide* with any special point of the preceding holohedry. The realization of the HCP structure is, as shown in the preceding section, the result of a complex secondary process in which *two physical constraints are taken into account.*

The first constraint is *the preservation of the interlayer distance between the (110) cubic planes in the HCP structure*, where they become the $(001)_h$ planes. This is a general property of the crystals of elements undergoing a BCC–HCP transformation, which are listed in Table 5.3. This table shows that, except for Ba and ^{3}He (which will be discussed in what follows), the variation of the interlayer distance is smaller than 3%, when going from one structure to the other. When this constraint is satisfied, the orthorhombic ($Cmmm$) symmetry *increases* to hexagonal ($P6/mmm, Z = 1$) as represented in Fig. 5.9(c).

In order to obtain the $P6_3/mmc$ structure, another constraint should be imposed, which consists in *considering the actual size of the atoms*, i.e., to exclude from the theoretical description of the transition process, shifts of atoms after they have entered into contact. Two types of contacts can be distinguished: the ones which occur, e.g., between pairs of atoms denoted (3,4), (5,6) or (1,8) in Fig. 5.8, which result from the primary (antiparallel shifting) mechanism, and

Table 5.3. Interlayer distances d (in Å) between the cubic (110) planes (column 3) and hexagonal (001) planes (column 5) for the elements listed in the first column. $(d_c - d_h)/d_c$ is the relative variation of d when going from the BCC to the HCP structure.

	HCP		BCC		
Elements	c_h	d_h	a_c	d_c	$(d_c - d_h)/d_c(\%)$
Ti	5.732	2.37	3.330	2.36	−0.42
Zr	5.150	2.58	3.620	2.56	−0.78
Hf	5.470	2.74	3.610	2.55	−7.45
Be	3.626	1.81	2.550	1.80	−0.56
Ca	6.440	3.22	5.380	3.10	−3.87
Ba	6.154	3.08	5.050	3.57	13.73
Fe	5.370	2.19	2.870	2.03	−7.88
Ti	5.530	2.77	3.882	2.75	−0.73
^{3}He	5.986	2.49	3.963	2.80	11.07
^{4}He	5.798	2.90	5.110	2.91	0.34
Gd	5.700	2.85	5.060	2.87	0.7
Tb	5.6936	2.85	5.020	2.84	−0.35
Dy	5.6475	2.82	3.980	2.81	−0.36
Yb	6.3859	3.19	5.440	3.14	−1.59
Y	5.7306	2.87	5.110	2.91	1.38

the ones which take place in the perpendicular direction, such as between the pairs (4, 8), (1, 3) or (6, 7) in Fig. 5.8. These latter contacts are the result of the secondary mechanism assumed in our model, namely of the compression (or decompression) along the four-fold axes, and are equivalent to collective atomic shifts, which are, for instance, along the Z axis: $\Delta z = |\boldsymbol{a}_c - \boldsymbol{c}_h|/2$, or in terms of the atomic radius: $\Delta z = (4R/\sqrt{3} - 2R)/2$.

Figure 5.10(a) illustrates the role of the contacts between the atomic spheres, which is *to create a geometrical barrier*, acting as a *mirror plane*, perpendicular to the direction of the shifts. *This barrier modifies the "map" of the special points of the lattice at which the symmetry of the structure may increase.* Figure 5.10(b) shows that it produces a reduction by 3/2 of the periodicity of the shifts, so that the unit length $a\sqrt{2}$ in Fig. 5.9(a) becomes $2a\sqrt{2}/3$. Consequently, the exponential terms $\exp\{\pm(4\pi i/a\sqrt{2}) \cdot \xi\}$ in Eq. (5.21) become $\exp\{\pm(6\pi i/a\sqrt{2}) \cdot \xi\}$, and the magnitude of the order-parameter

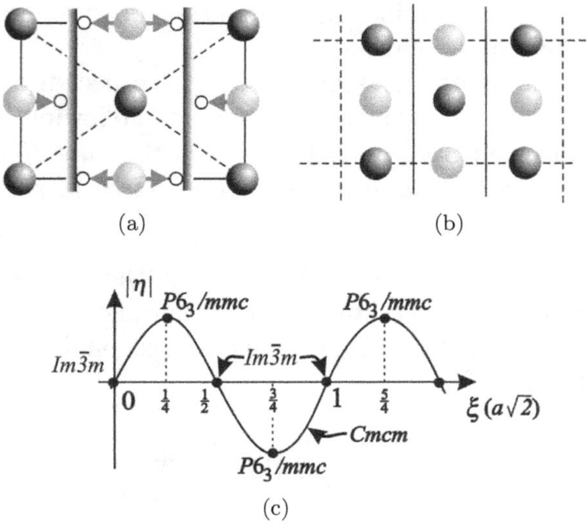

Fig. 5.10. (a) Creation of a geometrical barrier due to contacts between atoms. (b) Subsequent reduction of the periodicity of the shifts. (c) Periodicity of the order-parameter at the BCC–HCP transformation, following Eq. (5.23).

$|\boldsymbol{\eta}(\xi)|$ takes the form

$$|\boldsymbol{\eta}(\xi)| = \eta_0 \cdot \sin\frac{6\pi}{a\sqrt{2}}\xi, \tag{5.23}$$

represented in Fig. 5.10(c). In this figure, one gets periodically the initial BCC structure (for $\xi = 0, 1/2, 1, \ldots$, in $a\sqrt{2}/3$ units) and the HCP structure for $\xi = 1/4, 3/4, 5/4, \ldots$, in agreement with our crystallographic description.

5.3.4. *Alternative description of the BCC–HCP transition: The HCP parent phase*

The functional dependence $|\boldsymbol{\eta}(\xi)|$ represented in Fig. 5.10(c) deserves two observations:

(1) It has been obtained by modifying the periodicity associated with the "primary" antiparallel shifting mechanism, in order to account for necessary constraints: the preservation of the interlayer distance between definite families of planes, and the "hard-sphere" character of the atoms. These constraints are taken into

account by the secondary mechanism assumed in our reformulation of Burgers mechanism.

(2) The sinusoidal curve is *symmetric* with respect to the displacement axis *for the* $P6_3/mmc$ *structure.* Following the general scheme given in Sec. 4.6.2, this means that *the HCP structure may also play the role of the initial structure in the BCC–HCP transition.*

In order to verify this point, let us write the lattice function ψ, starting from the HCP structure. The antiparallel shifts, which lead to the BCC structure, take place along the hexagonal $[1\bar{1}0]$ directions. For an arbitrary shift $\pm\xi$, one has

$$\psi(\xi) = A\left\{\exp\left[\frac{4\pi i}{a\sqrt{3}}\left(-\frac{a\sqrt{3}}{12}+\xi\right)\right]\right.$$
$$\left.+\exp\left[\frac{4\pi i}{a\sqrt{3}}\left(\frac{a\sqrt{3}}{12}-\xi\right)\right]+\text{c.c.}\right\}. \qquad (5.24)$$

In the initial (HCP) phase: $\xi = 0$ and $\psi_{\text{HCP}} = 1$. The increment $\delta\psi(\xi) = \psi_{\text{HCP}} - \psi(\xi)$, which is proportional to the order-parameter magnitude, is thus:

$$\delta\psi = 2A\cdot\left[\sin\left(\frac{4\pi}{a\sqrt{3}}+\frac{\pi}{6}\right)-\frac{1}{2}\right]. \qquad (5.25)$$

The functional dependence (5.25), given in Fig. 5.11(a), shows an asymmetric behavior with respect to the ξ axis, as the maxima of the curve correspond to the tetragonal structure $I4/mmm\,(Z=1)$ (after taking into account the macroscopic deformation), and the minima, to the simple hexagonal structure of symmetry $P6/mmm\,(Z=1)$. The body-centered tetragonal symmetry increases to BCC when applying a decompression along the [110] hexagonal direction, which coincides with the Z axis. This *secondary* mechanism increases the angle between the two-fold axes of the hexagonal structure from $\pi/3$ to 0.392π. Let us stress that the $Im\bar{3}m$ space-group *figures among the symmetries reached for special points* of the $Cmcm$ holohedry, as shown in Fig. 5.11(b), where one can see that it coincides with specific values of the shifts along the axes of coordinates.

Two clarifying remarks have also to be made on the curve of Fig. 5.11(a). (1) The curve has been obtained by considering the

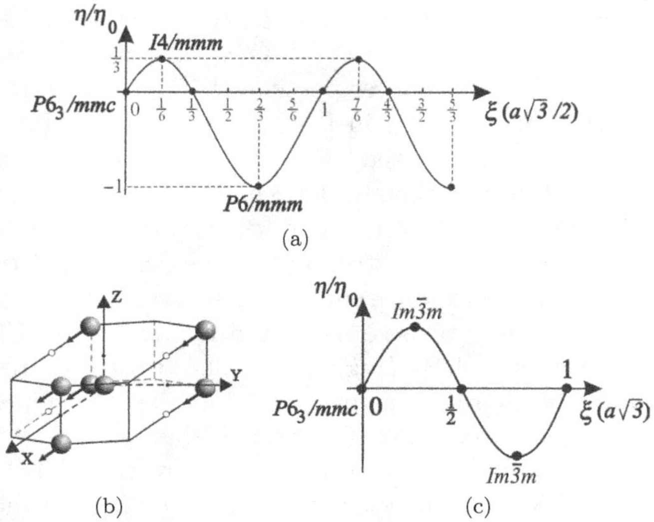

Fig. 5.11. (a) Functional dependence of the HCP–BCC mechanism following Eq. (5.25). (b) Holohedry corresponding to *Cmcm*. Open circles indicate the special points at which the orthorhombic symmetry increases to cubic. (c) Periodic dependence of the HCP–BCC order-parameter, following Eq. (5.26).

atoms as located at *points* of the lattice. This leads for the phase of $P6/mmm$, to a configuration in which the atoms of one layer are exactly over the atoms of the next layer, i.e., a situation which implies a considerable increase of the interlayer distance, or a noticeable deformation of the atoms. *As neither happens* in the close-packed model under consideration, one has again to introduce the geometrical constraints imposed by the size of the atoms and by the invariability of the interlayer distance when going from one structure to another. In particular, one can see in Fig. 5.6(b) that the realistic path followed by the atoms starting from the HCP positions is to be shifted along the directions $[120]_{\mathrm{HCP}}$, $[210]_{\mathrm{HCP}}$ or $[110]_{\mathrm{HCP}}$, i.e., to pass through valleys between touching atoms. *This excludes the phases of symmetry $P6/mmm$ to be formed*, and yields a simplified dependence of the order-parameter modulus, as compared to Eq. (5.25), which is

$$\eta = \eta_0 \cdot \sin \frac{8\pi}{3a\sqrt{3}}\xi. \qquad (5.26)$$

The corresponding curve given in Fig. 5.11(c) expresses a more realistic situation than the curve of Fig. 5.11(a). In particular, it is *symmetric* with respect to the other possible (BCC) initial structure.

(2) The two curves shown in Figs. 5.11(a) and 5.11(c), and the better compatibility of an initial HCP structure with the symmetry of the holohedry of the orthorhombic lattice, represented in Fig. 5.11(b), suggest that the HCP structure may constitute a more "natural" parent phase than the BCC structure in the HCP–BCC transition mechanism. In the phase diagram of ^{3}He represented schematically in Fig. 5.4(e), the HCP phase appears indeed *above* the BCC phase. In the phase diagram of Ba (Fig. 5.4(f)), the two structures appear at approximately the same temperature, but the HCP phase is stable at higher pressure. Table 5.3 reveals that for these two elements the interlayer distance varies from more than 10% between the HCP and BCC structures, denoting a "softening" of the atomic spheres.

The preceding facts suggest that, when the hard-sphere model does not strictly apply (as for ^{3}He or Ba), the HCP structure is preferably the *initial* structure, while on the contrary, when the hard-sphere assumption is more strictly verified, the BCC phase is the natural initial structure. This interpretation is consistent with the current scheme depicting the filling of space by atoms: with increasing temperature, the Debye–Waller factors increase and thicken the "soft" part of the electronic shell, favoring an interpenetration of the atomic spheres. In this case, a filling of space corresponding to the BCC configuration is sufficient, as the atoms are already interpenetrated. At lower temperature, the possibility of interpenetration decreases for atoms, and the filling of space follows more strictly the HCP close-packing configuration. In ^{3}He and Ba, which are "soft" atoms, the temperature has less influence on the filling of space, and the HCP structure recovers its role of parent structure.

5.3.5. *Thermodynamics of the BCC–HCP transition: Improper symmetry-breaking strains*

The dependences on critical shifts of the order-parameter magnitudes associated with the BCC–HCP and HCP–BCC transitions, as given by Eqs. (5.23) and (5.26), respectively, are *symmetric* with respect to the displacement axis, and thus correspond both to an effective

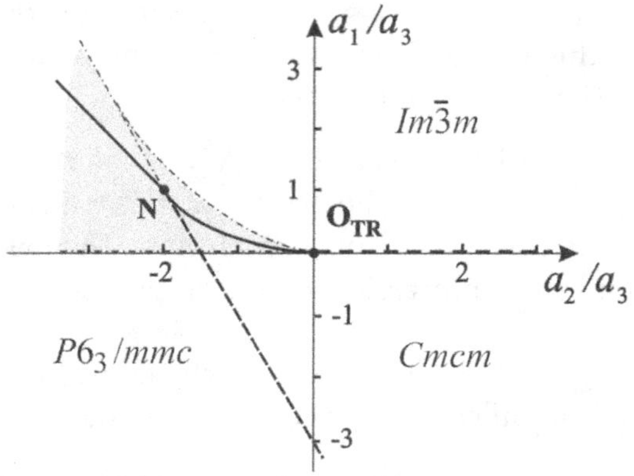

Fig. 5.12. Phase diagram containing the BCC and HCP phases, following Eq. (5.27). Full, dashed and dashed-dotted lines are, respectively, the first-order, second-order transitions and limit of stability lines. N is a three-phase point. O_{TR} is a tricritical point.

order-parameter expansion of the form

$$F(|\boldsymbol{\eta}(\xi)|) = F_0(P,T) + \frac{\alpha}{2}|\boldsymbol{\eta}(\xi)|^2 + \frac{\beta}{4}|\boldsymbol{\eta}(\xi)|^4 + \frac{\gamma}{6}|\boldsymbol{\eta}(\xi)|^6, \qquad (5.27)$$

which has to be expanded at least up to the sixth degree, in order to account for a region of stability including a direct BCC–HCP transition line. Equation (5.27) is identical to the expansion (4.9) discussed in Chapter 4, and thus the topology of the corresponding phase diagram, represented in Fig. 5.12, is identical to the phase diagram of Fig. 4.4 in Chapter 4. The thermodynamic path followed for the elements listed in Table 5.3 is far from the region of stability of the orthorhombic phase, which among elements is found, e.g., in uranium (see Sec. 5.6).

Expansion (5.27) reflects the symmetry of the primary order-parameter associated with the BCC–HCP mechanism. For a full phenomenological description of the transition, one needs to add to Eq. (5.27) a coupling term which accounts for the macroscopic deformation involved in the transition mechanism. The strain components which couple to the primary order-parameter can be found by decomposing the symmetrized *second* power of $\tau_4(\boldsymbol{k}_9)$ into the irreducible representations of the point-group $m3m$. One finds

$$[\tau_4(\boldsymbol{k}_9)]^2 \cong \mathrm{A_{1g}} + \mathrm{E_g} + \mathrm{F_{2g}}, \qquad (5.28)$$

where A_{1g}, E_g and F_{2g} are, respectively, the one, two, and three-dimensional irreducible representations of $m3m$, spanned by the following combinations of the strain-tensor:

(i) $e_{xx}+e_{yy}+e_{zz}$, which transforms as A_{1g}, is the rigidity, expressing the change in volume of the crystal at the transition.
(ii) (ζ_1, ζ_2) defined by Eq. (5.19), which have the symmetry of E_g, account for the compression or decompression along the four-fold axes.
(iii) $(e_{xyz}, e_{xz}, e_{xy})$, which span F_{2g}, are the shear strains describing an eventual modification of the interlayer distance.

In the BCC–HCP mechanism, one can neglect, in a first approximation, the change in volume and the variation of the interlayer distance. Besides, the equilibrium value of the tetragonal deformation of the unit cell is $\zeta_1 = \zeta \neq 0$, $\zeta_2 = 0$ (see Sec. 5.4). Accordingly, the effective coupling term is

$$F(\eta, \zeta) = \delta \cdot |\boldsymbol{\eta}(\xi)|^2 \cdot \zeta, \tag{5.29}$$

where δ is a negative coupling constant.

Let us stress that, although the form of Eq. (5.29) reflects the "improper" character of the secondary strain ζ, the role of ζ, as described in our mechanism, results in an *increase* (to $P6_3/mmc$) of the orthorhombic symmetry $I4/mmm$, at least when considering an initial BCC structure, and in the stabilization of the HCP phase. Thus, in contrast to the property of the secondary order-parameters in the second-order or weakly first-order structural transitions (see Chapter 1, Sec. 1.2.4) to be non-symmetry-breaking quantities, *the secondary strains in reconstructive transitions can be associated to a breaking of symmetry* (or more exactly to an "increase" in symmetry). In other words, they play the role of a distinct order-parameter, though they are induced by the primary mechanism. This apparent contradiction is resolved by the following observations:

(1) The coupling term (5.29) expresses a connection between the strains and the critical atomic shifts ζ. Thus, sufficiently large values of ζ contribute to the large specific values of ζ, which correspond to an increase of the orthorhombic symmetry.
(2) As shown in Table 5.4, the magnitude of the strains involved in the BCC–HCP transition in elements is generally larger than

Table 5.4. Magnitude of the spontaneous macroscopic strain e_{zz}, involved in the BCC–HCP transition, for the elements listed in column (a). Note the large values of e_{zz} found for ^{3}He and Ba.

Elements	HCP a_h (Å)	BCC a_c (Å)	e_{zz}
Ti	2.976	3.33	0.11
Zr	3.24	3.62	0.11
Hf	3.33	3.61	0.08
Be	2.291	2.55	0.1
Ca	3.94	5.38	0.1
Ba	3.901	5.05	0.23
Fe	2.705	2.87	0.06
Ti	3.48	3.882	0.1
^{3}He	3.046	3.963	0.23
^{4}He	3.555	5.11	0.14
Gd	3.629	5.06	0.11
Tb	3.601	5.02	0.1
Dy	3.5903	3.98	0.1
Yb	3.8799	5.44	0.13
Y	3.6474	5.11	0.11

10%, i.e., one order of magnitude larger than the strains usually involved in ferroelastic transitions [22].

The primary instability, which transforms as $\tau_4(\boldsymbol{k}_9)$, coincides with an optical phonon mode at the N-point of the BCC Brillouin zone. Except for Li and Na (see Sec. 5.5), no detectable temperature-dependent softening of this mode has been observed in crystals of the elements listed in Table 5.4, confirming, on the one hand, the absence of intrinsic softening predicted in our approach for direct reconstructive transitions to limit state, and on the other hand, that the transition occurs far from the region of stability of the orthorhombic phase. By contrast, a dip in the dispersion curve at, or close to, the N-point was clearly observed in Ti, Zr, Hf, Na, Ba and Fe [23], and also in the BCC phase of some other elements, such as Mo, Cr, Nb, W, K or Cs [23] in which the HCP phase has not been recognized till now.

The secondary strain ζ corresponds to the acoustical transverse mode Σ_4. It may have two types of dynamical effects: a *direct one*, such as a change in the slope of the Σ_4 mode with temperature (the softening of the $c_{11} - c_{12}$ elastic constant), or an *indirect effect* connected with the coupling to the primary instability that would result in a shift of the dip at the N-point observed in the dispersion phonon branch [0.5 0.5 0], related to the change of the lattice parameters below the transition.

At last, among the other properties deriving from our approach to the BCC–HCP transition, let us briefly describe the corresponding domain structure. Starting from a BCC structure, one should observe *six orientational domains* in the HCP phase, transforming into each other by the three-fold cubic axes lost at the transition, and two types of antiphase domains, corresponding to the loss of the cubic a_c and b_c lattice parameters. In the reversed HCP–BCC transition, no antiphase domains should be found in the cubic phase, as the transition is induced by a zone-center instability, but three orientational domains should be created, which can be deduced one from another by the six-fold rotations lost at the transition.

5.4. The BCC–FCC Transition

5.4.1. *The Bain deformation*

The BCC–FCC transition is almost as frequent, among elements, as the BCC–HCP transition. It is found in about 20 elements, in the phase diagram of which the BCC phase appears generally at high temperatures, below the melt. The stability of the FCC structure, which is closer packed than the BCC structure, increases with pressure, and in the phase diagram of K, Cs and Rb, FCC stability region appears only at high pressures, at approximately the same temperature than the BCC phase. Figures 5.13(a)–5.13(c), show the main types of phase diagrams of the elements in which the two phases are inserted. Like the BCC–HCP transition, the BCC–FCC transition is always found to be strongly first-order, and associated with large atomic displacements, of the order of the unit-cell parameters.

One can visualize the way a BCC lattice transforms into an FCC lattice via a deformation which stretches the cubic unit cell by the same extent along one of the four-fold axes (e.g., [001]) and compresses it by the same extent along the other four-fold axes. Such a macroscopic deformation, which is represented in Figs. 5.14(a) and

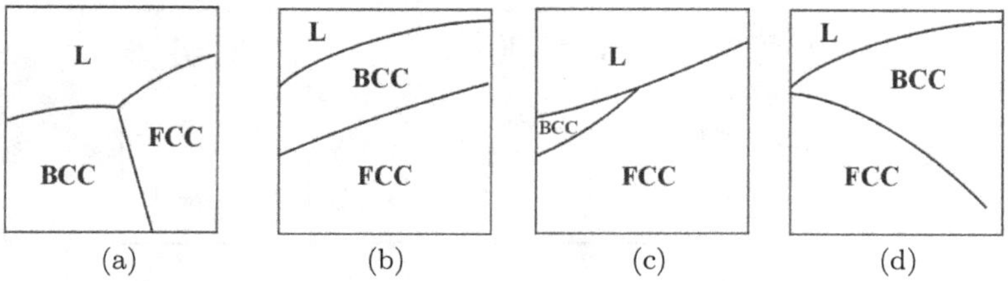

Fig. 5.13. Topology of the BCC–FCC transition line in some elemental crystals. The represented phase diagrams are found, e.g., in Rb (a), Mn (b), La (c) and Yb (d).

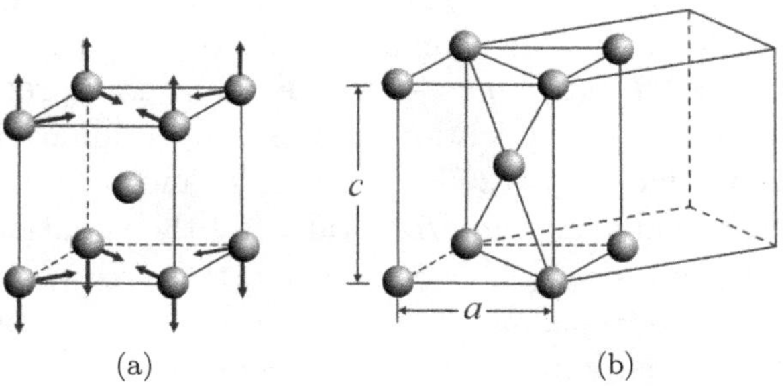

Fig. 5.14. Macroscopic deformation representing the Bain deformation: (a) corresponding atomic displacements; (b) connection between the BCC and FCC unit cells.

5.14(b), is called the *Bain deformation* [24]. Figure 5.14(b) shows the connection between the BCC and FCC ($Fm\overline{3}m$, $Z = 1$) unit cells, which yields the following orientation relationship between the corresponding, ($\boldsymbol{a}_c$, $\boldsymbol{b}_c$, $\boldsymbol{c}_c$) and ($\boldsymbol{a}_1^{\mathrm{F}}$, $\boldsymbol{a}_2^{\mathrm{F}}$, $\boldsymbol{a}_3^{\mathrm{F}}$), lattice parameters:

$$\boldsymbol{a}_1^{\mathrm{F}} \| \boldsymbol{a}_c, \boldsymbol{a}_2^{\mathrm{F}} \| (\boldsymbol{a}_c + \boldsymbol{b}_c + \boldsymbol{c}_c), \boldsymbol{a}_3^{\mathrm{F}} \| (\boldsymbol{a}_c + \boldsymbol{c}_c). \tag{5.30}$$

This translational relationship is associated with a wave-vector located at the center (Γ-point) of the BCC Brillouin zone. The spontaneous strains, which account for the Bain deformation, are the same combinations of the strain-tensor components, ζ_1 and ζ_2 given by Eq. (5.19), which play the role of a secondary deformation in Burgers mechanism. The corresponding two-dimensional irreducible representation $\mathrm{E_g}$ ($\tau_3(\boldsymbol{k}_{11})$ in Kovalev's notation [8]) has the following generating matrices:

$\{2_x, 2_y, \bar{1}\|0\}$	$\{2_{xy}\|0\}$	$\{3^{xyz}\|0\}$
$\begin{bmatrix}1 & 0\\0 & 1\end{bmatrix}$	$\begin{bmatrix}1 & 0\\0 & -1\end{bmatrix}$	$\frac{1}{2}\begin{bmatrix}-1 & -\sqrt{3}\\\sqrt{3} & -1\end{bmatrix}$

which lead to the following, possibly stable, low-symmetry groups: $I4/mmm\,(Z = 1)$ for $\zeta_1 = \zeta \neq 0$, $\zeta_2 = 0\,(\text{or}\,\zeta_2 = \zeta \neq 0,\ \zeta_1 = 0)$, and $Immm$ $(Z = 1)$ for $\zeta_1 \neq \zeta_2 \neq 0$. When, for the first solution, one has the specific ratio $c/a = \sqrt{2}$, where a and c are the tetragonal lattice parameters, the tetragonal space-group $I4/mmm$ *identifies as the face-centered cubic space-group* $Fm\bar{3}m$. The preceding ratio corresponds to the multiplication of the conventional BCC lattice parameters, by a factor $\sqrt{3}/2$ along X and Y, and by a factor $\sqrt{(3/2)}$ along Z. Accordingly the numerical values of the spontaneous strain components are $e_{xx} = e_{yy} = \sqrt{3}/2 - 1$ and $e_{zz} = \sqrt{(3/2)} - 1$, which give $\zeta = 1 - 1/\sqrt{2} \cong 0.293$. Besides, in order to keep the atoms in contact at the transition, one has to consider as a secondary order-parameter the rigidity $\delta = (e_{xx} + e_{yy} + e_{zz})/\sqrt{3}$, which measures the decrease in volume at the transition. One finds numerically: $\delta = 1 + 1/\sqrt{2} - \sqrt{3} \cong -0.025$.

Like the BCC–HCP transition, the Bain deformation has a *periodic character*, which can be seen by considering the angle θ between the diagonals of the (110) cubic planes, shown in Fig. 5.15(a). Thus, in the BCC lattice, $\theta \cong 70°32' \cong 0.392\pi$. For $\theta = 90° = \pi/2$,

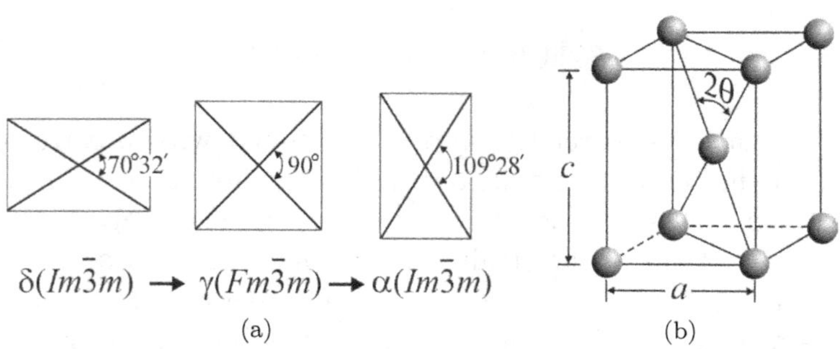

Fig. 5.15. (a) Periodic character of the Bain deformation. (b) Angle θ between the diagonals of the deformed cube, used as a variational parameter in Eq. (5.31).

the (110) BCC planes transform into the (010) FCC planes. Further increase of the angle θ up to $\theta \cong 109°28' \cong (1-0.392)\pi$ brings about the crystal again to the BCC symmetry, but with a different orientation, turned by $\pi/2$ with respect to the initial BCC structure. For other arbitrary values of θ, the lattice remains tetragonal (Fig. 5.15(b)). One can calculate the dependence of $\zeta(\theta) = 2[e_{xx}(\theta) - e_{zz}(\theta)]/\sqrt{6}$ by expressing e_{xx} and e_{zz} *independently* in function of the angle θ. One finds, using Fig. 5.15(b):

$$e_{xx} = \sqrt{\frac{3}{2}}\sin[2(\theta_0 + \Delta\theta)] - 1, e_{zz} = \sqrt{3}\cos[2(\theta_0 + \Delta\theta)] - 1, \quad (5.31)$$

where $\theta_0 = \frac{1}{2}\sin^{-1}\sqrt{3}/2 = \frac{1}{2}\cos^{-1}\sqrt{1}/3\sin^{-1}\sqrt{3}/2 \cong 27°22'$ is one-fourth of the large angle $[109°28' \cong (1-0.392)\pi]$ between the diagonals in the BCC unit cell, and $\Delta\theta$ is the variational angle. Figures 5.16(a) and 5.16(b) show the variations of e_{xx} and e_{zz} in function of $\Delta\theta$. One can see that *three* structures appear periodically for definite values of $\Delta\theta$: the BCC structure for $e_{xx} = e_{zz} = 0$, the FCC structure for maximal e_{xx} and e_{zz}, and a planar hexagonal ($P6/mmm$) structure. This latter unphysical atomic configuration results from the fact that the variation laws (5.31) do not hold for any value of θ (or $\Delta\theta$), as this parameter represents an absolute measure of the angle between the diagonals, and is thus restricted by the following constraints:

(1) The arguments of $e_{xx}(\Delta\theta)$ and $e_{zz}(\Delta\theta)$ should be limited, as they represent a macroscopic deformation of the crystal.
(2) After reaching the FCC close-packed structure, a further deformation of the unit cell along the $[100]_{\mathrm{BCC}} \| [100]_{\mathrm{FCC}}$ directions

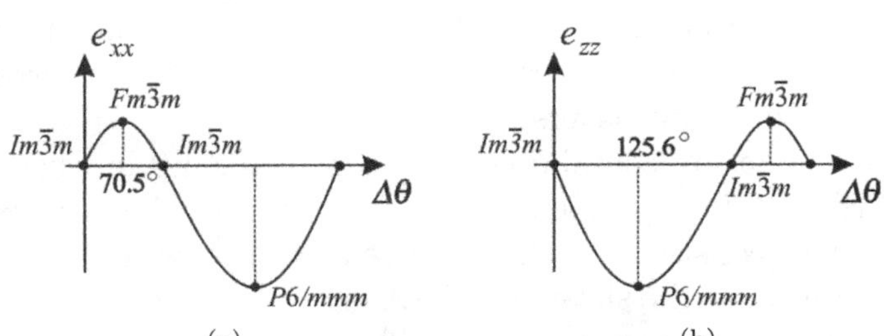

Fig. 5.16. Dependence of e_{xx} and e_{zz} on the variation of the diagonal angle $\Delta\theta$ at the BCC–FCC transition.

is impossible, at least if one assumes a hard-sphere model; as in the preceding directions, close-packed chains of atoms are formed with atoms in contact with each other.

In order to remove the preceding restrictions, *one should express the order-parameter as a function of the average displacements of the atoms.* This brings us to analyze the atomistic mechanism which underlies the Bain deformation.

5.4.2. *The Bain deformation revised*

In the shearing deformation model, currently proposed for the BCC–FCC transition [25] as equivalent to the Bain deformation mechanism, the structural stability of the BCC structure is lowered with respect to a homogeneous shear strain of the (011) planes in the $\pm[0\bar{1}1]$ directions, which gives rise to an FCC structure. An experimental support to this interpretation is the decrease of the shear modulus $c_{11}-c_{12}$ observed in a number of BCC metals when approaching the FCC phase. We can propose the following path, in reference to Figs. 5.17(a) and 5.17(b), which materializes at the atomistic level the preceding macroscopic shearing mechanism. (1) The layer containing the atoms labeled 7–12 is shifted by $a\sqrt{2}/6$ (where a is the cubic lattice parameter) in the [011] BCC direction, the atom in 0 position reaching the center of a triangle formed by the atoms denoted 8, 9 and 11; (2) the next layer, formed by atoms 1–6 will be subsequently shifted by $a\sqrt{2}/3$. This will place each atom initially in 0 position, at the center of the corresponding triangles of the following layer; (3) at last, a shifting of the third layer by $a\sqrt{2}/2$ will force the atoms located in this layer to occupy equivalent positions, i.e., the corner atoms (0, 0) will lie on a straight line passing by the centers of the triangles (8, 9, 11) and (2, 4, 5) and orthogonal to the three successive layers.

Hence, the layers containing the atoms labeled 0 in Figs. 5.17(a) and 5.17(b) will be equivalent crystallographically, i.e., the resulting structure will consist of a three-layered close-packed structure (an FCC structure), as the shears will have eliminated the repulsive effect of one layer with respect to the neighboring ones.

Let us stress that the above microscopic description of the Bain deformation is closely analogous to Burgers mechanism, as both

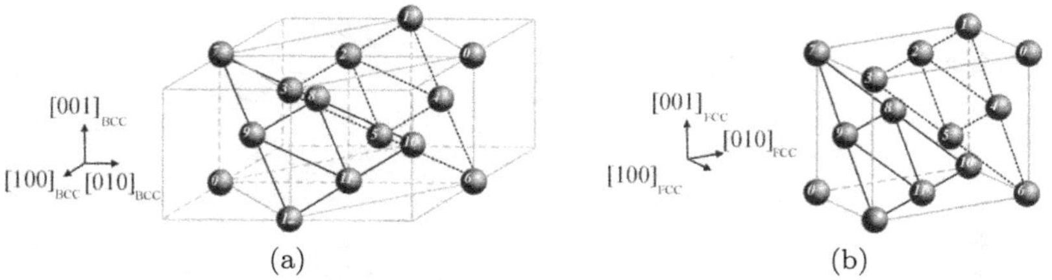

Fig. 5.17. Unit cell of (a) the BCC and (b) the FCC structures. The numbering of the atoms used in the description of the BCC–FCC transition mechanism is given in the text.

mechanisms preserve the interlayer distance, and thus the repulsive effect of one layer with respect to the others. The essential difference is that, in the BCC–HCP transition, two successive layers are assumed to shift in opposite directions, i.e., the displacements are associated with a soft mode at the surface of the BCC Brillouin zone, whereas for the BCC–FCC transition, *two successive layers are sheared in the same direction*, and thus correspond to a zone-center soft mode.

The periodicity of the atomic displacements associated with the BCC–FCC transition is shown in Fig. 5.18(a). The mechanism can be depicted in terms of antiparallel shifts of the $(\bar{1}10)$ atomic planes in the $\pm[110]$ directions. A zero shift is chosen at middle distance between two arbitrary planes, and the successive planes correspond to increasing shifts ξ_n enumerated $1, 2, 3, \ldots, n$. The shifts of the atoms in the nth plane are

$$\xi_n = \frac{c}{4}(2n+1) \cdot \tan\alpha, \tag{5.32}$$

where α is the angle between the Z axis and the direction formed by the shifted atoms (Fig. 5.18(a)), which characterizes the macroscopic deformation.

The lattice function expressing the structure for a shift ξ is

$$\psi(\xi) = A\left\{\exp\left(\frac{2\pi i}{c}\cdot\frac{c}{4}\right)\cdot\exp\left(\frac{4\pi i}{\tilde{a}}\xi\right) + \exp\left(-\frac{2\pi i}{c}\cdot\frac{c}{4}\right)\right.$$
$$\left.\cdot\exp\left(-\frac{4\pi i}{\tilde{a}}\xi\right) + \text{c.c.}\right\}, \tag{5.33}$$

with $\tilde{a} = (2a\sqrt{2})/3$. For $\xi = 0$ (the BCC structure) $\psi(0) = 0$, and the increment $\delta\psi(\xi)$ associated with the BCC–FCC transition can be

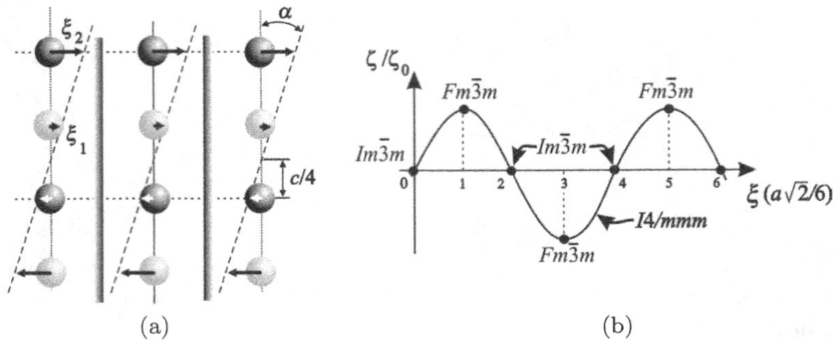

Fig. 5.18. (a) Periodicity of the antiparallel shifting mechanism at the BCC–FCC transition. (b) Periodic dependence of the order-parameter following Eq. (5.34).

simply expressed as

$$\delta\psi(\xi) = A \cdot \sin \frac{3\pi}{a\sqrt{2}}\xi. \tag{5.34}$$

The corresponding dependence of the order-parameter is shown in Fig. 5.18(b). It expresses a physically realistic scheme, as compared to Fig. 5.16(a) or 5.16(b), and holds for non-limited magnitude of ξ, as large antiparallel shifts are compatible with a limited macroscopic deformation and respect the constraints imposed by the close packing of atoms. In particular, the "layer" $P6/mmm$ phase found in the preceding curves is absent from Fig. 5.18(b).

5.4.3. *Phase diagram associated with the BCC–FCC transition*

In the sinusoidal curve shown in Fig. 5.18(b), the FCC phase is symmetric with respect to the ξ-axis. In our approach, this fact is interpreted as the possibility for the FCC structure to be the initial structure in the BCC–FCC transition. In the phase diagrams of elements [17], one effectively finds the FCC phase at higher temperatures than the BCC phase in ^{3}He (Fig. 5.13(c)). In iron (Fig. 5.13(a)), the FCC structure is located between two BCC (δ and α) phases [17], providing a striking confirmation of the interplay between the BCC and FCC structures shown in Fig. 5.18(b). In K, Cs and Rb, the BCC and FCC phases appear at approximately the same temperature, but

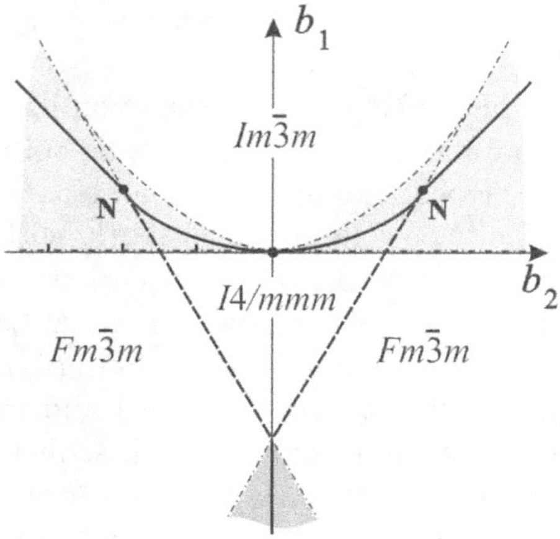

Fig. 5.19. Phase diagram associated with the BCC–FCC transition and with the potential (5.35). The symmetric character of the phase diagram results from a folding with respect to the a_1 axis of the right-hand part of the phase diagram of Fig. 5.3. Full, dashed and dashed-dotted lines are, respectively, the first- and second-order transition lines, and limit of stability lines. N are three-phase points. The two symmetric $Fm\bar{3}m$ regions correspond to anti-isostructural phases.

the region of stability of the FCC phase extends at higher pressures (Fig. 5.13(d)).

The form of the order-parameter expansion associated with the BCC–FCC transition is determined by the symmetry of the order-parameter (ξ_1, ξ_2) and by the equilibrium condition ($\xi_1 \neq 0$ or $\xi_2 \neq 0$), which gives the orientation of the vector η in the ε_2-space, showing that the FCC phase may correspond to two possible anti-isostructural configurations. It yields

$$F[T, P, \zeta(\xi)] = F_0(T, P) + b_1\zeta(\xi)^2 + b_2\zeta(\xi)^3 + b_3\zeta(\xi)^4, \qquad (5.35)$$

with $|\zeta(\xi)| = \zeta_0 \cdot \sin\frac{3\pi i}{a\sqrt{2}}\xi$. This *symmetric form* of ζ does not contradict the existence of a *cubic invariant* in Eq. (5.35), since it results from the limitations imposed on the dissymmetrical functions, represented in Figs. 5.16(a) and 5.16(b), in order to take into account the physical constraints inherent in the Bain deformation. Thus, the phase diagram of Fig. 5.19, which corresponds to Eq. (5.35), is obtained from the phase diagram of Fig. 5.3, by keeping only the right-hand side of the phase diagram of Fig. 5.3 (the physically acceptable part) and by a subsequent reflection with respect to the

a_1 "plane". It finally gives a *symmetric* topology with respect to the coefficient b_1 in Eq. (5.35).

Our description of the BCC–FCC transition illustrates the property of the order-parameter periodicity, to be fully expressed only in function of the critical atomic shifts, and not in function of a macroscopic variational parameter, consistently with the microscopic nature of the density-wave approach. Verifications of this description can be found: (1) in the existence of the tetragonal Landau intermediate state, which is found in a number of alloys ($I4/mmm$ is the symmetry of "martensite" in carbon steels [18]) and in some elements. For example, one has in plutonium the full sequence of structures: $\varepsilon(Im\bar{3}m) - \delta'(I4/mmm) - \delta(Fm\bar{3}m)$. The same sequence is also found in protactinium [17]. Three tetragonal domains are usually observed in the tetragonal structure, which should be also found in the FCC structure or in the BCC structure when coming for the FCC phase.

5.4.4. *The BCC–FCC phase diagram*

Our description of the BCC–HCP and BCC–FCC transitions shows that the HCP and BCC structures may both constitute the *initial* structure in the framework of our approach. Such a possibility disappears when the FCC and HCP phases coexist in the same phase diagram, far from the eventual region of stability of the BCC phase. This situation is found, e.g., for some elements, such as He, Co, Pb, Si, La, Ce, Pr, Nd or in compositions such as ZnS or SiC [2,17]. In this case, the FCC–HCP transition loses its displacive nature, in relation with the close packing of the two structures, and acquires an *ordering* character, the mechanism of which is described in Chapter 6.

The situation is different when the FCC and HCP structures coexist with the BCC phase. This is, e.g., the case in Fe and Tl, in the phase diagram of which the three phases merge at a triple point. Here, the transition mechanisms and corresponding phase diagram can be simply described by combining the Burgers and Bain deformation mechanisms, taking the BCC structure as the initial structure [20]. In order to investigate the region at which the three phases coexist, one should use the thermodynamic potential:

$$F(T, P, |\boldsymbol{\eta}|, |\zeta|) = F_0(T, P) + F_1[|\eta(\xi_1)|] + F_2[|\zeta(\xi_2)|] + F_3(|\boldsymbol{\eta}|, |\zeta|), \tag{5.36}$$

where $F_1[|\boldsymbol{\eta}(\xi_1)|]$ and $F_2[|\zeta(\xi_2)|]$ coincide, respectively, with Eqs. (5.27) and (5.35), and can be rewritten for the sake of homogeneity as

$$F_1(\eta) = \frac{a_1}{2}\eta^2 + \frac{a_2}{4}\eta^4 + \frac{a_3}{6}\eta^6 \text{ and } F_2(\zeta) = \frac{b_1}{2}\zeta^2 + \frac{b_2}{3}\zeta^3 + \frac{b_3}{4}\zeta^4.$$

ξ_1 and ξ_2 denote the independent shifts associated with the BCC–HCP and BCC–FCC transitions, respectively. F_3 expresses the coupling between the two order-parameters. From the symmetries of the order-parameters, one finds

$$F_3(\eta, \zeta) = \delta\zeta\eta^2 + (\gamma\zeta^2\eta^2)/2. \quad (5.37)$$

Minimization of F with respect to ξ_1 and ξ_2 yields the following equations of state:

$$\eta(a_1 + a_2\eta^2 + a_3\eta^4 + 2\delta\zeta + \zeta^2)\frac{\partial\eta}{\partial\xi_1} = 0, \quad (5.38)$$

$$(b_1\zeta + b_2\zeta^2 + b_3\zeta^3 + \delta\eta^2 + \gamma\zeta^2)\frac{\partial\zeta}{\partial\xi_2} = 0. \quad (5.39)$$

Restricting the description of these equations to the region of the phase diagram in which the three phases are close to the triple point, one obtains the phase diagram of Fig. 5.20(a). A striking feature of this diagram, which contains the two *limit*, HCP and FCC, phases, is that all the transition lines and limit of stability lines are *straight lines*, in agreement with the general properties established in Chapter 4 (Sec. 4.4.3). The equations corresponding to the lines of first-order transitions are

$$a_1 + a_2 + a_3 = 0 \text{ (BCC − HCP)}, b_1 + b_2 + b_3 = 0 \text{ (BCC − FCC)},$$

$$a_1 + a_2 + a_3 = b_1 + b_2 + b_3 \neq 0 \text{(FCC − HCP)}.$$

The preceding lines intersect at the triple point T, whose coordinates are in the (a_1, b_1) plane $(-a_2 - a_3; -b_2 - b_3)$. The equations giving the limit of stability lines are as follows:

(1) $a_1 = -(\gamma + \delta)$ (for the BCC phase into the HCP phase),
(2) $b_1 = -(3b_2 + b_3)$ (for the FCC into the HCP phase),

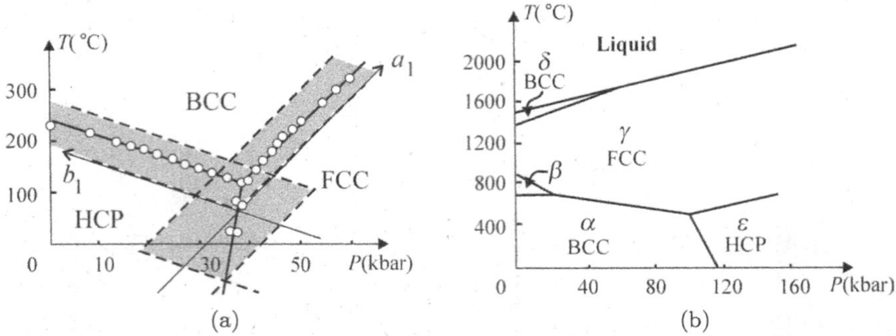

Fig. 5.20. (a) Phase diagram, close to the triple point, corresponding to the thermodynamic potential (5.36). The lines correspond to the best fit with the experimental points found by Jayaraman *et al.* [28] for thallium. (b) Pressure–temperature phase diagram of iron from Ref. [17].

(3) $b_1 = -\gamma$ (for the BCC into the FCC phase),
(4) $a_1 = (2a_2 + a_3)$ (for the HCP into the BCC phase).

The pressure–temperature phase diagrams of iron, schematized in Fig. 5.20(b), provide a good verification of the preceding theoretical diagram. At atmospheric pressure, iron crystallizes from the melt into the BCC structure (δ-Fe) and at lower temperature transforms into the austenite FCC phase (γ-Fe) and again into a BCC structure (α-Fe). With increasing pressure, a triple point is found [26] at $P_T \cong 97$ kbar and $T_T \cong 450°$C, at which the γ and α-phases meet the HCP ε-phase. The metastable regions of the BCC phase into the FCC and HCP phases, shown in Fig. 5.20(a), have been verified by Poniatovskii [27], who observed an overcooling of the BCC phase into FCC and HCP iron. The lines separating the phases in Fig. 5.20(b) are shown from an insufficient number of experimental points [2], in order to confirm conclusively their predicted straight-line character. In contrast, the phase diagram of thallium, measured by Jayaraman *et al.* [28] in the vicinity of the triple point, clearly confirms this property. In Fig. 5.20(a), the experimental points given by these authors are superimposed to the theoretical curves calculated for the transition lines and limit of stability lines. A linear transformation of the (a_1, b_1) plane into the pressure–temperature plane has been performed for comparison with experiment. One can verify in Fig. 5.20(a) that a very good fit is obtained between the theoretical lines and the experimental points.

It has already been pointed out that the δ and α BCC phases of iron can be interpreted as two differently oriented structures, predicted from the periodic dependence of the order-parameter shown in Fig. 5.15(a). No available data allow one to compare the different orientations assumed in our model for the δ and α structures. A diffraction spectrum of the two phases would provide such a verification since cycling across the transition should progressively lead to a replacement of two-fold positions (e.g., in the cubic direction [110]) by a four-fold diffraction pattern.

In conclusion, for iron and thallium, the FCC–HCP transition is interpreted as a displacive transition between two low-symmetry phases, which are both derived from a common parent BCC structure. This interpretation does not hold when the BCC phase is absent from the phase diagram, or does not have a region of coexistence with the FCC and HCP phases, as it will be shown in Chapter 6.

5.5. The Martensitic Transformations in Lithium and Sodium: Advanced Phenomenological Theory

Analysis of Secs. 5.2–5.4 shows that the crystal-geometrical approach to martensitic reconstructive phase transitions allows us unequivocally to identify corresponding primary order-parameters and their image symmetry. In a spirit of Landau-type theory, the approach considers reconstruction as a *uniform* transformation of a parent crystal structure to a distorted one. However, at this stage, it maps the corresponding atomic displacements as shifts of the dimensionless points. Although such a simplified picture allows to successfully consider the specific topology of the P–T phase diagrams, the stability domains of both non-Landau reconstructive phases and Landau ones, it is not fully realistic, and it results in missing important effects like, for instance, the stability of extremely complex structures recently discovered in many elemental crystals under high pressure [29, 30].

One can conclude that, by analogy to the classic problem of the theoretical crystal strength [31], and due to operating with a large value of the order-parameter, any advanced approach to reconstructive phase transitions should account for the possible appearance of *inhomogeneous* structures. A realistic thermodynamic model (i) should deal with a thermodynamic potential *density*, (ii) the latter should contain not only the primary order-parameter periodic contributions but also the contributions of the strain field, including those induced by the non-uniform order-parameter distribution.

We introduce the advanced phenomenological approach through the example of martensitic transformations in light alkaline metals.

5.5.1. *Experimental features of the martensitic transformations in Na and Li*

Phase diagrams of Na and Li (Fig. 5.21) were studied in detail by using synchrotron X-ray diffraction under high pressure in recent decades, but comprehensive single-crystal investigations of the metals at ambient pressure and low temperature were already carried out back in the 1980–1990s.

A number of neutron scattering investigations of sodium and lithium by Blaschko and coworkers [36–40] as well as Smith and coworkers [41–46] have added decisive finishing touches to the complex pictures of the martensitic transformations which take place at about 35 K in Na, and 77 K in Li. One of the essential points which has been under discussion is the structure of the low-symmetry phases that were initially described from X-ray powder data [47, 48] as faulted HCP structure coexisting with a large amount of untransformed BCC matrix. A neutron diffraction experiment on Li by McCarthy *et al.* [49] showed that powder data could not be well indexed by the double-layered HCP stacking. Overhauser proposed [50] that a nine-layered rhombohedral (9R) polytype structure could provide a better account for the preceding data. Subsequent elastic neutron scattering studies of Li single crystal [38, 44] confirmed

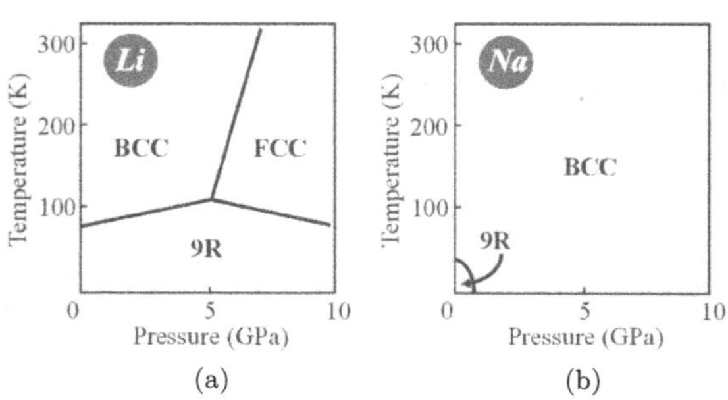

Fig. 5.21. Topology of the $P{-}T$ phase diagrams of Li and Na metals [32–35].

a faulted 9R structure. On heating above 80 K, an FCC structure appears, and remains exclusively with the BCC phase between 120 K and about 180 K. The existence of a perfect FCC structure on warming was independently confirmed in an high-resolution experiment [45], within the same temperature range, at 6.5 kbar and at atmospheric pressure.

Although Na was initially regarded as equivalent to Li [47, 48], successive studies showed that its low-temperature phase was different in some respects. Thus, on the one hand, Schwartz *et al.* [37] found no short-range ordered polytype sequence but long-range close-packed HCP, 9R and BCC sequences whose relative volume fractions depend on temperature and specimen characteristics. On the other hand, Berliner *et al.* [42] interpreted the crystallography of the low-temperature phase of Na as a complex mixture of almost hexagonal, rhombohedral polytypes, with short and long periods, forming a ladder of structures connected one to another by stacking faults. In both studies [37, 42], no FCC structure was reported.

Another important point of controversy concerns the lattice instabilities revealed by inelastic neutron scattering measurements on the Σ_4 phonon branch above the transitions. Although Smith *et al.* [42, 44, 46] claim that no evidence of transformation precursors can be found in Na and Li, Schwartz *et al.* [40] observe in Li a partial softening of the acoustic [110] phonon branch, polarized along $[1\bar{1}0]$, extending from $\boldsymbol{k} = 0.1$ [110] to the surface of the BCC Brillouin zone (the N-point): from 200 K to 100 K, the softening increases toward the zone boundary, whereas between 100 K and 80 K, a small dip is observed near $\frac{1}{3}[110]$. In Na, Blaschko and Krexner [36] find the same phonon branch to soften close to $\boldsymbol{k} = 0.42$ [110], a result confirmed by the recent work of Abe *et al.* [51].

5.5.2. *Thermodynamic models of phase transitions in sodium and lithium*

Let us show that a coherent theoretical description of most of the experimental features of the transformations in lithium and sodium can be deduced from the specific nature of the corresponding symmetry-breaking mechanisms. Three close-packed structures appear to coexist and to dominate alternately below the BCC phase: the HCP and 9R phases in Na, and the 9R and FCC phases in Li.

Let us show that the different sequences of phases observed on cooling and heating in these crystals can be understood by simple thermodynamic considerations, which involve couplings between the corresponding order-parameters. The order-parameters of the three structures are associated with the primary instabilities pertaining to the $\Gamma - \Sigma - \mathrm{N}$ line of the Brillouin zone (Fig. 5.22). They establish hexagonal ratios in (110) BCC-type layers and new translational symmetry in the perpendicular $[110]_{\mathrm{BCC}}$ directions. In addition, three types of strains are involved for a realistic description of the low-symmetry structures, as follows:

$$e_6 = e_{xy}; e_3 = (e_{xx}+e_{yy}-2e_{zz})/\sqrt{6}; e_1 = (e_{xx}+e_{yy}+e_{zz})/\sqrt{3}. \quad (5.40)$$

e_6 provides no contact between the atoms in the [001] BCC direction, but decreases the interlayer distance in the [110] direction (Figs. 5.23(a) and 5.23(b)). e_3 restores the close packing both inside the layers (e_{zz}) and between layers ($e_{xx} + e_{yy}$) (Fig. 5.23(c)). e_1 is the non-symmetry-breaking rigidity, which accounts for a change in volume at the transition.

(1) The BCC–HCP transition is associated with an instability located at the N-point of the BCC Brillouin zone (see Sec. 5.5.2), and its order-parameter expansion (5.27) can be rewritten under the compact form

$$F_1(\eta) = a_1\eta^2 + a_2\eta^4 + a_3\eta^6. \quad (5.41)$$

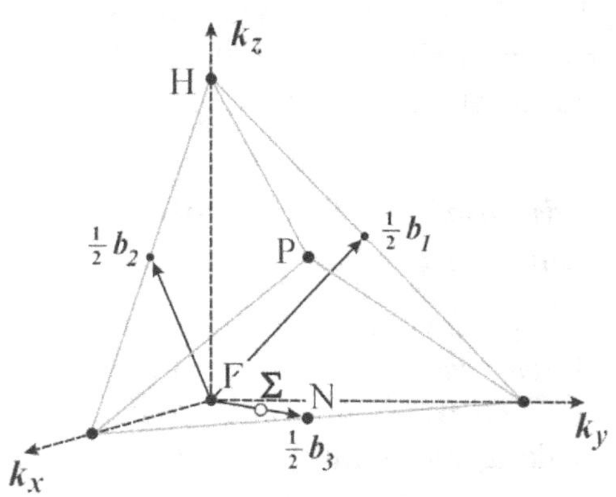

Fig. 5.22. Independent octant of the body-centered cubic Brillouin zone.

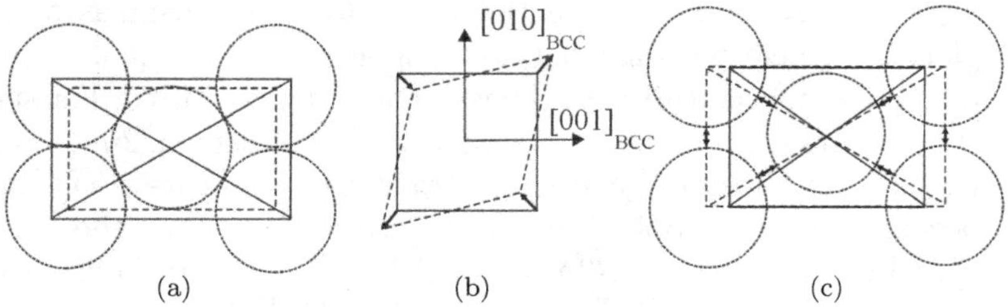

Fig. 5.23. Effect of the macroscopic strains e_3 (a) and e_6 ((b) and (c)) on the BCC unit cell. (a)–(c) show projections on the $(1\bar{1}0)$, (001) and $(1\bar{1}0)$ BCC planes, respectively. Dashed lines represent deformed states with respect to the undeformed BCC structure in solid lines.

The full thermodynamic potential includes, in addition, the coupling terms with the secondary strains:

$$F_1(\eta, e_1, e_3, e_6) = F_1(\eta) + \eta^2(\delta_1 e_3 + \delta_2 e_6 + \delta_3 e_1). \qquad (5.42)$$

Assuming a pure BCC–HCP transition mechanism in Na would give, e.g., the following numbers:

$$e_3 = -0.29015, e_6 = -0.1518, e_1 = 0.08625. \qquad (5.43)$$

(2) The BCC–FCC transition is induced by a primary instability at the center Γ of the BCC Brillouin zone, which is associated with the effective order-parameter expansion (5.35), in which we put $\zeta = e_3$:

$$F_2(e_3) = b_1 e_3^2 + b_2 e_3^3 + b_3 e_3^4. \qquad (5.44)$$

Considering the coupling of e_3 with e_6 and e_1 gives the full thermodynamic potential:

$$F_2(e_1, e_3, e_6) = F_2(e_3) + \mu_1 e_3^2 e_1 + \mu_2 e_3 e_6^2. \qquad (5.45)$$

(3) The BCC–9R transition has not been considered in the preceding sections. The current mechanism for this transformation was proposed by Wilson and de Podesta [52]. It consists in a shifting of unequal amounts in the $[110]_{BCC}$ direction of the atoms pertaining to nine successive BCC layers (Fig. 5.24(a)). However, when considering the displacement associated with the tensile strain

e_3, one can again describe the BCC–9R mechanism as a displacement field with antiparallel displacements of equal magnitude $a\sqrt{2}/6$, each third BCC layer remaining unshifted. For such critical shifts, the monoclinic structure of symmetry $C2m$ $(z=3)$ which is obtained for general arbitrary displacements along $[1\overline{1}0]$ becomes the rhombohedral 9R structure. Note that the corresponding space group $R\overline{3}m$ $(z=3)$ is not a subgroup of $Im\overline{3}m$ since its three-fold axis is parallel to the [110] direction and does not coincide with the three-fold axis of the BCC structure. In the preceding description, the symmetry-breaking mechanism is associated with a 12-dimensional IR of the $Im\overline{3}m$ space group denoted [9] $\tau_3(\boldsymbol{k}_4) = \Sigma_3$ with $\boldsymbol{k}_4 = 2/3(\pi/a,\, \pi/a, 0)$. The corresponding 12-component order-parameter $(\zeta_i)(i = 1, \ldots, 12)$ takes in the 9R phase the equilibrium values

$$\zeta_1 = \zeta_2 = \zeta \neq 0, \;\; \zeta_i = 0 \text{ for } i = 3\text{–}12. \tag{5.46}$$

The dependence of ζ on the critical displacements ξ along $[1\overline{1}0]$ has the same periodicity as $e_3(\xi)$:

$$\zeta(\xi) = \zeta_0 \cdot \sin[(3\pi/a\sqrt{2})\xi]. \tag{5.47}$$

Figure 5.24(b) represents the $\zeta(\xi)$ periodicity with the successive onset of BCC, monoclinic and 9R sequences of phases. Using the transformation properties of the ζ_i $(i = 1\text{–}12)$ by the IR $\tau_3(\boldsymbol{k}_4)$ and the equilibrium conditions (5.46), one can construct the order-parameter expansion

$$F_3[\zeta(\xi)] = F_{03}(T, P) + c_1\zeta(\xi)^2 + c_2\zeta(\xi)^4 + c_3\zeta(\xi)^6. \tag{5.48}$$

$F_3[\zeta(\xi)]$ is formally identical to $F_1(\eta)$ in Eq. (5.27). Accordingly, the same topology as shown in Fig. 5.12 will hold for the phase diagram resulting from the minimization of $F_3[\zeta(\xi)]$ with respect to ξ, the 9R and monoclinic $C2/m$ phases replacing, respectively, the HCP and $Cmcm$ phases in Fig. 5.12.

5.5.3. *Improper symmetry-breaking strains*

In the description of the BCC–HCP transformation mechanism, we have shown that, beyond the primary displacive mechanism, additional spontaneous strains have to be taken into account in order

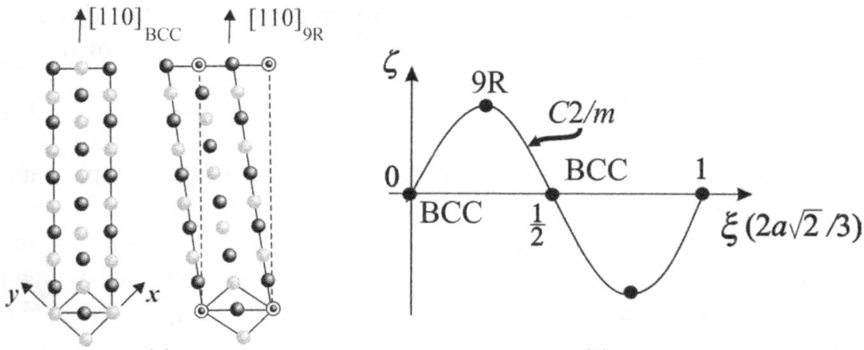

Fig. 5.24. (a) BCC–9R transformation mechanism following the description of Wilson and de Podesta (Ref. [52]). (b) Dependence of the order-parameter as a function of the critical displacements at the BCC–9R transformation, following Eq. (5.47).

to obtain the actual HCP structure. These strains couple to the order-parameter, and the corresponding coupling terms have to be added to the expansion $F_1[\eta(\xi)]$ given by Eq. (5.27). In the case of the BCC–HCP transformation, e.g., identification of the relevant strains is obtained [22] by decomposing the symmetrized second power of the irreducible representation $(\tau_4(\boldsymbol{k}_9))^2 = \mathrm{A_{1g}}+\mathrm{E_g}+\mathrm{F_{2g}}$ (see Eq. (5.28)), where $\mathrm{A_{1g}}$, $\mathrm{E_g}$ and $\mathrm{F_{2g}}$ are, respectively, one-, two- and three-dimensional irreducible representations of $m\overline{3}m$ point group in the standard (spectroscopic) notation. These irreducible representations are spanned by the following combinations of the strain-tensor components:

(i) $\mathrm{A_{1g}}$ corresponds to the rigidity $e_1 = (e_{xx}+e_{yy}+e_{zz})/\sqrt{3}$ expressing a possible change in volume at the transition.
(ii) $\mathrm{E_g}$ transforms as the tensile strains (ζ_1, ζ_2) (Eq. (5.19)) associated with the Bain deformation of the BCC unit cell ($\zeta_1 = e_3 \neq 0, \zeta_2 = 0$). Figure 5.23(a) illustrates the effect of e_3, which is to restore the close packing both inside the layers (e_{zz}) and between layers ($e_{xx} + e_{yy}$).
(iii) $\mathrm{F_{2g}}$ has the symmetry of the triplet (e_{yz}, e_{xz}, e_{xy}) describing a modification of the interlayer distance. In the HCP phase, $e_{yz} = e_{xz} = 0$, $e_{xy} = e_6 \neq 0$. Figures 5.23(b) and 5.23(c) show that e_6 provides no contact between the atoms in the $[001]_{\mathrm{BCC}}$ direction, but decreases the interlayer distance in the $[110]_{\mathrm{BCC}}$

direction. Therefore, the full thermodynamic potential describing the BCC–HCP transformation takes the effective form

$$F_1[\eta(\xi), e_i] = F_1[\eta(\xi)] + \eta(\xi)^2(\delta_1 e_3 + \delta_2 e_6 + \delta_3 e_1). \qquad (5.49)$$

At this point, it has to be emphasized that, in the standard approach to structural transitions [22], a coupling term of the form $\eta^2 e_i$ reflects the "improper" non-symmetry-breaking character of the spontaneous strain e_i. In our description of the BCC–HCP mechanism, it has been actually shown that e_3 results in an increase (to $P6_3/mmc$) of the orthorhombic symmetry *Cmcm*. Thus, in contrast to the property of secondary order-parameters of being non-symmetry-breaking quantities, the secondary strains at the BCC–HCP transformation take part in the breaking of the symmetry. More precisely, they induce an "increase" of the structural symmetry and play the role of a distinct order-parameter, though they are induced by the primary order-parameters. This apparent contradiction is resolved when noting that the coupling terms in Eq. (5.49) establish a connection between the macroscopic strains and the critical atomic shifts ζ_c. The large values of ζ_c occurring at the transformation give rise to large spontaneous strains. For example, assume a pure BCC–HCP mechanism in Na would give the following numbers: $e_3 = -0.29015, e_6 = -0.1518, e_1 = 0.08625$ (see Eq. (5.43)), which is more than one order of magnitude larger than the numbers usually found in ferroelastic transitions. The same order of magnitude is found for the "improper" strains involved in the BCC–HCP transformation in all the elemental crystals, e.g., the compression e_{zz} is larger than 10% in Ti, Zr or Yb, and reaches values above 20% in Ba or ^{3}He (see Table 5.3).

The property of improper strains to become symmetry-breaking quantities holds for all martensitic transformations of the reconstructive type, i.e., in which a group–subgroup relationship between the initial and final structures is lost. At the BCC–FCC transformation, where the tensile strain e_3 is the primary order-parameter, analogous considerations yield

$$F_2[e_3, e_1, e_6] = F_2[e_3(\xi)] + \mu_1 e_3^2 e_1. \qquad (5.50)$$

For the BCC–9R transformation, one finds

$$F_3[\zeta, e_i] = F_3[\zeta(\xi)] + \zeta^2(\nu_1 e_1 + \nu_3 e_6 + \nu_2 e_3). \qquad (5.51)$$

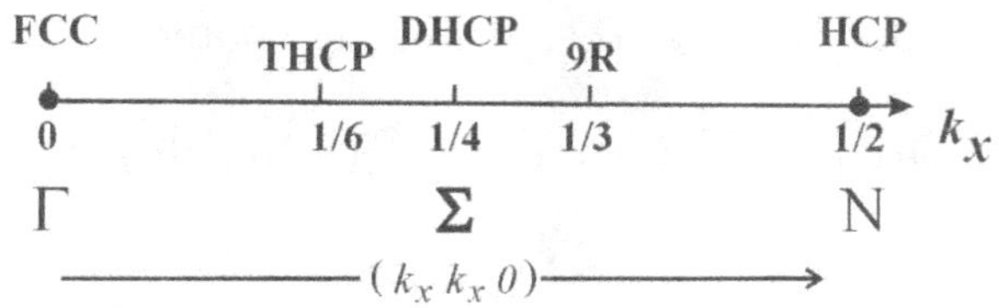

Fig. 5.25. Length of the k_x component of the wave vectors associated with the transformations from a BCC phase to an FCC ($k_x = 0$), HCP (1/2), 9R (1/3), DHCP (1/4), or THCP (1/6) structure, on the Σ-line $(k_x, k_x, 0)$ of the BCC Brillouin zone.

The same coupling invariants have been considered by Gooding *et al.* [53] in their model of the BCC–9R transformation. However, these authors did not assume a transcendental dependence of the primary order-parameter ζ on the critical displacements ξ, and overlooked, therefore, the symmetry-breaking character of the spontaneous strains corresponding to the maxima of $\zeta(\xi)$.

The primary order-parameters associated with the transformations from the BCC to the HCP, FCC and 9R structures establish hexagonal ratios between the lattice parameters in (110) BCC layers and new translational symmetries in the $[110]_{\text{BCC}}$ directions. The corresponding wave vectors are located on the $\Gamma_5 - \Sigma_4 - \text{N}_4$ line of the BCC BZ (Fig. 5.22).

Figure 5.25 shows that along this line, which coincides with the bisector $(k_x, k_x, 0)$ of the (x, y) plane in reciprocal space, the FCC, 9R and HCP structures are, respectively, obtained for $k_x = 0, k_x = 1/3$ and $k_x = 1/2$.

Note that for the values $k_x = 1/4$ and $k_x = 1/6$, one may also get double and triple HCP structures which are found in several lanthanide elements.

5.5.4. *The phase diagram of sodium and lithium*

The results obtained in Secs. 5.5.2 and 5.5.3 can be used as a basis for describing the phase diagrams of sodium and lithium. Let us first consider the experimental data reported by Schwarz *et al.* [37] for Na. The sequences of phases observed by these authors on cooling and heating involve the BCC, 9R and HCP phases. Using the effective forms of the thermodynamic potentials associated with the BCC–HCP (Eq. (5.49)) and BCC–9R (Eq. (5.51)) transformations, one can write the total effective thermodynamic potential as

$$\Phi_1[\eta(\xi_1), \zeta(\xi_2), e_i] = F_1[\eta(\xi_1), e_i] + F_3[\zeta(\xi_2), e_i] + \gamma_1 \eta^2(\xi_1)\zeta^2(\xi_2), \tag{5.52}$$

where ξ_1 and ξ_2 denote the respective displacements occurring at the BCC–HCP and BCC–9R transformation mechanisms. The γ_1 term is the lowest-degree coupling invariant (bi-quadratic) between the η and ζ order-parameters.

The general features of the phase diagrams obtained by minimizing Φ_1 with respect to ξ_1 and ξ_2 are shown in Fig. 5.26 within the plane (c_1, a_1) of the phenomenological coefficients of the quadratic invariants in F_1 and F_3. Since these coefficients depend linearly on temperature and pressure, the topology of the transition lines and phases in Fig. 5.26 is preserved in the corresponding temperature–pressure phase diagrams. Note that a negative value has been assumed for the coefficients a_2 and c_2. It implies that the intermediate Landau state in Fig. 5.12 is unstable. The figure reveals that besides the 9R ($\eta = 0, \zeta \neq 0$) and HCP ($\eta \neq 0, \zeta = 0$) phases, an additional phase corresponding to an 18-layer polytype structure (18R) can be stabilized for $\eta \neq 0$, $\zeta \neq 0$. This phase can be reached from the BCC phase either via the 9R or HCP phase (Fig. 5.26). Figure 5.27(a) focuses on the region of the phase diagram represented in Fig. 5.26 surrounding the triple point T_1. The phase sequences indicated by the thick arrows 1 and 2 coincide with the thermodynamic paths followed by Schwarz *et al.* [37] in their diffuse neutron scattering investigation, which are schematically reproduced in Fig. 5.27(b). Thus, on cooling one goes from BCC to (BCC+9R+HCP), whereas on heating one gets successively the phase sequence (BCC + 9R + HCP) → (HCP + BCC) → BCC. The difference in the phase sequences on cooling and heating corresponds to a standard hysteretic behavior usually expected at first-order transformations. A less usual behavior which relates to the specific transformation kinetics in Na and Li is observed in a second cooling cycle during which Schwarz *et al.* [37] find a reduction in the amount of the HCP phase with an almost pure 9R phase coexisting with the BCC matrix. Therefore, cycling across the transition shifts the thermodynamic paths toward the pure 9R region. Such a tendency is consistent with the neutron powder diffraction study on

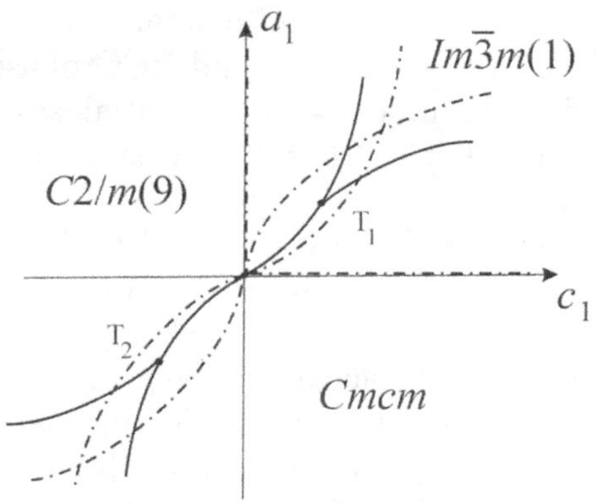

Fig. 5.26. Phase diagrams associated with the coupled order-parameter expansion defined by Eq. (5.52). The description of the phase diagrams is in the text. Full and dash-dotted lines are, respectively, the first-order transition and limit of stability lines. A strong positive coupling between the order-parameters η and ζ assumes ($\delta > 0$), whereas a_2 and c_2 are negative. T_1 and T_2 are triple points.

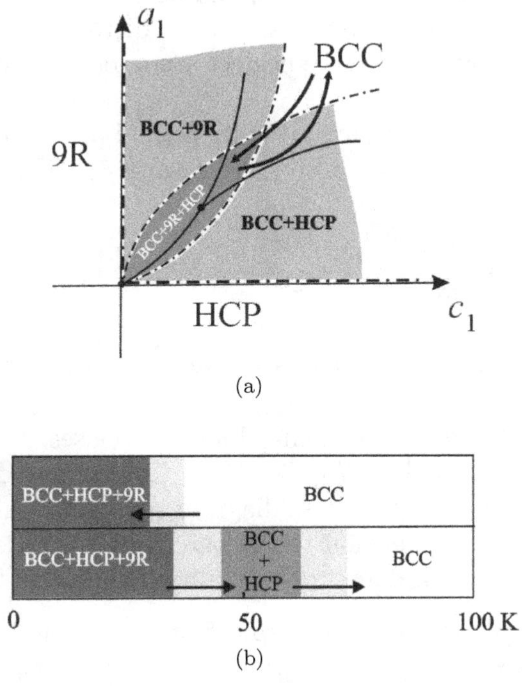

Fig. 5.27. (a) Region of the phase diagram of Fig. 5.26 surrounding the triple point T_1. The arrows indicate the sequences of phases found in sodium (Ref. [37]), which are represented in (b).

Na by Berliner *et al.* [46] who found a small fraction of HCP phase coexisting with large amounts of 9R and BCC phases.

The full effective thermodynamic potential associated with the coupled BCC–9R and BCC–FCC transformations can be written as

$$\Phi_2[\zeta(\xi_2), e_3(\xi_3), e_i] = F_3[\zeta(\xi_2), e_i] + F_2[e_3(\xi_3), e_i] + \gamma_2\zeta^2(\xi_2)e_3(\xi_2) + \gamma_3\zeta(\xi_3^2)e_3(\xi_3^2), \quad (5.53)$$

where ξ_2 and ξ_3 are the displacements, respectively, associated with the BCC–9R and BCC–FCC transformation mechanisms. The γ_2 and γ_3 terms are the lowest-degree couplings which have to be considered to insert the BCC, 9R and FCC phases within the same phase diagram. In contrast to the ordinary case of "improper" coupling (see Sec. 2.4.3), two different e_3 exist: (i) induced $e_3(\xi_2) \sim T$ and (ii) critical $e_3(\xi_3) \sim \sqrt{T}$. We take it into account attributing the γ_2 term, responsible for establishing a hexagonal metric in the nine-layered structure, exclusively to the corresponding ζ order-parameter, while γ_3 in Eq. (5.53) relates only to the "cubic" distortion of the intermediate tetragonal lattice, i.e., we neglect a non-critical contribution to e_3 induced by ζ.

The forms of F_3 and F_2 are given by Eqs. (5.35), (5.48), (5.50) and (5.51). Minimizing F_2 with respect to ξ_2 and ξ_3 yields the following equations of state:

$$\frac{\partial\Phi_2}{\partial\xi_2} = 2\zeta\frac{\partial\zeta}{\partial\xi_2}\cdot(c_1 + 2c_2\zeta^2 + c_3\zeta^4 + 3/2\cdot\gamma_2\zeta e_3) = 0,$$
$$\frac{\partial\Phi_2}{\partial\xi_3} = e_3\frac{\partial e_3}{\partial\xi_3}\cdot(2b_1 + 3b_2e_3 + 4b_3e_3^2 + 2\gamma_3\zeta^2) = 0. \quad (5.54)$$

Assuming unstable intermediate Landau phases (of symmetries $I4/mmm$ and $C2/m$), i.e., $|b_2| > b_2^N$ in Fig. 5.19 and $c_2 < c_2^N$ in Fig. 5.12, one obtains the phase diagram represented in Fig. 5.28 in the (b_1, c_1) plane. In addition to the parent BCC phase ($\eta = \zeta = 0$), three phases possess a domain of stability:

(1) The FCC phase which corresponds to $\zeta = 0$ and $\partial e_3/\partial\xi_3 = 0$. It is realized for the fixed strain $e_1^S(\xi_3^c)$, where $\xi_3^c = a\sqrt{2}/6$ (see Sec. 5.5.2).
(2) The 9R phase corresponding to $\zeta(\xi_2) \neq 0$ and $\partial\zeta/\partial\xi_2 = 0$. It is also realized for a fixed strain $e_2^S(\xi_2^c) \neq 0$ (see Sec. 5.5.2).

(3) A 27R polytype structure obtained for $\zeta(\xi_2) \neq 0, e_3(\xi_3) \neq 0$, which is the minimal common superstructure for the FCC and 9R structures. It involves tensile strains e_3 varying between e_1^S and e_2^S.

In other words, the stabilization of the 9R structure requires the existence of an improper (symmetry-breaking) secondary strain e_3 due to the specific form of the nonlinear coupling term γ_2. Since e_3 is also the primary order-parameter giving rise to an FCC phase, an intermediate 27R polytype structure may likewise possess a region of stability adjacent to the 9R phase.

The structure of the Na martensite disclosed by Schwarz *et al.* [40] and Berliner *et al.* [42] has been described by thermodynamic considerations involving the coupling between the mechanisms (order-parameters) assumed in our approach for the BCC–HCP, BCC–9R, and BCC–FCC transformations. The description stresses the important role played by the tensile strain e_3 in these transformations, which acts as a primary order-parameter or as a secondary symmetry-breaking quantity. It remains to understand why the thermodynamic paths partly differ in the two experimental studies. The most consistent explanation relates to the intrinsically faulted nature of the considered close-packed structures, which has been stressed by Schwarz *et al.* [40] and serves as a basis for the structural analysis of the Na martensite by Berliner *et al.* [42]. The amount of stacking faults increases on cycling across the martensitic transformation, i.e., from virgin to deteriorated samples. At given temperature and pressure, it shifts the phase fractions toward a larger amount of longer polytype sequences, e.g., from HCP to 9R as observed by Schwarz *et al.* [40]. Besides, the concentration of stacking faults pre- existing in virgin samples differs from one sample to another depending on sample preparation.

The phase diagram of lithium differs in a number of aspects from the sodium phase diagram. The neutron diffraction studies of Smith *et al.* [43, 44, 46] and Blaschko *et al.* [38, 39] and synchrotron X-ray studies [34, 35] analyze the Li martensite in terms of 9R structure coexisting with the BCC matrix. The two groups also find a long-range FCC structure, which appears only on heating both at atmospheric pressure and at 6.5 kbar [39, 44]. The diffuse scattering analysis of a wider region of reciprocal space by Schwarz and

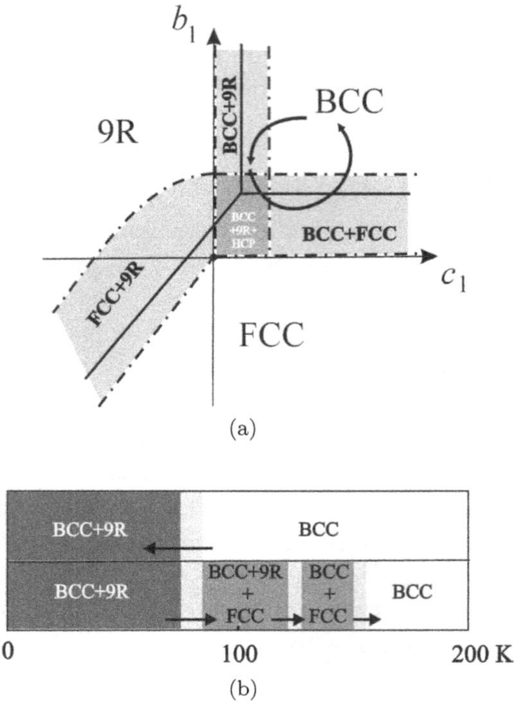

Fig. 5.28. Phase diagram associated with the coupled order-parameter expansion defined by Eq. (5.53). The description of the phase diagram is given in the text. The same convention as in Fig. 5.26 is used for the transition and limit of stability lines. The arrows indicate the experimental path possibly followed by Berliner *et al.* [42].

Blaschko [44] reveals that the long-range 9R phase coexists at low temperature with a faulted polytype structure, where short-range ordering tendencies of HCP, FCC and 9R phases are simultaneously present. Reference to the considerations developed for constructing the phase diagram of Fig. 5.28 shows that a long-range FCC structure adjacent to a long-range 9R structure and induced from the BCC phase cannot occur since the two structures require for their stabilization different fixed values of the tensile strain e_3.

Figure 5.28(a) represents the region surrounding the triple point. The arrows 1 and 2 reflect the phase sequences reported by Schwarz and Blaschko (Fig. 5.28(b)) on cooling and heating, respectively. On cooling, one gets the BCC $\rightarrow$ (BCC + 9R) sequence, whereas on heating, the (BCC + 9R + FCC) → (BCC + FCC) → BCC sequences are obtained.

It remains to understand the transformation mechanism which results in Li and Na in a low-temperature phase, reflecting two or three simultaneous instabilities located in different regions of the Σ-line in the BCC Brillouin zone. The simplest explanation refers to the actual topology of the pressure–temperature phase diagrams of these elements, which still display large unexplored high-pressure regions [17]. It assumes that, in both Li and Na, the low-temperature region in which the martensitic transformation takes place is located *not far from a triple point*, within the region of metastability of three, or even four, stable or potentially stable phases, as confirmed by pseudopotential and band calculations [33].

The pressure–temperature phase diagram of Li [17], shown in Fig. 5.21(a), reveals that the regions of stability of the BCC, FCC and martensite phases may merge at about 3 kbar. Besides, the calculations of Vaks *et al.* [32] predict a narrow region of stability for a HCP structure, close to the 9R region. In Na, the region of stability of the martensite phase surrounds the origin of the phase diagram (Fig. 5.21(b)), and the eventual stabilization of an FCC structure seems improbable for positive pressures and temperatures, although *ab initio* calculations [32] show its possible stability close to the 9R and HCP regions.

Two correlated series of facts suggest a more precise complementary interpretation of the transition mechanisms in Li and Na. Above the transitions, in the premartensitic region, experimental observations [54] indicate a complex nucleation process involving the existence of large microdomains, possessing different microstructures, i.e., 9R, FCC and HCP. This indetermination in the structure of the martensitic "embryos" is consistent with the unusually large extension of the softening region observed by Schwarz and Blaschko in Li [39] for the acoustic [110] phonon branch, which extends from 0.1 [110] to 0.5 [110] above 100 K, around 0.33 [110] below 100 K, and even close to the Γ point in a deteriorated crystal of Li. In Na, the extension of the softening region for the same phonon branch [36] is smaller, but still goes from 0.4 [110] to the N-point.

These observations allow the following speculative interpretation for the transition mechanism: the appearance of distinct microstructures which involve different strain fields [55] acting inhomogeneously on the transition order-parameters (due to the coupling between the two quantities expressed by Eqs. (5.49), (5.50) and (5.51)), give *a potentially incommensurate character to the low-temperature structure*, which is attested by the temperature dependence of the k-vector along the Σ-line, and by the incommensurate values of $\boldsymbol{k}$ in the softening regions [36, 39, 43]. However, the large shifts of the atoms which take place at the transition lead the system to *a lock-in transition*, at which *the k-vector locks simultaneously* at the lock-in values corresponding to the fractional numbers 0 (FCC), 1/3 (9R) and 1/2 (HCP), reflecting the various tendencies competing above the transition.

The preceding mechanism can be formalized using the phenomenological approaches to nucleation [56] and to incommensurate "strain-induced" phases [57] (Sec. 5.6). By contrast, the unusual behavior reported for the precursor instabilities, in the BCC phase of Li and Na, can be better understood in the framework of a more general description of the phonon spectrum of martensitic transformations (Sec. 5.7).

5.6. Strain-Induced Inhomogeneous States and Martensitic Transformations

A new, not fully understood phenomenon, was discovered during the last decade by intensively developing high-resolution single crystal synchrotron radiation diffraction under high pressure. In elemental crystals, within a wide pressure range, complex crystal structures containing hundreds of atoms in a unit cell were observed. The Ba IV structure which contains, following [58], 768 atoms per unit cell is a record to date. The general symmetry-based approach (Chapter 1 and Ref. [22]), however, predicts no such phases [59].

In this section, we extend our approach by accounting for the elastic properties of the crystal lattice. It enables us to reveal a universal scenario for the onset of incommensurate phases, induced by the reconstructive phase transitions. The low-temperature martensitic transformations in the alkali metals allow us to disclose their general thermodynamic origin.

It is worth recalling (see Sec. 2.3) that the incommensurate phases are special states of solids that differ from ordinary crystals by the lack of periodicity in one or more directions, whereas a long-range order exists unlike amorphous solids. The reasons for the appearance of the incommensurate phase and its properties are well described by the Landau theory of phase transitions [22].

5.6.1. *Incommensurate phases in the Landau phase transition theory*

A typical density of the Landau free energy used to consider an incommensurate phase is of the form

$$F(y) = \overline{F} + \frac{\alpha}{2}(\eta_1^2 + \eta_2^2) + \frac{\beta_1}{4}(\eta_1^2 + \eta_2^2)^2 + \frac{\beta_2}{4}[(\eta_1^2 - \eta_2^2)^2 - (2\eta_1\eta_2)^2] + \sigma\left(\eta_2\frac{d\eta_1}{dy} - \eta_1\frac{d\eta_2}{dy}\right) + \frac{\delta}{2}\left[\left(\frac{d\eta_1}{dy}\right)^2 + \left(\frac{d\eta_2}{dy}\right)^2\right]. \quad (5.55)$$

Here η_1 and η_2 are components of the order-parameter. The term with the coefficient σ is the Lifshitz invariant (Sec. 2.3). The invariant can exist only if the order-parameter has two components or more. In the case where the invariant is allowed by symmetry, there inevitably appears an incommensurate phase. If the length of the crystal in the y direction is L, the free energy per unit length is

$$F = \frac{1}{L}\int_0^L F(y)dy. \quad (5.56)$$

Remember that in the Landau theory, it is convenient to assume that the single coefficient α depends on the temperature varying linearly with it.

In the absence of the Lifshitz invariant, a minimum of F would correspond to a uniform solution $\eta_1 = \text{const}$ and $\eta_2 = \text{const}$ because of the last term in Eq. (5.55), which is always present (we assume that $\delta > 0$). Minimizing the free energy $F(y)$ yields two nontrivial solutions corresponding to two possible commensurate

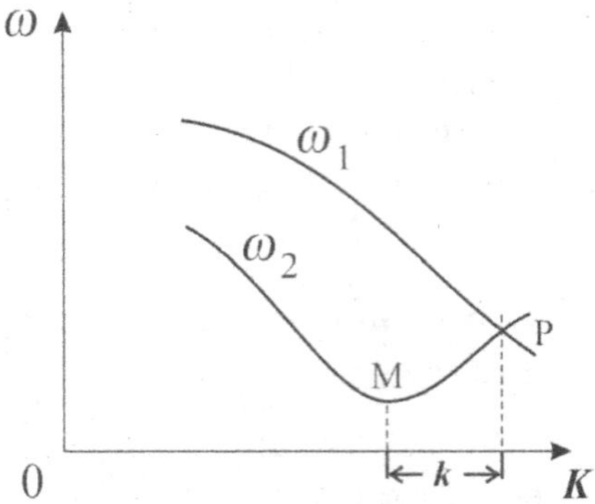

Fig. 5.29. Phonon branches near the Brillouin-zone boundary in the parent phase.

phases:

(i) $\eta_1 = \eta \neq 0, \eta_2 = 0$, or *vice versa*,

$$\eta^2 = -\frac{\alpha}{\beta_1 + \beta_2}, F = \overline{F} - \frac{\alpha^2}{4(\beta_1 - \beta_2)}. \quad (5.57)$$

(ii) $$\eta_1^2 = \eta_2^2 = -\frac{\alpha}{2(\beta_1 - \beta_2)}, F = \overline{F} - \frac{\alpha^2}{4(\beta_1 - \beta_2)}. \quad (5.58)$$

The first solution is energetically preferable if $\beta_2 < 0$, the second one if $\beta_2 > 0$.

At a commensurate–commensurate transition, the soft mode occurs at the Brillouin-zone boundary to which the wave vector $\boldsymbol{K} = \boldsymbol{b}_2/2$ corresponds, where $\boldsymbol{b}_2$ is the basic reciprocal-lattice vector directed along the y axis (Fig. 5.29).

The existence of the Lifshitz invariant at $\boldsymbol{K} = \boldsymbol{b}_2/2$ means a nonzero slope of the phonon branch at point P (Fig. 5.29). Two phonon branches meet at point P in this instance as long as all dispersion curves can be transferred to the first Brillouin zone. The lower branch is necessarily a minimum inside the zone, and as the temperature lowers, it is the frequency at point M that vanishes first, although $\alpha > 0$ (see Fig. 2.19). As a result, a superstructure with the wave vector $\boldsymbol{K} = \boldsymbol{b}_2/2 - \boldsymbol{k}$ arises in the crystal, whereas there is no reason for the ratio $\boldsymbol{k}/\boldsymbol{b}_2$ to be rational. Two mutually incommensurable periods, therefore, coexist in the crystal: the parent crystal lattice period and the period of the undulated displacive distortion.

Thus, one finds an incommensurate phase without any periodicity commensurate to the crystal lattice but with a definite long-range order.

Substituting Eqs. (5.57)–(5.58) into (5.55) and minimizing the functional F yield the following Euler–Lagrange equations:

$$\delta\frac{d^2\eta_1}{dy^2} + 2\sigma\frac{d\eta_2}{dy} - \alpha\eta_1 - \beta_1\eta_1(\eta_1^2 + \eta_2^2) - \beta_2\eta_1(\eta_1^2 - 3\eta_2^2) = 0, \tag{5.59}$$

$$\delta\frac{d^2\eta_2}{dy^2} + 2\sigma\frac{d\eta_1}{dy} - \alpha\eta_2 - \beta_1\eta_2(\eta_1^2 + \eta_2^2) - \beta_2\eta_2(\eta_2^2 - 3\eta_1^2) = 0. \tag{5.60}$$

Near the phase-transition point, where η_1 and η_2 are small, one can neglect the nonlinear terms in Eqs. (5.59) and (5.60). The resulting linear equations admit a solution

$$\eta_1 = \rho \cdot \cos(ky), \eta_2 = \rho \cdot \sin(ky), \tag{5.61}$$

with $\rho =$ const and

$$k = \frac{1}{\delta}\left[\sigma \pm (\sigma^2 - \alpha\delta)^{1/2}\right]. \tag{5.62}$$

The last equation shows that the maximal value of α at which a real value of k exists is $\alpha = \alpha_0 = \sigma^2/\delta$ while then $k = \sigma/\delta$. Therefore, at $\alpha = \alpha_0 > 0$, there occurs a phase transition into an incommensurate phase. Equations (5.58) and (5.59) do not lend themselves to analytical solutions in the general case. The equations were solved numerically [60,61], and there exists a strict analytical solution valid for an especial relation between the coefficients in Eq. (5.55) [62]. The obtained solutions show that, as the temperature lowers, the value of k decreases and tends to zero. When k approaches zero, there occurs a phase transition to one of the commensurate phases described by Eq. (5.57) or (5.58). This phase transition, referred to as a *lock-in transition*, has specific features in comparison with an ordinary second-order transition. Near the lock-in transition, the incommensurate phase becomes domain-like. The structure of the incommensurate phase inside the domains factually coincides with that of the corresponding low-temperature commensurate phase, whereas distinctions take place only in the domain walls, called the discommensurations. If the lock-in transition is continuous, it happens when the distance between the discommensurations becomes infinite, in which case the crystal will contain only one or two domains with the commensurate periodicity of phase.

It is worth recalling that all modes in the crystal are coupled, which gives rise to secondary order-parameters which should be taken into account in the free energy, and whose role augments as the temperature lowers and the order-parameter increases. In the case of an incommensurate phase, higher-order derivatives of the order-parameters and their high-degree terms can have remarkable impact if the temperature is not close to the phase-transition point.

5.6.2. *Strain-induced effects at the BCC–9R transition*

Let us apply our unified phenomenological approach to the martensitic reconstructive phase transition BCC–9R, which occurs in Li ($T_M \sim 77K$), Na ($T_{\mathrm{M}} \sim 35\,\mathrm{K}$), and presumably, K metals [17].

The primary symmetry-breaking BCC–9R mechanism relates to the 12-dimensional irreducible representation $\tau_3(\boldsymbol{k}_4)$, with $\boldsymbol{k}_4 = 0.33(\pi/a, \pi/a, 0)$ (Sec. 5.5.2 and Ref. [63]). One finds that the corresponding thermodynamic potential contains "mixed" fifth-degree invariants, which couple the order-parameter and space derivatives ∇_i (irreducible representation $\mathrm{F_{1u}}$) to shear strain-tensor components u_{jk} ($\mathrm{F_{2g}}$) or to the symmetry-adopted $\varepsilon_1 = (u_{xx} - u_{yy})/\sqrt{2}$ and $\varepsilon_2 = (u_{xx} + u_{yy} - 2u_{zz})/\sqrt{6}$ ($\mathrm{E_g}$):

$$\begin{aligned} I_1^c = u_{xy} \cdot \nabla_z[(\eta_1^3 - 3\eta_1^2\eta_2 + \eta_2^3 + \eta_3^3 - 3\eta_3^2\eta_4 + \eta_4^3) \\ + (\eta_5^3 - 3\eta_5^2\eta_6 + \eta_6^3 + \eta_7^3 - 3\eta_7^2\eta_8 + \eta_8^3)] + \cdots . \qquad (5.63) \end{aligned}$$

In order to demonstrate their generic structure, one of four invariants is displayed in Eq. (5.63), the others appear in an effective form in what follows. Only the first characteristic terms are shown in Eq. (5.63), the others can be produced by permuting subscript indexes.

In the literature, the identification of reconstructive phase transitions so far mainly underlined the break of the group–subgroup relationship between symmetries of the parent and non-Landau phases. However, little attention has been paid to the less-evident fact that the *purely geometrical* conditions exist, and they control, in the non-Landau phase, the values of the order-parameter and spontaneous strain components, so that their magnitudes are independent of external parameters (temperature, pressure, etc.) [9]. Therefore, the effective degree of the invariants of type (5.63) in the thermodynamic potential lowers, and this latter dramatically increases the impact of high-order mixed gradient terms.

The symmetry change $Im\overline{3}m$ to $R\overline{3}m$, as it was shown in Sec. 5.2, requires for the 12-component order-parameter ($\eta_i, i = 1 - 12$) satisfying in the 9R-phase condition:

$$\eta_1 = \eta_2 = \eta_0 \neq 0 \,\text{and}\, \eta_j = 0 \quad \text{for } j = 3 - 12, \tag{5.64}$$

where the order-parameter equilibrium value η_0 =const. Taking into account that the spontaneous strain components u_{xy}^0 and ε_2^0 are also constant in 9R [63], one immediately realizes that the corresponding coupling invariants assume the form similar to the proper Lifshitz invariant:

$$I_{1c}^{\text{eff}} = u_{xy}^0 \cdot \eta_0^2 \cdot (\nabla_z \eta_1 + \nabla_z \eta_2),$$
$$I_{2c}^{\text{eff}} = \eta_0^2 \cdot [\sqrt{3}\varepsilon_1(\nabla_x \eta_1 + \nabla_x \eta_2) + \varepsilon_2^0(\nabla_y \eta_1 + \nabla_y \eta_2)]. \tag{5.65}$$

We stress that, in contrast to Landau-type phase transitions, the above reduction does not lead to a necessary small size of the renormalized coupling constants generally. Then, one can write down an *effective* thermodynamic density in the form ($\eta_1 = \eta_2 = \eta$):

$$F[\eta, u_{ik}] = F_1[\eta] + F_2(u_{ik}) + F_3[\nabla\eta] + F_4(\eta, u_{ik}\nabla\eta). \tag{5.66}$$

The $F_1[\eta(\Delta)] = a_1\eta^2 + a_2\eta^4 + a_3\eta^6$ term is a non-equilibrium contribution of the primary order-parameter. The order-parameter $\boldsymbol{\eta}$ is a periodic function of critical displacements Δ of corresponding atomic layers along the $[\overline{1}10]_{\text{BCC}}$ direction, e.g., $\eta(\Delta) = \eta_0 \cdot \sin(3\pi\Delta/a\sqrt{2})$ [63]. F_2 is the elastic strain energy; it includes important improper terms coupling the primary order-parameter $\{\eta_i\}$ to shear strain-tensor components $I_s^{\text{eff}} \approx (\eta_1^2 + \eta_2^2) \cdot u_{xy}$, and to its diagonal components

$I_e^{\text{eff}} \approx (\eta_1^2+\eta_2^2)\cdot(e_1+e_2)$. $F_3[\nabla\eta]$ accounts for the conventional order-parameter gradient contribution. Finally, F_4, being proportional to I_{jc}^{eff} (5.65), is a coupling energy of the non-uniform order-parameter distribution and a strain field. It should be specially noted that the proper Lifshitz gradient invariant does not appear in Eq. (5.62) due to the symmetry restrictions. Nevertheless, the above symmetry-adopted terms linear in the order-parameter gradient ensure appearance of the stable incommensurate phases. The latter can be driven either by spontaneous strains or by the deviatoric stresses due to the non-hydrostatic compression conditions.

Already, the equilibrium conditions (5.64) and the form of the inhomogeneous energy contributions (5.65) allow us to predict some crystal-geometrical characteristics for the strain-induced incommensurate phase. Equation (5.64) identifies the atomic plane $(110)_{\text{BCC}}$ as becoming a hexagonal close-packed $(001)_{\text{9R}}$, and directs new three-fold axes along $[110]_{\text{BCC}}\|[001]_{\text{9R}}$. The coupling terms in F_2 predict that the spontaneous strains u_{xy} and ε_{1-2} should be induced in a new structure. The former, according to I_{1c}^{eff}, may impose, in the rhombohedral phase, incommensurate structure modulation along the $[001]_{\text{BCC}}$ direction, while the latter (see I_{2c}^{eff}) does the same in the $[110]_{\text{BCC}}\|[001]_{\text{9R}}$ direction.

To identify the incommensurate structure in detail, one has to solve not simply algebraic but rather the partial differential equations of state coming from the functional (5.63), with certain boundary conditions. Generally very difficult, the problem can be solved in some special cases. In particular, if the strain distribution is approximately uniform, a triple sine-Gordon equation should be solved.

Its simplest solutions describing one-dimensional incommensurate phases read

$$x = \int_{x_0}^{\Delta} \frac{du}{\sqrt{u(1-u^2)(E+a_1u-a_2u^2-a_3u^3)}}, \qquad (5.67)$$

where x_0 and E are the integration constants.

Possible incommensurate phases are given by inverting the hyperelliptic integral $x = x(\Delta)$. Generally, it yields several periodic solutions, each defined by the couple of adjacent real roots of the denominator zeroes in Eq. (5.67) keeping the positive radical. The period of each incommensurate phase structure is given by the numerical value of Eq. (5.67), where the high- and low-integration limits are equated to the chosen roots. Minimizing the potential (5.66), one finds the relevant solution $\Delta_0(x)$. It depends on the phenomenological parameters a_1 and E, and the magnitude of the uniform strain component coupled to the order-parameter gradient. The component plays a role of the chemical potential for solitons in incommensurate phase, and it controls the stability limits of the latter. Note that the onset and evolution of the strain-induced incommensurate phase at reconstructive phase transition shows a clear analogy to the formation of soliton superstructures at the Landau-type phase transition when the proper Lifshitz invariant is allowed [64]. The only new mathematical feature is that for reconstructive phase transition the full set of periodic solutions contains few distinct branches corresponding to different root couples, the solutions are hyperelliptic integrals. These are rare in problems in physics, while elliptic ones are common. Concerning the physics, the difference is much more remarkable as the existence of such branches reflects the possibility of phase transitions between different incommensurate phases. The analysis of such sequential phase transitions observed, e.g., in Eu [65] could be an exciting topic in itself.

Another scenario comes into play if, after elimination of inhomogeneous strains, the sign of the renormalized coefficient at the order-parameter quadratic gradient invariant becomes negative. Then, in the vicinity of a continuous transition to the incommensurate phase, the corresponding linearized equations of state assume the same form as those near the Lifshitz point at the Landau-type phase transition. Nevertheless, the fully developed incommensurate structure is different since the thermodynamic potential at reconstructive phase transition is more complex than the ones describing the conventional Landau incommensurate phases.

It is worth comparing our approach to the already existing models of inhomogeneous intermediate states observed at reconstructive phase transitions. In the framework of traditional crystal-geometrical models of strongly discontinuous martensitic transformations, the origin of anomalous satellites was attributed to the stacking faults accommodating the distortions of twinned martensite plates to the austenite matrix [66]. Polynomial expansion by strains is often used as a thermodynamic potential for crystals undergoing martensitic transformations (see, e.g., [67]). It corresponds actually to the case of proper ferroelasticphase transition of the Landau type. In this case, inhomogeneous structures are considered to be induced by defects. Following a similar strategy, i.e., inventing an analog of improper ferroelasticphase transition, a model was used in [53, 68] whereby a phonon mode, corresponding to atomic layer shuffling at martensitic transition, is a primary order-parameter which is coupled to the secondary uniform strains. One notices that both approaches are missing essential features of reconstructive phase transition and their order-parameters. The former approach fully neglects the contribution of the actual order-parameter in the thermodynamic potential (5.66), and the free energy is minimized only with respect to the strain-field components. Obviously, it allows one to determine the mesoscopic structure morphology in the phase coexistence region only if the contribution of the elastic energy is dominant in the potential. The latter approach suffers because of the crude approximation of the displacement field $\Delta(x)$ by a phonon in a Brillouin-zone point with rational coordinates, defined by a commensurate structure of the martensite phase. This is irrelevant to the case, as generally the spatial dependence of $\Delta(x)$ minimizing periodic thermodynamic potential at reconstructive phase transition cannot be described as a "frozen" phonon. In particular, the fact has been confirmed for Li experimentally [69]. Moreover, the unit-cell parameters of the Li lattice in austenite ($Im\bar{3}m$) and martensite ($R\bar{3}m$) phases allow a practically perfect coherent habit–plane interface between the parent BCC structure and an *unfaulted* 9R phase. It would be logical to conclude that a simple crystal-geometrical approach is an adequate tool, while martensitic transformation could occur via sidewise growth of the perfect martensite plates. However, instead of the latter, a mosaic of irregular martensite segments arises showing the presence of highly faulted or disordered polytype structures. It should be mentioned that the authors of Ref. [67] used *high-purity* samples, while in experiments with Li-based solid solutions, no similar close-packed polytypes were observed [70]. It appears to reveal the inherent character of the polytype-formation mechanism, while impurity suppresses the process. Thus, it is natural to believe that it is the existence of the above improper Lifshitz invariants that cause the instability of 9R with respect to complex polytype structures.

General symmetry requirements imposed on order-parameters which imply the mixed coupling term $\{\eta_i^n \cdot u_{ik} \cdot \nabla\eta_j\}$ should be derived following the standard group-theoretical procedures (see, e.g., Ref. [22]). Compared to the symmetry restrictions on the Lifshitz invariants, selection rules for the new gradient terms are less restrictive. Indeed, the symmetry properties of the conventional Lifshitz invariants $\{\eta_i \nabla \eta_k\}$ are described by the tensorial product of the antisymmetrized square $\{\eta_i^2\}$ of the order-parameter representation and of the vector representation spanned by the coordinates x_i. For such a term to be invariant, the antisymmetrized square of the order-parameter should contain the vector representation V: $\{\eta_i^2\} \supset \mathrm{V}$. In the above case of reconstructive PTs, the gradient operator ∇, spanning V, transforms to $u_{\mathrm{ik}}\nabla$, and the corresponding representation is a direct product of the one, $[U_{\mathrm{ik}}]$, spanned by the components of the symmetric second-rank tensor and V. The above condition transforms to $\{\eta_i^2\} \supset \mathrm{V} \otimes [U_{\mathrm{ik}}]$. The product, in the general case, is a *reducible* representation, and $\{\eta_i^n\}$ $(n \geq 2)$ can contain any of its irreducible parts.

For example, in the cubic point group $m\bar{3}m$, irreducible representation $\mathrm{F_{1u}}$ is the vector one, and it has to be present in the expansion of $\{\eta_i^n\}$. However, the operator $u_{\mathrm{ik}}\nabla$ transforms as $\mathrm{F_{2g}} \otimes \mathrm{F_{1u}} = \mathrm{A_{2u}} \oplus \mathrm{E_u} \oplus \mathrm{F_{1u}} \oplus \mathrm{F_{2u}}$, and any of four latter representations, but not exclusively $\mathrm{F_{1u}}$, can be in $\{\eta_i^n\}$. It is worth noting that the above condition formally forbids the existence of the coupling invariants, but does not totally forbid the onset of an incommensurate phase, since the latter can arise due to energetic but not symmetry reasons.

Generalizing, one can predict, for reconstructive phase transitions whose order-parameters allow high-order invariant $\{\eta_i^n \cdot u_{\mathrm{ik}} \cdot \nabla \eta_j\}$, a tendency for incommensurate phase to occur under the application of large stresses, or in a coexistence range of strain-dominant martensitic phase, or its onset in plastically distorted crystals. It is promoted by the increase, in the above cases, of impact of the mixed terms responsible for the formation of strain-induced inhomogeneous structures. The widespread extremely complex structures recently discovered in elemental crystals under high pressure (see review papers [29, 30]) seem to be good examples of the trend.

Summarizing, it is worth underlining that the approach developed in the section is *model-free*. It is *symmetry-based* and addresses the fundamental properties of phase transitions to a non-Landau state. Both the periodic dependence of order-parameter $\eta(\Delta)$ and the existence of effective improper Lifshitz invariants in the corresponding thermodynamic potential are controlled exclusively by the crystal symmetry. Due to weak symmetry restrictions for mixed gradient invariants, relevant to reconstructive phase transitions, the strain-induced incommensurate phases seem to be more probable than classical Landau ones.

5.7. Phonons and Martensitic Reconstructive Transformations

Let us first calculate the contribution of the secondary strains into the height of the energy barrier at a reconstructive transformation using, as an illustrative example, the BCC–HCP transformation. The form of the energy barriers in each relevant region of the phase diagram of Fig. 5.12 is shown in Fig. 5.30(a). Figure 5.30(b) represents the energy barrier along the BCC–HCP transition line. This line is given by the equation $\partial\eta/\partial\xi = 0$, which yields $\eta = \eta_0$ from Eq. (5.23), and by the equality of the effective potentials $F_1(\eta = 0) = F_1(\eta = \eta_0)$, where $F_1(\eta)$ is defined by Eq. (5.27). Therefore, the equation of the

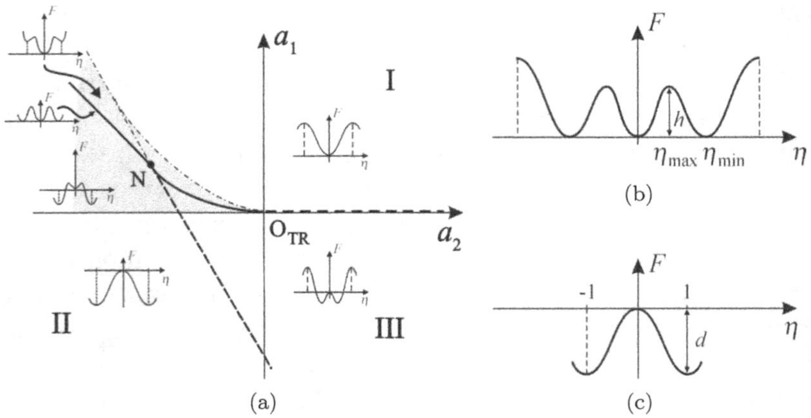

Fig. 5.30. (a) Profile of the energy barrier in each relevant region of the phase diagram of Fig. 5.12. (b) Energy barrier along the BCC–HCP reconstructive transition line. h is the height of the barrier. (c) Depth d of the potential well in the HCP phase.

BCC–HCP transition line is

$$\frac{a_1}{a_3} = -\frac{a_2}{a_3}\eta_0^2 - \eta_0^4, \tag{5.68}$$

which corresponds to a straight line in the (a_1, a_2) plane. As it is convenient to assume $\eta_0 = 1$, the coordinates of the minima of the energy barrier in Fig. 5.30(b), associated with nonzero values of the order-parameter, are

$$\eta_{\rm min}^2 = -\frac{a_2}{2a_3}, \tag{5.69}$$

whereas the coordinates of the closest maxima are

$$\eta_{\rm max}^2 = -\frac{a_2}{6a_3}. \tag{5.70}$$

From Eqs. (5.68) and (5.69), we can deduce the height h of the barrier

$$h = F_1(\eta_{\rm max}) - F_1(\eta_{\rm min}) = -\frac{a_2^3}{54a_3^2}, \tag{5.71}$$

which is a positive quantity since $a_2 < -2a_3\eta_0^2$ in the region of stability of the HCP phase. In order to evaluate the influence of the secondary strains e_3, e_6 and e_1 on h, one has to use the full effective form of $F_1[\eta(\xi), e_i]$ given by Eq. (5.49) to which must be added the elastic energy associated with the strains e_3, e_6, and e_1:

$$F_1^{\rm el}(e_i) = \frac{1}{2}(c_{11} - c_{12})e_3^2 + \frac{1}{2}c_{44}e_6^2 + \frac{1}{2}(c_{11} + 2c_{12})e_1^2, \tag{5.72}$$

where the $c_{ij}(i = 1\text{–}6, j = 1\text{–}6)$ are elastic constants. Minimization of $F_1[\eta(\xi), e_i] + F_1^{\rm el}(e_i)$ with respect to the e_i yields the equilibrium values of the spontaneous secondary strains as functions of the effective order-parameter η:

$$e_3^e = -\frac{\delta_1\eta^2}{c_{11} - c_{12}}, e_6^e = -\frac{\delta_2\eta^2}{c_{44}}, e_1^e = -\frac{\delta_3\eta^2}{c_{11} + 2c_{12}}. \tag{5.73}$$

Introducing the preceding expressions of the e_i in $F_1[\eta, e_i]$ yields the renormalized order-parameter expansion

$$F_1[\eta(\xi), e_i] = a_1\eta^2 + \tilde{a}_2\eta^4 + a_3\eta^6, \tag{5.74}$$

where $\tilde{a}_2 = a_2 - \frac{\delta_1^2}{c_{11}-c_{22}} - \frac{\delta_2^2}{c_{44}} - \frac{\delta_3^2}{c_{11}-2c_{22}}$ is smaller than a_2.

Accordingly, the renormalized height of the energy barrier is

$$\tilde{h}(e_i) = -\frac{\tilde{a}_2^3}{54a_3^2}, \tag{5.75}$$

which is lower than h. Hence, *the secondary strains favor the transformation from the* BCC *to the* HCP *phase.* This conclusion holds more generally for phase transformations of the reconstructive type and is consistent with the result obtained by Gooding *et al.* [71] in their model of the BCC–9R transformation.

At variance with a presupposed view that strongly first-order transitions imply strong anharmonicity of the lattice, let us now show that *martensitic reconstructive transformations may correspond to a jump between two harmonic potential wells.* In a phenomenological Landau-type approach, an inharmonic behavior can be related to two main different situations.

(i) A single order-parameter is associated with the transition. The anharmonicity is here expressed by the influence of the order-parameter invariants of degrees higher than two in the thermodynamic potential and is reflected in the potential well by the existence of nonequidistant energy levels (frequencies) corresponding to transitions to different possibly stable states. A second-order transition is induced by a continuous variation of the potential profile through the flat bottom regime corresponding to a soft mode frequency $\omega_{\mathrm{sm}} = 0$. The anharmonicity is weak in this case since it is essentially governed by the quadratic term in the potential. A first-order transition corresponds to a jump between two wells, and the description of the changes in the potential profile requires inharmonic equations. However, this does not necessarily imply that the critical phonon displays anharmonicity, as one has also to compare the soft mode frequency ω_{sm} and the height h of the energy barrier. If $\omega_{\mathrm{sm}} << h$, the crystal will remain in the fully harmonic regime across the transition. If ω_{sm} is of the order of h, an inharmonic regime should be observed for the critical phonon. Note that the phonon frequency should be more sensitive to temperature in this latter case.

(ii) More than one order-parameter is required for the transition mechanism. The anharmonicity is then reflected also by the coupling invariants between the order-parameters. The situation is more complex since the energy levels in the potential well possess nonzero widths, and the soft modes can be associated with different wave vectors. However, similar conclusions can be drawn as in the single-order-parameter case considering the relative magnitude of the soft mode frequencies and of the height of the energy barrier.

Let us illustrate the preceding considerations by some numbers extracted from concrete examples of reconstructive BCC–HCP transformations. In Ba, the value found for the jump in entropy [72] which is expressed in our model by $\Delta = -(\partial a_1/\partial T) \cdot \eta^2(T_1)$, where T_1 is the BCC–HCP transition temperature, gives

$$a_1 = -1.084\,\mathrm{cal/mol/K} \text{ and } a_2 = -0.084\,\mathrm{cal/mol/K}. \tag{5.76}$$

Assuming the normalizing value $a_3 = 1$, one gets an estimate of the barrier height $h = 5\,\mathrm{meV/atom}$. Since the order-parameter keeps a constant value $\eta = \eta_0 = 1$ in the HCP phase, the depth of the potential well (Fig. 5.30(c)) is $d = F_1(1) = a_1 + a_2 + a_3$. Its minimal value is reached on the second-order transition line, which separates, in Fig. 5.30(a), the HCP and intermediate orthorhombic phases. The equation of this line is $a_1 = -2a_2 - 3a_3$, which gives $d_{\mathrm{min}} = -a_2 - 2a_3 \cong -25\,\mathrm{meV/atom}$. Note that these estimates are in good agreement with the numbers found by Chen *et al.* [73] from first principles total energy calculations ($h \cong 4\,\mathrm{meV/atom}$, $d_{\mathrm{min}} \cong 27.2\,\mathrm{meV/atom}$) confirming the consistency of our approach.

Comparing h and d_{min} shows that the potential depth is about five times larger than the potential barrier. This ratio and the experimental value found for the frequency of the $T_1(N)$ phonon mode [73] in Ba, which is 3.1 meV, suggest that the atoms are in a reasonably harmonic situation at the BCC–HCP transformation. In Zr [72], $\Delta = -1.37\,\mathrm{cal/mol/K}$, from which one can deduce by analogous estimates $h \cong 15.5\,\mathrm{meV/atom}$ and $d_{\mathrm{min}} \cong -69\,\mathrm{meV/atom}$, which is of the same order as the value $d_{\mathrm{min}} \cong -45\,\mathrm{meV/atom}$ obtained by pseudopotential calculations [74]. The depth of the well is again about five times larger than the barrier height. It reflects a strong harmonic regime when compared to the soft mode frequency [74], which is 5.17 meV. A similar conclusion can be drawn [15] for Ti in which the entropy jumps [51] at the BCC–HCP transformation:

$\Delta = -0.866$ cal/mol/K corresponding to $h = 53.5$ meV/atom, while the frequency of the soft mode [74] is ~ 5 meV.

Coming back to the phase diagram of Fig. 5.30(a), let us now discuss in more detail the soft mode behavior when approaching the HCP phase from the BCC phase assuming that the transformation corresponds to a jump between two harmonic wells. In the following discussion, the BCC, HCP- and intermediate orthorhombic phases are, respectively, denominated I, II and III, and a_1 is assumed to symbolize the temperature axis.

For $a_2 > 0$, the second-order I–III transition takes place for $a_1 \sim \omega_{\text{sm}}^2 = 0$. The zero-frequency line $a_1 = 0$ prolongates, for $a_2 < 0$, onto the limit of stability between phase I and phase III. In the region $-3a_3/2 < a_2$, the I–III phase transition becomes first-order, and the hardening of the soft mode reflects the increasing magnitude of the structural distortion of phase I in phase III. The maximum distortion is reached on the continuous transition line between phase III and phase II.

Since phase II corresponds to a constant distortion of phase I structure, the soft mode frequency ω_{sm} reaches its maximum at the three-phase point N and then keeps a saturated value $\omega_{\max}$ across the reconstructive I–II transition. A rough estimate of the softening of the phonon mode when approaching phases II and III from phase I is the ratio $(\Delta\omega/\omega)^2 \cong \delta/\delta_{\max}$, where δ and $\delta_{\max}$ represent the projections on the a_1 axis of the thermodynamic paths shown in Fig. 5.31, i.e., δ is the projection of the path between the limit of stability lines I–II and II–III, while $\delta_{\max}$ is the maximum value of δ reached for $a_2 = 0$. Using the equations of the limit of stability lines and the difference of their coordinates along the a_1 axis (Fig. 5.31), one finds

$$\delta = \frac{a_2^2 + 8a_2a_3 + 12a_3^2}{4a_3}, \delta_{\max} = \frac{(a_2 + 3a_3)^2}{3a_3},$$

$$\text{and } \frac{\delta}{\delta_{\max}} = \frac{3}{4} + \frac{3a_3(2a_2 + 3a_3)}{4(a_2 + 3a_3)^2}, \tag{5.77}$$

which varies from zero at the point L ($a_1 = 3a_3, a_2 = 3a_3$), where the stability lines I–II and II–III merge, to 1 at the origin ($a_1 = a_2 = 0$). In other words, no softening should take place above the L-point ($a_2 > 3a3$, $\Delta\omega/\omega = 0$), whereas a maximum softening ($\Delta\omega/\omega = 1$)

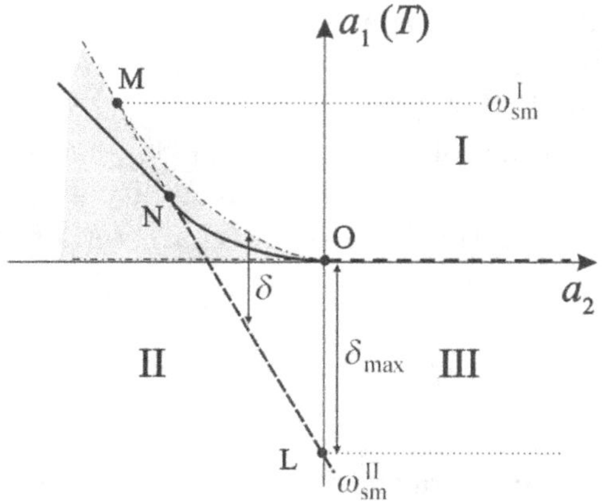

Fig. 5.31. Soft mode frequencies ω_{sm} along the transition lines separating the BCC phase from the HCP and intermediate phases. δ and δ_{max} are used for an estimate of the partial softening (5.76).

should occur at the origin. However, such behavior corresponds to an ideal limit situation (e.g., full harmonic regime, single order-parameter with no coupling to secondary variables). For real reconstructive transitions, it has to be interpreted in an extended and reciprocal way, namely, a *slight softening* of the phonon mode should be observed at a martensitic reconstructive transformation for thermodynamic paths close to the region of stability of the intermediate phase (III), whereas far from this region (for large negative values of a_2 in Fig. 5.31), *no softening should be observed.*

In summary, a number of suggestions have been pointed out in this section to explain the features of the martensitic transformations in Li and Na: coexistence of structures related to different order-parameters in metastability regions of the phase diagram, critical influence of secondary strains, harmonicity of the lattice vibrations at strongly first-order transitions, inherent incommensurate character of the martensite phase. These suggestions find more striking confirmations in the experimental results reported for the martensitic transformations in a number of alloys, in which the Σ_4 instability line is involved, the parent structure being either BCC (β-phase alloys) or of the CsCl type. Thus, an incommensurate modulation has been found between the CsCl high-temperature structure and the low-temperature 7M martensite phase of NiAl alloys [75–78]. The transition corresponds to another point of the Σ-line, around

$\boldsymbol{k} = \frac{1}{7}\left[\frac{2\pi}{a}, \frac{2\pi}{a}, 0\right]$, at which a softening of the transverse acoustic branch is clearly observed [76]. Incommensurate modulations have been also reported in TiNi(Fe) [79–81] and VR_4 alloys [82]. The theoretical interpretations of these data [83–86] stress the importance of the coupling between the primary instability and the macroscopic strains, but overlook the limit state character of the martensitic phase.

References

[1] S. K. Sikka, Y. K. Vohra and R. Chidambaram, *Progress in Materials Science* **27** (1982) 245.
[2] E. Tonkov, *Phase Diagrams of the Elements Under Pressure* (Nauka, Moscow, 1979).
[3] D. J. Cometto, G. L. Houze Jr. and R. F. Hehemann, *Trans. Metall. Soc. AIME* **233** (1965) 30.
[4] J. Silcock, *Acta Met.* **6** (1958) 481.
[5] Y. Bagariatskii, T. Tagunova and G. Nosova, *Dokl. Acad. Nauk SSSR* **105** (1955) 1225; *Acta Cryst.* **14** (1961) 1087.
[6] S. Sass, *J. Less Common Met.* **28** (1972) 157; *Acta Met.* **17** (1969) 813.
[7] D. De Fontaine, N. Paton and J. Williams, *Acta Met.* **19** (1971) 1153.
[8] O. V. Kovalev, *Irreducible Representations of the Space-Groups* (Gordon and Breach, New York, 1965).
[9] P. Toledano and V. Dmitriev, *Reconstructive Phase Transitions in Crystals and Quasicrystals* (World Scientific, Singapore, 1996).
[10] V. P. Dmitriev, S. B. Rochal, Yu. M. Gufan and P. Toledano, *Phys. Rev. Lett.* **60** (1988) 1958.
[11] D. De Fontaine and O. Buck, *Phil. Mag.* **27** (1973) 967.
[12] H. Cook, *Acta Met.* **21** (1973) 1445; **22** (1974) 239; H. Cook and W. Pardee, *Acta Met.* **25** (1977) 1043.
[13] S. C. Moss, D. T. Keating and J. D. Axe, *Solid State Commun.* **13** (1973) 1465; *Phys. Rev. Lett.* **35** (1975) 530.
[14] C. Stassis, J. Zaretsky and N. Wakabayashi, *Phys. Rev. Lett.* **41** (1978) 1726.
[15] W. Petry *et al.*, *Phys. Rev. B* **43** (1991) 10933; A. Heiming *et al.*, *Phys. Rev. B* **43** (1991) 10948; J. Trampenau *et al.*, *Phys. Rev. B* **43** (1991) 10963.
[16] S. L. Sass and B. Borie, *J. Appl. Crystallogr.* **5** (1972) 236.
[17] D. A. Young, *Phase Diagrams of the Elements* (University of California Press, Berkeley, 1991).

[18] Z. Nishiyama, *Martensitic Transformations* (Academic Press, New York, 1978).
[19] W. G. Burgers, *Physica* **1** (1934) 561.
[20] V. P. Dmitriev, Yu. M. Gufan and P. Toledano, *Phys. Rev. B* **44** (1991) 7248.
[21] H. T. Stokes and D. M. Hatch, *Isotropy Subgroups of the 230 Crystallographic Space Groups* (World Scientific, Singapore, 1989).
[22] J. C. Toledano and P. Toledano, *The Landau Theory of Phase Transitions* (World Scientific, Singapore, 1987), Chap. 3.
[23] Landolt-Bornstein Tables: *Metals: Phonon States, Electron States and Fermi Surfaces* (Springer-Verlag, Berlin, 1983), Vols. 13a, b, c.
[24] E. C. Bain, *Trans. AIME* **70** (1924) 25.
[25] H. Warlimont and L. Delaey, *Martensitic Transformations in Copper-Silver and Gold Based Alloys* (Pergamon, Oxford, 1974), Chap. 3.
[26] H. M. Strong, R. E. Tuft and R. E. Hanneman, *Metal. Trans.* **4** (1973) 2657.
[27] E. G. Ponyatovskii, *Kristallografiya* **4** (1959) 257.
[28] A. Jayaraman, W. Klement, R. C. Newton and G. C. Kennedy, *J. Phys. Chem. Solids* **24** (1963) 7.
[29] O. Degtyareva, *High Pressure Res.* **30** (2010) 343.
[30] M. McMahon and R. Nelmes, *Z. Krystallogr.* **219** (2004) 742.
[31] J. P. Hirt and J. Lothe, *Theory of Dislocations* (McGraw-Hill, New York, 1968).
[32] V. Vaks, M. I. Katsnelson, V. G. Koreshkov *et al.*, *J. Phys.: Condens. Matter* **1** (1989) 5319.
[33] C. Guillaume, E. Gregoryanz, O. Degtyareva *et al.*, *Nature Phys.* **7** (2011) 211.
[34] A. M. Schaeffer, W. Cai, E. Olejnik *et al.*, *Nat. Comm.* **6** (2015) 8030.
[35] G. J. Ackland, M. Dunuwill, M. Martinez-Canales *et al.*, *Science* **356** (2017) 1255.
[36] O. Blaschko and G. Krexner, *Phys. Rev. B* **30** (1984) 1667.
[37] W. Schwarz, O. Blaschko and I. Gorgas, *Phys. Rev. B* **46** (1992) 14448.
[38] G. Ernst, C. Artner, O. Blaschko and G. Krexner, *Phys. Rev. B* **33** (1986) 6465.
[39] W. Schwarz and O. Blaschko, *Phys. Rev. Lett.* **65** (1990) 3145.
[40] W. Schwarz, O. Blaschko and I. Gorgas, *Phys. Rev. B* **44** (1991) 6785.
[41] H. G. Smith, R. Berliner, J. D. Jorgensen and J. Trivisonno, *Phys. Rev. B* **43** (1991) 4525.
[42] R. Berliner, H. G. Smith, J. R. D. Copley and J. Trivisonno, *Phys. Rev. B* **46** (1992) 14436.
[43] H. G. Smith, *Phys. Rev. Lett.* **58** (1987) 1228.

[44] H. G. Smith, R. Berliner, J. D. Jorgensen, M. Nielsen and J. Trivisonno, *Phys. Rev. B* **41** (1990) 1231.
[45] H. G. Smith, R. Berliner and J. Trivisonno, *Phys. Rev. B* **49** (1994) 8547.
[46] R. Berliner, O. Fajen, H. G. Smith and R. L. Hitterman, *Phys. Rev. B* **40** (1989) 12086.
[47] C. S. Barrett and O. R. Trauz, *Trans. Am. Inst. Min. Metall. Pet. Eng.* **175** (1948) 579.
[48] C. S. Barrett, *Acta Crystallogr.* **9** (1956) 671.
[49] C. M. McCarthy, C. W. Tomson and S. A. Werner, *Phys. Rev. B* **22** (1980) 575.
[50] A. W. Overhauser, *Phys. Rev. Lett.* **53** (1984) 65.
[51] H. Abe, K. Ohshima, T. Suzuki, S. Hoshino and K. Kakurai, *Phys. Rev. B* **49** (1994) 3739.
[52] J. A. Wilson and M. de Podesta, *J. Phys. F: Met. Phys.* **16** (1986) L121.
[53] R. Gooding and J. A. Krumhansl, *Phys. Rev. B* **38** (1988) 1697.
[54] H. Abe and K. Ohshima, *J. Phys. Soc. Jpn.* **67** (1998) 2952.
[55] W. Pichl, *Scripta Metall.* **31** (1994) 1593.
[56] A. A. Bulbich, V. P. Dmitriev, O. A. Zhelnova and P. E. Pumpyan, *Sov. Phys.: JETP* **71** (1991) 619.
[57] A. L. Korzhenevskii and V. Dmitriev, *J. Phys.: Condens. Matter* **27** (2015) 375401.
[58] I. Loa, R. J. Nelms, L. F. Lundegard and M. I. McMahon, *Nat. Mater.* **11** (2012) 627.
[59] A. Arakcheeva, M. Bykov, E. Bykova *et al.*, *IUCrJ* **4** (2017) 152.
[60] Y. Ishibashi and V. Dvořak, *J. Phys. Soc. Jpn.* **44** (1978) 32.
[61] H. Shiba and Y. Ishibashi, *J. Phys. Soc. Jpn.* **44** (1978) 1592.
[62] V. A. Golovko, *Sov. Phys.: JETP* **60** (1984) 625.
[63] O. Blashko, V. Dmitriev, G. Krexner and P. Toledano, *Phys. Rev. B* **59** (1999) 9095.
[64] Yu. A. Izyumov and V. Syromyatnikov, *Phase Transitions and Crystal Symmetry* (Kluwer, Dordrecht, 1990).
[65] E. Hackmann, V. Kargamanova, J. Kunz and C. Lämmerzahl, *Phys. Rev. D* **78** (2008) 124018.
[66] A. G. Khachaturyan, S. M. Shapiro and S. Semenovskaya, *Phys. Rev. B* **43** (1991) 10832.
[67] S. R. Shenoy, T. Lookman, A. Saxena and A. R. Bishop, *Phys. Rev. B* **60** (1999) R12537.
[68] R. J. Husband, I. Loa, K. A. Munro *et al.*, *Phys. Rev. B* **90** (2014) 214105.
[69] M. Krystian and W. Pichl, *Phys. Rev. B* **62** (2000) 13956.

[70] Ch. Maier, R. Glas, O. Blaschko and W. Pichl, *Phys. Rev. B* **51** (1995) 779.
[71] R. G. Gooding, Y. Y. Ye, C. T. Chou, K. M. Ho and B. N. Harmon, *Phys. Rev. B* **43** (1991) 13626.
[72] E. Yu. Tonkov, *High Pressure Phase Transformations* (Gordon and Breach, Philadelphia, 1992).
[73] Y. Chen, K. M. Ho and B. N. Harmon, *Phys. Rev. B* **37** (1988) 283.
[74] K. M. Ho, C. L. Fu and B. N. Harmon, *Phys. Rev. B* **29** (1984) 1575.
[75] S. M. Shapiro, J. Z. Larese, Y. Noda, S. S. Moss and L. E. Tanner, *Phys. Rev. Lett.* **57** (1986) 3199.
[76] S. M. Shapiro, B. X. Xang, G. Shirane, Y. Noda and L. E. Tanner, *Phys. Rev. Lett.* **62** (1989) 1298.
[77] S. M. Shapiro, B. X. Yang, Y. Noda, L. E. Tanner and D. Schayvers, *Phys. Rev. B* **44** (1991) 9301.
[78] Y. Noda, S. M. Shapiro, G. Shirane, Y. Yamada and L. E. Tanner, *Phys. Rev. B* **42** (1990) 10397.
[79] S. M. Shapiro, Y. Noda, Y. Fujii and Y. Yamada, *Phys. Rev. B* **30** (1984) 4315.
[80] S. K. Satija, S. M. Shapiro, M. B. Salamon and C. M. Wayman, *Phys. Rev. B* **29** (1984) 6031.
[81] P. Moine, J. Allain and B. Reuker, *J. Phys. F: Met. Phys.* **14** (1984) 2517.
[82] Y. Yamada, *J. Phys. Soc. Jpn.* **59** (1990) 2517.
[83] I. Folkins and B. M. Walker, *Phys. Rev. B* **40** (1989) 255.
[84] R. J. Gooding and J. A. Krumhansl, *Phys. Rev. B* **39** (1989) 1535.
[85] Y. Yamada, Y. Noda, M. Takimoto and K. Furukawa, *J. Phys. Soc. Jpn.* **54** (1985) 2940.
[86] Y. Yamada, Y. Noda and K. Fuchizaki, *Phys. Rev. B* **42** (1990) 10405.

Chapter 6

Ordering-Type Reconstructive Phase Transitions

6.1. Introduction

In the introductory example of order–disorder transition considered in Chapter 4 (Sec. 4.2.4), it has been shown that the loss of group–subgroup relationships between the disordered and ordered phases could only be realized when assuming a fractional concentration of the initial structure. It resulted in the creation of regularly distributed vacancies in the ordered phase, and subsequently to an increase of symmetry of the actual distribution of atoms with respect to the symmetry of the distribution function of the sites. It is needless to say that the preceding situation is not the more realistic one, as condensed matter has a strong tendency to form fully occupied close-packed atomic structures. Hence, structures with regularly distributed vacancies constitute exceptions, and order–disorder transitions do not have, in general, a reconstructive character: they are, in most cases, second-order or slightly first-order, and fulfill the group–subgroup relationship criterion [1, 2].

There exists, however, another type of ordering transitions, which is compatible with close-packing atomic configurations, and is, in some respect, *inherent* to such configurations, namely *transitions between close-packed structures*. Indeed, in the simplest hexagonal close packing of identical atoms, the stacking of identical layers which form polytypes and the transitions between them serve as a useful illustration of our logic.

6.1.1. *Polytype structures*

The International Union of Crystallography defines the notion of polytypism as follows: "An element or compound is polytypic if it occurs in several different structural modifications, each of which may be regarded as built up by stacking layers of identical structure and composition, and if the modifications differ only in their stacking sequence. Polytypism is a special case of polymorphism: the two-dimensional translations within the layers are (essentially) preserved whereas the lattice spacings normal to the layers vary between polytypes and are indicative of the stacking period" [3]. In close packing of identical hard spheres, any successive layer (sometimes termed as a "modular layer") relates to the underlying one by the *stacking vectors*

$$\boldsymbol{s}_1 = \frac{1}{3}\boldsymbol{a}_h + \frac{2}{3}\boldsymbol{b}_h + \boldsymbol{e};\ \boldsymbol{s}_2 = -\frac{1}{3}\boldsymbol{a}_h - \frac{2}{3}\boldsymbol{b}_h + \boldsymbol{e}, \tag{6.1}$$

where $\boldsymbol{a}_h$ and $\boldsymbol{b}_h$ denote hexagonal basic translation vectors within the stacking layers (Fig. 6.1) and $\boldsymbol{e}$ is a vector perpendicular to $\boldsymbol{a}_h$ and $\boldsymbol{b}_h$, its modulus is equal to the minimal interlayer distance $|\boldsymbol{e}| = d_{[001]}$ [4].

Crystal structures of some metals, alloys and inorganic compounds can be described in terms of close packing of equal spheres. In inorganic compounds and alloys, "small" atoms or ions occupy voids of close-packed "large" spherical atoms. Nature provides us with a rich variety of polytypic structures. The most known textbook examples are zinc sulfide (ZnS), silicon carbide (SiC), cadmium iodide (CdI_2) and gallium selenide (GaSe) (see Ref. [5]). These compounds demonstrate pronounced polytypism, each crystallizing in structures with different sequences of identical layers. It indicates actually a low energy of interlayer interaction, and therefore, easy possibility of changing the sequences in more ordered or disordered structures. This fact favors naturally the formation of *stacking faults* in ordered structures, and conversely, the formation of *disordered polytype structures*, which are random sequences (statistically disordered) of hexagonal closepacked layers. The existence of disordered polytypes is well established in materials such as ZnS or SiC [3]. Their ordering yields ordered polytypic structures which are formed by short sequences of stacked planes, as the HCP (2H, 4H, 6H) and FCC (3R) structures,

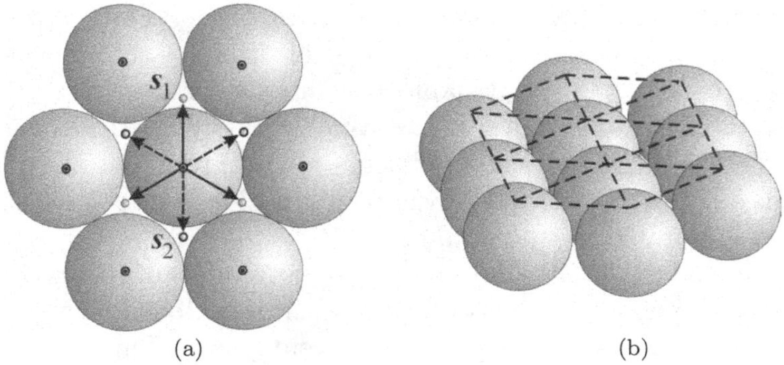

Fig. 6.1. Close packing of hard spheres. (a) Projection of the layer on the (001) plane. Centers of the large spheres are in A position; small gray dots represent the B position and small double circles denote the C positions. (b) Layer of equiradii closepacked spheres.

or by middle (7R, 9R, ...) or longer sequences (27R, 45R, ...) of hexagonal planes.

6.1.2. *Disordered polytype as a parent structure*

The problem of occurrence or the stability of disordered polytype structures in crystals has two aspects. On the one hand, a formal crystal–geometric model of polytypic phase transitions is not forbidden to use as a parent structure, a virtual one-dimensionally disordered polytype, considering it as an archetype structure — an approach typical in crystal-chemical and mineralogical classifications of crystal families. On the other hand, such an approach is missing energy arguments, and therefore, is not fully justified from the point of view of condensed matter physics. Although finding highly-disordered polytype structures in many inorganic compounds and minerals is a strong argument, some doubts on its global stability remain. Finally, statistical physics provided us with necessary justifications investigating a "crystal problem" applied to polytypism.

The "crystal problem" in statistical physics, which remains largely unsolved, is the problem of understanding whether the crystal state is a ground state for bulk matter or a disordered one ("non-crystalline" in its terms) may play this role (see, e.g., [6] and references therein). Polytypism is a limited version of the crystal problem, i.e., a three-dimensional structure is viewed as a well-defined crystalline in two dimensions but varies its order in the third one. Thus, statistical physics, in the framework of the classical discrete-state problem in one dimension [7], gives reliable conclusions on possible stacking sequences in real polytypes. It says that any polytype family should not be assumed to have ordered (crystalline) ground state [6]. The conclusion provides a solid physical ground for crystal–geometrical models elaborated in this chapter.

Along the preceding line, there exist a number of reconstructive transitions, the mechanisms of which can be described by an ordering process from a disordered polytype parent phase. This is, e.g., the case for the FCC–HCP transition which takes place in Co, He, Pr or Nd, i.e., when the FCC and HCP phases are not both adjacent to a parent BCC phase in the pressure–temperature phase diagram [8]. The HCP–FCC transitions in the preceding elements display a weakly first-order character [9–11] excluding the possibility of large displacements that would be required in a displacive reconstructive mechanism. In metal cobalt and carbon, a reordering mechanism between the phases is suggested by the observation of a strong tendency to form disordered polytype structures above the regions of stability of the FCC (or diamond) and HCP phases [12, 13]. However, in most of the examples discussed in this chapter, a disordered polytype structure is *assumed* to be the parent structure, i.e., it is a *latent* phase, as it provides the best crystal–geometrical correlation between the observed phases surrounding a reconstructive transition. In this case, the choice of an ordering mechanism, instead of a displacive one, is based also on strong experimental indices, such as the existence of regular distributions of stacking faults [14] or clear-cut indication of atomic ordering as in some phases of ice [15].

From the point of view of our approach, there exist precise additional criteria which help to recognize an ordering-type reconstructive transition as the absence of intermediate Landau-type phases, which appear in displacive mechanisms but are forbidden in ordering mechanisms. Let us note in this respect, that the diffusion process usually assumed for ordering transitions has a partially correlated character for reconstructive transitions, due to the in-layer close packing of atoms. Another indicative property is the existence of different types of stacking faults which result, in our interpretation, from the distribution of domain walls between antiphase domains.

In the *purely* ordering type of reconstructive transitions discussed in this chapter, the density-wave formalism does not play an important role, as we deal only with a variation of the waves' *amplitudes.* By contrast, when both ordering and displacive mechanisms are involved, such as in the graphite–diamond transition, the description of the transition in terms of density waves is essential for the understanding of the transition process.

As in Chapter 5, we will start from the simplest situation found among ordering-type reconstructive transitions: the FCC–HCP transition mechanism (and its variants), which possesses among these transitions the same universal character as Burgers mechanism among displacive-type reconstructive transitions.

6.2. The FCC–HCP Reconstructive Transition

6.2.1. *Crystallographic description and transition order-parameter*

Among the elements, the FCC–HCP transition is found in ^{3}He, ^{4}He, Co, Fe, Tl, Am and in most lanthanides [8]. The FCC and HCP structures represent the simplest close-packing configurations of hard sphere atoms. Their unit cells are formed by *three* and *two* stacked hexagonal layers, respectively, which are shown in Figs. 6.2(a) and 6.2(b). Each layer is shifted with respect to another by $a_h\sqrt{3}$ in the [120] HCP direction, where a_h is the basic vector of the (rhombohedral) FCC unit cell, in hexagonal coordinates.

For a better insight of the FCC and HCP structures, let us underline that, although a close packing of atoms represented by hard spheres occupying a minimal volume may be realized in several ways [5, 16], in real crystals close packing always corresponds to a *layered* configuration, which gives the possibility of isolating planes of atoms packed in the closest manner. These planes are stacked according to some rules of filling up space, and they represent hexagonal packing of spheres, each of which is in contact with six nearest neighbors (position A in Fig. 6.1(a)). In the centers of the triangles formed by neighboring atoms exist geometrically equivalent sites B and C (Fig. 6.1(a)). A spatial close packing is realized when each of the successive layers occupies the free spacings left by the preceding layer,

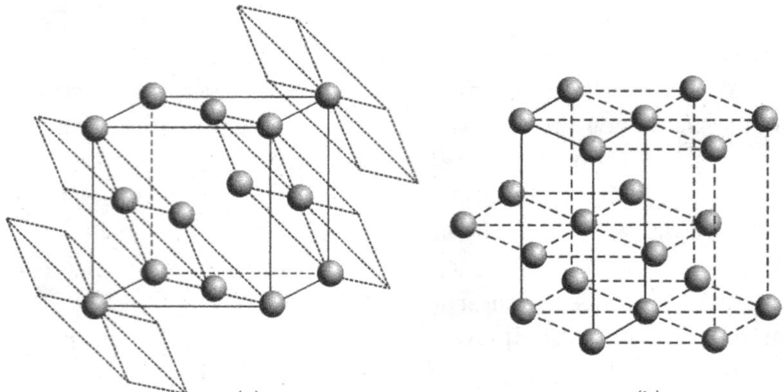

Fig. 6.2. (a) Three-layered FCC structure. (b) Two-layered HCP structure.

related to positions of the B or C type. The stacking order of the layers determines the close-packed structure.

Figure 6.1(b) shows a layer of equiradii (R) close-packed spheres. Within the layer, the close packing condition is $a_h = b_h = 2R$, where a_h and b_h are the hexagonal lattice parameters. For stacked layers, one has an additional condition reflecting the close packing in the third dimension, namely: $c_h = a_h\sqrt{(8/3)}$ for a two-layer stacking (HCP) structure, $c_h = a_h\sqrt{6}$ for a three-layer FCC structure, etc.

The basic vectors of the HCP unit cell shown in Fig. 6.2(b) can be expressed as functions of the basic vectors of the FCC rhombohedral (primitive) unit cell as

$$\boldsymbol{a}_h = \boldsymbol{b}_c - \boldsymbol{a}_c, \boldsymbol{b}_h = \boldsymbol{a}_c - \boldsymbol{c}_c, \boldsymbol{c}_h = 2(\boldsymbol{a}_c + \boldsymbol{b}_c + \boldsymbol{c}_c)/3. \tag{6.2}$$

From Eq. (6.2), or directly from the structures shown in Fig. 6.2, one can easily deduce that *the maximal substructure*, common to the HCP and FCC structures, is composed by a *monolayer hexagonal structure*. It is also easy to verify that this substructure, which will be denoted L, has the symmetry $P6/mmm$ and a unit-cell volume $V_{\mathrm{L}} = V_{\mathrm{H}}/6 = V_{\mathrm{C}}/3$, where V_{H} and V_{C} are the volumes of the HCP and FCC primitive unit cells, respectively. The L-unit cell, which is represented within the FCC and HCP structures in Figs. 6.3(a) and 6.3(b), is *filled by 1/3 atom*, i.e., it corresponds to an occupancy (a concentration) $x = 1/3$ for the L-structure. This fractional number must be understood as follows: in a given L-monolayer, the atoms

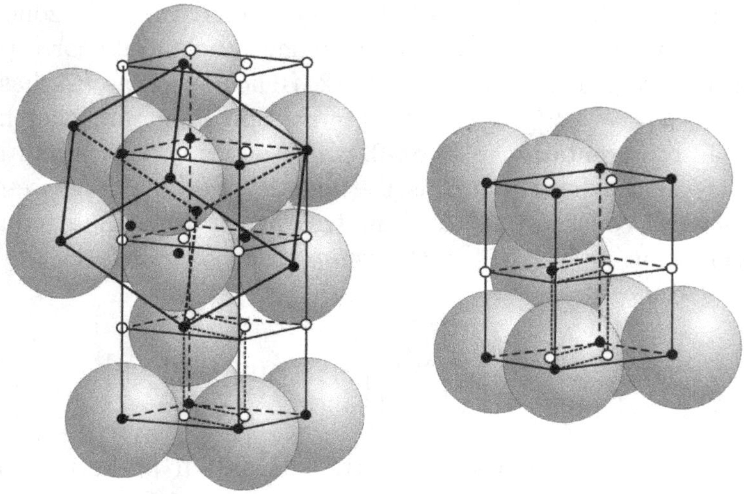

(a) (b)

Fig. 6.3. Unit cell of the disordered polytype structure within (a) the FCC structure, (b) the HCP structure.

occupy the positions 1(a):(000). Only one among the three (A, B and C) positions in Fig. 6.1(a) is occupied in the monolayer. In the next layer, the atoms cannot be again in position A, but only in position B *or* C, let us say B. In the following layers, they will thus occupy the position A or C, and so forth. In other words, the L-layer stacking corresponds to a statistically disordered polytype structure in which the A, B and C sites are equivalent: the 1(a) positions are occupied with equal probabilities by atoms and vacancies.

We will now assume that the disordered polytype L-structure is the *initial structure* for our description of the FCC–HCP reconstructive transition. With that goal, let us write the relationships between the basic vectors of the HCP and FCC unit cells, and the basic vectors ($\boldsymbol{a}_{\mathrm{L}} = \boldsymbol{s}_1$, $\boldsymbol{b}_{\mathrm{L}} = \boldsymbol{s}_2, \boldsymbol{c}_{\mathrm{L}} = \boldsymbol{e}$ from Eq. (6.1)) of the L-structure:

$$\boldsymbol{a}_h = 2\boldsymbol{a}_c = \boldsymbol{b}_{\mathrm{L}}, \boldsymbol{b}_h = \boldsymbol{a}_{\mathrm{L}} + 2\boldsymbol{b}_{\mathrm{L}}, \boldsymbol{c}_h = 2\boldsymbol{c}_{\mathrm{L}}; \tag{6.3}$$

$$\boldsymbol{a}_c = \boldsymbol{a}_{\mathrm{L}} + \boldsymbol{b}_{\mathrm{L}} + \boldsymbol{c}_{\mathrm{L}} = \boldsymbol{b}_{\mathrm{L}}, \boldsymbol{b}_c = -\boldsymbol{a}_{\mathrm{L}} + \boldsymbol{c}_{\mathrm{L}}, \boldsymbol{c}_c = -\boldsymbol{b}_{\mathrm{L}} + \boldsymbol{c}_{\mathrm{L}}. \tag{6.4}$$

From Eqs. (6.3) and (6.4), one can deduce the wave-vectors expressing the breaking of translational symmetry at the *virtual* L–HCP and L–FCC transitions. One finds, respectively, $\boldsymbol{k}_{15} = \frac{1}{3}(\boldsymbol{a}_{\mathrm{L}}^* + \boldsymbol{b}_{\mathrm{L}}^*) + \frac{1}{2}\boldsymbol{c}_{\mathrm{L}}^*$, and $\boldsymbol{k}_{10} = \frac{1}{3}(\boldsymbol{a}_{\mathrm{L}}^* + \boldsymbol{b}_{\mathrm{L}}^* + \boldsymbol{c}_{\mathrm{L}}^*)$, where $\boldsymbol{a}_{\mathrm{L}}^*$, $\boldsymbol{b}_{\mathrm{L}}^*$ and $\boldsymbol{c}_{\mathrm{L}}^*$ are the reciprocal lattice vectors of the L-hexagonal Brillouin zone. $\boldsymbol{k}_{15}$ coincides with the H-point of the preceding zone and its corresponding invariance group is $\hat{G}(\boldsymbol{k}_{15}) = \bar{6}m2$. Hence, the star $\boldsymbol{k}^*_{15}$ has two branches: $\boldsymbol{k}^1_{15} = -\boldsymbol{k}^2_{15} = (4\pi/3a, 0, \pi/c)$. Since, on the other hand, the small representation $\hat{\tau}_1$ of $\bar{6}m2$, which describes the permutation of atoms in the L-structure, is the identity irreducible representation, one can construct the 2×2 matrices generating the irreducible representation $\tau_1(\boldsymbol{k}^*_{15})$ of the $P6/mmm$ space group as

$$\begin{matrix} \{6\,|\,0\}\{\bar{1}\,|\,0\} & \{1\,|\,3\boldsymbol{c}_{\mathrm{L}}\} & \{1\,|\,\boldsymbol{a}_{\mathrm{L}}\} & \{1\,|\,4\boldsymbol{a}_{\mathrm{L}}\} \\ \begin{bmatrix} 0 & 1 \\ 1 & 0 \end{bmatrix} & \begin{bmatrix} -1 & 0 \\ 0 & -1 \end{bmatrix} & \begin{bmatrix} \varepsilon & 0 \\ 0 & \varepsilon^2 \end{bmatrix} & \begin{bmatrix} 1 & 0 \\ 0 & 1 \end{bmatrix} \end{matrix} \tag{6.5}$$

with $\varepsilon = \exp(2\pi i/3)$. Using the transformation properties of the matrices (6.5), one gets the possible symmetries of the phases corresponding to the two-component order-parameter (η_1, η_2) associated with $\tau_1(\boldsymbol{k}^*_{15})$. Three symmetries are found: (I) $P6/mmm$ $(V \times 6)$ for $\eta_1 = \eta_2 \neq 0$; (II) $P6_3/mmc$ $(V \times 6)$ for $\eta_1 = -\eta_2 \neq 0$; and (III) $P\bar{6}m2$ $(V \times 6)$ for $\eta_1 \neq \eta_2 \neq 0$. The second solution corresponds to the six-fold unit cell and space-group symmetry of the HCP structure. Thus, the virtual L $\rightarrow$ HCP transition is associated with the direction of the two-component order-parameter in the ε_2 space, given by $\eta_1 = -\eta_2 = \eta \neq 0$. Let us emphasize that this transition is of the standard Landau type, the $P6_3/mmc\,(Z = 2)$ symmetry being group–subgroup related to the "latent" L-symmetry $(P6/mmm\,(Z = 1/3))$.

From a structural point of view, the L–HCP transition involves a splitting of the initial one-fold site 1(a) into three two-fold positions shown in Fig. 6.3(a): 2(b):(000,00 1/2), 2(c):±(1/3 2/3 1/4), and 2(d):±(1/3 2/3 3/4). As the concentration in the initial structure is $x = 1/3$, after the transition one of the preceding positions has to be occupied by two $(1/3 \times 6)$ atoms in the ordered HCP structure. As a result of close packing, the hard sphere system of atoms can occupy either position 2(c) or 2(d), which are crystallographically equivalent (Fig. 6.3(b)). Therefore, the ordered crystal has a two-layered hexagonal close-packed structure with the standard ratio for the unit-cell parameters $|\boldsymbol{c}_h|/|\boldsymbol{a}_h| = a_h\sqrt{(8/3)} \cong 1.63$.

The wave-vector $\boldsymbol{k}_{10}$, which is associated with the L–FCC translational relationship (6.4), is located on *the edge* of the hexagonal

Brillouin zone of the L-structure, along the K–H line. As its invariance group is $\hat{G}(\boldsymbol{k}_{10}) = 3m$, the star $\boldsymbol{k}_{10}^*$ has *four* branches:

$$\boldsymbol{k}_{10}^1 = \left(\frac{4\pi}{3a}, 0, \frac{2\pi}{3c}\right) = -\boldsymbol{k}_{10}^2, \boldsymbol{k}_{10}^3 = \left(\frac{4\pi}{3a}, 0, -\frac{2\pi}{3c}\right) = -\boldsymbol{k}_{10}^4.$$

From the identity irreducible representation of $3m$ (which describes the permutation of atoms in the L-structure), one can construct the generating 4×4 matrices associated with $\tau_1(\boldsymbol{k}_{10}^*)$ as

$$\begin{array}{ccccc} \{6\,|\,0\} & \{\bar{1}\,|\,0\} & \{1\,|\,3\boldsymbol{c}_h\} & \{1\,|\,\boldsymbol{a}_h\} & \{1\,|\,4\boldsymbol{a}_h\} \\ \begin{bmatrix} 0&0&0&1\\0&0&1&0\\0&1&0&0\\1&0&0&0 \end{bmatrix} & \begin{bmatrix} 0&1&0&0\\1&0&0&0\\0&0&0&1\\0&0&1&0 \end{bmatrix} & \begin{bmatrix} 1&0&0&0\\0&1&0&0\\0&0&1&0\\0&0&0&1 \end{bmatrix} & \begin{bmatrix} \varepsilon&0&0&0\\0&\varepsilon*&0&0\\0&0&\varepsilon&0\\0&0&0&\varepsilon* \end{bmatrix} & \begin{bmatrix} \varepsilon&0&0&0\\0&\varepsilon*&0&0\\0&0&\varepsilon*&0\\0&0&0&\varepsilon \end{bmatrix} \end{array} \tag{6.6}$$

from which we can deduce the possible symmetries induced by $\tau_1(\boldsymbol{k}_{10}^*)$ and the equilibrium values of the corresponding four-component order-parameter, denoted $(\zeta_1, \zeta_2, \zeta_3, \zeta_4)$, which are summarized in Table 6.1.

One can see in Table 6.1 that two solutions involve a threefold multiplication of the L-unit cell, in agreement with Eq. (6.4). They possess the space groups $R3m$ and $R\bar{3}m$. This latter solution actually coincides with the FCC symmetry due to the specific crystal–geometrical conditions fulfilled by the L-unit cell, which are:

Table 6.1. Stable states associated with the irreducible representation $\tau_1(\boldsymbol{k}_{10})$. Columns (b), (c) and (d) give, for the seven states numbered in column (a), the equilibrium values of the four-component order-parameter, the corresponding space group, and the multiplication of the L-unit cell, $\varepsilon = \exp\{2i\pi/3\}$.

(a)	(b)	(c)	(d)
1	$\zeta\zeta\zeta\zeta$	$P6/mmm$	9
2	$\zeta_1\zeta_2\zeta_3\zeta_4$	$P3m$	9
3	$\zeta_1\zeta_1\zeta_2\varepsilon^2\zeta_2$	$P\bar{3}m$	9
4	$\zeta\zeta\ 0\ 0$	$R\bar{3}m$	3
5	$0\ 0\ \zeta_1\zeta_2$	$R3m$	3
6	$\zeta_1\zeta_2\zeta_1\zeta_2$	$P\bar{6}m2$	9
7	$\zeta_1\zeta_2\zeta_2\varepsilon^2\zeta_2$	$P6mm$	9

(1) a concentration $x = 1/3$ of the L-structure, (2) a ratio $|\boldsymbol{c}_\mathrm{L}|/|\boldsymbol{a}_\mathrm{L}| = \sqrt{2}$, corresponding to a close-packed structure.

In order to understand why the fulfillment of the preceding conditions lead to a coincidence of the $R\overline{3}m$ and $Fm\overline{3}m$ symmetries, let us clarify some crystallographic aspects of the $P6/mmm \to R\overline{3}m$ mechanism. In this mechanism, the initial 1(a) position splits into the 1(a) and 2(c):$\pm(1/3\,1/3\,1/3)$ positions. Since atoms occupy only one of the 1(a) positions in the rhombohedral structure (Fig. 6.3(b)), they are localized at lattice nodes but not in sites located inside the unit cell. This appears in Figs. 6.4(a) and 6.4(b), where the position denoted X, e.g., is empty. As a consequence, additional four-fold axes are created in the structure, which are connected with the six-fold axis of the latent structure by $\{6_z|\boldsymbol{c}_\mathrm{L}\}_\mathrm{L}$ acting as $\{4_z|0\}_\mathrm{c}$. This correspondence is illustrated in Fig. 6.4(a), which shows that a four-fold axis exists only if atoms of the X-type are absent. If, on the contrary, such atoms are present, the four-fold rotations would bring the X atom in position Y, which is an empty position.

The coincidence $R\overline{3}m \leftrightarrow Fm\overline{3}m$ is also determined by the close-packing condition $|\boldsymbol{c}_\mathrm{L}|/|\boldsymbol{a}_\mathrm{L}| = \sqrt{2}$ which implies a $\pi/3$-angle between the vectors $\boldsymbol{a}_c$, $\boldsymbol{b}_c$ and $\boldsymbol{c}_c$. It is notorious that a rhombohedral lattice possessing equal angles α between its basic vectors has the FCC cubic symmetry for $\alpha = \pi/3$. Let us stress that the preceding close-packing condition (which corresponds to $|\boldsymbol{c}_h|/|\boldsymbol{a}_h| \cong 1.63$) is well

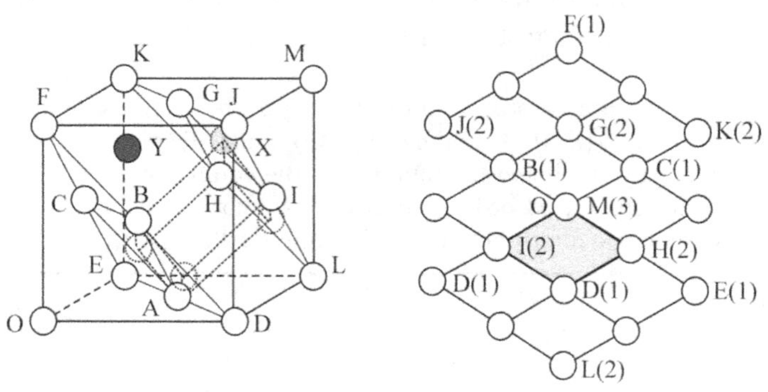

(a) (b)

Fig. 6.4. The figures show the connection existing between the concentration of the latent phase unit cell and the $R\overline{3}m \to Fm\overline{3}m$ increase in symmetry. In (a), a four-fold axis exists only if the atoms of the X-type are absent. (b) Projection on the (111) cubic plane of the FCC structure.

Table 6.2. Fulfillment of the close-packing condition for elemental crystals undergoing the FCC–HCP transition.

	a_h(Å)		c_h(Å)		
	Hexagonal (exp.)	Cubic (recalculated from experimental)	Hexagonal (exp.)	Cubic (recalculated from experimental)	c_h/a_h
Tl	3.457	3.392	5.525	5.518	1.52
Pb	3.265	3.514	5.387	5.717	1.65
Fe	2.461	2.589	3.952	4.412	1.61
La	3.770	3.770	6.080	6.133	1.61
Ce	3.673	3.444	5.901	5.602	1.61
Pr	3.672	3.465	5.918	5.636	1.61
Nd	3.658	3.408	5.899	5.544	1.61
Yb	3.880	3.895	6.386	6.336	1.64
He	3.450	3.010	5.519	4.897	1.62
Co	2.505	2.516	4.089	4.093	1.63
Sm	3.618	3.621	5.830	6.880	1.61
Am	3.468	3.475	5.620	5.652	1.62

satisfied for the elements undergoing the FCC–HCP transition, as shown in Table 6.2.

In summary, the L–FCC transition is associated to the non-vanishing components $\zeta_1 = \zeta_2 = \zeta \neq 0$ of a four-component order-parameter, and the FCC–HCP transition can thus be interpreted as *a transition between two "low-symmetry" ordered phases*, induced by different ordering processes from a common disordered (latent) poly-typic phase of hexagonal symmetry $P6/mmm$. Thus, the transition has the symmetry of a reducible six-dimensional order-parameter, with only four effective components $\eta_1 = \eta_2, \zeta_1 = \zeta_2$. Let us note that the corresponding reducible representation $\tau_1(\boldsymbol{k}_{10}) \oplus \tau_1(\boldsymbol{k}_{15})$ leads to three types of low symmetry states: those induced separately by (η_1, η_2) and by the (ζ_i), which have been listed above, and those resulting from the coupling of the two irreducible order-parameters. These additional states, in which components of the two order-parameters are non-zero, are given in Table 6.3. They involve an 18-fold multiplication of the L-unit cell, and will be shown, in the next section, to correspond to an intermediate region of the phase diagram located between the two preceding types of states.

Table 6.3. Stable states resulting from the coupling of the irreducible representations $\tau_1(\boldsymbol{k}_{10})$ and $\tau_1(\boldsymbol{k}_{15})$. Column (b) gives the equilibrium values of the order-parameter components, whereas columns (c) and (d) provide the corresponding space groups and multiplication of the L-unit cell. $\varepsilon = \exp\{2i\pi/3\}, \omega = \exp\{i\pi/3\}$.

(a)	(b)	(c)	(d)
1	$\zeta_1\zeta_2\zeta_2\zeta_1\eta_1\eta_1$	$P6mm$	18
2	$\zeta_1\zeta_2\zeta_1\zeta_2\eta_1\eta_2$	$P\overline{3}m$	18
3	$\zeta_1\zeta_2\zeta_2\zeta_1\eta_1 - \eta_1$	$P6_3mc$	18
4	$\zeta_1\zeta_1\varepsilon * \zeta_1\varepsilon\zeta_2\eta_1\omega * \eta_1$	$P6_3/mmc$	18
5	$\zeta_1\zeta_1\varepsilon * \zeta_1\varepsilon\zeta_2\eta_1\varepsilon\eta_1$	$P6/mmm$	18
6	$\zeta_1\zeta_1\zeta_2\varepsilon * \zeta_2\eta_1\omega * \eta_1$	$P\overline{3}m1$	18
7	$\zeta_1\zeta_1\zeta_2\varepsilon * \zeta_2\eta_1\varepsilon\eta_2$	$P\overline{3}m1$	18
8	$\zeta_1\zeta_2\zeta_3\zeta_4\eta_1\eta_2$	$P3m$	18

6.2.2. *Thermodynamic description: The phase diagrams of helium*

As a verification of the thermodynamic implications of the FCC–HCP transition mechanism described above, let us construct the theoretical diagram predicted for the latent L, FCC and HCP phases, and compare it to the experimental pressure–temperature phase diagram obtained for ^{4}He [17, 18].

The integrity basis corresponding to the six-dimensional reducible representation $\tau_1(\boldsymbol{k}_{10}) \oplus \tau_1(\boldsymbol{k}_{15})$ is given in Table 6.4. It contains 15 invariants. However, when taking into account the equilibrium relations corresponding to the two-component (η_1, η_2) and four-component $(\zeta_1, \zeta_2, \zeta_3, \zeta_4)$ order-parameters associated with the L–HCP and L–FCC transitions, respectively, which are $\eta_1 = \eta_2 = \eta \neq 0$, and $\zeta_1 = \zeta_2 = \zeta \neq 0$, $\zeta_3 = \zeta_4 = 0$, one gets the following effective potential:

$$F(\eta, \zeta) = a_1\zeta^2 + a_2\zeta^3 + a_3\zeta^4 + b_1\eta^2 + b_2\eta^4 + c\zeta^2\eta^2, \qquad (6.7)$$

which has been truncated to the fourth degree. The corresponding equations of state

$$\zeta(2a_1 + 3a_2\zeta + 4a_3\zeta^2 + c\eta^2) = 0, \qquad (6.8)$$

$$\eta(2b_1 + 4b_2\eta^2 + 2c\zeta^2) = 0, \qquad (6.9)$$

Table 6.4. Integrity basis corresponding to the six-dimensional reducible representation $\tau_1(\boldsymbol{k}_{10}) \oplus \tau_1(\boldsymbol{k}_{15})$.

Degree	Invariants
2	$I_1 = \zeta_1\zeta_2 + \zeta_3\zeta_4; I_2 = \eta_1\eta_2$
3	$I_3 = \zeta_1^3 + \zeta_2^3 + \zeta_3^3 + \zeta_4^3$
4	$I_4 = \zeta_1\zeta_3\eta_2^2 + \zeta_2\zeta_4\eta_1^2; I_5 = \zeta_1\zeta_2\zeta_3\zeta_4$
5	$I_6 = \zeta_1\zeta_2(\zeta_3^3 + \zeta_4^3) + \zeta_3\zeta_4(\zeta_1^3 + \zeta_2^3);$
	$I_7 = \eta_1^2(\zeta_1^2\zeta_4 + \zeta_2^2\zeta_3) + \eta_2^2(\zeta_3^2\zeta_2 + \zeta_4^2\zeta_1)$
6	$I_8 = \zeta_1^3\zeta_3^3 + \zeta_2^3\zeta_4^3; I_9 = \zeta_1\zeta_3\eta_1^4 + \zeta_2\zeta_4\eta_2^4;$
	$I_{10} = \zeta_1^2\zeta_3^2\eta_1^2 + \zeta_2^2\zeta_4^2\eta_2^2; I_{11} = \eta_1^6 + \eta_2^6; I_{12} = \zeta_1^3\zeta_4^3 + \zeta_2^3\zeta_3^3$
7	$I_{13} = \zeta_2\zeta_4\eta_1^2(\zeta_2^3 + \zeta_4^3) + \zeta_1\zeta_3\eta_2^2(\zeta_1^3 + \zeta_3^3);$
	$I_{14} = \eta_1^4(\zeta_4^2\zeta_1 + \zeta_3^2\zeta_2) + \eta_2^4(\zeta_1^2\zeta_4 + \zeta_2^2\zeta_3)$
8	$I_{15} = \eta_1^2(\zeta_1^2\zeta_4^4 + \zeta_2^4\zeta_3^2) + \eta_2^2(\zeta_1^4\zeta_4^2 + \zeta_2^2\zeta_3^4)$

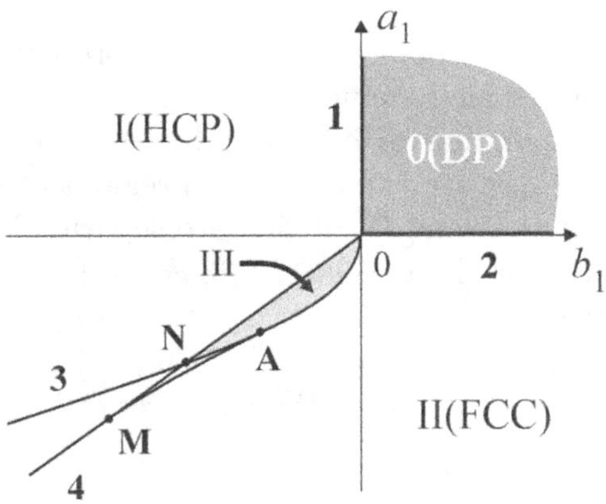

Fig. 6.5. Phase diagram corresponding to the potential (6.7) when $\Delta = 4a_3b_2 - c^2 < 0$ and $c > 0$. Only the limit of stability lines are represented. The diagrams correspond to a positive value of ζ.

can be solved following the examples discussed in Chapter 2 (Sec. 2.4). It leads to different phase diagram topologies depending on the signs of c, a_2 and $\Delta = 4a_3b_2 - c^2$. Figure 6.5 shows the phase diagram corresponding to $c > 0$ and $\Delta < 0$. It contains four phases denoted 0, I, II, III: 0 is the L-phase obtained for $\eta = 0$, $\zeta = 0$; I is the HCP phase, which verifies $\eta \neq 0$, $\zeta = 0$; II is the FCC phase ($\eta = 0, \zeta \neq 0$); and III is a six-layered structure (triple HCP) stabilized for $\eta \neq 0, \zeta \neq 0$.

In Fig. 6.5, the boundaries of the L-phase with respect to the HCP and FCC phases correspond to $b_1 = 0$ (line $0-1$) and $a_1 = 0$ (line $0-2$), respectively. The limit of stability of the HCP phase with respect to the FCC phase is also a straight line $a_1 = cb_1/2b_2$ (line $0-4$), and the value of the HCP order-parameter within phase I is given by $\eta^2 = -b_1/2b_2$. The boundaries of the FCC phase II are determined by the equations

$$\zeta(3a_2 + 6a_3\zeta) > 0, \text{ and } b_1 + c\zeta > 0. \tag{6.10}$$

Therefore, for positive values of ζ, one has the situation assumed in Fig. 6.5, in which the boundaries of phase II correspond to the parabolic branch 3–A–0, and to the $a_1 = 0$ line. In this case, the stability region between the lines 3–A–0 and 0–4 contains the domain of stability of phase III, and the line AM represents the limit of stability of phase III with respect to phase II.

The topology of the phase diagram is more complicated if the value of the FCC order-parameter ζ changes sign. Figures 6.6(a) and 6.6(b) (corresponding to $c > 0$ and $c < 0$, respectively) show that, in this case, there are two disjuncted regions of stability for the FCC phase, denoted II and II′, which are separated by a forbidden (unstable) interval for the ζ values: $(-3a_2/4a_3) < \zeta < 0$, where a_2 and a_3 are assumed to be positive (i.e., taking $a_2 < 0$ is equivalent to changing the sign of ζ in our considerations). Figure 6.7 shows the phase diagram corresponding to the same conditions $c > 0$, $\Delta < 0$,

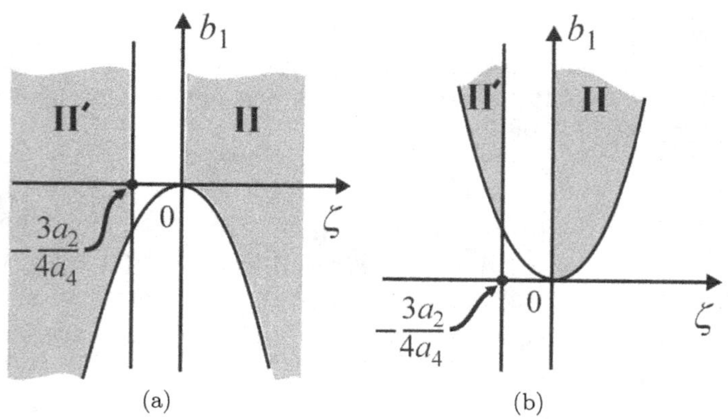

Fig. 6.6. Regions of stability of the FCC phase as a function of the ζ order-parameter for: (a) $c > 0$ and (b) $c < 0$.

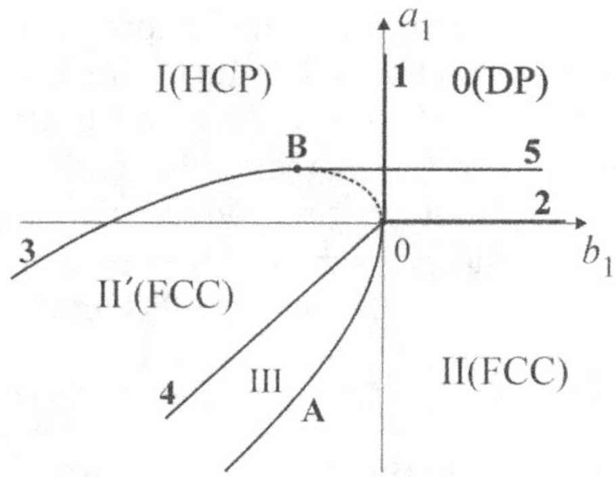

Fig. 6.7. Phase diagram corresponding to the potential (6.7), for $\Delta < 0$ and $\Delta > 0$, taking into account two regions of stability for ζ.

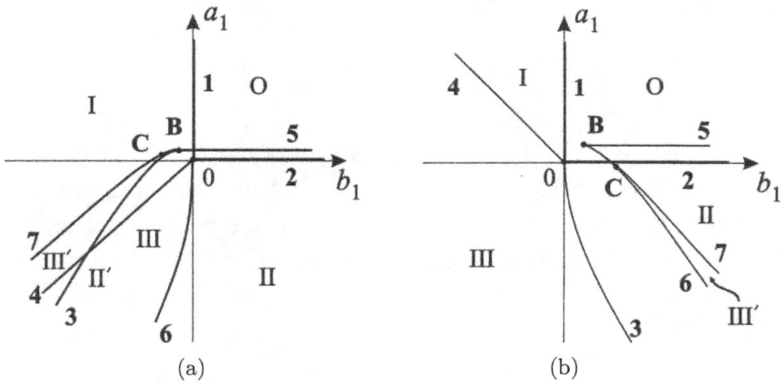

Fig. 6.8. Phase diagrams corresponding to the potential (6.7) for: (a) $\Delta > 0$, $c > 0$, and (b) $\Delta < 0$, $c < 0$.

but with two regions of stability of the FCC phase: II ($\zeta > 0$) and II′ [$\zeta < (-3a_2/4a_3)$], and limited by the parabola $b_1 = \pm c\zeta^2$. One can see that the boundaries of phase II′ extend to another portion of the parabola (3–B) and to the B–5 line. Here, phase III is located between phases II′ and II, in the region limited by the lines 4–0 and 6–A–0. Figures 6.8(a) and 6.8(b) assume, respectively, the conditions ($\Delta > 0, c > 0$) and ($\Delta > 0, c < 0$). One can verify that, for such conditions, phase III also splits into separate stability regions determined by different signs of ζ.

We can verify the reality of the preceding theoretical phase diagrams by comparison with the experimental phase diagrams obtained in a number of elements. The experimental points in Fig. 6.9(a) have been obtained by Franck [17] and Franck and Daniels [18] for the FCC–HCP transition in ^{4}He, on cooling and heating up to 9 kbar. They can be compared to the metastability region located near the N-point in Fig. 6.5. Figure 6.9(b) reproduces the surrounding of this region, together with the pressure and temperature axes, used for the linear transformation of the three coefficients a_1, b_1 and a_2:

$$a_1 = \gamma_1 P + \delta_1 T + \varepsilon_1, b_1 = \gamma_2 P + \delta_2 T + \varepsilon_2, a_2 = \alpha a_1 + \beta b_1 + B. \quad (6.11)$$

It yields the analytical equations of the 3–N line as follows:

$$a_1 = (3\beta c\tau^3 - 4a_3\tau^2 - 3B\tau) \cdot (2 + \alpha\tau)^{-1}, b_1 = c\tau^2, \quad (6.12)$$

where τ is a parameter. The 4–N line has the linear equation

$$a_1 = \frac{cb_1}{2b_2}. \quad (6.13)$$

Figure 6.9(a) shows the best fit obtained by the least squares method for Eqs. (6.12) and (6.13), with the experimental points. The numerical values taken by the adjusting constants, assuming a dimensionless thermodynamic potential (6.7) and the normalizing

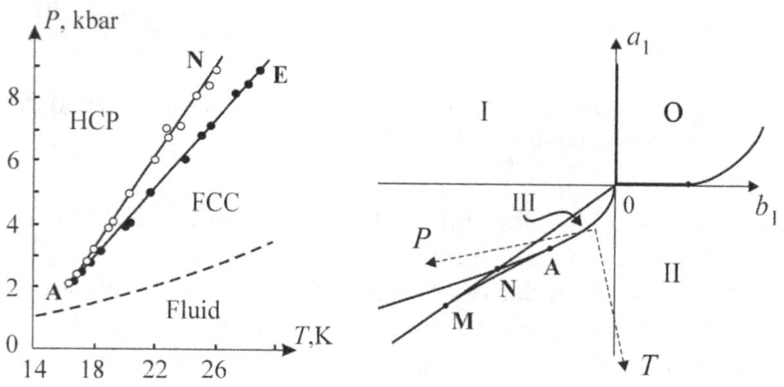

Fig. 6.9. The FCC–HCP transition in ^{4}He. (a) Experimental points from Ref. [17] and the best fit obtained from our calculations. The points are represented by small open circles on cooling and small solid circles on heating. (b) Details of the phase diagram of Fig. 6.5 in the region of measurements. Dashed lines are the pressure–temperature coordinates used for the fit.

condition $b_2 = 1$, are: $\alpha = -0.5$, $\beta = 0.556$, $a_3 = 3.286$, $B = -0.648$, $c = 3.76$, $\varepsilon_1 = 4.89$, $\varepsilon_2 = -2.6$, $\delta_1 = -0.471$/K, $\delta_2 = 0.159$/K, $\gamma_1 = -0.327$/bar, $\gamma_2 = -0.942$/kbar. These values allow to evaluate the distance of the N-point from the melting line, which is about 0.25 K, and the largest width of phase III, parallel to the T-axis, which is less than 0.15 K. The disordered polytype region in Fig. 6.9(b) lies outside the experimental region of stability of the ^{4}He crystal.

The decrease of the hysteresis region between the FCC and HCP phases, with decreasing temperature and pressure, as shown in Fig. 6.9(a), illustrates the fact that reconstructive transitions of the reordering type can be close to second-order, i.e., they can take place with a small change in the probability of occupancy of the atoms between the ordered structures. This decrease cannot be attributed to specific microscopic properties of ^{4}He, and to the Nernst principle, as our model is general and free from any microscopic assumptions on the transition mechanism.

6.3. Martensitic Transformation in Cobalt

6.3.1. *Introduction*

The FCC–HCP transformation in Co is currently designated as martensitic [19] due to its diffusionless character, considerable thermal hysteresis and the typical nucleation and growth processes. It shows a shape memory-type effect which concerns the relative orientation of the FCC and HCP structures when cycling across the transition [20]. Furthermore, the transformation temperature T_m as well as the amount and extent of transformed structures is drastically affected by various factors, namely, the external stresses or the alloying with soluble atoms (such as Fe and Ni) or with metallic and nonmetallic elements of limited solubility range. There exist, however, a number of features which make the transition in Co different than the martensitic transformations found, for example, in BCC-based metals and alloys. Thus, the transformation, which takes place at $T_m \cong 695$ K at ambient pressure is weakly first-order, as attested by small changes in enthalpy ($\Delta h = 113$ cal/mol on heating) and volume ($\Delta V/V \cong 3.33 \times 10^{-3}$) and by a sharp singularity of the specific heat [21]. On the other hand, no significant softening of the relevant phonon branches could be observed [22–26] in the low- (HCP) or high-temperature (FCC) phases but only a slight decrease of the c_{44} hexagonal shear constant [[24, 26] when approaching T_m from the HCP phase. There is also a strong tendency to disordering of the structure which is reflected in the dependence with temperature of the width of some diffraction lines [27] and anomalous diffuse scattering along certain directions in reciprocal space, which can both be interpreted in terms of stacking faults [28]. In the region of coexistence of the two structures around T_m, the disorder is also manifested in pure cobalt by the appearance of a modulated structure and in cobalt alloys by the stabilization of high-order polytypes.

6.3.2. *Specific features of the transformation in Co*

The HCP (α or ε) – FCC (β or γ) transformation in Co takes place around $T_m \cong 695\,\mathrm{K}$ at ambient pressure more than 1000°C below the melting transition ($T_f = 1768\,\mathrm{K}$) [8]. With increasing pressure above 70 GPa, the HCP–FCC transition line remains almost parallel to the melting line [29]. Despite the loss of group–subgroup relationships between the symmetries of the phases, the transition in Co has a weakly discontinuous (first-order) character. This is attested by the small jump in enthalpy ($\Delta h = 113\,\mathrm{cal/mol}$ on heating) [30] and volume ($\Delta V/V \cong 3.33 \times 10^{-3}$) [21] involved at T_m and by a relatively small hysteresis of about 20 K at ambient pressure [30]. It is also reflected in the close relationship between the structural features of the two phases, namely, (1) the distance between the close-packed planes varies only by about 0.3% at T_m, both structures displaying the almost ideal close-packed rate $c/a \cong 1.623$, (2) the same atomic coordination exists for the first and second nearest neighbors, and (3) the FCC and HCP lattices reversibly connect with the epitaxial relationships $(111)_{\mathrm{FCC}}//(001)_{\mathrm{HCP}}$ and $[11\bar{2}]_{\mathrm{FCC}}//[120]_{\mathrm{HCP}}$.

The main distinctive feature of the transformation in Co is the strong asymmetry of the FCC → HCP and HCP → FCC thermodynamic paths which manifests in the following properties:

(i) The average value of the transformation enthalpy is different on heating (113 cal/mol) and cooling (84 cal/mol) [30]. (ii) The HCP → FCC transformation is always complete, but the reverse FCC → HCP transformation is incomplete; even at room temperature, weak reflections of the FCC structure are still present. (iii) The DHCP ε' phase found at high pressure [31] is stabilized on quenching the FCC phase but not on heating the HCP phase. (iv) The disorder behavior of HCP and FCC Co differ greatly [32]. The FCC phase just above T_m is well ordered. In contrast, a stacking disorder is always present in the HCP phase even during the early stages of the transformation. This is attested by the observation of diffuse streaks along the $[10\zeta]_{HCP}$ direction which are not detected in the FCC structure [32]. Analysis of the Debye–Waller factor reveals no anomaly in the HCP phase but an increase of the atomic square amplitude in the FCC phase on approaching T_m from above [26]. The FCC and HCP precursor regimes are also different [26]. Independently from the cubic domains, the c_{44} hexagonal elastic constant decreases some 50 K below T_m by above 27% [22], but the corresponding FCC constant $(c_{11} - c_{12} + c_{44})/3$ shows no equivalent [26].

Other intriguing features of the transformation in Co are as follows:

(1) The form of the specific heat anomaly which shows a sharp increase from both sides of T_m with almost the same slope [33], at variance with the current shape of c_p found at first-order transformations (see Sec. 4.4.1).
(2) The absence of soft mode behavior when approaching T_m from the FCC or HCP sides despite the c_{44} anomaly, which is accompanied by an increase of the internal friction [34] and a simultaneous decrease of the entire branch of transverse acoustic vibration corresponding to the mode. In the following subsections, we give a unified description of the preceding properties in the framework of the theoretical model developed in Sec. 6.2.

6.3.3. *The phonon spectrum of cobalt and the elastic anomaly*

Investigations of the dynamical properties of cobalt [22–26] have been performed in order to verify the eventual existence of a soft mode in connection with the martensitic nature of the transformation and with the suggestion [35] that its mechanism could be triggered by a small decrease of phonon energy related to small displacements. However, no softening behavior has been found when approaching the transition either from the HCP [27] or FCC [22, 26] sides.

This is consistent with the purely reordering character of the transformation assumed in our approach since the average atomic positions remain fixed across ordering-type transitions, and the jumps of atoms between the sites are supposed to be uncorrelated. This picture is still reinforced by the reconstructive character of the transition, which, as shown in Sec. 4.5.1, implies no critical fluctuations.

The absence of temperature dependence of the relevant phonon branches constitutes, however, only a partial confirmation of reordering-type mechanism. A more precise confirmation of the mechanism can be found in the frequency dependence $\omega(\boldsymbol{k})$ on the reduced $\boldsymbol{k}$-vector of the phonon spectra as it can be used to verify the coincidence of the structures surrounding the transformation in the directions of space preserved by the reconstruction of the lattice, i.e., one expects to find a coincidence of the phonon branches for the two structures in these directions and a softening of $\omega(\boldsymbol{k})$ at special points corresponding to the translational connection of the two structures. Let us show that an indirect confirmation of the underlying existence of a latent polytypic phase can be found in the phonon spectra reported for cubic [22, 23, 26] and hexagonal [24, 25] cobalt.

In order to disclose the structural relationship between the FCC and HCP structures in reciprocal space, let us represent the connections between the Brillouin zones of the three FCC, HCP and L structures. Figures 6.10(a)–6.10(c) show, respectively, the FCC and HCP Brillouin zones and their embedding within the L-hexagonal Brillouin zone. Figure 6.10(c) reveals that the coinciding symmetry directions for the three zones are $\Gamma\mathrm{A_L} = \Gamma\mathrm{L_C} = 2\Gamma\mathrm{A_H}$ and $\Gamma\mathrm{M_L}//\Gamma\mathrm{K_H}//\Gamma\mathrm{K_C}$ where the indices L, C and H refer to the L, FCC and HCP structures. Figures 6.11(a) and 6.11(b) show the phonon spectra obtained for the HCP and FCC phases of Co in the preceding directions. Figure 6.11(c) represents the superposition of the two spectra, the full circles corresponding to hexagonal cobalt [25] and the open circles to cubic cobalt [22, 27]. We can verify the following: (i) In the direction $\mathrm{GA_L}$, there is a perfect coincidence for the full and open circles which lie on the same phonon branches in agreement with the doubling of the basic translation in this direction. (ii) Along the ΓKM reciprocal space direction, there is a good agreement for two phonon branches, whereas for the third branch, the two structures give different curves.

This discrepancy can be explained by the fact that the experiments on FCC and HCP cobalt were performed at temperatures

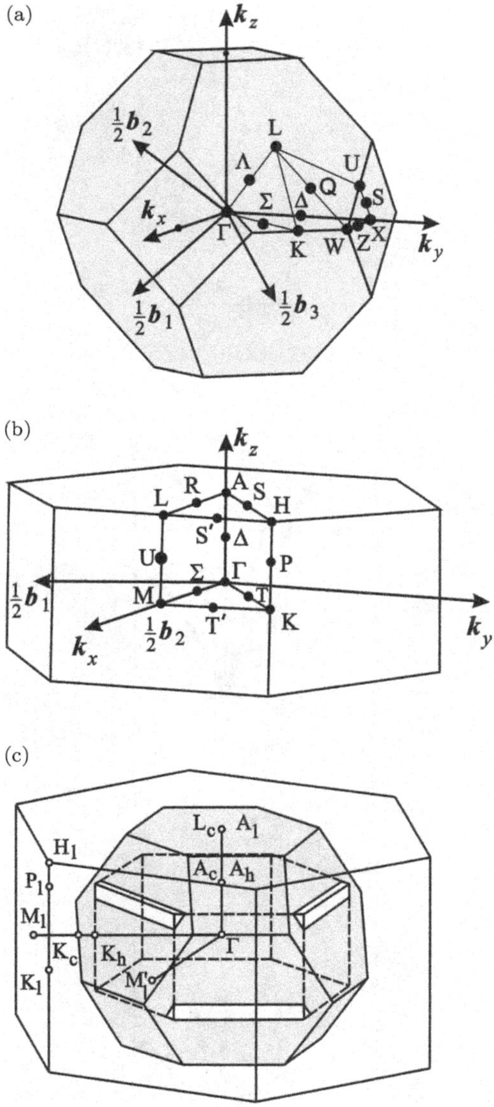

Fig. 6.10. (a) FCC Brillouin zone. (b) HCP Brillouin zone. (c) Embedding of the FCC and HCP Brillouin zones in the L-phase Brillouin zone.

differing by about 700 K [22, 23, 25], and by the observation by Frey *et al.* [24] that the slope of this branch increases when the temperature decreases. This explanation is supported by the convergence of the two curves when approaching the transition temperature T_m.

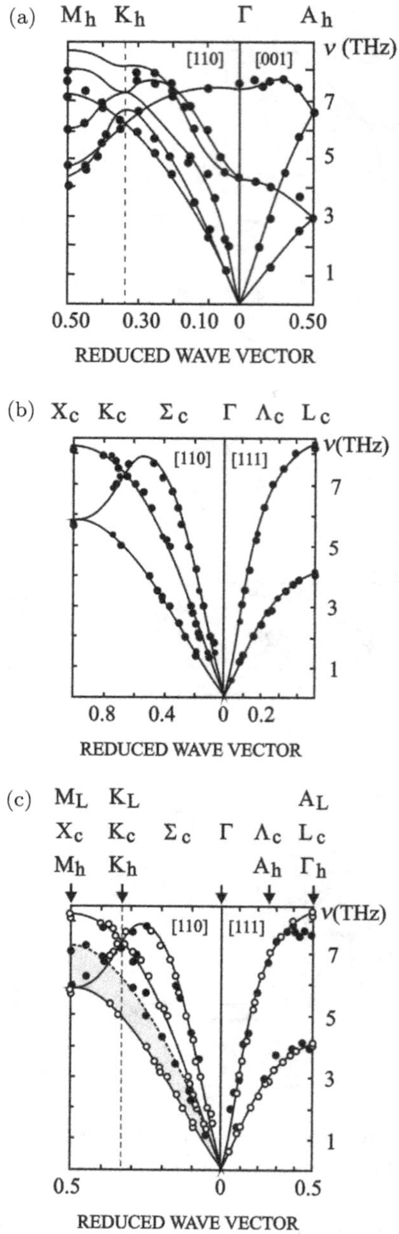

Fig. 6.11. (a) Phonon spectrum of hexagonal cobalt from Ref. [25]. (b) Phonon spectrum of cubic cobalt from Refs. [22,23]. (c) Superposition of the two preceding spectra. In (c), full and open circles correspond to hexagonal and cubic cobalt, respectively. Gray area shows difference between acoustic branches in cubic and hexagonal phases.

Note also that the hexagonal spectrum [25] was obtained on pure cobalt, while the cubic spectrum was measured on $Co_{0.92}Fe_{0.08}$ [22, 23]. (iii) The crossing of two branches at the K_h point in the hexagonal spectrum (Fig. 6.11(a)) is related to the fact that at this point two small IRs of the $mm2$ point group degenerate in a two-dimensional IR of the point group $\overline{6}m2$. However, the crossing at K_h of the two branches in the cubic spectrum (Fig. 6.11(b)), despite the fact that this point is not at the surface of the FCC Brillouin zone, constitutes a verification of the existence of the L-structure since K_h coincides (Fig. 6.11(c)) with the surface point K_L of the L-structure: At K_h, the cubic structure is reminiscent of the "parent" disordered L-structure.

The phonon spectrum may also serve as a test to check the influence of macroscopic deformations which are involved in reconstructive transformation mechanisms as such deformations modify the distance between equivalent sites in the initial and final structures (i.e., they change the *metric* of the structures). The optical phonon curves will thus display shifts with respect to their location in the undeformed structure, whereas acoustic phonon branches undergo changes in their slopes. In this respect, the small dip of the optical branch reported by Frey *et al.* [24] in the hexagonal phase in the direction ΓA_h at $2(0, 0, \pi/c)/3$ on approaching T_m can be related to the coupling of the primary optical instability to the secondary shear strain e_4, which as noted in Sec. 2.1 is required for the formation of the FCC structure. One has effectively $\frac{2}{3}c_h^* = \frac{1}{3}c_L^*$, which corresponds to the wave-vector $\boldsymbol{k}_{10}$ associated with the L–FCC virtual transition.

The preceding anomalous dip observed in the HCP phase represents a precursor indication of the shear strain $e_4 = e_{yz}$ arising spontaneously in the FCC structure. The corresponding acoustic instability will consist in a decrease in the slope of the hexagonal elastic constant c_{44}. c_{44} actually diminishes about 50 K below T_m from the value 0.73×10^{12} to $0.53 \times 10^{12}\,\text{dyn/cm}^3$ at T_m, i.e., $\Delta c_{44}/c_{44} \cong 27\%$. This has been first observed by Frey *et al.* [24] and confirmed by Strauss *et al.* [26]. The elastic constant anomaly is consistent with a dip of the Debye–Waller factor measured by Bokshtein *et al.* [36] and with the increase of the internal friction related by Bidaux *et al.* [34] to shear modes parallel to the hexagonal planes. The fact that in our approach no spontaneous strain is needed as a secondary order-parameter for the formation of the HCP phase is confirmed by the absence of elastic constant anomalies within the FCC phase on approaching T_m and especially of $c' \sim c_{11} - c_{12} + c_{44}$.

6.3.4. *Asymmetry of the interphase region*

The FCC–HCP transformation has been assumed in Sec. 6.2.1 to take place between two differently ordered structures corresponding to the distinct set of effective order-parameter components: $(\eta, -\eta)$ for the HCP phase and $(\zeta, \zeta, 0, 0)$ for the FCC phase. As shown for reconstructive transformations between SiO_2 polymorphs (see Sec. 6.5 and Ref. [37]), when two adjacent phases are associated to distinct subgroup symmetries of a parent phase, their interphase is asymmetric. Furthermore, if one considers a thick phase boundary, i.e., not reduced to one or two layers, there exists a surface S_m inside the phase boundary which coincides with a minimal deformation. Therefore, the volumes of the phases on each side of S_m will be different and possess the respective symmetries of the phases. On the other hand, since at the surface S_m one shifts from one to another set of order-parameter components $(\eta, -\eta) \to (\zeta, \zeta, 0, 0)$, the symmetry of the S_m surface will correspond to a cross-section of the parent phase.

The preceding picture applies to the FCC–HCP transformation in Co. It is consistent with the difference of FCC and HCP volumes involved in the region of coexistence of the two phases, which is much larger on the HCP side than on the FCC side. The S_m surface in Co has the simple hexagonal symmetry $P6/mmm$ of the disordered polytype structure, and the phase boundary between the HCP and FCC interphase volumes should be macroscopically distorted by the shear strain e_{xy} required for the formation of the FCC close packing but not for the HCP structure.

6.3.5. *Specific heat anomaly*

Since the transformation occurs between two differently ordered structures, the order-parameter moduli corresponding to the HCP and FCC phases fulfill the conditions

$$\frac{\eta}{\eta^k_{\max}} = \frac{\zeta}{\zeta^k_{\max}} = 1, \tag{6.14}$$

where $\boldsymbol{k}$ is the wave-vector corresponding to the ΓA direction in the hexagonal Brillouin zone and to the ΓL direction in the cubic Brillouin zone. Therefore, the temperature dependence of the effective order-parameter associated with the direct FCC–HCP mechanism will display a step-like behavior represented in Fig. 6.12(a). Such behavior is typical of reconstructive transitions between fully ordered phases.

From the conditions (6.12), one can deduce that the specific heat $c_p = -T(\partial^2 F/\partial T^2)_P$, where F is given by Eq. (6.7), has the same expression on both sides of T_m: $c_P^{\mathrm{HCP}} = c_P^{\mathrm{FCC}} = -T(\partial^2 F_0/\partial T^2)_P$. Therefore, $(\Delta c_p)_{T=Tm} = 0$, i.e., no jump of the specific heat takes place at the transition. This is in contrast with the upward jump found for Δc_p on cooling at first- and second-order transitions between group–subgroup related phases [31]. On the other hand, assuming $a_1 = a_0(T - T_{c1})$ and $b_1 = b_0(T - T_{c2})$ in Eq. (6.6) where a_0 and b_0 are positive constants, one finds a finite discontinuity of the entropy $S = -\partial F/\partial T$ at T_m, which on cooling is given by

$$S_{\mathrm{FCC}} - S_{\mathrm{HCP}} = b_0(\eta^k_{\max})^2 - a_0(\zeta^k_{\max})^2. \qquad (6.15)$$

Specific heat $c_p(T)$ will display a narrow δ-shaped peak at T_m as shown in Fig. 6.12(b). The two preceding properties of the specific heat ($\Delta c_p = 0$ and δ-shaped peak at T_m) have been shown (Sec. 4.4.2) to constitute a typical signature of reconstructive transformations. Such properties are well illustrated by the experimental curve found for $c_p(T)$ in cobalt [33], which is reproduced in Fig. 6.12(c).

6.3.6. *Stacking faults and domain structure*

From our approach to the FCC–HCP transition, one can deduce that the FCC and HCP structures are intrinsically faulted. There are two different origins for the stacking faults. One type of stacking fault which is symmetry induced and independent of temperature results from the existence of antiphase and orientational domains, which occur at the L–HCP and L–FCC transitions. Another type of (temperature-dependent) stacking faults relates to the ordering character of the order-parameters. Let us first analyze the stacking faults associated with the domain texture of the phases. At the

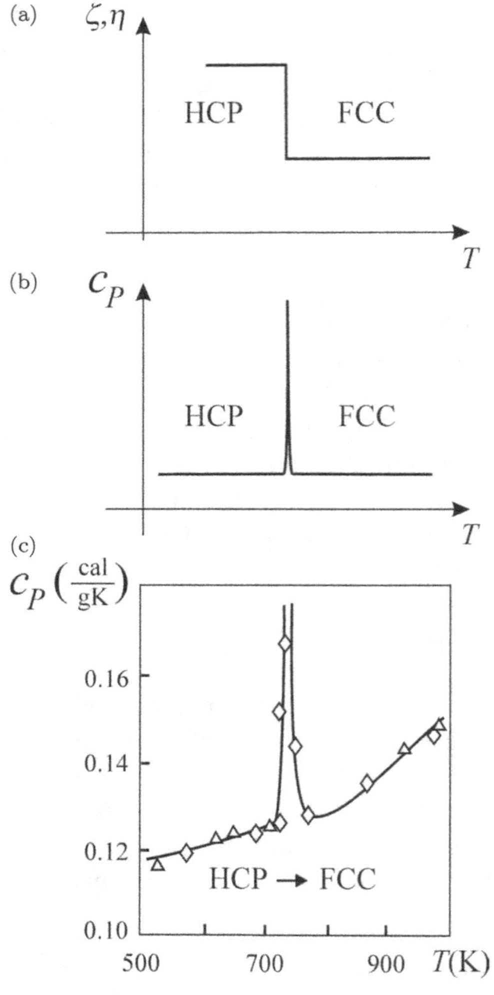

Fig. 6.12. Temperature dependence of (a) the order-parameter and (b) the specific heat c_p across a transition between fully ordered HCP and FCC states. (c) Specific heat anomaly measured at the FCC–HCP transformation in cobalt from Ref. [33].

L–HCP transition, the point group symmetry ($P6/mmm$–$P6_3/mmc$) is not modified but one has a six-fold decrease of the translational symmetry expressed by Eq. (6.2). Therefore, antiphase domains are created, which transform into one another by the translations lost at the transition. Table 6.5 lists the equilibrium values of the two-component order-parameter (η_1, η_2) for each of the six types of

Table 6.5. Antiphase domains corresponding to the L–HCP transition. (b) Equilibrium values of the two-component order-parameter (η_1, η_2) for each of the six types of antiphase domains numbered in column (a). (c) Sequence of layers corresponding to each type of domain. (d) Symmetry operations transforming domain 1 into other domains. $\omega = \exp\{i\pi/3\}, \varepsilon = \exp\{2i\pi/3\}$.

(a)	(b)	(c)	(d)
1	$\eta \quad -\eta$	*CBCB...*	$\{1\|0\}$
2	$-\eta \quad \eta$	*BCBC...*	$\{1\|\boldsymbol{c}_{\mathrm{L}}\}$
3	$\varepsilon\eta \quad \omega\eta$	*ABAB...*	$\{1\|\boldsymbol{a}_{\mathrm{L}}\}$
4	$\omega * \eta \quad \varepsilon * \eta$	*BABA...*	$\{1\|(\boldsymbol{a}_{\mathrm{L}} + \boldsymbol{c}_{\mathrm{L}})\}$
5	$\varepsilon * \eta \quad \omega * \eta$	*ACAC...*	$\{1\|(-\boldsymbol{a}_{\mathrm{L}})\}$
6	$\omega\eta \quad \varepsilon\eta$	*CACA...*	$\{1\|(\boldsymbol{a}_{\mathrm{L}} + \boldsymbol{c}_{\mathrm{L}})\}$

antiphase domains (column (b)) as well as the possible sequences of layers corresponding to each type of domain (column (c)). For example, the first domain $(\eta - \eta)$ implies a layer sequence of the type ... *CBCB*... whereas the domain $(-\eta\eta)$ yields the sequence ...*BCBC*.... The symmetry operators which transform the antiphase domain denoted 1 into the other antiphase domains are listed in column (d). The domain texture in the HCP phase must be compatible with the atomic packing. This excludes certain contacts between the domains that would generate sequences of the type *AA*, *BB* or *CC*. Therefore, the number of allowed contacts is limited. Using the standard notation [38] which labels *h* a hexagonal layer having an identical surrounding (e.g., *B* in *ABA*) and *c* a hexagonal layer having a nonsymmetric surrounding (e.g., *B* in *ABC*), one finds two sorts of possible contacts ...*hhhchhh*... and ...*hhhcchhh*..., both corresponding to *deformation* stacking faults [38].

At the L–FCC transition, the underlying change in the point group symmetry $(6/mmm \rightarrow \overline{3}m)$ and the corresponding threefold multiplication of the L-unit cell lead to two different types of domains: (i) *two* orientational domains transforming into one another by the lost six-fold rotation and (ii) *three* antiphase domains. Table 6.6 lists the six different types of domains and the corresponding equilibrium values of the four-component order-parameter (ζ_i). In column (c) of Table 6.6, one finds the layer sequences associated with each domain, e.g., ...*ACBACB*... for $(\zeta, \zeta, 0, 0)$

Table 6.6. Antiphase and orientational domains corresponding to the L–FCC transition. (a) Numbering of the domains. (b) Equilibrium values of the four-component order-parameter $(\zeta_1, \zeta_2, \zeta_3, \zeta_4)$ for each domain. (c) Layer sequences associated with each domain. (d) Symmetry operations transforming the first domain into the others. $\varepsilon = \exp\{2i\pi/3\}$.

(a)	(b)	(c)	(d)
1	$\zeta \;\; \zeta \;\; 0 \;\; 0$	$ACBACB\ldots$	$\{1\|0\}$
2	$\varepsilon\zeta \;\; \varepsilon*\zeta \;\; 0 \;\; 0$	$BACBAC\ldots$	$\{1\|\boldsymbol{b}_{\mathrm{L}}\}$
3	$\varepsilon*\zeta \;\; \varepsilon\zeta \;\; 0 \;\; 0$	$CBACBA\ldots$	$\{1\|(-\boldsymbol{a}_{\mathrm{L}})\}$
4	$0 \;\; 0 \;\; \zeta \;\; \zeta$	$ABCABC\ldots$	$\{m_z\|0\}$
5	$0 \;\; 0 \;\; \varepsilon\zeta \;\; \varepsilon*\zeta$	$CABCBA\ldots$	$\{m_z\|\boldsymbol{a}_{\mathrm{L}}\}$
6	$0 \;\; 0 \;\; \varepsilon*\zeta \;\; \varepsilon\zeta$	$BCABCA\ldots$	$\{m_z\|2\boldsymbol{a}_{\mathrm{L}}\}$

and $\ldots BACBAC\ldots$ for $(\varepsilon\zeta, \varepsilon^*\zeta, 0, 0)$. Column (d) indicates the symmetry operators which transform the first domain into the others.

As for the L–HCP transition, the contacts between the domains in the FCC structure must preserve the atomic close packing. At variance with the HCP structure, one finds two types of stacking faults: (1) Deformation-type stacking faults resulting from the contact between, e.g., the domains denoted 1 and 3 in Table 6.5. It gives the layer stacking $\ldots ACBABACB\ldots$ corresponding to $\ldots ccchhccc\ldots$; (2) Twining (growth) stacking faults due, e.g., to the contacts between the domains denoted 1 and 4. The layer sequence is in this case $\ldots ACBABCABC\ldots$ corresponding to $\ldots ccchccc\ldots$.

Accordingly, the domain structure and the structure of the domain walls lead to intrinsically faulted HCP and FCC structures, the distribution of stacking faults being symmetry induced. These stacking faults are independent from temperature, i.e., they cannot be annihilated by external fields (aging, annealing, etc.) and give rise to a single domain, since we deal with a probabilistic mechanism with no conjugated field. Another type of temperature-dependent defects can be found in the close-packed FCC and HCP structures originating in the nonmaximal character of the order-parameters, which is inherent in the assumed ordering-type mechanism. In the disordered polytype structure, each close-packed layer corresponds to a stacking fault. The ordering process leading to the HCP and FCC structures can be characterized by the number

$$\Delta = 1 - \frac{N_d}{N}, \tag{6.16}$$

where N is the total number of close-packed layers and N_d is the number of stacking faults. N_d/N represents the concentration of stacking faults, and one has $\Delta = 0$ in the disordered polytype structure and $\Delta = 1$ in the ideal close-packed structures. Intermediate states correspond to $0 < \Delta < 1$. The value of Δ at a given temperature and pressure is determined by the number of defects. Far from the transition within the close-packed phases, the asymptotic value of Δ will reflect the symmetry-induced type of stacking faults. Close to the transition, Δ accounts as well for the temperature-dependent defects. In the following subsection, we formalize such considerations in the framework of a phenomenological description of the ordering of close-packed structure, which is inspired by a model of segregation in complex fluids [39]. This will give a more realistic picture of the parent polytype structure assumed in our approach.

6.3.7. *Degree of order in the FCC and HCP structures*

The antiphase and orientational domains associated with a transition from a disordered polytype structure to HCP and FCC structures are listed in Tables 6.5 and 6.6. It has been stressed (Sec. 6.3.6), in our description of the ordering process to close-packed structures, that the preceding transitions should occur within periodically stacked domains which form the partially ordered state taking place below the melt. Therefore, the resulting deformation and growth stacking faults preexist in the HCP structures independently from the martensitic transformation between the two structures. When going from one structure to the other, domains are created originating in the loss of symmetry operations related to the absence of group–subgroup relationship between the symmetries of the structures.

At the FCC–HCP transformation, four types of orientational domains are produced, which can be deduced from one another by four-fold cubic rotations or equivalently by the four [111] cubic planes. In the HCP structures, two neighboring domains differ by 70.5° as found in Refs. [20] and [40]. Figures 6.13(a) and 6.13(b)

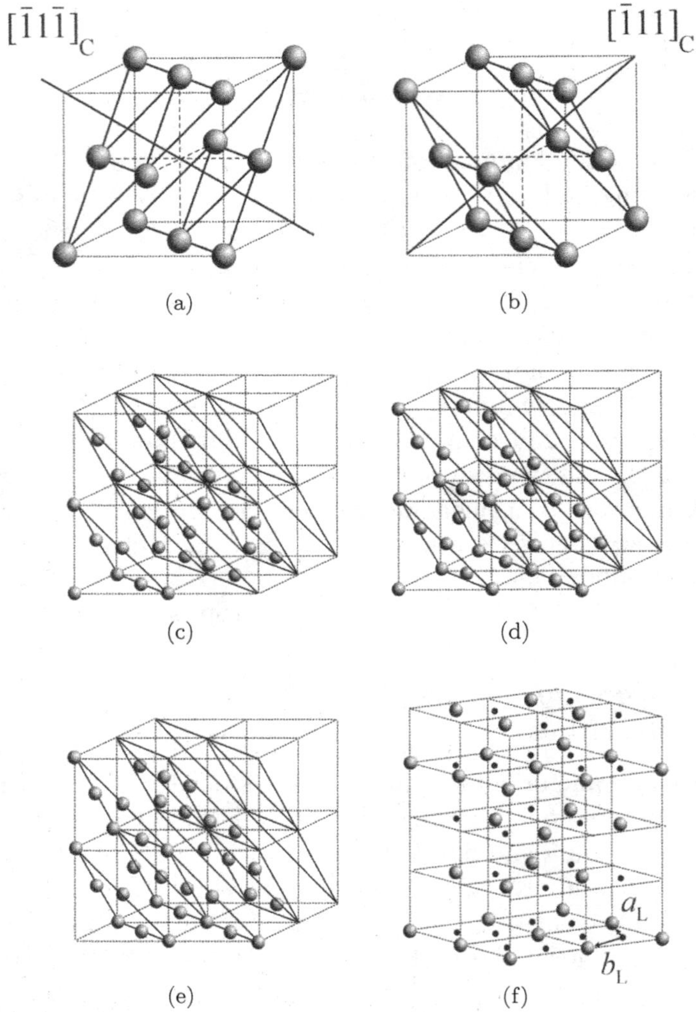

Fig. 6.13. (a) and (b) Two among the four cubic orientational domains within the HCP structure formed by *BCBCBC*... sequences of hexagonal close-packed planes in the cubic framework and corresponding to the [0001] hexagonal direction. (c)–(e) Three antiphase domains in the HCP structure associated with one cubic domain and corresponding to *ABAB*... (c), *ACAC*... (d), and *CBCB*... (e) sequences. (f) One of the three types of antiphase domains formed at the HCP–FCC transformation and corresponding to the *ABCABC*... sequence of planes.

represent two among the cubic domains formed by $BCBCBC\ldots$ sequences of hexagonal close-packed planes in the cubic framework and corresponding to the hexagonal [0001] direction. In addition, each cubic variant gives rise to three types of antiphase domains since the FCC translations are lost in the HCP lattice as shown by Eq. (6.1), at the corresponding cell doubling transition. Three antiphase domains corresponding to $ABAB\ldots$, $ACAC\ldots$ and $CBCB\ldots$ sequences are represented in Figs. 6.13(c)–6.13(e). The full set of 12 (antiphase and orientational) domains have been observed by Bibring and Sebilleau [40] and Nelson and Altstetter [20], whereas the variants associated to only three cubic directions were reported by Gaunt and Christian [41] and Babkevich *et al.* [42].

The reverse HCP−FCC transformation creates also three types of antiphase domains, which shift the atoms along the $\{2\bar{1}\,\bar{1}0\}_{\mathrm{L}}//\{\bar{1}010\}_{\mathrm{H}}$ and $\{\bar{1}2\bar{1}0\}_{\mathrm{L}}//\{0\bar{1}10\}_{\mathrm{H}}$ directions by $|a_a|/\sqrt{3}$. Figure 6.13(f) shows one of such cubic domains corresponding to the sequence of hexagonal planes $ABCABC\ldots$. Two other domains ($BCABCA\ldots$ and $CABCAB\ldots$) can be derived from the corresponding shifts of the $ABCABC\ldots$ structure. Since the cubic phase corresponds to a sheared rhombohedral $R\bar{3}m$ symmetry, two types of orientational domains may be produced which transform into one another by the (001) hexagonal plane. However, only one variant is generally observed after the transformation from the HCP phase [28].

As shown in Sec. 6.3.6, these different symmetry-induced variants should produce temperature independent (growth or deformation) stacking faults at variance with the temperature-dependent stacking faults inherently associated with a nonmaximal value of the ordering order-parameter. The detailed investigation by Frey and Boysen [32] based on elastic neutron scattering data on single crystals concludes that the HCP and FCC phases of Co are disordered by a smaller amount than reported in previous studies on powder samples [27, 43], i.e., 2.5% for the HCP phase and 0.5% for the FCC phase. The degree of disorder in the HCP phase is essentially due to the growth faults and is temperature independent when approaching the transformation being also not affected by aging of the sample across the transformation. In contrast, the FCC phase, below and above the transformation, is well ordered with no evidence of growth faults. However, the observation [20, 41], that a crystal of cobalt cycled through the transformation and remaining below 600°C gives on cooling the same single variant in the HCP phase, is in favor of the existence above the transformation of symmetry-induced (deformation) stacking faults.

Summarizing, the reordering process between the HCP and FCC cobalt phases, assumed in the model, takes place from a less-ordered HCP to a more-ordered FCC structure. In the two phases in the region surrounding the transformation, the observed (small) disorder corresponds to symmetry-induced stacking faults, which are predominantly of the growth type in the HCP phase and of the deformation type in the FCC phase. The residual random disorder associated with the ordering of the two close-packed structures below the melt is negligible.

6.3.8. *The transformation mechanisms between the FCC and HCP structures in cobalt*

Let us show that the order-parameter symmetries, assumed in Sec. 2.1 for the ordering mechanisms leading to the FCC and HCP structures, allow one to account qualitatively for the observations reported for the FCC–HCP and HCP–FCC transformations in the series of studies by Karnthaler and his co-workers [44–47] on cobalt and Co–Ni alloys as well as for the previously proposed mechanisms [48–58]. In Sec. 2.1, it has been shown that the FCC structure results from specific crystal–geometrical conditions fulfilled by the three-layered rhombohedral structure of symmetry $R\overline{3}m$ induced in the ordering process, which leads to an enlargement of the rhombohedral symmetry to cubic. Therefore, the FCC–HCP and reverse HCP–FCC transformations proceed via an underlying intermediate structure.

As already noted in Sec. 6.2.1, the $Fm\overline{3}m - R\overline{3}m$ symmetry change corresponds to a "pseudo-proper" ferroelastic transition [59] at which the spontaneous strain $e_4 = e_{yz}$ couples bilineary to the primary ordering parameter ζ. The shear stress s_{yz} conjugated to e_{yz} is parallel to the close-packed layers and induces as a secondary effect a deformation coinciding with the strain component e_{xx} between two hexagonal planes where the atoms are, e.g., in A and C positions. Such deformation results in a stretching of, let us say, the A atoms, which, due to the geometrical constraints, produces a jump-like shifting of the atoms from their A positions to the B positions as represented in Fig. 6.14. In this figure, one can see that this shifting gives rise to the onset of a glissile Shockley partial dislocation (also called "partial") characterized by one among the three equivalent Burgers vectors $\boldsymbol{p}_i$ of the $(\boldsymbol{a}_{\mathrm{H}}/3)$ $1\bar{1}00$ type where $\boldsymbol{a}_{\mathrm{H}}$ is the hexagonal lattice vector. Note that the onset of a partial in a layer formed by A atoms does not deform the layer with atoms in C positions, since the A atoms are located in the hollows of the C layer. Only the atoms of C type located above and below the dislocation line can be slightly shifted (but not necessarily) in the [111] cubic direction. The row of these out-of-plane atoms, therefore, appears as a nucleus for a dislocation in the C layer. This process, being repeated, leads to the growth of a hexagonal lamella in the cubic structure. Besides, the set of partials in successive planes appears as a sharp interphase front parallel to the $(10\bar{1}0)$ hexagonal plane (i.e., the cubic (210) plane) (Fig. 6.14).

The preceding atomistic mechanism is consistent with the dislocation mechanisms assumed in most models of the FCC–HCP transformation [50–58] and with the direct electron microscopy observations [60]. An almost similar mechanism leading to the onset of a partial dislocation in a close-packed hexagonal plane may be

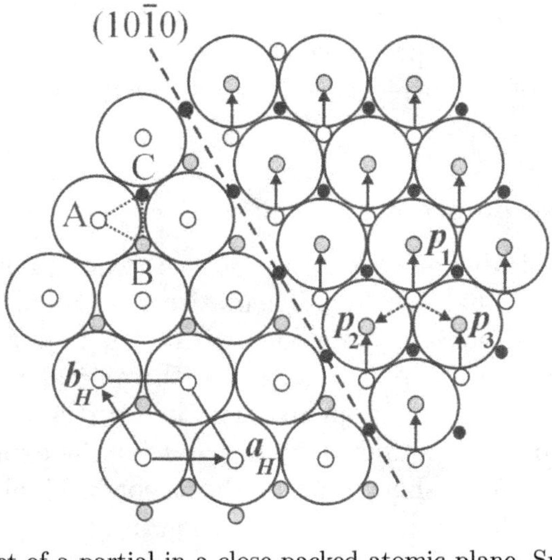

Fig. 6.14. Onset of a partial in a close-packed atomic plane. Small open circles represent positions A, small full circles are positions C. Gray small circles are positions B. The dashed line represents a glissile Shockley partial dislocation characterized by the $\boldsymbol{p}_i$ Burgers vector and belonging to the $(10\bar{1}0)$ hexagonal plane.

invoked for the reversed HCP–FCC transformation although it is less directly grounded on the corresponding order-parameter symmetry. The $P6/mmc \to R\overline{3}m$ symmetry change is an "improper" ferroelastic transition [61], i.e., it corresponds to the spontaneous onset of the elastic constant $c_{14} = c_{xxyz}$ as the result of an improper coupling of the order-parameter with c_{14} of the form $\eta^2 \cdot c_{14}$. Since the physical quantity conjugated to c_{14} is the stress product $s_{xx}s_{yz}$, it may lead again to the coupled strains e_{xx} and e_{yz}, which produce the stretching of hexagonal layers and the resulting partial dislocation shown in Fig. 6.14. Hence, the atomistic mechanism leading to partial dislocations in the hexagonal layers can be only considered as a secondary effect for the HCP–FCC thermodynamic path. A more consistent formation of the HCP–FCC transformation mechanism should be made in terms of an accumulation of stacking faults nucleating initial FCC lamellas of a few atoms in the HCP phase, which then grow and expand to form the FCC structure. Such mechanism is supported by the faulting disorder found in the HCP structure (Sec. 6.3.7), which is essentially symmetry induced (antiphase domains) as discussed in Sec. 6.3.6, and does not correspond to a random distribution of stacking faults as assumed by Fujita and Veda [54] or Pandey and Lele [58]. Note, however, that a small amount of disorder in the stacking fault distribution is predicted in our approach due to the nonmaximal character of the ordering parameter Δ expressed by Eq. (6.15).

6.3.9. *Phase diagram of cobalt*

Recently, two studies have focused on the experimental phase diagram of cobalt [31,62], which is schematized in Fig. 6.15(a). One can see that the HCP phase is embedded within the region of stability of FCC cobalt. When comparing with the theoretical phase diagrams of Figs. 6.5–6.8, it clearly appears that the fourth-degree expansion given by Eq. (6.6) is insufficient to account for the experimental features. Figure 6.15(b) shows one of the theoretical phase diagrams associated with the sixth-degree expansion in η:

$$F(T,P,\zeta,\eta) = F_0(T,P) = a_1\zeta^2 + a_2\zeta^3 + a_3^4 + b_1\eta^2 + b_2\eta^4 + b_3\eta^6 + c\zeta^2\eta^2, \quad (6.17)$$

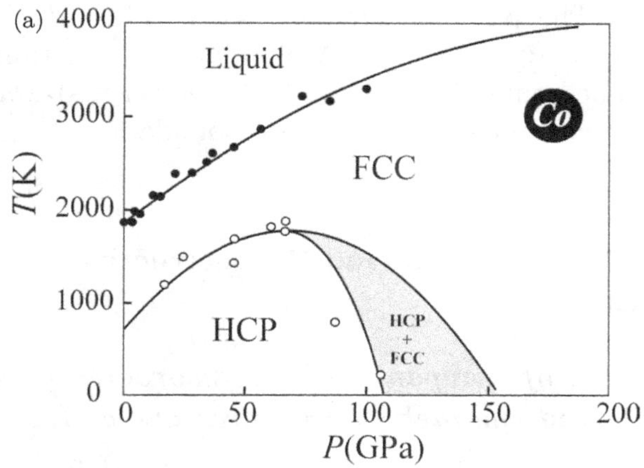

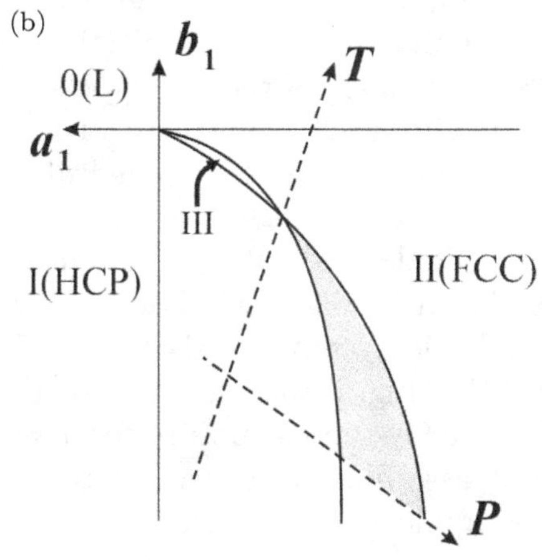

Fig. 6.15. (a) Experimental phase diagram of cobalt from Refs. [31, 62]. The shadowed area corresponds to the region of coexistence between the HCP and FCC phases. (b) Theoretical phase diagram in the (a_1, b_1) plane, associated with the free-energy expansion defined by Eq. (6.16). The dashed arrows represent the pressure and temperature axes in reference to the phase diagram of (a).

which differs from the theoretical phase diagram of Figs. 6.5–6.8 by the experimental property that the limit of stability line between the HCP and FCC phases is curved toward the HCP phase as suggested by experimental phase diagram. In order to obtain a suitable fitting with the experimental curves, one needs to take into account a linear dependence of three coefficients $(a_1, a_2 \text{ and } b_1)$ on temperature and pressure. The phase diagram shown in Fig. 6.15(b) assumes $b_3 > 0$ and $\Delta_1 = b_2^2 - 3b_1b_3 > 0$. A more precise quantitative model (with determined numerical values of the coefficients) would require more experimental points for the phase boundaries.

6.4. The Graphite–Diamond Reconstructive Transformation

6.4.1. *Degree of occupancy of the disordered polytype cell, and symmetry of ordered cubic structures*

In the phenomenological approach to the FCC–HCP transition, an essential parameter appears to be the degree of occupancy x of the disordered polytype unit cell by the atoms (the concentration): for a concentration 1/3 (or smaller) of the polytype structure, the symmetry induced by $\tau_1(\boldsymbol{k}_{10})$ is $Fm\bar{3}m$, whereas for larger concentrations, let us say 1/2, the symmetry is rhombohedral $R\bar{3}m$. However, the dependence on x of the preceding "low symmetry" structure is not straightforward, and it depends also on the *geometry* of the latent unit cell, as it is shown in Fig. 6.16. For $BaTiO_3$, an angle of $\pi/2$ between the cubic lattice vectors leads to the $Pm\bar{3}m$ simple cubic structure for $x = 1/12$ (Fig. 6.16(b)), while for the same angles, one gets the octahedral ReO_3-type structure for $x = 1/4$ (Fig. 6.16(c)) or for $x = 1/3$ but assuming an ordered latent structure, i.e., two different crystallographic positions are occupied in the latent unit cell (Fig. 6.16(d)).

Figure 6.16(e) shows that for the same geometrical conditions fulfilled in the FCC–HCP model, namely a ratio $c_L/a_L = \sqrt{2}$, or equivalently an angle of $\pi/3$ between $\boldsymbol{a}_c$, $\boldsymbol{b}_c$, $\boldsymbol{c}_c$, *but for a concentration x=2/3, one obtains the rhombohedral graphite structure* ($R\bar{3}m, Z = 2$) [63]. This observation will be used in the present section, as a basis for the description of the (hexagonal) graphite–(cubic) diamond transition mechanism. More precisely, we will show that this mechanism can be understood as a transition between two ordered phases, which are deduced from a parent polytypic disordered phase, by combined ordering and shifting mechanisms for the diamond structure, and by ordering plus decompression for the graphite structure [64].

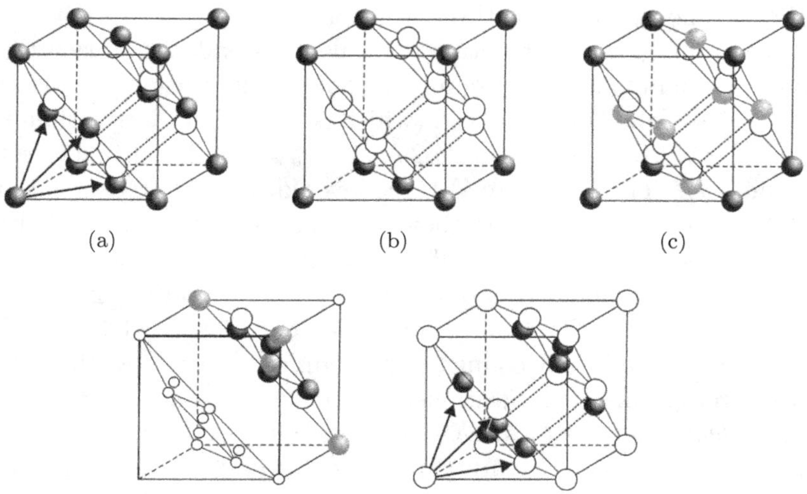

(a) (b) (c)

(d) (e)

Fig. 6.16. Relation between the symmetry of the cubic structure and the crystallo-geometrical parameters (angle between the cubic lattice vectors, concentration of the L-unit cell). Comments on (a)–(e) are in the text.

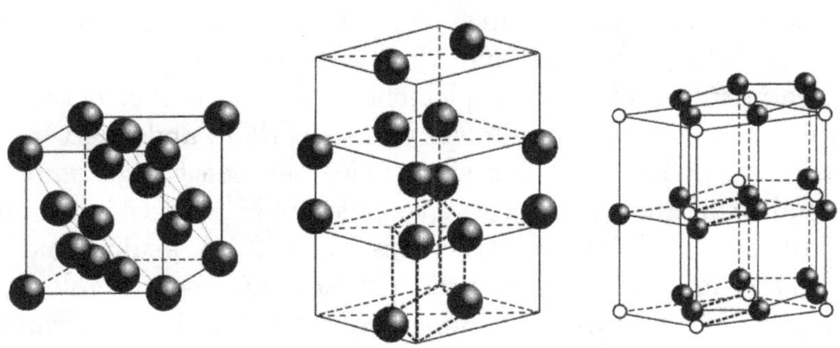

(a) (b) (c)

Fig. 6.17. (a) and (b) Structure of diamond in (a) cubic and (b) hexagonal settings. (c) Graphite structure.

6.4.2. *Crystal–geometrical properties of hexagonal graphite and cubic diamond*

The diamond structure corresponds to the space group $Fd\bar{3}m$ with two atoms in the primitive (rhombohedral) unit cell and eight atoms in the conventional cubic cell (Fig. 6.17(a)). The cubic parameter is $a_c = 3.567$ Å, and the minimal distance between atoms is 1.445 Å [65].

Although the packing of carbon atoms is compact, one cannot speak of close packing, as the structure is covalently bonded. Figure 6.17(b) shows the diamond structure in hexagonal configuration. The lattice parameters of the hexagonal cell (which is three times larger than the rhombohedral cell) are $a_d = 2.52$ Å and $c_d = 6.3$ Å, with six atoms per unit cell. The structure is three-layered, as the FCC structure, but *the layers are not flat*: they are composed by atoms, half of which are shifted up, and half shifted down, with respect to the average atomic planes, by about 0.26 Å. The c_d/a_d ratio is 2.50 which differs by less than 2% from the ideal FCC ratio of $\sqrt{6} \cong 2.45$. So, if we do not take into account the preceding deviations with respect to the average in-plane positions, *the structure of diamond appears as complementary to the FCC structure*: the occupied positions in diamond correspond to empty position in an FCC structure, and *vice versa*.

The structure of graphite, shown in Fig. 6.17(c), has the space group $P6_3/mmc$ with four atoms per unit cell. The hexagonal parameters are $a_g = 2.46$ Å and $c_g = 6.7076$ Å [66]. Hence, one has less than 2% difference between the a_g and a_d parameters: the (001) hexagonal planes of the two structures (or the (111) cubic plane) are nearly identical. Along the same line, the two-layered graphite structure can also be interpreted as complementary to the HCP structure: the occupied positions in graphite are empty in HCP, and *vice versa*.

Aside from the mentioned similarities between the graphite and diamond structures, there exists a considerable difference in the interlayer distance for the two structures: $d_d = 2.1$ Å in diamond and $d_g = 3.39$ Å in graphite. This property plays an essential role in the description of the graphite–diamond transition mechanism, which is proposed in the next section [64].

6.4.3. *The graphite–diamond transition mechanism*

Figures 6.18(a) and 6.18(b) represent, respectively, one unit cell of hexagonal graphite and a tripled unit cell of rhombohedral graphite. The maximal substructure common to these two structures is represented in the two figures: it is formed by a monolayer structure of symmetry $P6/mmm$, with a 2/3 occupancy of the unit cell, i.e., it is complementary to the maximal substructure common to the HCP and FCC structures, and also represents a statistically disordered

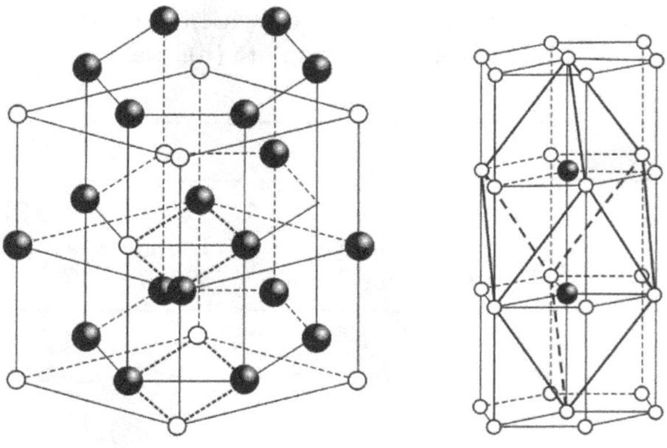

Fig. 6.18. (a) One unit cell of hexagonal graphite. (b) Tripled unit cell of rhombohedral graphite. The L-maximal substructure to the two preceding structures is represented in dashed line.

polytype structure. As in the preceding sections, we will use this latent structure as a parent phase (denoted L) for the graphite and diamond phases.

Columns (b) and (c) of Table 6.7 summarize the irreducible representations and mechanisms, which are, respectively, involved in the L $\to$ graphite and L $\to$ diamond transitions. One can see that the same wave-vectors $\boldsymbol{k}_{15} = \frac{1}{3}(\boldsymbol{a}_{\mathrm{L}}^* + \boldsymbol{b}_{\mathrm{L}}^*) + \frac{1}{2}\boldsymbol{c}_{\mathrm{L}}^*$ (the H point) and $\boldsymbol{k}_{10} = \frac{1}{3}(\boldsymbol{a}_{\mathrm{L}}^* + \boldsymbol{b}_{\mathrm{L}}^* + \mathbf{c}_{\mathrm{L}}^*)$ of the hexagonal L-Brillouin zone, which have been used for the FCC–HCP transition, take place for the graphite and diamond structures.

From Table 6.7, one can give the following description:

(1) The hexagonal graphite structure results from an ordering mechanism, $\mathrm{L}(P6/mmm, Z = 2/3) \to P6_3/mmc$ $(Z = 4)$, associated with a two-component order-parameter (η_1, η_2) transforming as $\tau_1(\boldsymbol{k}_{15})$. For the equilibrium values $\eta_1 = -\eta_2 = \eta$, a hexagonal six-fold structure of symmetry $P6_3/mmc$, with four atoms per unit cell, is formed, which represents a compressed structure with respect to hexagonal graphite, as the ratio $c_{\mathrm{h}}/a_{\mathrm{h}} \cong 1.63$ (since $c_{\mathrm{L}}/a_{\mathrm{L}} = \sqrt{2}$) differs from $c_{\mathrm{g}}/a_{\mathrm{g}} = 2.73$. Accordingly, an additional decompression along Z is necessary, corresponding to an "improper" strain e_{zz}, which transforms as the totally symmetric irreducible representation $(\tau_1(\boldsymbol{k}_{16}))$ of the $P6/mmm$

Table 6.7. Mechanisms involved in the transition from the latent structure ($P6/mmm, Z = 2/3$) to the hexagonal graphite (b), diamond (c) and rhombohedral graphite (d) structures.

(a)	(b)	(c)	(d)
Structure mechanism	Hexagonal graphite ($P6_3/mmc$, $Z = 4$)	Diamond ($Fd\overline{3}m, Z = 2$)	Rhombohedral graphite ($R\overline{3}m, Z = 4$)
Ordering $\tau_1(\boldsymbol{k}_{15})$	■		
Ordering $\tau_1(\boldsymbol{k}_{10})$		■	■
Shifting $\tau_1(\boldsymbol{k}_{10})$		■	
Decompression $\tau_1(\boldsymbol{k}_{16})$	■		■

space group, at the center (Γ-point) of the hexagonal Brillouin zone. Let us stress that while e_{zz} is not a symmetry-breaking order-parameter (i.e., it couples to η via the "improper" coupling $\eta^2 e_{zz}$), it cannot be considered as a "secondary" order-parameter, as it produces an enlargement of the interlayer distance by a factor 1.67.

(2) The diamond structure is obtained by correlated ordering *and* shifting mechanisms from the L-structure. The two mechanisms transform as the same four-component order-parameter ζ_i (i = 1–4), which has the symmetry of $\tau_1(\boldsymbol{k}_{10})$, for the same equilibrium values $\zeta_1 = \zeta_2 = \zeta \neq 0$, $\zeta_3 = \zeta_4 = 0$, i.e., they are bilinearly coupled in a "pseudoproper" coupling. The ordering process leads to a three-layered rhombohedral structure of symmetry $R\overline{3}m(Z = 2)$, which has a tripled unit cell with respect to the L structure. However, this phase plays the role of an intermediate Landau state in the framework of our approach. To realize the diamond structure, one needs additional antiparallel shifts of the carbon atoms along the [111] rhombohedral directions (the [001] hexagonal directions), which transform the flat layer stacking into the framework-type diamond structure. For specific displacements of magnitude $c_L/8 \cong 0.26$ Å, the atoms reach the centers of regular tetrahedra, and the site symmetry increases from $3m$ to $\overline{4}3m$, while the overall symmetry increases from $R\overline{3}m$ to $Fd\overline{3}m$. Note that, in agreement with our description of displacive reconstructive mechanisms, the $Fd\overline{3}m$ symmetry

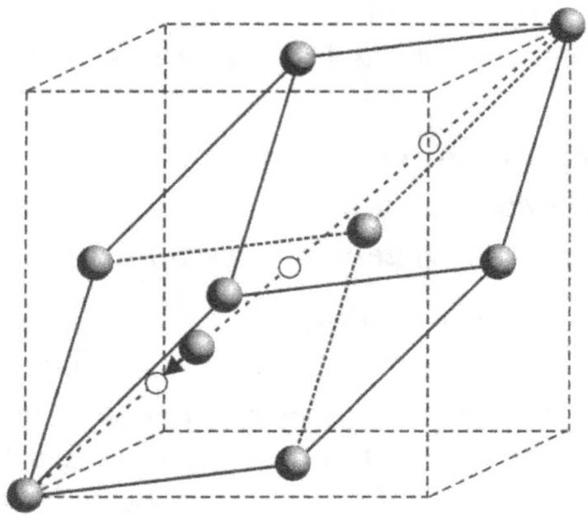

Fig. 6.19. Holohedry of $R\overline{3}m(Fm\overline{3}m)$ and special point corresponding to the increase in symmetry to $Fd\overline{3}m$.

corresponds to a special point of the $Fm\overline{3}m$ holohedry (i.e., the holohedry of the rhombohedral lattice with special geometrical constraints) as shown in Fig. 6.19.

Column (d) of Table 6.7 indicates the mechanism which leads to another polymorphic form of carbon: rhombohedral graphite, which has been found by Lipson and Stokes [63]. It is obtained by the same ordering mechanism ($\tau_1(\boldsymbol{k}_{10})$) than diamond, but requires an additional decompression $e_{zz}(\tau_1(\boldsymbol{k}_{16}))$ as hexagonal graphite. The disordered wurtzite-type structure of hexagonal diamond, synthesized by Bundy and Kasper [67], involves different combined mechanisms of ordering and shifts. Let us emphasize that in our presentation of the mechanisms, we have chosen arbitrarily to start from a "compressed" latent structure for the L-polytype. Starting from a "decompressed" polytype structure would lead to describing the transitions to the hexagonal and rhombohedral graphite structures, by simple ordering mechanisms. By contrast, the "L $\rightarrow$ diamond" transition mechanism would require an additional compression e_{zz}.

To conclude, let us stress that an intensive effort has been made, during the last decades, to understand the nature of the graphite–diamond transition mechanism, and also to reproduce this transition to a metastable state in laboratory. Our description, which provides a crystallographic description of the mechanism, and the symmetry of the corresponding order-parameter, suggests that the transition should take place more easily through the energetically advantageous path, passing by the rhombohedral graphite structure.

6.4.4. *Changes in the chemical bonds at the graphite–diamond transition: Physical nature of the order-parameter*

Following Buerger's classification [68], the graphite–diamond transition is a typical representative example of "bond-type" transformation, i.e., a reconstructive transition involving a disruption of the bonds between nearest-neighbor atoms. In order to illustrate this diagnostic, and to connect it to our proposed mechanism, let us investigate the modifications of chemical bonding, which occur at the graphite–diamond transition. With that goal, we can start from the well-known facts that the carbon atoms have the electronic ground state $1s^2 2s^2 2p_x 2p_y$ and that in the chemical bonding, the excited states $1s^2 2s 2p_x 2p_y 2p_z$ also take part, by hybridization of the s electrons of one atom with the p electrons of its neighbors [69]. The participation of the p-orbitals gives their covalent character to the bonds, and the sp^2 hybridization which predetermines the *layered character* of the graphite structure is described by linear combinations of the atomic functions [69]:

$$\psi_{sp}^1 = \frac{1}{\sqrt{3}}(\varphi_{2s} + \sqrt{2}\varphi_{2p_x}); \quad \psi_{sp}^2 = \frac{1}{\sqrt{6}}(\sqrt{2}\varphi_{2s} + \varphi_{2p_x} + \sqrt{3}\varphi_{2p_y});$$

$$\psi_{sp}^3 = \frac{1}{\sqrt{6}}(\sqrt{2}\varphi_{2s} + \sqrt{2}\varphi_{2p_x} + \sqrt{3}\varphi_{2p_y}). \tag{6.18}$$

The p_z-orbitals are symmetry independent from the p_x- and p_y-orbitals, and their effect on the hybridization of orbitals determines the interlayer bonding, and thus the interlayer distance. In contrast, the effect of the p_x- and p_y-orbitals fixes the interatomic distance *within* the layers.

In diamond, one has three equivalent pairs of atoms producing a hybridization of the sp^3 type, which corresponds to a linear combination of the basic functions [69]:

$$\psi_{sp}^1 = \frac{1}{2}(\varphi_{2s} + \varphi_{2p_x} + \varphi_{2p_y} + \varphi_{2p_z});$$

$$\psi_{sp}^2 = \frac{1}{2}(\varphi_{2s} + \varphi_{2p_x} - \varphi_{2p_y} - \varphi_{2p_z});$$

$$\psi_{sp}^3 = \frac{1}{2}(\varphi_{2s} - \varphi_{2p_x} + \varphi_{2p_y} - \varphi_{2p_z});$$

$$\psi_{sp}^4 = \frac{1}{2}(\varphi_{2s} - \varphi_{2p_x} - \varphi_{2p_y} + \varphi_{2p_z}). \tag{6.19}$$

From Eq. (6.19), one can deduce that the electron density distribution has maxima in the directions of the apexes of tetrahedra, a fact which determines a *framework type* of structure. Besides, as the s-basis functions have a spherical distribution, the basis functions (6.18) have transformation properties coinciding with that of the p-orbitals, which transform as *bivectors* [69]. We can, therefore, image the effect of the electron density distribution as a set of bivectors localized in the positions of the carbon atoms.

According to the preceding considerations, one can represent the chemical bonds in the parent polytype structure, as shown in Figs. 6.20(a) and 6.20(b). Due to the orientational disorder and to the six-fold symmetry of the L-structure, the bivectors form a six-fold planar star, each branch of the star being oriented in the direction of the nearest neighbor in-layer position. In addition, bivectors will be directed along the six-fold axis. Thus, there will be two sets of orthogonal bivectors forming the basis of the irreducible representation $\tau_1(\boldsymbol{k}_{15})$. Figure 6.20(c) shows the equilibrium configuration of ordered bivectors in hexagonal graphite. One can verify that no mixture of p_z- and $(p_x - p_y)$-orbitals is needed.

The corresponding sp^2 hybridization determines the layered structure of hexagonal graphite. By contrast, the ordered diamond structure, shown in Fig. 6.20(d), corresponds to an sp^3 type of hybridization, in which the mixed p_x-, p_y- and p_z-orbitals form the basis of the irreducible representation $\tau_1(\boldsymbol{k}_{10})$ and result in a non-layered structure of the framework type. Let us stress, in this respect, that the transition which transforms the disordered polytype into the diamond structure can be equivalently interpreted by three distinct physical mechanisms: an ordering or a shifting mechanism, or a change in the electronic distribution. All three mechanisms possess the symmetry of the irreducible representation $\tau_1(\boldsymbol{k}_{10})$, and only a complete set of measurements, not simply symmetry-based analysis, may allow to disclose which of them plays the leading role.

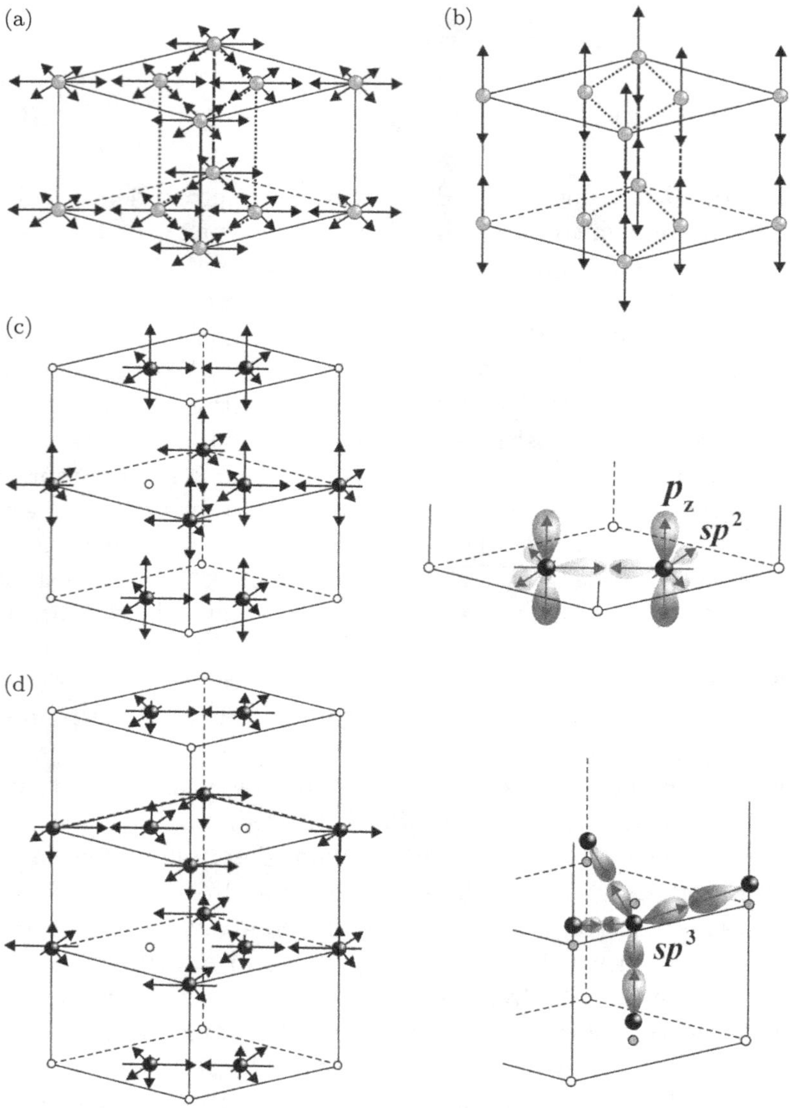

Fig. 6.20. Distribution of disordered p-orbitals represented by bivectors: (a) p_x and p_y, (b) p_z in the L-structure. (c) Ordered p_x and p_y and sp^2 hybridized orbitals in hexagonal graphite. (d) Ordered and sp^3 hybridized orbitals in cubic diamond.

6.5. Polytypic Structures in Zinc Sulfide

The wurtzite–sphalerite type of reconstructive transition occurs in a number of sulfides (MnS, ZnS, CdS, $Ga_2S_3, \ldots$) and selenides (MnSe, CdSe,...). Among these compounds, ZnS is known for the variety of its polytypic structures [3]. When cooling ZnS from the melt, its high-temperature wurtzite-type structure ($P6_3mc, Z = 2$) undergoes a transition at about 1024°C to a cubic sphalerite-type structure ($F\bar{4}3m, Z = 2$), which is never fully accomplished, the final product being a polytype "microtwinned" structure, containing about 10% of stacking faults [70].

The disordered polytype latent phase model formulated for the FCC–HCP transition applies with only slight modification to the transition in ZnS. Figures 6.21(a) and 6.21(b) represent the hexagonal and cubic structures of ZnS. In both structures, the large anions S are close packed and occupy positions 2(b) : (1/3 2/3 0) in the hexagonal wurtzite structure, whereas in the sphalerite cubic structure, they are in positions 1(a). The relatively small atoms Zn are, respectively, in positions 2(b): (1/3 2/3 3/8) and (1/12 1/12 1/12) in the two preceding structural modifications. In the unit cells shown in Figs. 6.16(a) and 6.16(b), the maximal substructure common to the hexagonal and cubic structures is represented. It is formed by a monolayer structure of symmetry $P6_3mc$, in which the anions are in position 1(a), and the cations in positions (0 0 3/4). As the concentration of this substructure is $x = 1/3$, i.e., the two ions Zn and S are randomly distributed, occupying one among three possible positions

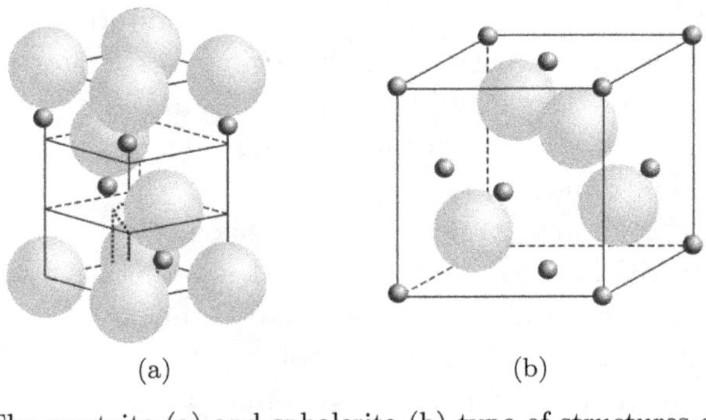

(a) (b)

Fig. 6.21. The wurtzite (a) and sphalerite (b) type of structures of ZnS. Small black spheres represent Zn, large gray spheres represent S.

Table 6.8. Experimental values for the lattice parameters of the compounds listed in the first column. The calculated values use the ideal ratios (6.17) and (6.18).

	Sphalerite	L-polytype	Wurtzite			
	a_h^{exp}(Å)	$a_{\mathrm{L}}^{\mathrm{calc}}$(Å)	a_h^{exp} (Å)	a_h^{calc}(Å)	c_h^{exp}(Å)	c_h^{calc}(Å)
MnS	5.610	2.29	3.996	3.996	6.446	6.48
ZnS	5.406	2.21	3.811	3.823	6.234	6.25
CdS	5.892	2.38	4.132	4.124	6.730	6.73
Ga_2S_3	5.182	2.12	3.685	3.664	6.028	5.97
MnSe	5.830	2.38	4.130	4.122	6.730	6.73
CdSe	6.050	2.47	4.300	4.278	7.070	6.97

in each layer, we deal with a disordered polytypic structure, which plays the same role as the L-phase in the FCC–HCP transition.

The same relationship (6.2) exists between the wurtzite and sphalerite basic unit-cell vectors, and the relations (6.3) and (6.4) also hold between, respectively, the cubic or hexagonal unit cells and the "latent" polytypic unit cell. These relations, which derive from the close-packed character of the phases, imply the following ratios between the lengths of the lattice parameters:

$$a_{\mathrm{c}} = a_{\mathrm{L}}\sqrt{6}, a_{\mathrm{h}} = a_{\mathrm{L}}\sqrt{3} \quad \text{and} \quad c_{\mathrm{h}} = 2c_{\mathrm{L}} = 2a_{\mathrm{L}}\sqrt{3}, \tag{6.20}$$

and *between* the hexagonal and cubic lattice parameters:

$$a_{\mathrm{h}} = a_{\mathrm{c}}/\sqrt{2} \quad \text{and} \quad c_{\mathrm{h}} = 2a_{\mathrm{c}}/\sqrt{3}. \tag{6.21}$$

Table 6.8 shows that, for most of the compounds undergoing the wurtzite–sphalerite transition, the experimental values reported for their lattice parameters fulfill the preceding ratios. It means that the FCC–HCP model applies without adding any important secondary macroscopic deformation.

These preliminary considerations show that the wurtzite–sphalerite transition can be described as a transition between two low-symmetry differently ordered structures, both derived from a common parent disordered polytypic structure. The group-theoretical formalism remains the same as for the FCC–HCP transition, but *the symmetries of the structures are lowered.* Thus, the two- and four-dimensional irreducible representations (denoted

$\tau_1(\boldsymbol{k}_{15})$ and $\tau_1(\boldsymbol{k}_{10})$) of the parent space group $P6_3mc$ induce, respectively, the $P6_3mc(Z = 1/3) \rightarrow P6_3mc(Z = 2)$ six-fold wurtzite structure, and the $P6_3mc(Z = 1/3) \rightarrow F\bar{4}3m(Z = 1)$ three-fold sphalerite structure. The effective equilibrium values of the corresponding two- and four-component order-parameters are $(\eta, -\eta)$ and $(\zeta, \zeta, 0, 0)$, respectively, the form of the thermodynamic potential (6.7) and of the phase diagrams of Figs. 6.5–6.8 remains unchanged.

Note that the virtual transition (non-polar → polar) between the two hexagonal polytype forms $P6_3/mmc(Z = 1/3) \rightarrow P6_3mc(Z = 1/3)$ corresponds to a shifting along the six-fold axis of magnitude $c_L/4$, which brings the Zn cations in the center of tetrahedra formed by S anions. The same shifting mechanism lowers the FCC and HCP symmetries to $F\bar{4}3m(Z = 1)$ and $P6_3mc(Z = 2)$. Note also that ZnS provides a clear-cut verification of the description of stacking faults given in Sec. 6.3.6 as one observes the two sorts of stacking faults, (deformation and twinning types) under the form of various polytypic sequences, on the one hand, and of microdomains, on the other hand [71, 72]. These symmetry-induced stacking faults never disappear completely, and tend asymptotically to 10% of the sphalerite low-temperature structure [70]. By contrast, the temperature-dependent type of defects, related in our approach to the ordering parameter (5.15), is increased by cycling across the transition, and on the contrary, reduced by annealing far from the transition temperature [71, 72].

6.6. Combined Ordering-Displacive Mechanisms: Reconstructive Transitions in SiO_2

6.6.1. *Introduction*

Minerals, *despite* the complexity of their structures but also *due to* this complexity, constitute a preeminent field of application for the advanced phenomenological approach. Indeed, with rare exceptions, the polymorphic transitions found in minerals are discontinuous and even reconstructive transitions [73, 74]. However, the corresponding transformation mechanisms are, in majority, unknown partly because the density of the polymorphic structures notably increases with increasing pressure, and thus the transitions appear as having an

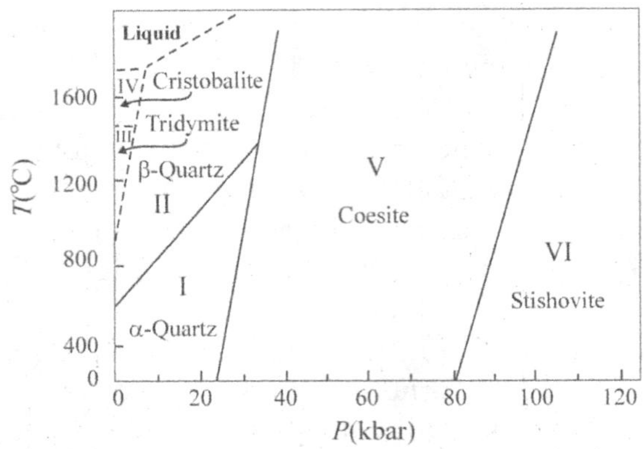

Fig. 6.22. Partial pressure–temperature phase diagram of SiO_2.

implicit morphotropic character, with a discontinuity in the topology of the structures. In this respect, the symmetry-based approach which assumes a parent structure with variable (fractional) concentrations x and combines ordering and displacive mechanisms seems to be well adapted for the description of the different phases in a given mineral. We will illustrate this proposition by discussing one of the most classical examples of phase transitions in minerals: the discontinuous transitions between the SiO_2 structures.

Figure 6.22 shows the pressure–temperature phase diagram of SiO_2, which contains at least *10* different phases, with well-established structures. Table 6.9 summarizes their denominations, symmetries, lattice parameters and densities. One can see in this table that from the lowest pressure (cristobalite, tridymite) to the highest-pressure forms (stishovite, $CaCl_2$-type), the density increases by a factor of 1.85. It suggests to adopt a parent structure which possesses different (fractional) concentrations.

6.6.2. *BCC parent structure for the SiO_2 polymorphs*

6.6.2.1. *Stishovite*

Let us start from the stishovite high-pressure form of SiO_2, which is stable in a wide range of pressures above 80 kbar [77]. In its rutile-type structure (space-group $P4_2/mnm, Z = 2$) represented in Fig. 6.23, the Si atoms occupy the corners and center (positions 2(a)) of the tetragonal unit cell. The oxygens have an octahedral

Table 6.9. The stable and metastable phases of SiO_2 [75, 76].

Number	Denomination	Space group	Lattice parameters (Å)	ρ (g/cm^3)
I	α-Quartz	$P3_121(Z=3)$	$a=4.4, c=5.29$	2.655
II	β-Quartz	$P6_222(Z=3)$	$a=4.44, c=5.45$	2.655
III	β-Tridymite	$P6_3/mmc(Z=4)$	$a=5.03, c=8.22$	2.37
IV	β-Cristobalite	$Fd\bar{3}m(Z=2)$	$a=7.05$	2.33
IV′	α-Cristobalite	$P4_12_12(Z=4)$	$a=4.97, c=6.92$	2.33
V	Coesite	$C2/c(Z=16)$	$a=b=7.17, c=12.38,$ $\gamma=120°$	3.01
VI	Stishovite	$P4_2/mnm(Z=2)$	$a=4.179, c=2.665$	4.29
VII	(α-PbO_2)-type	$Pbcn(Z=4)$	$a=4.181, b=4.617,$ $c=3.824$	4.26
VIII	Keatite	$P3_121(Z=12)$	$a=7.456, c=8.604$	2.50
IX	$CaCl_2$-type	$Pnnm(Z=2)$	$a=4.30, b=4.77,$ $c=4.50$	4.35

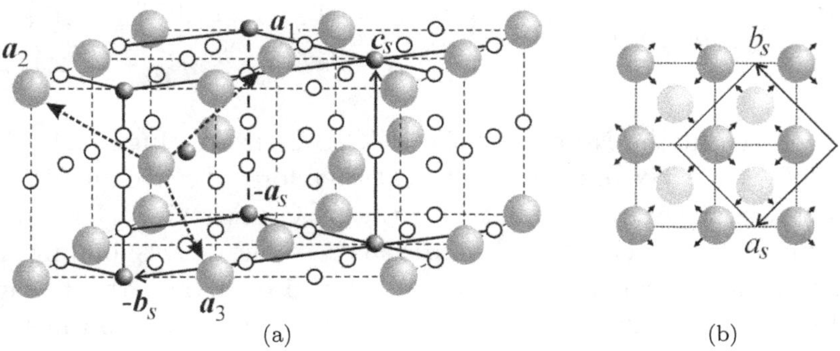

Fig. 6.23. (a) Stishovite rutile-type structure (solid lines) embedded in four conventional BCC unit cells. Large gray circles are oxygens. Full and open small circles are silicons and vacancies, respectively. (b) Shifting of the oxygen atoms in the (001) BCC plane at the BCC–stishovite transition. Dark- and clear-gray circles represent oxygens located at $z=0$ and $z=1/2$. The arrows symbolize the displacements associated with the two effective order-parameter components.

coordination [78] and lie on the diagonals of the faces perpendicular to the $\boldsymbol{c}$ axis. This corresponds to a slightly deformed BCC lattice shown in Fig. 6.23(a), one stishovite unit cell being formed by four BCC unit cells, with the following relationship between the basic translations:

$$\boldsymbol{a}_{\mathrm{st}} = \boldsymbol{a}_1 + \boldsymbol{a}_2 + 2\boldsymbol{a}_3, \boldsymbol{b}_{\mathrm{st}} = \boldsymbol{a}_1 - \boldsymbol{a}_2, \boldsymbol{c}_{\mathrm{st}} = \boldsymbol{a}_1 + \boldsymbol{a}_2, \tag{6.22}$$

where $\boldsymbol{a}_1$, $\boldsymbol{a}_2$ and $\boldsymbol{a}_3$ are the primitive BCC lattice vectors.

In the BCC sublattice, the oxygens are ordered (position 1(a), $x = 1$), whereas the Si atoms are randomly distributed in octahedrally coordinated positions 3(b), one site among six being filled. Therefore, one can describe the stishovite structure as the result of a hypothetical phase transition from a partially disordered BCC structure. The corresponding transition mechanism involves an ordering of the silicons in positions 2(a), while the 4(f) and 4(c) tetragonal positions remain vacant. Correlatively, the oxygens are shifted in the [110] and [1$\bar{1}$0] tetragonal directions by about $0.08|\boldsymbol{a}_{\mathrm{st}}| = 0.33$Å, as shown in Fig. 6.23(b).

The breaking of the translational symmetry expressed by Eq. (6.21) corresponds to the wave-vectors $\boldsymbol{k}_9 = \frac{1}{2}(\boldsymbol{a}_1^* + \boldsymbol{a}_2^*)$, where

$$\boldsymbol{a}_1^* = \left(0, \frac{2\pi}{a}, \frac{2\pi}{a}\right), \quad \boldsymbol{a}_2^* = \left(\frac{2\pi}{a}, 0, \frac{2\pi}{a}\right), \quad \boldsymbol{a}_3^* = \left(\frac{2\pi}{a}, \frac{2\pi}{a}, 0\right), \tag{6.23}$$

NO located at the N-point of the BCC BZ surface (Fig. 6.24). The star of $\boldsymbol{k}_9$ has six branches [79], and therefore, the irreducible representation $\tau_4(\boldsymbol{k}_9)$ associated with the BCC–stishovite transition is six-dimensional. The equilibrium values of the corresponding six-component order-parameter are $(\eta, -\eta, 0, 0, 0, 0)$, i.e., the tetragonal stishovite symmetry is determined by two effective nonvanishing components of the order-parameter. Assuming a cubic lattice parameter of $a = 2.9$ Å, which coincides with the best fit with respect to the different SiO_2 structures as deduced from our approach, the transition mechanism yields the calculated stishovite unit-cell parameters $a_{st}^c = 4.1$ Å and $c_{st}^c = 2.9$ Å. Compared with the experimental values [74] ($a_{st}^{\exp} = 4.179$ Å, $c_{st}^{\exp} = 2.665$ Å), it implies taking into account an additional compression along $\boldsymbol{z}$ of about 2% and a decompression along $\boldsymbol{x}$ and $\boldsymbol{y}$ of $\approx 8\%$.

The same BCC parent structure which has been used to deduce the stishovite structure will now be shown to constitute *the maximal substructure common to all the SiO_2 polymorphs provided different occupancies of the BCC lattice by the oxygen atoms are assumed.*

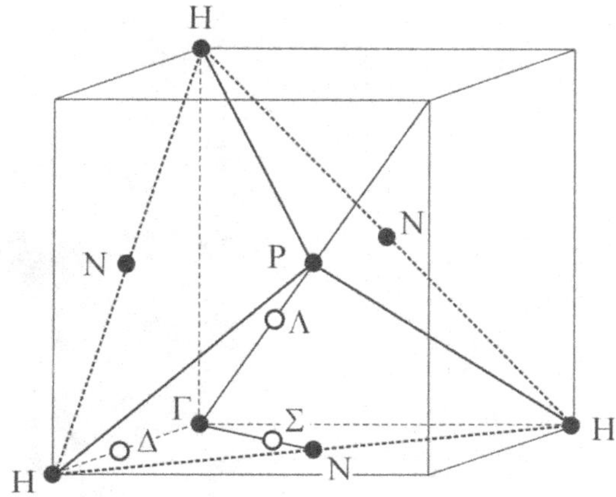

Fig. 6.24. Points and lines of the BCC Brillouin zone involved in the transition mechanisms from the BCC parent structure to the SiO_2 structures.

Accordingly, it will be used for describing successively the coesite, quartz, tridymite, cristobalite, $CaCl_2$-type and α-PbO_2 columbite-type forms of SiO_2.

6.6.2.2. *Coesite*

The embedding of the BCC unit cell within the monoclinic structure of coesite [80] ($C2/c(Z = 16)$) is shown in Fig. 6.25. In this unit cell, the oxygens occupy randomly 2 among 3 of the 1(a) positions ($x = 2/3$), whereas the Si atoms fill 1 among 9 of the 3(b) positions. Two independent mechanisms are needed to realize the coesite structure starting from a BCC structure: (1) ordering of both the O and Si atoms, which leads to a six-fold rhombohedral unit cell having the lattice parameters $\boldsymbol{a}_r = \boldsymbol{a}_1 - \boldsymbol{a}_2$, $\boldsymbol{b}_r = -\boldsymbol{a}_1 + \boldsymbol{a}_3$, $\boldsymbol{c}_r = 2(\boldsymbol{a}_1 + \boldsymbol{a}_2 + \boldsymbol{a}_3)$ and (ii) displacements of some of the oxygen positions with the rhombohedral unit-cell parameters

$$\boldsymbol{a}_{\text{coe}} = 2(\boldsymbol{a}_1 - \boldsymbol{a}_2),\ \boldsymbol{b}_{\text{coe}} = 2(-\boldsymbol{a}_1 + \boldsymbol{a}_3),\ \boldsymbol{c}_{\text{coe}} = 4(\boldsymbol{a}_1 + \boldsymbol{a}_2 + \boldsymbol{a}_3). \quad (6.24)$$

In this mechanism, the oxygens in monoclinic position 4(a) are unshifted, while those in positions 8(f) are shifted along the [111] BCC direction by about 0.15 Å. The remaining oxygens, in positions 4(a) and 8(f), are displaced along the diagonals of the BCC cube by $\cong 1.25$ Å.

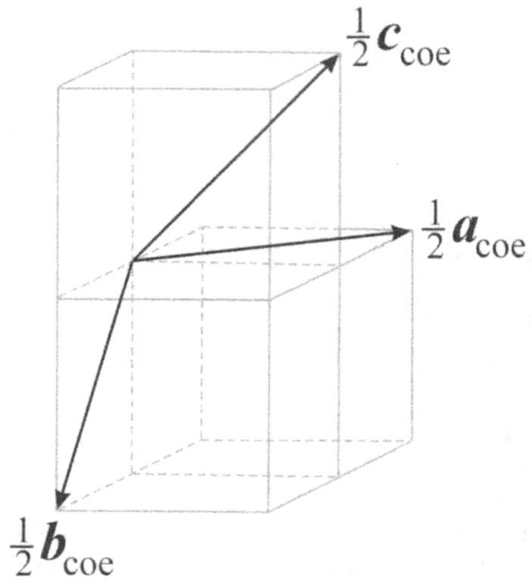

Fig. 6.25. Half of the coesite primitive translations and three BCC cells.

The combined ordering and displacive mechanisms associated with the BCC–coesite transition requires two distinct, eight- and six-dimensional irreducible representations of the $Im\bar{3}m$ space group: $\tau_1(\boldsymbol{k}_7) \oplus \tau_4(\boldsymbol{k}_9)$ where $\boldsymbol{k}_7 = \frac{1}{6}(\boldsymbol{a}_1^* + \boldsymbol{a}_2^* + \boldsymbol{a}_3^*)$ is located inside the BCC BZ along the Λ-line (Fig. 6.24). The equilibrium values of the corresponding coupled order-parameters are, respectively, $(\eta, 0, 0, 0, \eta, 0, 0, 0)$ and $(0, \xi, 0, 0, 0, 0)$. The calculated lattice parameters resulting from the assumed transition mechanism are $a^c_{\text{coe}} = b^c_{\text{coe}} = 8.2$ Å, $c^c_{\text{coe}} = 10.05$ Å, and $\gamma = 120°$. The difference with the experimental numbers [80] ($a^{\text{exp}}_{\text{coe}} = b^{\text{exp}}_{\text{coe}} = 7.17$ Å, $c^{\text{exp}}_{\text{coe}} = 12.38$ Å) is large, since one needs compressions of about 13% along $\boldsymbol{a}$, $\boldsymbol{b}$ and $\boldsymbol{c}$.

6.6.2.3. *β-Quartz*

The same occupancy ($x = 2/3$) of the BCC lattice by oxygen in coesite is found in the β-quartz structure [73] ($P6_3/mmc, Z = 3$) as shown in Fig. 6.26. Here, the combined ordering and displacive mechanism associated with the virtual BCC–β-quartz transition can be more simply formulated starting from an FCC structure, formed by hexagonal close-packed layers, which is obtained by a Bain deformation of the BCC unit cell. It consists in a shear strain $(2e_{zz} - e_{xx} - e_{yy})$ which stretches the BCC unit cell along one of the four-fold axis and compresses it by the same extent along the other four-fold axis (see

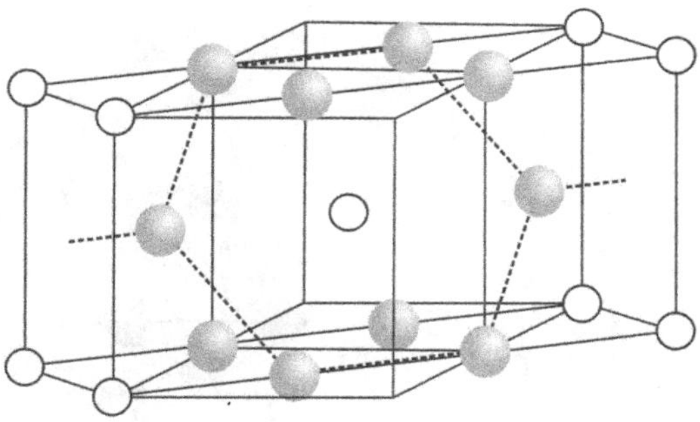

Fig. 6.26. Deformed hexagonal planes of β-quartz within the orthorhombic configuration obtained in the BCC–β-quartz mechanism. Thin lines represent the FCC structure. Full spheres and open circles are oxygens and vacancies, respectively.

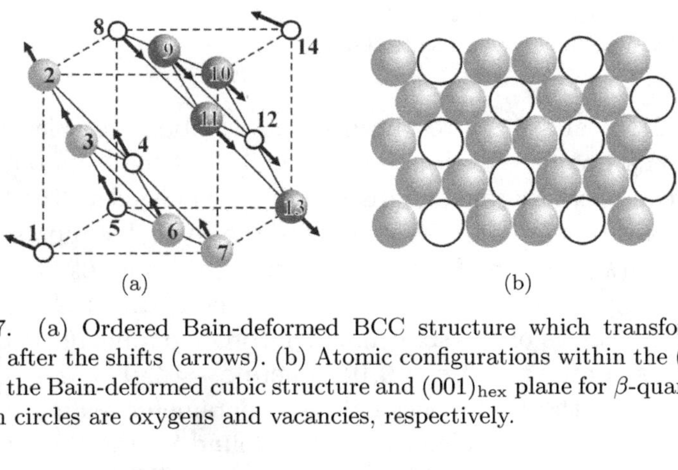

Fig. 6.27. (a) Ordered Bain-deformed BCC structure which transforms into β-quartz after the shifts (arrows). (b) Atomic configurations within the $(111)_{\mathrm{FCC}}$ plane for the Bain-deformed cubic structure and $(001)_{\mathrm{hex}}$ plane for β-quartz. Gray and open circles are oxygens and vacancies, respectively.

Secs. 5.4.1 and 5.4.2). The correspondence between the BCC, FCC and quartz unit cells is shown in Figs. 6.26 and 6.27(a). The relationship between the β-quartz and BCC primitive translations is

$$\boldsymbol{a}_q = -(\boldsymbol{a}_2 + 2\boldsymbol{a}_3), \boldsymbol{b}_q = 2\boldsymbol{a}_2 + \boldsymbol{a}_3, \boldsymbol{c}_q = 3\boldsymbol{a}_1 + \boldsymbol{a}_2 + 2\boldsymbol{a}_3. \quad (6.25)$$

The preceding figures show the ordering mechanism which brings the oxygen atoms in positions 2(e), the vacancies keeping the 1(a) positions, it leads to an orthorhombic structure ($Immm, Z = 3$), which involves a three-fold multiplication of the BCC unit cell with $\mathbf{a}_{\mathrm{orth}} = \boldsymbol{a}_1^F + \boldsymbol{a}_3^F$, $\boldsymbol{b}_{\mathrm{orth}} = -(\boldsymbol{a}_2^F + \boldsymbol{a}_3^F)$, $\mathbf{c}_{\mathrm{orth}} = 2\boldsymbol{a}_1^F + \boldsymbol{a}_2^F + \boldsymbol{a}_3^F$, where $\boldsymbol{a}_1^F$, $\boldsymbol{a}_2^F$ and $\boldsymbol{a}_3^F$ are the FCC lattice vectors. The β-quartz structure

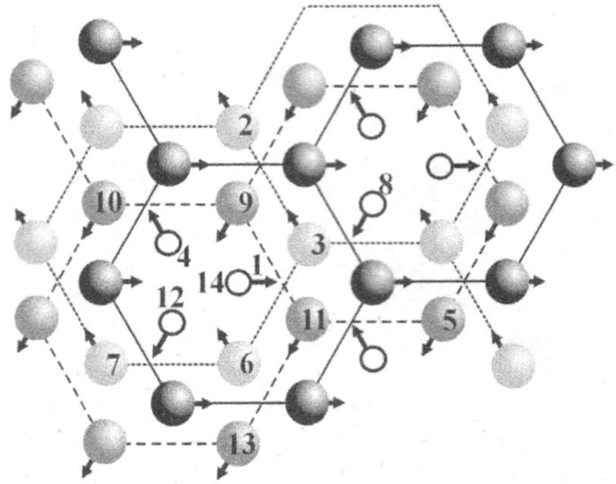

Fig. 6.28. Projection of the three-layered β-quartz structure on the $(0\bar{1}2)_{\text{BCC}} \| (001)_{\text{hex}}$ planes. Black, dark-gray and clear-gray circles are oxygens at $z = 0, 1/3$ and $2/3$, respectively. Open circles are vacancies. The arrows show the displacement field giving rise to β-quartz.

(Figs. 6.27(b) and 6.28) results from an additional displacement of the oxygens along the diagonals of the FCC unit cell by about 0.42 Å.

The complex ordering and displacive mechanisms for the BCC–β-quartz transition have the symmetry of the reducible representation $\tau_1(\boldsymbol{k}_8) \oplus \tau_4(\boldsymbol{k}_4)$, where $\boldsymbol{k}_8 = \frac{1}{3}(\boldsymbol{a}_1^* + \boldsymbol{a}_2^* + \boldsymbol{a}_3^*)$ and $\boldsymbol{k}_4 = \frac{1}{3}\boldsymbol{a}_3^*$ correspond, respectively, to the Λ and Σ lines of the BCC BZ. The equilibrium values of the six-component order-parameter transforming as $\tau_1(\boldsymbol{k}_8)$ are $(\eta, 0, 0, \varepsilon^*\eta, 0, 0)$, where $\varepsilon^* = 2\pi i/3$. The effective components of the 12-dimensional order-parameter related to $\tau_4(\boldsymbol{k}_4)$ are $(\eta, -\eta, 0, \ldots)$. The calculated values found for the β-quartz unit-cell parameters are $a_q^c = 4.81$ Å, $c_q^c = 6.48$ Å. With respect to the experimental values [73] ($a_q^{\text{exp}} = 5.45$ Å, $c_q^{\text{exp}} = 4.99$ Å), it implies considering a large compression of about 30% along the $\boldsymbol{y}$ axis and an analogous decompression along $\boldsymbol{z}$.

6.6.2.4. *Tridymite*

The occupancy of the BCC lattice associated with the tridymite structure [81] ($P6_3/mmc, Z = 4$) is $x = 1/2$, as one has one oxygen for two sites in position 1(a) and one silicon for six sites in position 3(b) (Fig. 6.29(a)). The BCC–tridymite transition mechanism can be decomposed into two steps: an ordering mechanism

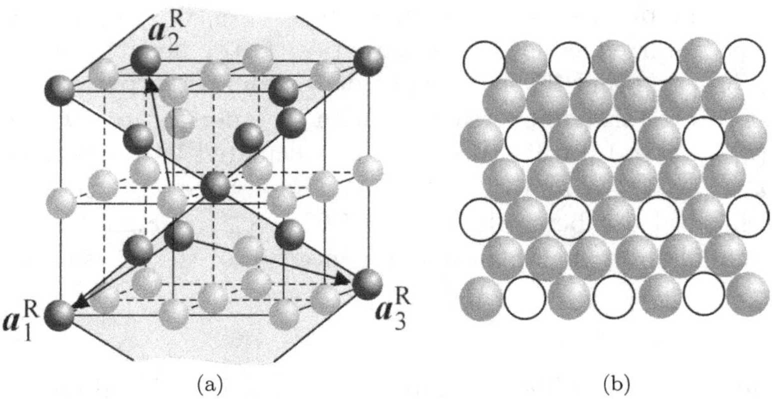

Fig. 6.29. (a) Connection between the BCC conventional cell (thin lines) and the rhombohedral atomic configuration taking place at the BCC–tridymite transition mechanism. The arrows are the basic vectors of the rhombohedral unit cell. Full and gray circles are a mixture of oxygens and vacancies (see text). (b) Atomic configuration within the $(001)_{\text{hex}}$ plane for tridymite. Gray and open circles are oxygens and vacancies, respectively.

which brings the oxygens in positions 2(c) and 6(g), and the silicons in position 4(f). The resulting ordered structure corresponds to an in-layer rhombohedral configuration represented in Fig. 6.29(a). In this figure, the dark-gray atoms represent oxygens and the clear-gray ones figure a mixture of vacancies and oxygens. The formation of the tridymite structure requires an additional Burgers-type mechanism (Secs. 5.3.1 and 5.3.2, Ref. [82]) which orders the light-gray atoms, and shifts in the antiparallel $\pm[1\bar{1}0]$ directions the atoms belonging to one between two (110) planes. The tridymite structure is obtained for the specific shifts of magnitude $a\sqrt{2}/12 \cong 0.34$ Å, for which the in-layer atomic configuration becomes hexagonal (Fig. 6.29(b)). The connection between the tridymite and BCC unit cells is

$$\boldsymbol{a}_{tr} = 2\boldsymbol{a}_1, \boldsymbol{b}_{tr} = 2\boldsymbol{a}_2, \boldsymbol{c}_{tr} = 2(\boldsymbol{a}_1 + \boldsymbol{a}_2 + \boldsymbol{a}_3). \tag{6.26}$$

It corresponds to a six-fold multiplication of the BCC unit cell.

The six-fold multiplication of the BCC unit cell, expressed by Eq. (6.25), which characterizes the BCC–tridymite transition, corresponds to two irreducible representations of the $Im\bar{3}m$ space group: a six-dimensional irreducible representation $(\tau_1(\boldsymbol{k}_9))$ associated with the first-step ordering mechanism described in Sec. 6.2.1, with the equilibrium order-parameter values $(0, \eta, 0, -\eta, 0, -\eta)$, and a 12-dimensional irreducible representation $(\tau_4(\boldsymbol{k}_4))$ reflecting the second step (ordering and displacements) of the Burgers-type mechanism leading to the tridymite structure, with an order-parameter symmetry $(\eta, -\eta, 0, 0, 0, 0, 0, 0, 0, 0, 0, 0)$. The calculated lattice parameters are $a_{tr}^c = 5.023$ Å and $c_{tr}^c = 8.20$ Å. These values are close to the experimental numbers [81] ($a_{tr}^{\rm exp} = 5.03$ Å, $c_{tr}^{\rm exp} = 8.22$ Å), i.e., no substantial additional deformation is needed for the transition.

6.6.2.5. *β-Crystobalite*

The ideal β-cristobalite structure [83] $(Fd\bar{3}m, Z = 2)$ and the underlying BCC lattice, which is filled by $x = 1/2$ oxygen, are shown in Fig. 6.30. The connection between the corresponding unit cells is

$$\boldsymbol{a}_{cr}^{\beta} = 2(\boldsymbol{a}_1 + \boldsymbol{a}_2), \quad \boldsymbol{b}_{cr}^{\beta} = 2(\boldsymbol{a}_1 - \boldsymbol{a}_2), \quad \boldsymbol{c}_{cr}^{\beta} = 2\boldsymbol{a}_3. \tag{6.27}$$

The mechanism of the BCC–β-cristobalite virtual transition consists essentially in an ordering process, which brings the oxygens in 4(c) and 4(d) positions. It results in a tetragonal structure $(I4_1/amd, Z = 2)$, which is transformed in ideal β-cristobalite by a Bain-type deformation, i.e., a decompression along the c-tetragonal axis, and a compression along the perpendicular two-fold axes. This deformation (Secs. 5.4.1 and 5.4.2) brings the oxygens in the FCC positions 4(c), whereas the Si atoms are distributed within SiO_4

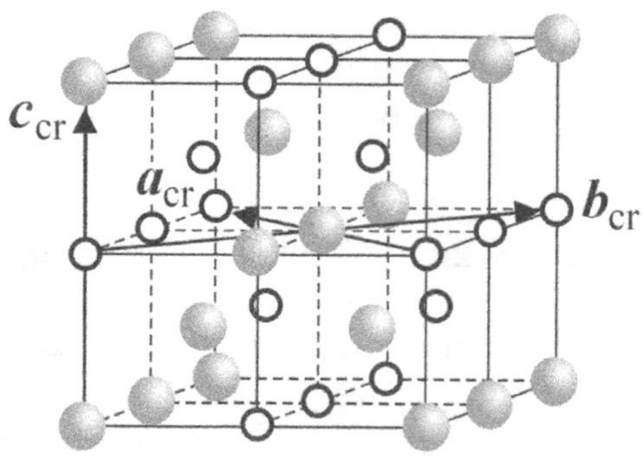

Fig. 6.30. Basic vectors of the β-cristobalite unit cell within the BCC lattice, before the Bain-type deformation. Full and open circles are oxygens and vacancies, respectively.

tetrahedra according to the "diamond law", i.e., they are in positions 2(a). A "nonideal" tetragonal structure ($I\bar{4}2d, Z = 2$) was proposed for β-cristobalite [84], which can be simply deduced from the FCC-structure type, by displacing the oxygens from the 4(c) to the 4(d) positions, the Si atoms keeping their 2(a) positions. In the same way, the α-cristobalite structure ($P4_12_12, Z = 4$) can also be directly deduced from the structure of β-cristobalite, by a tilting of the SiO_4 tetrahedra, which brings the oxygens from the positions 4(c) and 8(b) and the silicons from 2(a) to 4(a). The connection between the α and β unit cells is given by

$$\boldsymbol{a}_{cr}^{\alpha} = \boldsymbol{c}_{cr}^{\beta}, \boldsymbol{b}_{cr}^{\alpha} = \boldsymbol{a}_{cr}^{\beta}, \boldsymbol{a}_{cr}^{\alpha} = \boldsymbol{a}_{cr}^{\beta} + \boldsymbol{b}_{cr}^{\beta} + \boldsymbol{c}_{cr}^{\beta}. \qquad (6.28)$$

The BCC–β-cristobalite transition involves the same six-dimensional irreducible representation $\tau_1(\boldsymbol{k}_9)$ as the BCC–tridymite and BCC–coesite transitions, but with a different set of equilibrium values of the order-parameter components $(0, 0, -\eta, \eta, \eta, \eta)$. The calculated lattice parameters are $a_{cr}^{c} = b_{cr}^{c} = 8.2\,\text{Å}$ and $c_{cr}^{c} = 5.8\,\text{Å}$, which compared to the experimental values [83] ($a_{cr}^{\text{exp}} = 7.05\,\text{Å}$) require to consider the "reversed" Bain deformation mechanism restoring the FCC cristobalite structure. This deformation involves a compression along $\boldsymbol{a}$ and $\boldsymbol{b}$ of about 13% and a decompression along $\boldsymbol{c}$ of $\cong$ 21%. The tetragonal structure proposed for β-cristobalite is induced by the three-dimensional zone-center irreducible representation ($\tau_8(\boldsymbol{k}_{11})$) of the $Fd\bar{3}m$ space group. Another irreducible representation of $Fd\bar{3}m[\tau_4(\boldsymbol{k}_{10})]$ allows us to describe the β–α cristobalite transition, where $\boldsymbol{k}_{10}$ coincides with the X-point of the FCC BZ surface.

6.6.2.6. *$CaCl_2$-type SiO_2*

The orthorhombic form of SiO_2 ($Pnnm, Z = 2$), which has been obtained using a 350-kbar shock wave [85], corresponds to a slight distortion of the stishovite structure, with the oxygen displaced in general positions in the (x, y) plane, with unequal displacements in Fig. 6.23(b).

Although the $CaCl_2$-type structure of SiO_2 can be simply obtained from the stishovite parent structure at a zone-center irreducible representation $[\tau_5(\boldsymbol{k}_{17})]$ of the tetragonal BZ, which is commonly involved in rutile-type structures, one can also describe this structure as resulting directly from the BCC structure. The same irreducible representation $\tau_4(\boldsymbol{k}_9)$ associated with the BCC–stishovite transition leads to $CaCl_2-SiO_2$ but for different equilibrium values of the six-component order-parameter $(\eta_1, \eta_2, 0, 0, 0, 0)$. The calculated lattice parameters ($a^c_{CC} = 4.06$ Å, $b^c_{CC} = 2.9$ Å) differ substantially from the experimental values [85] ($a^{\exp}_{CC} = 4.09$ Å, $b^{\exp}_{CC} = 3.79$ Å, $c^{\exp}_{CC} = 4.50$ Å) and require us to consider the improper strains: $e_{xx} = e_{yy} = 7.3\%$, $e_{zz} = -17.9\%$ and $e_{xy} = 8.8\%$.

6.6.2.7. *α-PbO₂-type SiO₂*

The α-PbO_2–columbite-type SiO_2 ($Pbcn$, $Z = 4$) was synthesized by Liu *et al.* [86] at pressures greater than 350 kbar and temperatures greater than 1000°C. The connection between its hexagonal-deformed FeN-type unit cell [87] and the BCC structure assumed in our approach is shown in Fig. 6.31. The relationship between the orthorhombic and BCC lattice vectors is

$$\boldsymbol{a}_{\text{col}} = \boldsymbol{a}_1 + \boldsymbol{a}_2 + \boldsymbol{a}_3, \boldsymbol{b}_{\text{col}} = 2(\boldsymbol{a}_1 + \boldsymbol{a}_2), \boldsymbol{c}_{\text{col}} = -\boldsymbol{a}_1 + \boldsymbol{a}_2. \quad (6.29)$$

At the virtual BCC–α-PbO_2 transition, the oxygens, which are fully ordered in the BCC structure, reach the 8(d) positions. The silicons, which are initially disordered in 3(b) octahedral interstitials, order in 4(c) positions. This ordering process corresponds to two combined mechanisms: (i) a Burgers-type mechanism consisting of an antiparallel displacement of the oxygens along the [110] and $[1\bar{1}0]$ BCC directions, which produces a hexagonal arrangement of the oxygens within the (110) BCC planes; (ii) an ordering of the Si atoms in 2(a) hexagonal positions.

The mechanism involved in the BCC–$\alpha-PbO_2$-type SiO_2 transition is associated with the same irreducible representation $\tau_4(\boldsymbol{k}_9)$ than the BCC–stishovite transition. The additional ordering of the Si atoms in hexagonal positions corresponds to the $\tau_2(\boldsymbol{k}_8)$ irreducible representation with the following equilibrium values of the six-component order-parameters $(\eta, 0, 0, \eta, 0, 0)$. The calculated orthorhombic lattice parameters are $a^c_{\text{PbO}} = c^c_{\text{PbO}} = 4.06$ Å and

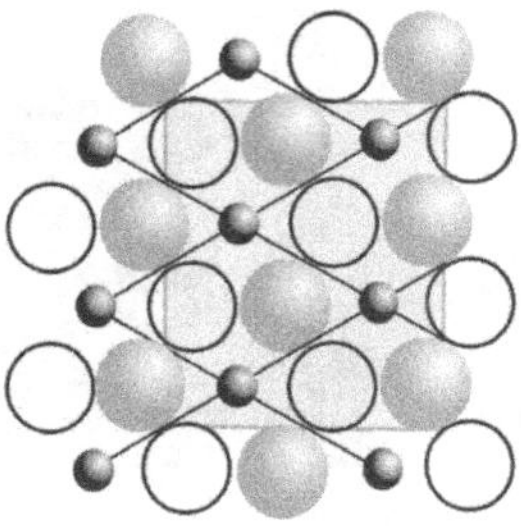

Fig. 6.31. Connection between the α-PbO_2 and hexagonal unit cells.

$b^c_{\mathrm{PbO}} = 5.8$ Å, which compared to the experimental values ($a^{\mathrm{exp}}_{\mathrm{PbO}} = 4.181, c^c_{\mathrm{PbO}} = 3.842$ Å and $b^{exp}_{\mathrm{PbO}} = 4.617$ Å) require compressions along $\boldsymbol{x}$, $\boldsymbol{y}$ and $\boldsymbol{z}$ ($e_{xx} = e_{yy} = 2.9\%, e_{zz} = 17.9\%$) and decompression along the $\boldsymbol{x}$, $\boldsymbol{y}$ direction ($e_{xy} = -4.4\%$).

6.6.3. *BCC parent structure: Degree of occupancy and density*

Figure 6.32 depicts the connections, assumed in our approach, between the SiO_2 polymorphs and their common virtual parent structure. The possibility of a high-pressure BCC configuration of SiO_2 and H_2O molecules has been already suggested by a number of authors [88–90] as representing the closest AB_2 packing which may be obtained when increasing the pressure. It also represents the most realistic configuration that can be assumed for AB_2 tetrahedrally coordinated close-packed clusters (at variance with HCP of FCC packings), which may form within the melt when approaching solidification. A BCC cluster realizes, in the simplest way, the tetrahedral and octahedral distributions of chemical bonds which constitute the basic motifs in covalently-bonded AB_2 polymorphs.

It is shown in Fig. 6.32 that there is a simple connection between pressure P, density ρ and the degree of occupancy x of the BCC structure by oxygen atoms, and the correlated influence of these parameters on the structural changes in SiO_2. For the lowest pressure phases (cristobalite, tridymite), one has the lower density of $\sim 2.35\,\mathrm{g/cm}^3$ and the minimal degree of occupancy $x = 1/2$. When

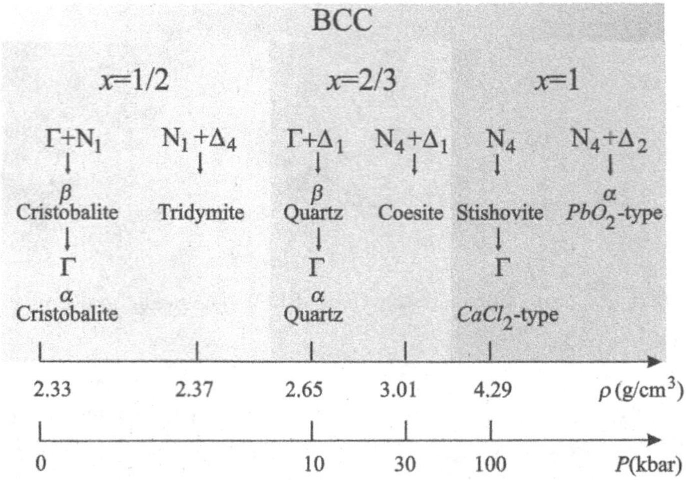

Fig. 6.32. Connection between the BCC parent structure and the SiO_2 polymorphs described in the present section. The points of the BCC Brillouin zone involved in the virtual transition mechanisms are indicated by the current Greek symbols, as well as the respective occupancies of the BCC unit cell. The densities and pressures corresponding to each structure are given at the bottom of the figure.

increasing pressure by a few kbar but below about 80 kbar, the density increases to $2.65 - 3\,\text{g/cm}^3$ and the degree of occupancy jumps to $x = 2/3$. For pressure above 80 kbar, the density reaches its highest values of $\sim 4.3\,\text{g/cm}^3$, and one has a maximal occupancy of $x = 1$ for the oxygens in the parent structure. This close correlation between P, ρ and x denotes an implicit morphotropic character for the sequence of reconstructive transitions which take place along the P-axis: tridymite $\rightarrow$ cristobalite $\rightarrow$ quartz $\rightarrow$ coesite $\rightarrow$ stishovite $\rightarrow \ldots$, i.e., the degree of occupancy x of the BCC structure plays the role of concentration in open systems such as solid solutions.

It is worth noting that the existence of vacancies in the BCC lattice, assumed in the model, should favor the amorphization process, since it allows an uncorrelated aggregation of SiO_2 clusters, which may lead in the intermediate region of the phase diagram ($x = 2/3$) to the formation of microcrystals possessing differently ordered and deformed structures with respect to the parent BCC structures.

Such microcrystals have been observed by electron microscopy in the amorphization process of quartz [91]. Accordingly, the α-SiO_2 phase can be interpreted as an assembly of microcrystalline regions involving different types of coordination, which is incompatible with a crystalline order.

6.6.4. *BCC parent structure: The Japanese twins of quartz*

An indication in favor of a parent BCC structure for quartz can be found in the existence of the so-called Japanese twins, observed in β-and α-quartz [73]. Following our description of the BCC–β-quartz transition, the orientation of the six-fold axes in β-quartz should correspond to cubic directions such as [120] and $[0\bar{1}2]$ (Fig. 6.33), which transform one into another by the cubic 4^x axis, and to the direction [201] obtained by a four-fold rotation 4^z. Furthermore, the structure of the domain walls separating the Japanese twins should reflect the structure of the parent BCC phase, i.e., it should correspond to a Bain-deformed section of the BCC structure having on average a planar orthorhombic symmetry. To determine the structure of the domain wall between the domains $1 \sim (\eta, -\eta, 0, 0, 0, 0, 0, 0, 0, 0, 0, 0)$ and $2 \sim (0, 0, 0, 0, 0, 0, 0, 0, \eta, -\eta, 0, 0)$, one can refer to the mechanism proposed in Sec. 6.6.2.3 for the BCC–β-quartz transition.

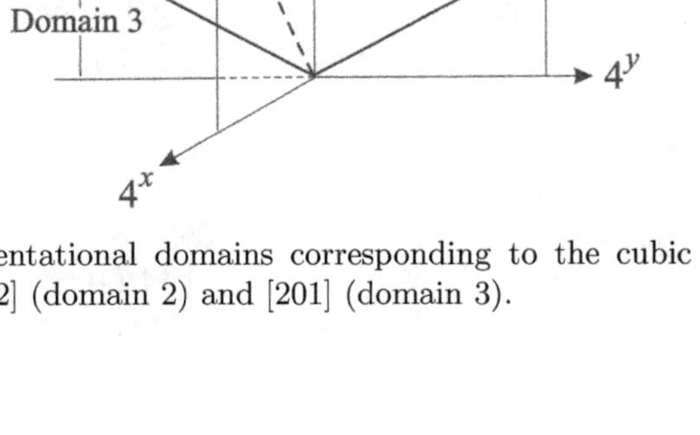

Fig. 6.33. Orientational domains corresponding to the cubic directions [021] (domain 1), $[0\bar{1}2]$ (domain 2) and [201] (domain 3).

This mechanism involves two distinct, 12- and 6-dimensional, order-parameters having the symmetries $\tau_3(\boldsymbol{k}_4)$ and $\tau_1(\boldsymbol{k}_8)$, respectively. The 12-dimensional order-parameter undergoes at the middle surface S_m, which corresponds to a minimal deformation, a cross-over from the equilibrium values $(\eta, -\eta, 0, 0, 0, 0, 0, 0, 0, 0, 0, 0)$ and $(0, 0, 0, 0, 0, 0, 0, 0, \eta, -\eta, 0, 0)$ when going from one to the other domain. Therefore, it vanishes on S_m and does not affect the form of this surface. By contrast, the six-dimensional order-parameter goes from the value $(\eta, 0, 0, \varepsilon^*\eta, 0, 0)$ to $(\varepsilon^*\eta, \eta, 0, 0, 0, 0)$. Hence, on the S_m surface, η_1 and η_4 exhibit a jump of $\pm(\eta - \varepsilon^*\eta)$. Consequently, S_m will have the symmetry of a *tetragonal* cross-section of the parent BCC structure, corresponding to the equilibrium values of the order-parameter $[(\eta - \varepsilon^*\eta), 0, 0, (\varepsilon^*\eta - \eta), 0, 0]$. It means that on the S_m surface the distortion component M_{yz} changes its sign, whereas M_{yy} and M_{zz} exhibit jumps of $\pm(\eta_1 - \eta_4)$ without changing sign [92]. Accordingly, the S_m surface, as well as the domain wall between the domains 1 and 2, should be deformed along the $\boldsymbol{y}$ and $\boldsymbol{z}$ axes.

References

[1] N. G. Parsonage and L. A. K. Staveley, *Disorder in Crystals* (Clarendon Press, Oxford, 1978).
[2] *Order-Disorder Transformations in Alloys*, ed. H. Warlimont (Springer Verlag, Berlin, 1974).
[3] A. Guinier, G. B. Bokij, K. Boll-Dornberger *et al. Acta Cryst. A* **40** (1984) 399.
[4] K. Dornberger-Schiff, *Acta Cryst.* **9** (1956) 593.
[5] A. R. Verma and P. Krishna, *Polymorphism and Polytypism in Crystals* (John Wiley Inc., New York, 1966).
[6] D. P. Varn and G. S. Canright, *Acta Cryst. A* **57** (2001) 4.
[7] G. Canright and G. Watson, *J. Stat. Phys.* **84** (1996) 1095.
[8] D. A. Young, *Phase Diagrams of the Elements* (University of California Press, Berkeley, 1991).
[9] J. C. Dugdale and J. P. Frank, *Phil. Trans. R. Soc. (Lond.) A* **257** (1964) 1.
[10] A. Jayaraman, *Phys. Rev. A* **139** (1965) 690.
[11] G. J. Piermarini and C. E. Weir, *Science* **144** (1964) 69.
[12] O. S. Edwards and H. Lipson, *Proc. R. Soc. Lond., Ser. A* **180** (1942) 268.

[13] K. Lonsdale, *Nature* **155** (1945) 572; C. V. Raman, *Nature* **156** (1945) 22.
[14] M. T. Sebastian and P. Krishna, *Random, Non-Random and Periodic Faults in Crystals* (Gordon and Breach, New York, 1994).
[15] P. V. Hobbs, *Ice Physics* (Clarendon Press, Oxford, 1974).
[16] N. V. Belov, *Structure of Inorganic Crystals and Metallic Phases* (AN SSSR, Moscow, 1947).
[17] J. P. Frank, *Phys. Rev. B* **22** (1980) 4315.
[18] J. P. Frank and W. B. Daniels, *Phys. Rev. Lett.* **44** (1980) 259.
[19] Z. Nishiyama, *Martensitic Transformations* (Academic Press, New York, 1978).
[20] J. O. Nelson and C. J. Altstetter, *Trans. AIME* **230** (1964) 1577.
[21] A. Goldsmith, *Handbook of Thermophysical Properties of Solid Materials. I. Elements* (Pergamon Press, Oxford, 1961), p. 142.
[22] S. M. Shapiro and S. Moss, *Phys. Rev. B* **15** (1977) 2726.
[23] E. C. Svensson, B. M. Powell, D. B. Woods and W. Teuchert, *Can. J. Phys.* **57** (1979) 253.
[24] F. Frey, W. Prandl, J. Schneider, C. Zeyen and K. Ziebeck, *J. Phys. F: Met. Phys.* **9** (1979) 603.
[25] N. Wakabayashi, R. H. Schermand and H. G. Smith, *Phys. Rev. B* **25** (1982) 5122.
[26] B. Strauss, F. Frey, W. Petry, J. Trampenau, K. Nikolaus, S. M. Shapiro and J. Bossy, *Phys. Rev. B* **54** (1996) 6035.
[27] O. S. Edwards and H. Lipson, *Proc. R. Soc. Lond., Ser. A* **180** (1942) 268.
[28] O. Blaschko, G. Krexner, J. Pleschiutschnig, G. Ernst, C. Hitzenberger, H. P. Karnthaler and A. Korner, *Phys. Rev. Lett.* **60** (1988) 2800.
[29] T. Onozuka, S. Yamaguchi, M. Hirabayashi and T. Wakiyama, *J. Phys. Soc. Jpn.* **37** (1974) 687.
[30] 30. R. Adams and C. Altstetter, *Trans. AIME* **242** (1968) 139.
[31] C. S. Yoo, P. Soderlind and H. Cynn, *J. Phys.: Condens. Matter* **10** (1998) L311.
[32] F. Frey and H. Boysen, *Acta Crystallogr. A* **37** (1981) 819.
[33] A. Goldsmith, *Handbook of Thermophysical Properties of Solid Materials. I. Elements* (Pergamon Press, Oxford, 1961), p. 233.
[34] J. E. Bidaux, R. Schaller and W. Benoit, *Acta. Metall.* **37** (1989) 803.
[35] J. A. Krumhansl and R. J. Gooding, *Phys. Rev. B* **39** (1989) 3047.
[36] B. S. Bokshtein, Yu. B. Voitkovskii, G. S. Nikol'skii and I. M. Razumovskii, *Sov. Phys.: JETP* **37** (1973) 283.
[37] V. P. Dmitriev, P. Toledano, V. I. Torgashev and E. K. H. Salje, *Phys. Rev. B* **58** (1998) 11911.

[38] H. Jagodinski, *Acta Crystallogr.* **7** (1954) 300.
[39] V. Dmitriev, P. Toledano, A. M. Figueiredo Neto and I. V. Lebedyuk, *Phys. Rev. E* **59** (1999) 771.
[40] H. Bibring and F. Sebilleau, *Rev. Met.* **52** (1955) 569.
[41] P. Gaunt and J. W. Christian, *Acta Metall.* **7** (1959) 529.
[42] A. Yu. Babkevich, F. Frey and H. Kahlert, *Philos. Mag. A* **72** (1995) 1341.
[43] S. Kajiwara, *Jpn. J. Appl. Phys.* **9** (1970) 385.
[44] T. Wartz and H. P. Karnthaler, *Philos. Mag. A* **73** (1996) 365, and references therein.
[45] C. Hitzenberger, H. P. Karnthaler and A. Korner, *Acta Metall.* **33** (1985) 1293.
[46] C. Hitzenberger, H. P. Karnthaler and A. Korner, *Acta Metall.* **36** (1988) 2719.
[47] T. Waitz and H. P. Karnthaler, *Phase Transit.* **67** (1999) 695.
[48] B. I. Nikolin, A. Yu. Babkevich, T. V. Izdkovskaya and S. N. Petrova, *Acta Metall. Mater.* **41** (1993) 513.
[49] S. A. Demin, A. A. Nekrasov and A. I. Ustinov, *Acta Metall. Mater.* **41** (1993) 2091.
[50] J. W. Christian, *Proc. R. Soc. Lond., Ser. A* **206** (1951) 61.
[51] Z. S. Basinski and J. W. Christian, *Philos. Mag.* **44** (1953) 791.
[52] A. Seeger, *Z. Metallkd.* **44** (1956) 247; **47** (1956) 653.
[53] W. Bollmann, *Acta Metall.* **9** (1961) 972.
[54] H. Fujita and S. Veda, *Acta Metall.* **20** (1972) 759.
[55] G. B. Olson and M. Cohen, *Metall. Trans. A* **7** (1976) 1897.
[56] S. Mahajan, M. L. Green and D. Brasen, *Metall. Trans. A* **8** (1977) 283.
[57] J. Singh and S. Ranganathan, *Phys. Status Solidi A* **73** (1981) 243.
[58] D. Pandey and S. Lele, *Acta Metall.* **34** (1986) 415.
[59] J.-C. Toledano and P. Toledano, *Phys. Rev. B* **21** (1980) 1139.
[60] T. Waitz and H. P. Karnthaler, *Acta Metall.* **45** (1997) 837.
[61] J.-C. Toledano and P. Toledano, *Phys. Rev. B* **16** (1976) 386.
[62] C. S. Yoo, H. Cynn, P. Soderlind and V. Iota, *Phys. Rev. Lett.* **84** (2000) 4132.
[63] H. Lipson and A. R. Stokes, *Nature* **149** (1942) 328.
[64] V. P. Dmitriev, S. B. Rochal, Yu. M. Gufan and P. Toledano, *Phys. Rev. Lett.* **62** (1989) 2495.
[65] W. H. Bragg, *Proc. Phys. Soc. Lond.* **33** (1921) 304; W. Ehrenberg, *Z. Kristallogr.* **63** (1926) 320.
[66] J. D. Bernal, *Proc. R. Soc. Lond., Ser. A* **106** (1924) 749; O. Hassel and H. Mark, *Z. Physik* **25** (1924) 317.
[67] F. P. Bundy and J. S. Kasper, *J. Chem. Phys.* **46** (1967) 3437.

[68] M. J. Buerger, in *Phase Transformations in Solids*, eds. E. Smoluchowski, J. Mayer and W. A. Weil (Wiley, New York, 1951), p. 183.
[69] L. Pauling, *The Nature of the Chemical Bond* (Cornell University Press, Ithaca, 1960), Chap. 4.
[70] M. P. Kulakov and S. Z. Shmurak, *Phys. Stat. Sol. A* **59** (1980) 147.
[71] O. Brafman and I. T. Steinberger, *Phys. Rev.* **143** (1966) 501.
[72] B. A. Abdikamalov, M. P. Kulakov, V. Sh. Shekhtman and S. Z. Shmurak, *Sov. Phys.: Solid State* **17** (1975) 2262.
[73] L. Bragg and G. F. Claringbull, *Crystal Structures of Minerals* (Bell and Sons, London, 1965).
[74] Lin-Gun Liu and W. A. Basset, *Elements, Oxides, Silicates* (Oxford University Press, New York, 1986).
[75] R. B. Sosman, *The Phase of Silica* (Rutgers University Press, New Brunswick, 1965).
[76] L. G. Liu, in *High Pressure Research in Geophysics*, eds. S. Akimoto and M. H. Manghnani (CAPJ, Tokyo, 1982), p. 349.
[77] L. G. Liu, W. A. Bassett and T. Takahashi, *J. Geophys. Res.* **79** (1974) 1160.
[78] E. C. T. Chao, J. J. Fahey and J. Littler, *J. Geophys. Res.* **67** (1962) 419.
[79] O. V. Kovalev, *Irreducible Representations of the Space Groups* (Gordon and Breach, New York, 1965).
[80] T. Zoltai and M. J. Buerger, *Z. Kristallogr.* **111** (1959) 129.
[81] J. E. Fleming and H. Lynton, *Phys. Chem. Glasses* **1** (1960) 148.
[82] P. Toledano and V. Dmitriev, *Reconstructive Phase Transitions* (World Scientific, Singapore, 1996), Chapter 3.
[83] R. W. G. Wyckoff, *Z. Kristallogr.* **62** (1925) 189.
[84] W. Nieuwenkamp, *Z. Kristallogr.* **92** (1935) 82.
[85] K. T. Park, K. Terakura and V. Matsui, *Nature* **336** (1988) 670.
[86] L. G. Liu, W. A. Bassett and J. Sharry, *J. Geophys. Res.* **78** (1978) 2301.
[87] V. N. German, M. A. Podurets and R. Trunin, *Sov. Phys.: JETP* **37** (1973) 107.
[88] H. Sowa, *Z. Kristallogr.* **184** (1988) 257.
[89] R. H. Hazen, W. L. Finger, R. J. Hemley and H. K. Mao, *Solid State Commun.* **72** (1989) 507.
[90] V. Dmitriev, S. Rochal and P. Toledano, *Phys. Rev. Lett.* **71** (1993) 553.
[91] K. J. Kingma, C. Meade, R. J. Hemley *et al.*, *Science* **259** (1993) 666.
[92] V. Dmitriev, P. Toledano, V. Torgashev and E. K. H. Salje, *Phys. Rev. B* **58** (1998) 11911.

Chapter 7

Phase Transitions from Isotropic State: Crystallization

7.1. Introduction

It is generally accepted that despite regular occurrence in materials crystallization or melting is the hardest to analyse phase transition. Already in his pioneering paper on symmetry-based phenomenological approach to phase transitions [1], L. Landau discussed crystallization process from isotropic liquid to periodic structures. However, in Ref. [1] it was more of motivating introduction than rigorous consideration of the problem. The principal reason is that crystallization typically occurs via *strong first-order* phase transition, while the phenomenological Landau theory operates with a free energy in the form of truncated series expansion by *small-value* order-parameters. Nevertheless, it was suggested that if somewhere crystallization process shows no signs of strong first-order transformation (e.g., a transformation is discontinuous, but its enthalpy is small), one can use either the thermodynamic model restricted to the fourth-degree expansion containing a cubic invariant (crystallization of symmetric units described by a scalar order-parameter) [1] or an expansion model containing only even-degree terms (crystallization of asymmetric units, pseudo-scalar order-parameter) but accounting for strong heterophase fluctuations [2].

In order to identify the applicability limits of the phenomenological theory of crystallization, existence of two crystalline types were assumed: (a) *conventional crystals* where atoms are localized in sites of a regular lattice and (b) *coherent crystals* characterized by a smooth periodic density modulation [3]. The latter, type (b), covers a broad class of materials possessing thermotropic liquid-crystalline phases which transform from the isotropic to periodically modulated state via weak first-order phase transitions. By contrast, for known simple liquids, including molten metals and alloys, the crystallization is a strong discontinuous transition, and the corresponding crystals belong to the (a) type. The group of materials integrating both crystal types constitute so-called lyotropic liquid crystals (alternative term, lyotropic complex fluids), which are organized molecular systems formed by amphiphilic organic molecules in the presence of a solvent, most often water.

In this chapter, we consider the applicability of the phenomenological approach to various types of crystallization processes, and discuss variants of the theory relevant to the corresponding cases.

7.2. Weak Crystallization in Classic Landau Theory

First stage of the crystallization process in isotropic liquid assumes to consist of onset of short-wavelength density modulations. The normalized amplitude of the modulation $\rho = \rho_{\text{short}}/\rho_0$ can be regarded as the order-parameter of the crystallization phase transition [1]. $\rho_{\text{short}} = (\rho_0 - \rho_{\text{d}})$ defines the difference between liquid state ($\rho_{\text{short}} = 0$) and crystalline solid state ($\rho_{\text{short}} \neq 0$). The order-parameter ρ is a scalar, thus, corresponding Landau potential $F(\rho)$ (non-equilibrium free energy) contains a cubic invariant. Consequently, the crystallization must be a first-order phase transition, which is fully consistent with the experimental situation. Since the classic Landau theory is valid for *small* order-parameters, only phase transitions from isotropic liquid to a *slightly density-modulated* anisotropic state can be considered. These transformations are termed *weak crystallization*. A comprehensive review of the results obtained for weak crystallization in the frame of the Landau theory can be found in Refs. [4, 5].

In brief, the symmetry group of an isotropic state is the extended Euclidean group $\tilde{E}_3$ which consists of all *continuous* proper and improper rotations and translations in the 3D-space [6]. It is worth recalling that in continuously degenerated systems fluctuations are not negligible at any temperature even in the anisotropic state [7]. This is why fluctuations become a compulsory element of the weak crystallization model.

Fully in the spirit of the density-wave approach (see Chapter 4), the average $\langle\rho\rangle$ is considered as a sum of an infinite number of Fourier harmonics, but without losing in generality, we can single out only reasonably a few principal ones. Thus,

$$\langle\rho(\boldsymbol{r})\rangle = 2\cdot \mathrm{Re}\sum_{j=1}^{N} a_j \cdot \exp(i\boldsymbol{q}_j\cdot\boldsymbol{r}),$$

where N is the number of principal harmonics, and a_j are their amplitudes. Absence, in general, of a fixed reference length scale in the isotropic liquid state makes logical considering the fluctuations of both density-wave amplitude and its wave vector. Then free-energy density $f(\rho)$ associated with the order-parameter ρ and its fluctuations, in Fourier space, reads as a standard Landau-potential expansion

$$\begin{aligned} f(\rho) = \frac{F(\rho)}{V} &= \sum_{\boldsymbol{q}} \omega(\boldsymbol{q})\rho(\boldsymbol{q})\rho(-\boldsymbol{q}) \\ &- \sum_{\boldsymbol{q}_1,\boldsymbol{q}_2,\boldsymbol{q}_3} \lambda_1(\boldsymbol{q}_1,\boldsymbol{q}_2,\boldsymbol{q}_3)\rho(\boldsymbol{q}_1)\rho(\boldsymbol{q}_2)\rho(\boldsymbol{q}_3) \\ &+ \sum_{\boldsymbol{q}_1,\boldsymbol{q}_2,\boldsymbol{q}_3,\boldsymbol{q}_4} \lambda_2(\boldsymbol{q}_1,\boldsymbol{q}_2,\boldsymbol{q}_3,\boldsymbol{q}_4)\rho(\boldsymbol{q}_1)\rho(\boldsymbol{q}_2)\rho(\boldsymbol{q}_3)\rho(\boldsymbol{q}_4), \end{aligned} \tag{7.1}$$

where V is the system volume, and $|\boldsymbol{q}_i| = |\boldsymbol{q}|$ [5]. Since $\rho(\boldsymbol{q}_i)$ are the Fourier components of the field ρ, and $f(\rho)$ is a real functional, summation in Eq. (7.1) should be performed over $\sum_i \boldsymbol{q}_i = 0$, i.e., vectors $\{\boldsymbol{q}_i^{\mathrm{eq}}\}$ in the reciprocal space have to be arranged as regular equilateral polygons. Namely, the cubic term is only non-zero for $\{\boldsymbol{q}_i\}$ forming equilateral triangles, the quadratic term obtains either from combining $\pm\boldsymbol{q}_i$ with $\pm\boldsymbol{q}_j$, or it can appear when four $\boldsymbol{q}_i$ form a non-coplanar quadrilateral.

The order-parameter ρ is the short-wavelength variable, thus coefficients in the Landau expansion (7.1) can depend on wave vectors $\boldsymbol{q}$. Interatomic interactions set a preferred length scale for crystallization, and this scale is introduced into second-order term. The latter term should be small near the phase transition point in the vicinity of a sphere $|\boldsymbol{q}| = q_0$. The corresponding coefficient takes the form of

$$\omega(\boldsymbol{q}) = t_n + b(\boldsymbol{q} - \boldsymbol{q}_0)^2, \tag{7.2}$$

where t_n is normalized temperature $t_n = \alpha \cdot (T - T_C)/T_C, T_C$ is a stability limit temperature, and α and b are phenomenological constants.

In the vicinity of the sphere $|\boldsymbol{q}| = q_0$ just mentioned above, the critical wave vectors $|\boldsymbol{q} - \boldsymbol{q}_0| << q_0$, and the quadratic term in Eq. (7.2) can be approximated as

$$(\boldsymbol{q} - \boldsymbol{q}_0^2) \cong \frac{(\boldsymbol{q}^2 - \boldsymbol{q}_0^2)^2}{4\boldsymbol{q}_0^2}.$$

Then, the Landau potential (7.1) transforms to

$$F = \int d^3r\{a_1\rho^2 + \frac{\beta}{8\boldsymbol{q}_0^2}[(\nabla^2 + \boldsymbol{q}_0^2)\rho]^2 - b_1\rho^3 + a_2\rho^4\}. \tag{7.3}$$

The Landau potential in the form of Eq. (7.3) relevant to the weak crystallization clearly shows two contributions: (i) a gradient term accounting for order-parameter fluctuations and (ii) a non-equilibrium part of the free energy. Minimizing the latter, one finds equilibrium array of $\boldsymbol{q}$-vectors and equilibrium values of the order-parameter components.

Comparing standard thermodynamic models presented in Chapter 2 (mean-field theory) and the weak crystallization theory for isotropic systems, two points are worth noting. First is the similarity of the non-fluctuational contribution to the free energy of the weak crystallization model, minima of which define the crystal geometry of solid state, and the standard Landau model (Sec. 1.3.3.3). Thus, the corresponding weak first-order phase transition occurs in the vicinity of the Landau point ($a_1 = b_1 = 0$), which is an isolated point of a second-order phase transition on the line of first-order transformations (see [1] and Sec. 1.3.3.2). Second point is the difference in the description of the effects induced by order-parameter fluctuations. We already underlined earlier that in isotropic systems order-parameter fluctuations can play a significant role even far from the transition point. Equation (7.3) mathematically expresses this physical hypothesis. In contrast to the crystalline parent state, where the gradient term is proportional to ∇^2 (see Sec. 5.6.1), the fluctuation effects in the isotropic state are accounted for by a fourth-degree gradient term proportional to ∇^4. It confirms the conclusion that even small fluctuations may significantly affect a weak crystallization process.

Determination of all possible structures which can be crystallized as a weak first-order transition from liquid is a non-trivial task in the weak crystallization theory. Indeed, the corresponding parent symmetry group has continuous symmetry, the order-parameter ρ has infinite number of components or, equivalently, infinite number of both $\boldsymbol{q}_i$-vectors identifying possible crystalline lattices and Fourier harmonics describing possible structures. Nevertheless, if the condition $\langle \rho \rangle \ll 1$ is valid, i.e., weak crystallization occurs in the vicinity of the Landau point, for the Landau functional (7.3), four phases can be identified: (1) isotropic liquid as a parent phase; (2) one-dimensionally modulated layered structure remaining isotropic in the perpendicular plane; (3) two-dimensionally modulated column-type structure; and (4) body-centered cubic structure [5].

Summarizing, one concludes that the weak crystallization theory actually considers weak first-order phase transitions between the *liquid-like* (uniform in space) and *solid-like* (non-uniform with characteristic wave vector $\boldsymbol{q}_0$) states, i.e., Landau (non-limit) phases in our notations; no limit (non-Landau) states are included. Its distinctive feature, following Refs. [4,5], is the important role of fluctuations of the order-parameter ρ, accounted for by the large phase volume. Fluctuation effects are strong due to the continuous symmetry of the parent phase and small amplitudes of density modulation of the distorted phase, and its liquid-like nature. It generates a number of qualitative effects, namely fluctuations make the crystallization phase transition a first-order phase transition even in the absence, in the Landau expansion, of the term cubic in ρ.

In the phase diagram of the weakly crystallized system, there arise both periodic phases of different symmetry as well as smectic, columnar and even quasicrystalline phases. The really important results the weak crystallization theory obtains in explaining the remarkable diversity of liquid-crystalline phases, emerging in a relatively narrow temperature range in real compounds. Apart from this, a consequence of the theory is the anomalous behavior of various physical characteristics, and, in the first place, the large values of the viscosity coefficients, which are one of the principal experimental characteristics of liquid crystals. Moreover, the theory makes it possible to establish links between the results of different types of experiments: calorimetric, dynamic and X-ray diffraction and scattering.

7.3. Liquid–Solid Phase Transition in the Advanced Phenomenological Approach

Liquid state directly borders on a rich variety of regular spatially organized structures in phase diagrams of elemental metals and alloys, inorganic and organic compounds. These include one-dimensional (sheet-like), two-dimensional (column-type) and three-dimensional ordered (cubic, hexagonal, rhombohedral, etc.) crystalline structures and mesophases. In the following sections, we apply the advanced symmetry-based phenomenological approach describing the formation of periodic structures from isotropic liquid, melt or molecular solution. Keeping in mind the variety of possible isotropic media (simple liquid, molecular mixture, etc.) as well as the periodic structures falling both in the categories of conventional crystals, identified by the strong localization of atoms in the lattice nods, and coherent crystals, characterized by non-zero probability for atoms and molecules to remain in intersite positions, we use in the following sections a more general term "*segregation-type transition*" instead of "crystallization" or "ordering".

7.3.1. *Phenomenological theory of segregation*

7.3.1.1. *Symmetry of the parent phase and the order-parameter space*

Let us start from the full-space symmetry of the phases. The symmetry group of an isotropic state is the extended Euclidean group $\tilde{E}_3$, which is a semi-direct product of the continuous three-dimensional translation group R_3, and orthogonal group $O(3)$ consisting of all continuous proper and improper rotations in the three-dimensional space [6]. We are considering the segregation-type transformation which appears as initial density distribution $\rho_0(\boldsymbol{r}) = \text{const}$ to be modulated periodically with the undulation function $\delta\rho(\boldsymbol{r}) = \rho_d(\boldsymbol{r}) - \rho_0(\boldsymbol{r})$. This periodic deviation of density $\delta\rho$ can be expanded, as in the case of discrete groups, using $\Phi_{\boldsymbol{k}_j^m}(\boldsymbol{r})$ basis functions of a relevant irreducible representation of $\tilde{E}_3$ group:

$$\delta\rho(\boldsymbol{r}) = \sum_{\boldsymbol{k}_j,m} \eta_{\boldsymbol{k}_j^m} \cdot \Phi_{\boldsymbol{k}_j^m}(\boldsymbol{r}), \tag{7.4}$$

where $\eta_{\boldsymbol{k}_j^m}$ are the order parameter components and $\boldsymbol{k}_j$ are the reciprocal space vectors, m specifies an irreducible representation. The functions

$$\Phi_{\boldsymbol{k}_j^m}(\boldsymbol{r}) = \Phi_{\boldsymbol{k}_j^m}(r,\theta,\varphi) = e^{i\boldsymbol{k}_j\boldsymbol{r}} \cdot Y_l^m(\theta,\varphi), \tag{7.5}$$

span infinite-dimensional irreducible representations of $\tilde{E}_3$ group, which will be denoted hereafter by $D_{\pm/}^{m,\boldsymbol{k}_j}$. The lower index indicates symmetry properties with respect to the inversion center. In order to construct irreducible representations $D_{\pm}^{m,\boldsymbol{k}_j}$, the method of induced representation can be used in which irreducible representations (small representations) of the invariant point group $G_{\boldsymbol{k}_j} = C_{\infty_v}$ of the $\boldsymbol{k}_j$ vector that is isomorphous to $SO(2)$ group induce irreducible representations for $\tilde{E}_3$ space group [6].

The "unit cell" of the parent isotropic state is reduced to a single subunit (atomic, molecular or micellar center of mass), which is invariant under all symmetry operations of $\tilde{E}_3$. Since the ordering process can be fully specified by the scalar parameter which is the probability density variation, and due to the invariance of the structure just mentioned, the $\{\eta_{\boldsymbol{k}_j^m}\}$ necessarily transforms as the irreducible representation denoted $D_{+}^{0,\mathbf{kj}}$, i.e., the irreducible representation induced by the identity (totally symmetric) small irreducible representation of $C_{\infty v}$.

The variational free energy $F(T;P;\eta_{\boldsymbol{k}_j}^m)$ is invariant by the symmetry transformations of the parent structure. Thus, it can be expanded in Taylor series by invariant homogeneous polynomials of order-parameter components. The general form of such a polynomial is

$$I_n = \sum_s C_s \cdot \eta_1 \cdots \eta_{1n_s} = \sum_s C_s \cdot e^{i(\boldsymbol{k}_1+\cdots+\boldsymbol{k}_{n_s})\boldsymbol{r}} \cdot |\eta_{\boldsymbol{k}_1}| \cdots |\eta_{\boldsymbol{k}_{n_s}}|. \tag{7.6}$$

Since the free energy $F(T;P;\eta_{\boldsymbol{k}_j^m})$ must be a function of modulus $|\eta_{\boldsymbol{k}_j}|$ but not their orientation, exponential multipliers in Eq. (7.6) have to be equal to unity. The $\{\boldsymbol{k}_j\}$ vector star consists of an infinite number of arms, and for any n_s, the corresponding low-symmetry star can be chosen such that $(\boldsymbol{k}_1+\cdots+\boldsymbol{k}_{ns})=0$ and the above conditions are satisfied. This means that, considering the segregation process, the variational free-energy $F(T;P;\eta_{\boldsymbol{k}_j^0})$ expansion contains invariant polynomials of all degrees, excluding the linear one.

In order to find the stable anisotropic phases, one needs to minimize the variational free energy (Landau potential) $F(T;P;\eta_{\boldsymbol{k}_j^m})$ with respect to the OP components:

$$\frac{F(T,P,\eta_{\boldsymbol{k}_j^m})}{\partial\eta_{\boldsymbol{k}_j^m}}=0. \tag{7.7}$$

Despite the infinite number of equations in Eq. (7.7), the problem can be solved *geometrically* in the order-parameter space ε for which the set of OP components $\{\eta_{\boldsymbol{k}_j^m}\}$ forms the basis [8, 9]. Remember that according to the standard procedure of the symmetry-based phenomenological approach, first, we should solve a model-free *angular problem*, consisting in minimizing the free energy and identifying the symmetry of distorted structures (Sec. 1.3). It is worth recalling that the $\delta\rho$ variation of probability density associated with a phase transition can be considered as a vector in the ε-space, and the components of $\{\eta_{\boldsymbol{k}_j^m}\}$ vector are the values of the order-parameter which minimize the thermodynamic potential. Minimization conditions (7.7) determine in ε-space $\{\eta_{\boldsymbol{k}_j^m}\}$ vectors (isotropy manifolds) which are invariant with respect to different subgroups of the order-parameter symmetry group (invariance groups), formed by the matrices of the relevant irreducible representation (image group) (see Sec. 1.3.2).

Matrices of the relevant $D_{+}^{0,\boldsymbol{k}_j}$ corresponding to the rotation subgroup of $\tilde{E}_3$ group are unity ones permuting the arms of $\{\boldsymbol{k}_j\}$ vector star. Matrices corresponding to the elements of translation subgroup are diagonal ones, the elements of these matrices are exponentials $\exp(i\boldsymbol{k}_j\boldsymbol{a}_i)$. For any crystal group, a homomorphous subgroup of the image group can be found, and the corresponding invariant (stationary) vector can be determined. One can use for this end the obvious idea that only for invariant components of $\{\eta_{\boldsymbol{k}_j^0}\}$ exponential multipliers will be equal to unity ($\exp(i\boldsymbol{k}_j\boldsymbol{a}_i) = 1$). Space arrangement of equilibrium $\{\boldsymbol{k}_{0j}\}$ and $\{\boldsymbol{a}_{0i}\}$ stars must be consistent also with the point group of the considered crystal class.

7.3.1.2. *Invariance groups in ε-space and isotropy groups in segregated states*

The vector directions which correspond to the minima of the free energy $F(\eta_{\boldsymbol{k}_j^m})$ are along the directions of maximal symmetry in the order-parameter space ε. Let us consider a few invariance groups – maximal subgroups of the order-parameter image group, the symmetry of isotropy groups G_d, and structures of the corresponding modulated liquid-like states.

Depending on the array of invariant vectors, i.e., on the number of nonvanishing independent $\boldsymbol{k}_j$, and on the respective equilibrium values of the $\eta_{\boldsymbol{k}_j}$, different segregated phases can be stabilized below the isotropic solution, whose symmetries correspond to subgroups G of $\tilde{E}_3$. The highest symmetry in the ε-space has a single-parametric pair of antiparallel vectors $\langle \eta_{-\boldsymbol{k}_j} = \eta_{+\boldsymbol{k}_j} \neq 0,\ \eta_{\boldsymbol{k}_i} = 0 \text{ for } i \neq j \rangle$. The corresponding order-parameter induces a one-dimensionally modulated phase of symmetry $G = D_{\infty h}$, which is described by a single pair of wave vectors $|\pm\boldsymbol{k}_j| = \pm\pi/a_j$ (Fig. 7.1(a)), where a_j is the period along the normal to the isotropic planes.

Two-dimensional cylindrical mesophases require two pairs of wave vectors: $\pm\boldsymbol{k}_1$ and $\pm\boldsymbol{k}_2$ (with $|\boldsymbol{k}_1| = |\boldsymbol{k}_2|$). One gets hexagonal ($G = 6m$) or square ($G = 4m$) mesophases for, respectively, an angle γ between $\boldsymbol{k}_1$ and $\boldsymbol{k}_2$ of $2\pi/3$ or $\pi/2$.

In the three-dimensionally modulated space, three independent basis vector stars $\{\boldsymbol{k}_j^0\}$ determine the cubic symmetry of the following periodic structures (Figs. 7.1(b)–7.1(d)):

$$\langle A\rangle:\ \pm\,\boldsymbol{k}_{[100]};\quad \langle B\rangle:\ \pm\,\boldsymbol{k}_{[111]};\quad \langle C\rangle:\ \pm\,\boldsymbol{k}_{[110]}. \tag{7.8}$$

Consequently, the only relevant set of order-parameter components are non-zero and equal to each other:

$$\eta^0_{\langle \boldsymbol{k}j\rangle} = \cdots = \eta^0_{\langle \boldsymbol{k}j'\rangle}. \tag{7.9}$$

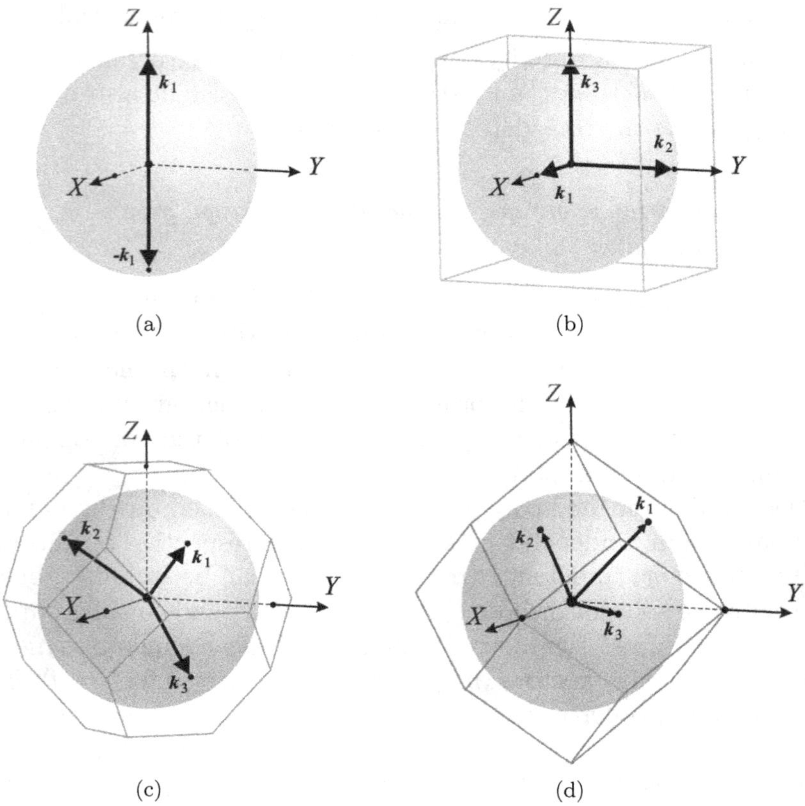

Fig. 7.1. Wave-vector stars corresponding to the order-parameter condensates: (a) one-dimensionally modulated phase of symmetry $G = D_{\infty h}$, (b) primitive cubic lattice $\langle A \rangle$, (c) face-centered cubic lattice $\langle B \rangle$, (d) body-centered cubic lattice $\langle C \rangle$ (see Eq. (7.8)).

One can see that, in direct space, 6-arm $\langle A \rangle$-star identifies the primitive cubic lattice, 8-arm $\langle B \rangle$ (body diagonals) — the face-centered cubic lattice, and 12-arm $\langle C \rangle$ (face diagonals) — the body-centered cubic lattice.

Along the same geometric minimization line, one finds, for the free energy $F(\eta^m_{\boldsymbol{k}_j})$, two solutions possessing icosahedral symmetry I_h and consisting of 12 sets of non-zero equal order-parameter components each, whose density modulation 6-arm wave-vector stars are

$$\boldsymbol{k}_1 = \langle 0\,1\,\tau \rangle, \boldsymbol{k}_2 = \langle 1\,\tau\,0 \rangle, \boldsymbol{k}_3 = \langle \tau\,0\,1 \rangle,$$
$$\boldsymbol{k}_4 = \langle \tau\,0\,-1 \rangle, \boldsymbol{k}_5 = \langle 0\,-1\,\tau \rangle, \boldsymbol{k}_6 = \langle -1\,\tau\,0 \rangle, \qquad (7.10)$$

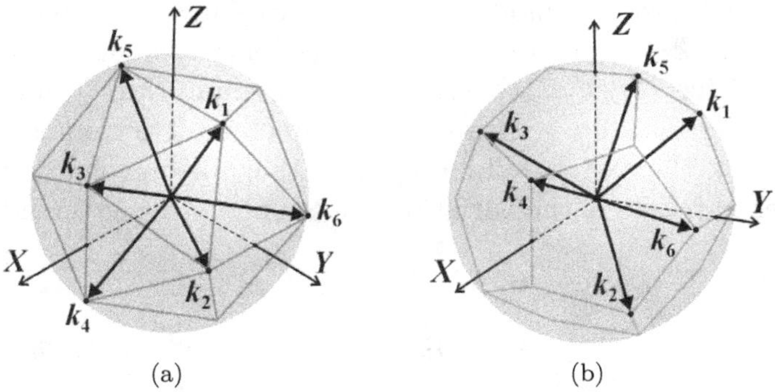

Fig. 7.2. (a) Icosahedral array of the wave-vectors (7.10). (b) Dodecahedral array of the wave-vectors (7.11).

and

$$\begin{aligned} \boldsymbol{k}_1 &= \langle -1\,1\,1 \rangle, \boldsymbol{k}_2 = \langle 1\,1\,-1 \rangle, \boldsymbol{k}_3 = \langle 1\,-1\,1 \rangle, \\ \boldsymbol{k}_4 &= \langle \tau\,0\,1/\tau \rangle, \boldsymbol{k}_5 = \langle 0\,1/\tau\,\tau \rangle, \boldsymbol{k}_6 = \langle 1/\tau\tau\,0 \rangle, \end{aligned} \qquad (7.11)$$

where $\tau = (1+\sqrt{5})/2$ is the golden ration. The star (7.10) describes propagation of icosahedral density modulation waves (Fig. 7.2(a)), while the star (7.11) identifies dodecahedral modulation (Fig. 7.2(b)), both forming aperiodic structures in three-dimensional crystal space but perfectly periodic ones in a six-dimensional superspace.

7.3.1.3. *Thermodynamic models*

In the case of one-dimensional density modulation (Fig. 7.1(a)) with nonvanishing order-parameter components fulfilling the equilibrium condition $\eta_{+\boldsymbol{k}_j} = \eta_{-\boldsymbol{k}_j} = \eta_{\boldsymbol{k}}$, the model can be reduced to the effective free-energy density

$$F_1(\eta_{\boldsymbol{k}}) = a_1\eta_{\boldsymbol{k}}^2 + b_1\eta_{\boldsymbol{k}}^3 + a_2\eta_{\boldsymbol{k}}^4 + \cdots. \qquad (7.12)$$

Two-dimensional cylindrical modulated phases require two pairs of wave-vectors: $\pm\boldsymbol{k}_1$ and $\pm\boldsymbol{k}_2$ (with $|\boldsymbol{k}_1| = |\boldsymbol{k}_2|$). The corresponding effective free-energy density is, therefore,

$$\begin{aligned} F_2(\eta_{\boldsymbol{k}_1}, \eta_{\boldsymbol{k}_2}) &= a_1(\eta_{\boldsymbol{k}_1^2} + \eta_{\boldsymbol{k}_2^2}) + b_1(\eta_{\boldsymbol{k}_1}^3 - 3\eta_{\boldsymbol{k}_1}\eta_{\boldsymbol{k}_2}^2) \\ &\quad + a_2(\eta_{\boldsymbol{k}_1}^4 + \eta_{\boldsymbol{k}_2}^4) + \cdots. \end{aligned}$$

It was analyzed in detail in Chapter 2, Sec. 2.2.1.2.

Three-dimensional modulated structures involve three pairs of independent $\boldsymbol{k}$ vectors $(\pm\boldsymbol{k}_1, \pm\boldsymbol{k}_2, \pm\boldsymbol{k}_3)$ for mesophases displaying a cubic lattice (Figs. 7.1(b)–7.1(d)). The equilibrium values of the corresponding six-component order parameters determine the actual symmetry of the segregated mesophases. For unequal components of the order parameter, the variational free-energy density reads

$$\begin{aligned} F_3(\eta_{\boldsymbol{k}_1}, \eta_{\boldsymbol{k}_2}, \eta_{\boldsymbol{k}_3}) &= a_1 I_1 + b_1 I_2 + a_2 I_3 \\ &= a_1(\eta_{\boldsymbol{k}_1}^2 + \eta_{\boldsymbol{k}_2}^2 + \eta_{\boldsymbol{k}_3}^2) + b_1 \eta_{\boldsymbol{k}_1}\eta_{\boldsymbol{k}_2}\eta_{\boldsymbol{k}_3} \\ &\quad + a_2(\eta_{\boldsymbol{k}_1}^4 + \eta_{\boldsymbol{k}_2}^4 + \eta_{\boldsymbol{k}_3}^4) + \cdots . \end{aligned} \tag{7.13}$$

It may describe (see Sec. 2.2.1.4) eight different phases depending on the equilibrium values of the $\eta_{\boldsymbol{k}_i}$ [8–10]: (I) $\eta_{\boldsymbol{k}_1} = \eta_{\boldsymbol{k}_2} = \eta_{\boldsymbol{k}_3} = 0$; (II) $\eta_{\boldsymbol{k}_1} \neq 0$, $\eta_{\boldsymbol{k}_2} = \eta_{\boldsymbol{k}_3} = 0$; (III$^\pm$) $\pm\eta_{\boldsymbol{k}_1} = \eta_{\boldsymbol{k}_2} = \eta_{\boldsymbol{k}_3} \neq 0$; (IV$^\pm$) $\pm\eta_{\boldsymbol{k}_1}$, $\eta_{\boldsymbol{k}_2} = \eta_{\boldsymbol{k}_3}$ $(\eta_{\boldsymbol{k}_1} > \eta_{\boldsymbol{k}_2})$; (V) $\eta_{\boldsymbol{k}_1}$, $\eta_{\boldsymbol{k}_2} = \eta_{\boldsymbol{k}_3}$ $(\eta_{\boldsymbol{k}_1} < \eta_{\boldsymbol{k}_2})$ and (VI) $\eta_{\boldsymbol{k}_1}$, $\eta_{\boldsymbol{k}_2}$, $\eta_{\boldsymbol{k}_3}$.

One can see that phase I in our model corresponds to the isotropic state. Phase II possesses $D_{\infty h}$ symmetry with translational symmetry broken in one direction and corresponds to a lamellar-type structure. The III$^\pm$ are cubic phases (different lattices are not distinguishable). Solutions IV$^\pm$, in the chosen model, correspond to three-dimensional uniaxial phases, but one cannot distinguish between tetragonal and rhombohedral symmetry. Phases V and VI are orthorhombic and monoclinic, respectively.

The locations of the phases pertaining to the phase diagram in the orbit space (I_1, I_2, I_3) are shown in Fig. 7.3. The phase diagram in the orbit space (manifold of orbits, see Chapter 1) not only reveals

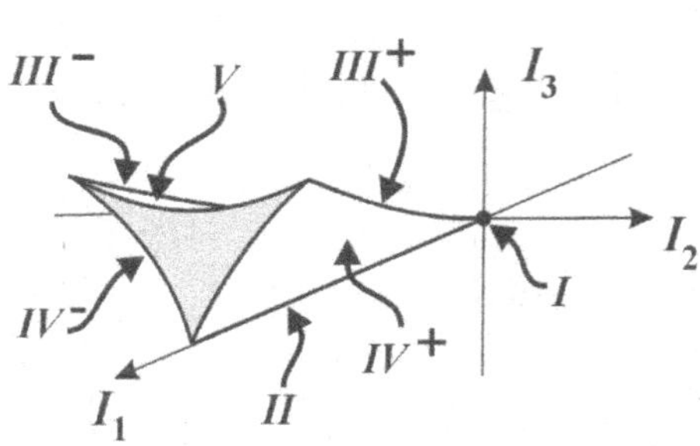

Fig. 7.3. Topology of phases in the orbit space (I_1, I_2, I_3). Location of the phases corresponding to the image-group defined by Eq. (7.13).

the full set of possible stable states for the corresponding image-group but also provides an indication of the topological properties of the phases in the real (e.g., temperature–pressure) phase diagram. The phase I (nonparametric solution) is located at the origin point $I_1 = I_2 = I_3 = 0$. The stability regions of one-parametric solutions II and III$^{\pm}$ are represented by the edges of a pyramid, the faces correspond to the two-parametric solutions IV$^{\pm}$ and V with the body filled by the three-parametric phase VI. The number characterizing the reduction of dimensionality of the phase stability region in the orbit space is known to be equal to the dimensionality of the corresponding interphase border. From Fig. 7.3, one can deduce that, when undergoing a phase transition from the isotropic state (origin point in the phase diagram, zero-parametric solution) to a phase associated with single-parametric solution of equations of state (II or III$^{\pm}$), the dimensionality is reduced by one. This means that phases II and III$^{\pm}$ can be reached *directly* from the phase I across a first-order transition *line* in contrast to the phases IV^{+} and V which can touch phase I only at an *isolated point* in the two-dimensional (e.g., temperature–pressure) phase diagram. This means that only lamellar-type and cubic or icosahedral phases (among those mentioned above) can be expected to border an isotropic phase in the phase diagram [11].

Two points have to be stressed. First, one should remember that low-symmetry phases discussed above are actually modulated liquid-like (partially segregated) states but not conventional crystals (fully segregated states). Second, although applying a model-free image-group technique to find minima of free energy, we cannot obtain solutions corresponding to the conventional crystal state as the latter ($|\eta_{\mathbf{k}_j}| = 1$) simply do not exist in the classic linear Landau theory. In Chapter 4, we showed that such "non-Landau" (limit) solutions appear due to a transcendental functional dependence of the phenomenological (averaged) order-parameter on local (atomic) parameters.

7.3.2. *Partially and fully segregated states*

Considerations developed in the preceding section provide a general framework for determining the order-parameter symmetries for the phase transition from the isotropic liquid state to partially segregated liquid-like modulated structures. It allows solving, on the model-free basis, some of the problems faced by the weak crystallization theory. However, they still are insufficient for describing transitions to the fully segregated states. The probability for objects to be in the inter-site space is still non-zero. In Chapter 4, we already considered the possibility of reconstructive (breaking the group–subgroup relation) transformations of the ordering type, which occur to the non-Landau (limit $\eta = 1$) phases. Here, in the fluctuation-affected description, let us use another approach deriving a functional form of the phenomenological order-parameter by considering the deviations of the local order-parameters from the average value, by analogy with thermal fluctuations (which they, in fact, are), as we did it in Chapter 4, Sec. 4.5.2 [12].

Let us consider again $\eta_{\boldsymbol{k}_j}$ as long-range order-parameters and $\xi_{\boldsymbol{k}_j}$ as variation of short-range order-parameters, i.e., probability for the segregation transformation. It describes density-wave modulation in the direction defined by the $\boldsymbol{k}_j$ vectors. The explicit form of the $\eta_{\boldsymbol{k}_j}(\xi_{\boldsymbol{k}_j})$ function for the given segregated phase (for the given set of n non-zero order-parameter components) can be obtained by minimizing, with respect to $\eta_{\boldsymbol{k}_j}$, the thermodynamic potential

$$\Phi(\eta_{\boldsymbol{k}_j}) = \int_{V_\xi} \left\{ F[\eta_{\boldsymbol{k}_j}(\xi_{\boldsymbol{k}_j})] + \sum_{j=1}^{n} g_j (\nabla \eta_{\boldsymbol{k}_j})^2 \right\} d^n \xi_{\boldsymbol{k}_j}, \tag{7.14}$$

where the integral is over a volume V_ξ in the ξ-order-parameter space. $F[\eta(\xi)]$ is the free-energy density and the Ginzburg g_i gradient invariants account for the deviations of the values of $\eta_{\boldsymbol{k}_j}$ from the $\xi_{\boldsymbol{k}_j}$. For the sake of simplicity, we can treat an effective order-parameter and neglect external inhomogeneous fields. Using one of the effective order-parameter expansions $F(\eta_{\boldsymbol{k}})$ given by Eqs. (7.8) and (7.9) yields the equation of state

$$g \frac{d^2 \eta_{\boldsymbol{k}}}{d\xi_{\boldsymbol{k}}^2} = a_1 \eta_{\boldsymbol{k}} + \frac{3}{2} b_1 \eta_{\boldsymbol{k}}^2 + 2a_2 \eta_{\boldsymbol{k}}^3 + \cdots . \tag{7.15}$$

When the right-hand expansion is restricted to the third power, Eq. (7.15) coincides with the general elliptic equation, which can be solved exactly [13]. More generally, the bifurcation from the solution $\eta_{\boldsymbol{k}}(0) = 0$, which corresponds to the non-segregated isotropic state, to the solution $\eta_{\boldsymbol{k}} = \eta_{\boldsymbol{k}_j}(\xi_{\boldsymbol{k}_j}) \neq 0$, corresponding to a partially or fully segregated state, can be obtained by linearizing Eq. (7.10) around the value $\eta_k(0) = 0$ [14]. One gets the second-order linear differential equation

$$g\frac{d^2\eta_k^*}{d\xi_k^2} = a_1^*\eta_k^*, \tag{7.16}$$

where η_k^* is the eigenfunction corresponding to the eigenvalue a_1^* at which the solution $\eta_{k_j}(\xi_{k_j}) \neq 0$ branches off the solution $\eta_k(0) = 0$. Taking into account the boundary conditions defining the initial isotropic state ($\eta_k^*(0) = 0$) and fully segregated state ($\eta_k^*(1) = 1$), one finds the asymptotically exact solution [14] in the vicinity of the point $a_1 = a_1^*$, which is expressed as

$$\eta_{k_j}(\xi_{k_j}) = \eta_k^{\max}\left|\sin\frac{\pi}{2}\xi_{k_j}\right|, \tag{7.17}$$

where the amplitude $\eta_k^{\max}$ is to be determined by the nonlinear terms in Eq. (7.14).

7.3.3. *Phase diagram of a segregating system*

Introducing the function $\eta_{k_j}(\xi_{k_j})$ given by Eqs. (4.39) or (7.17) in the effective free-energy density $F(\eta_k)$ (Eq. (7.11)) restricted to the fourth degree in η_k, we obtain a phase diagram reflecting all essential topological details. Thus, by minimizing the functional $F[\eta_j(\xi_i)]$ with respect to ξ_i variable, one gets

$$\frac{\partial F}{\partial \eta_j}\cdot\frac{\partial \eta_j}{\partial \xi_i} = \left(\sin\frac{\pi}{2}\xi\right)\left(2a_1 + 3b_1\sin\frac{\pi}{2}\xi + 4a_2\sin^2\frac{\pi}{2}\xi\right)\cdot\left(\cos\frac{\pi}{2}\xi\right) = 0, \tag{7.18}$$

which yields three possible stable phases:

(1) ISO: $\sin\frac{\pi}{2}\xi = 0, \xi = 0$ – parent isotropic state;
(2) FS: $\cos\frac{\pi}{2}\xi = 0, \xi = \pm 1$ – fully segregated states;
(3) PS: $\sin\frac{\pi}{2}\xi = \frac{-3b_1 \pm \sqrt{9b_1^2 - 32a_1a_2}}{8a_2}$ – partially segregated states.

Figure 7.4 shows, in the plane of phenomenological coefficients $(a_1/a_2, b_1/a_2)$, the phase diagram associated with the equation of state (7.18).

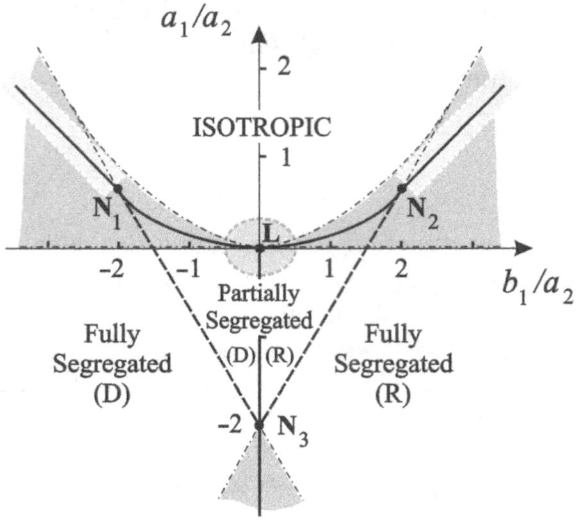

Fig. 7.4. Phase diagram associated with effective one-component order-parameter expansion (7.12). Full, dashed and dotted-dashed lines are, respectively, discontinuous, continuous transitions and limit of stability lines. The regions of phase coexistence are shadowed. N_1 and N_2 are triple points, N_3 is a four-phase point. L is the Landau point, in the vicinity of which the weak-crystallization theory is valid (light-gray area). The same color shadows strong-crystallization lines.

Thus, the partially and fully segregated phases, which display identical symmetries, correspond to distinct regions of the phase diagram. These regions are separated by the lines of *continuous topological transitions*, which are determined by the property that the probability η_k reaches its maximum (saturation) value $\eta_k = 1$. The two, partially and fully segregated, phases, can be reached from the isotropic solution across the lines of first-order transitions which merge at two symmetric three-phase points N_1 and N_2. More generally, the phase diagram of Fig. 7.4 is symmetric with respect to the a_1 axis ($b_1 = 0$), or equivalently, with respect to the change in the sign of η_k: each partially or fully segregated phase possesses a symmetric analog corresponding to an opposite sign for η_k. Defining negative values of ξ_k as the normalized probability for objects to be in the segregated regions leads to the interpretation that two symmetric phases, with respect to the a_1/a_2 axis in the phase diagram of Fig. 7.4, correspond, respectively, to a *direct* (D) structure

($\eta_k > 0$, A_nB_m) and its *reversed* analog ($\eta_k < 0$, A_mB_n). In terms of structural phase transitions, these are anti-isostructural phases (see Sec. 2.2.2). Figure 7.4 shows that the directed–reversed phase transition is discontinuous, the corresponding first-order transition line being limited by a three-phase point L (isolated Landau point of the second-order phase transition) and by a four-phase point N_3, at which partially and fully segregated, direct and reversed, partially segregated phases merge. Note that when taking into account the higher-degree terms in F or considering multidimensional order-parameters (Eqs. (7.12) and (7.13)), the phase diagram of Fig. 7.4 becomes asymmetric with respect to the a_1/a_2 axis or more complex, but its essential features remain unchanged, i.e., the direct and reversed mesophases still correspond to the left- and right-hand side regions of the phase diagram, although their locations become non-symmetric with respect to the a_1/a_2 axis. Thus, despite the restricted character of the thermodynamic model (7.12), we obtained an adequate picture of transformations from the isotropic state. The validity domain of the weak-crystallization theory ($\rho \ll 1$ in Eq. (7.4), i.e., $b_1/a_2 \ll 1$) surrounds the Landau L-point, while direct liquid–solid transformation occurs beyond points N_1 and N_2; the corresponding lines of strongly first-order phase transition, according to the general features of non-Landau phases (Sec. 4.4.3), are straight lines:

$$a_1 = \pm b_1 - a_2. \tag{7.19}$$

In order to make the theory of segregation-type phase transitions complete, we should consider the crystallization of objects different from isotropic, whose order-parameter is not a trivial scalar. In the free-energy terms, it means treating the thermodynamic potential expansion $F[\eta(\xi)]$ which contains only even-degree invariants. We can refer to the results of Sec. 4.2.1.3 where the corresponding phenomenological model was analyzed in detail.

Thus, the effective free-energy density for a segregating system reads

$$F[\eta(\xi)] = a_1 \sin^2\left(\frac{\pi}{2}\xi\right) + a_2 \sin^4\left(\frac{\pi}{2}\xi\right) + a_3 \sin^6\left(\frac{\pi}{2}\xi\right), \tag{7.20}$$

where $\eta(\xi)$ takes the form of Eq. (7.18). Minimizing the functional $F[\eta(\xi)]$ with respect to the essential variable ξ yields three possible states:

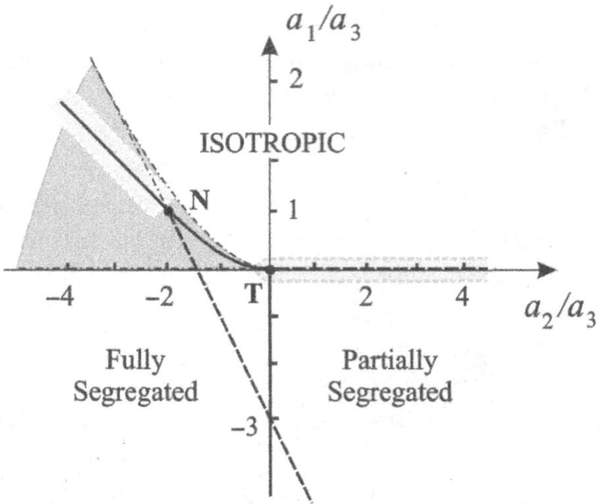

Fig. 7.5. Phase diagram representing the order-parameter expansion (7.20) in the plane of phenomenological coefficients ($a_3 > 0$). Solid and dashed lines are discontinuous and continuous transition lines, respectively. Dashed-dotted lines are limit of stability lines. T is a tricritical point, N is a three-phase point. The phase coexistence region is shadowed in dark-gray, light-gray regions shadow the strong crystallization line ($a_2 < 0$) and weak crystallization domain ($a_2 > 0$).

(1) ISO: $\sin\frac{\pi}{2}\xi = 0, \xi = 0$ – parent isotropic state;
(2) FS: $\cos\frac{\pi}{2}\xi = 0, \xi = 1$ – fully segregated states;
(3) PS: $\sin\frac{\pi}{2}\xi = \frac{-a_2 \pm \sqrt{a_2^2 - 3a_1a_3}}{3a_3}$ – partially segregated states.

The corresponding phase diagram is shown in Fig. 7.5, in the plane of the phenomenological ratios $(a_1/a_3, a_2/a_3)$. In contrast to the preceding model, two solutions of the equation of state η_0 and $-\eta_0$ describe not different (direct and reversed) phases but two domains of the same phase. Second-order transition line meets a first-order one at the tricritical point T ($a_1 = 0, a_2 = 0$), and a triple point N merges three transition lines: (i) the first-order transition line between isotropic (liquid) state and fully segregated one (conventional crystal), (ii) the first-order transition line from the isotropic to partially segregated (modulated liquid-like) state, and (iii) the line of continuous topological transitions between partially and fully segregated states. For clear understanding, we mapped on the phenomenological phase diagram of Fig. 7.5 the domain of the strong crystallization and a region neighboring the second-order transition line where the weak-crystallization theory was applied (see Ref. [2]).

7.3.4. *Segregated states in atomic and molecular systems*

In order to relate experimentally observed states to phases predicted in the framework of the advanced phenomenological theory, let us consider examples of structures and materials by their classes.

7.3.4.1. *Simple metals*

Elemental metals provide us with a single example set of transformations, which is the direct crystallization of metal melt to conventional crystalline structures — the liquid–solid phase transformation. No reliable observation of density modulation waves in a liquid metal has been reported in the literature so far. Nevertheless, metals demonstrate to us an important thermodynamic regularity: the theory predicts that solid–liquid–solid transition line is a straight line (see Eq. (7.19)), at least in coordinates of phenomenological coefficients, and experimentally mapped melting lines do match the prediction.

It is convenient, in the Landau theory, to consider only the phenomenological coefficients in the lowest-degree terms to be linearly dependent on intensive external thermodynamic parameters temperature T and pressure P:

$$\begin{aligned} a_1 &= \alpha_1(T - T_{\mathrm{C}}) + \beta_1(P - P_{\mathrm{C}}), \\ b_1 &= \alpha_2(T - T_{\mathrm{C}}) + \beta_2(P - P_{\mathrm{C}}). \end{aligned} \tag{7.21}$$

This is a strong assumption as one finds no atomistic or thermodynamic arguments reducing such dependencies to simple linear functions. However, if we find that the agreement between the theory and experiment is sufficiently good even under such strong limitations, the arguments in favor of the theory validity become valuable.

Considerable efforts have been made in the last decades toward mapping melting curves of simple metals under high temperature and high pressure. Reliable data were collected using advanced techniques of synchrotron X-ray diffraction and X-ray absorption spectroscopy

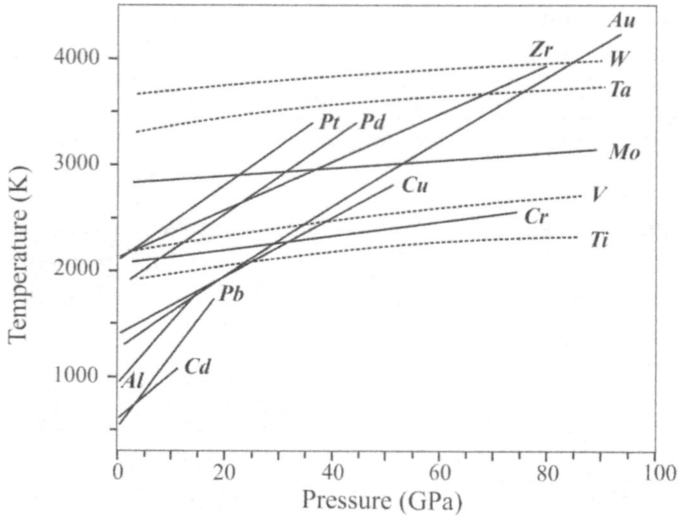

Fig. 7.6. Comparison of experimental data of pressure-dependent melting temperature of W, Ta, Mo, V, Ti, Cr from Ref. [15]; Pb, Al, Cd [16]; Cu, Pt, Pd [17]; Zr [18] and Au [19]. Solid lines are straight, while dotted lines are slightly curved.

in diamond anvil cells. Figure 7.6 shows sketched experimental data obtained for some elemental metals and available in the literature. One immediately concludes on its excellent agreement with the theory prediction. Indeed, the most part of mapped melting lines are straight lines, and only a few (Ti, V, W and Ta) slightly deviate from the linear $T(P)$ function. This demonstration is especially important since, on the one hand, we consider strongly the first-order phase transition (actually, maximally strong as the order-parameter jumps in the transformation point from zero to its maximal value), but on the other hand, the theoretical prediction for the corresponding transition to the non-Landau phase is model-free and does not depend on the maximal degree of the free-energy expansion.

7.3.4.2. *Lyotropic liquid crystals*

The class of lyotropic liquid crystals, another term for which is lyotropic complex fluids, exhibits the most reach variety diversity of partially and fully segregated structures observed in a certain compound: from crystalline structures with positional order along one, two or three dimensions to disordered concentrated micellar solutions. Lyotropic liquid crystals are molecular systems formed by amphiphilic molecules in the presence of a solvent. Full view of experimental facts and theoretical interpretations can be found in Refs. [20, 21] containing collections of review chapters, and in the monograph on the topic [22].

Amphiphilic molecules have both a polar and a non-polar part well defined in their molecular structures. Very strong hydrophobic–hydrophilic interaction between molecules and solvent may originate the formation of segregated superstructures. In order to minimize the interaction energy, amphiphilic molecules tend to build up closed aggregates hiding their hydrophobic organic tails from the water environment. Although the latter process is not included in the segregation scenario considered in the preceding sections, it fully fits into the general symmetry-based scheme: primary molecular agglomerates merge into more complex spatially arranged structures without breaking the general phase symmetry.

In very diluted regimes, the amphiphilic molecules in the solvent remain isolated, but when a critical micellar concentration is achieved, they self-assemble in spherical micelles. Such an aggregation process is controlled by the chemical factors and does not change the symmetry of the system. The isotropic *micellar solution* can be referred to as a parent phase for the segregation-type phase transitions (Fig. 7.7(a)). Its symmetry is $\tilde{E}_3 = R_3 \rtimes O(3)$.

The next phase in the generic sequence is the one-dimensionally modulated lamellar phase possessing the symmetry $R_2 \rtimes D_{\infty h}$ and normally bordering the isotropic phase. A partially segregated micellar solution (Fig. 7.7(b)) is followed by the fully segregated one (see Ref. [23] and references therein) in which micelles merge into continuous lamellas (Figs. 7.7(c) and 7.7(d)).

Again, the merging of densely packed micelles to continuous lamellas (Figs. 7.7(c) and 7.7(d)) is controlled by the chemical factors, and a symmetry-based theory is incapable to distinguish them — both correspond to the non-Landau solution of the equation of state (7.18). Possible uniaxial distortion of the micelles and their nematic-type orientational ordering are discussed in the next section.

The two-dimensional density modulation process (Fig. 7.8(a)) ends up in the hexagonal non-Landau phase where the surfactant molecules are aggregated to form long rods (Fig. 7.8(b)), which are packed onto a two-dimensional hexagonal lattice, having a high

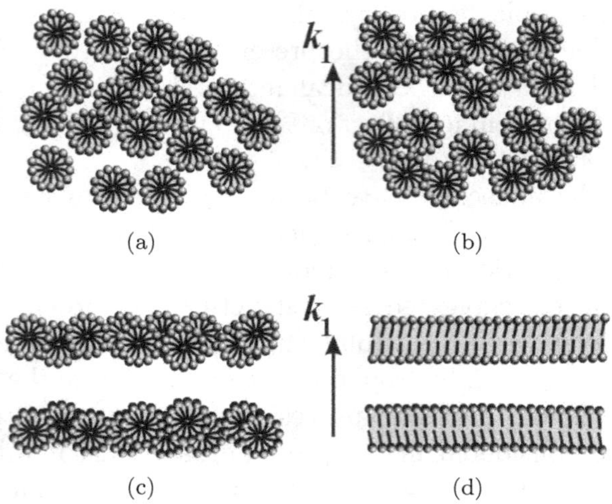

Fig. 7.7. Step-by-step evolution of micelle-solvent mixture: (a) isotropic micellar phase; (b) partially segregated and (c) fully segregated pseudo-lamellar micellar phase; (d) fully segregated lamellar structure.

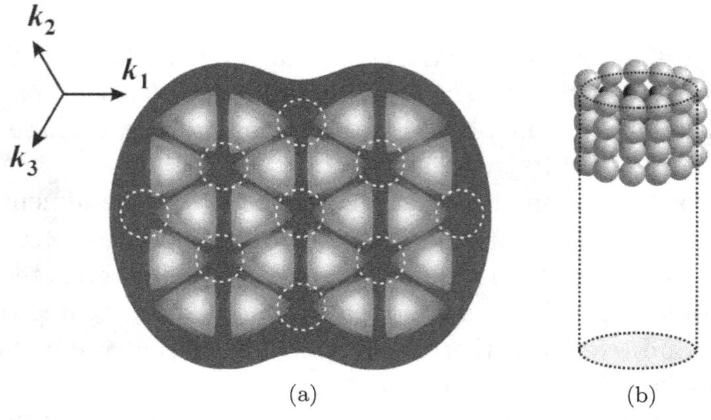

Fig. 7.8. (a) Plane cross-section of the hexagonal two-dimensionally density-modulated mixture; dotted circles show regions of maximal micellar density. (b) Schematic structure of molecular cylinders in the fully segregated state (side view).

degree of long-range positional order in two dimensions. Its symmetry is $R_1 \rtimes D_{6h}$.

The rich variety of the structures and space-groups of three-dimensional translationally-ordered lyotropic cubic liquid-crystalline phases even allows one to discuss the cubic polymorphism for

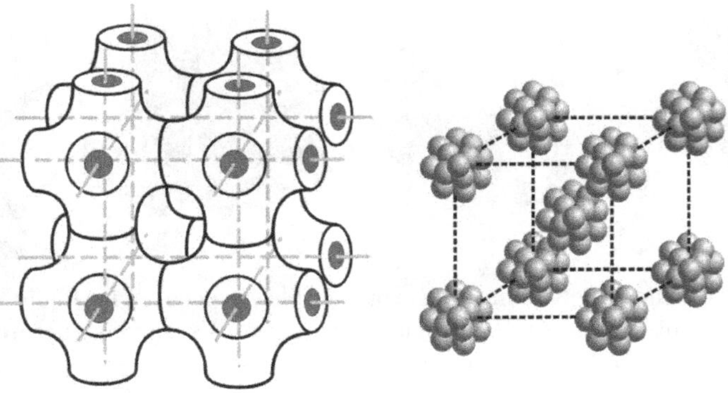

(a) (b)

Fig. 7.9. Cubic structures of lyotropic liquid crystals. (a) Bicontinuous structure; straight lines is a skeletal graph showing primitive cubic lattice. (b) Body-centered cubic micellar structure.

different complex fluids [20]. Although the cubic phases exhibit distinct structures, there are two fundamentally different classes: (i) the bicontinuous cubic, consisting of continuous bilayers of amphiphilic molecules that divide space into two interwoven networks of water and surfactant (Fig. 7.9(a)); and (ii) the micellar structures composed of discrete aggregates of "oil in water" or "water in oil" type (Fig. 7.9(b)). The latter class is conventional crystals and can be unambiguously identified as a non-Landau (limit) state of a three-dimensional fully segregated system.

In the partially segregated cubic phases, micelles concentrate along the lines linking the volumes surrounding the high-symmetry points in the structure [24]. These latter belong, in the cubic space groups, to the ($x00$), ($xx0$) and (xxx)-type Wyckoff positions [25]. Similar to the scenarios of one- and two-dimensional modulation, the high concentration of micelles provokes their merger, forming continuous passages and conserving, therefore, non-zero probability for amphiphilic molecules to be in segregated regions [24]. Thus, we obtain full set of structure types predicted in the framework of the phenomenological theory of segregation.

The phenomenological theory of segregation-type transformations predicts the existence of inverted (reverse) micellar structures which display exchanged configurations as anti-isostructural states with respect to the corresponding direct mesophases [26].

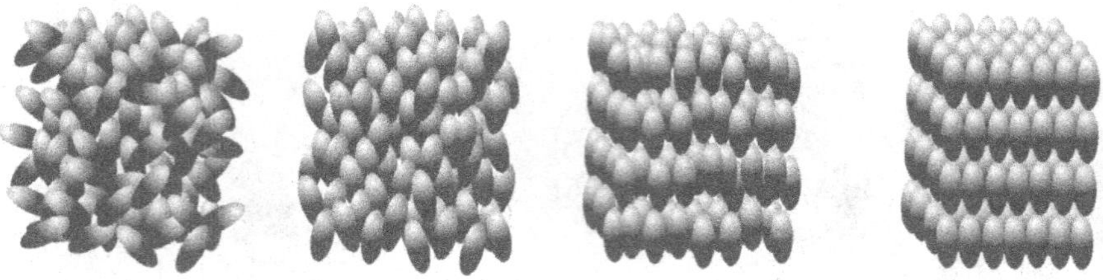

(a) (b) (c) (d)

Fig. 7.10. Schematic arrangement of rod-like molecules in liquid-crystalline phases: (a) isotropic state; (b) nematic structure; (c) smectic structure with soft diffuse layers; (d) fully segregated smectic phase.

The bicontinuous and micellar cubic structures as well as the hexagonal cylindrical structures indeed are known to exist either as organization of aggregates of amphiphiles in continuous aqueous environment or as water droplets in a continuous medium of amphiphiles. In the first case, they are classified as direct structures, and we discussed them above; in the second case – inverted (reversed) structures, completing the list of structure types experimentally observed in lyotropic liquid crystals [21].

7.3.4.3. *Thermotropic liquid crystals*

Phase transitions in the mesogen compounds exhibiting various molecular liquid-crystalline states which occur between the crystalline solid state and the isotropic liquid at distinct temperature ranges are the field of the most successful application of the weak-crystallization theory. The thermotropic liquid crystals can roughly be grouped into two main types, depending on the geometry of their anisotropic molecules. The liquid crystals derived from the rod-shaped molecules (Fig. 7.10) are referred to as calamities. This class of compounds is well studied and they are extremely useful for practical applications. Another type, the mesophases formed from disk-like molecules, are called discotics (Fig. 7.11). Namely, the anisotropy of molecules on the macroscopic level leads to the appearance between solid and liquid states of mesophases possessing a strong anisotropy of its various characteristics. A review of the structures and properties of the thermotropic liquid crystals may be found, e.g., in the classic monographs by de Gennes [27] and Chandrasekhar [28].

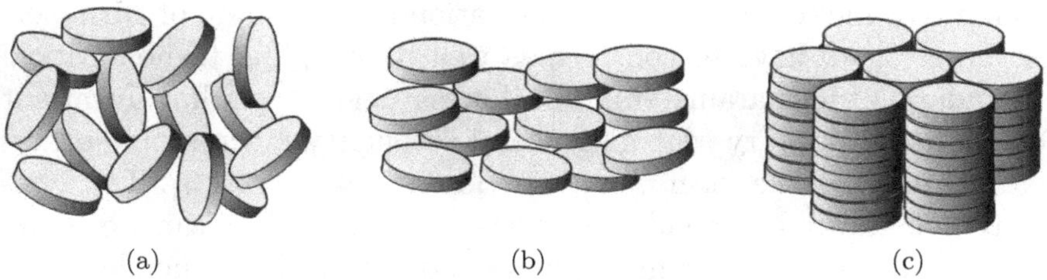

Fig. 7.11. Schematic arrangement of disk-like molecules in (a) isotropic state, (b) discotic nematic, and (c) discotic disordered hexagonal columnar phase.

The natural parent phase for thermotropic liquid-crystalline structures is the isotropic liquid where the centers and orientation of the molecules are randomly distributed (Figs. 7.10(a) and 7.11(a)), its symmetry group is $\tilde{E}_3 = R_3 \rtimes O(3)$.

The simplest type of ordered liquid crystals is nematics, yet it is the most widely used type. In nematic state, the molecules start to orient in a common direction (Figs. 7.10(b) and 7.11(b)) lowering symmetry to $R_3 \rtimes D_{\infty h}$. Actually, nematic phase is an anisotropic liquid, and no density modulation occurs at the isotropic liquid–nematic phase transition. The order-parameter for the transition to the nematic phase is a symmetrical traceless second-rank tensor [27, 29]. Its components form the basis for the $D_+^{2,0}$ irreducible representation characterized by the reciprocal-space vector $\boldsymbol{k} = 0$. Moreover, in the case of lyotropic mixtures, the corresponding order-parameter has double action. It induces the uniaxial contraction or extension of spherical micelles transforming the latter to oblate or prolate rotation ellipsoids, and simultaneously orients them along the common direction. Since both mechanisms have identical symmetry, the transition can be classified as "pseudo-proper" (Sec. 2.4.3). Oblate and prolate forms of micelles correspond to different signs of the order-parameter and represent anti-isostructural low-symmetry phases (see Secs. 2.2.1.2 and 2.2.2).

Propagation of density waves along the direction characterized by the reciprocal-space vector $\pm\boldsymbol{k}_1 \neq 0$ (Fig. 7.1(a)) results in lamellar (Fig. 7.7(c)) or smectic (Fig. 7.10(c)) structure. Their symmetry is $R_2 \rtimes D_{\infty h}$. Uniaxial symmetry of the segregated state induces a pseudo-nematic behavior of the system. Indeed, the corresponding primary segregation order-parameter components transform into the irreducible representation $D_+^{0,\boldsymbol{k}_j}$. However, in order to get the full scenario of the segregation transformation in the system of anisometric objects, we have to consider as well its coupling to non-critical secondary order-parameters. The lowest degree coupling invariant between the primary segregation and secondary macroscopic order-parameters has the form $\eta_i^2 X_j$ (see Secs. 1.2.4 and 2.4.3). The symmetry of induced (secondary) order-parameter $\{X_j\}$ can be found by decomposing the symmetrized second power of irreducible representation spanned by the primary order-parameter components $\{\eta_i\}$ [9, 10]. For relevant $D_+^{0,\boldsymbol{k}_j}$, we have

$$[D_+^{0,\boldsymbol{k}_j}]^2 = D_+^0 + D_+^1 + D_+^2 + D_+^3 + \cdots + \left\{ D_+^{m,(\boldsymbol{k}_j+\boldsymbol{k}_{j'})} \right\}, \quad (7.22)$$

where $D_+^l \equiv D_+^{l,0}$ are the irreducible representations of the $O(3)$ point group, and the irreducible representations in parentheses are related to the second harmonics of density modulation function $\delta\rho$. The mechanism corresponding to the totally symmetric irreducible representation, D_+^0, as it is responsible for the micelle volume variation, is relevant only to the lyotropic systems (the associated function is the trace of asymmetrical second-rank tensor). D_+^1 is a pseudo-vectorial irreducible representation and the corresponding mechanism results in ferromagnetic-type ordering that has meaning only for molecules possessing diamagnetic properties. The components of a symmetrical second-rank tensor form the basis for the D_+^2 irreducible representation and at least two mechanisms, as we discussed above, are related to this irreducible representation in liquid crystals. Higher-rank symmetric tensors spanning irreducible representations D_+^3, $D_+^4, \ldots$, etc. describe more complex tensorial ordering mechanisms. Crystallization to the smectic phase does not require primary nematic-type alignment, and therefore, (i) macroscopic mechanism of molecular or micellar orientational ordering is a natural component of the segregation scenario, and (ii) the transformation can occur directly from the isotropic state. According to the form of coupling, improper nematic-type ordering in the segregating system has to vary linearly with temperature changing ($\eta \sim T$), while in proper nematic phase, it is proportional to $\eta \sim \sqrt{\mathrm{T}}$.

Two-dimensional density modulation, controlled by the same order-parameter $D_+^{0,\boldsymbol{k}_j}$, leads to forming molecular columnar structures in the thermotropic systems. One of them, vertically disordered discotic hexagonal structure possessing the symmetry $R_1 \rtimes D_{6h}$, is shown in Fig. 7.11(c).

Summarizing our overview of structures crystallized in the systems of various nature, we conclude that both states, Landau and non-Landau, can be identified in various solidifying materials.

7.4. Solidification to Aperiodic Structures

7.4.1. *Quasicrystalline order*

Our natural habitude to consider any regular arrangement of atoms in terms of exclusively three-dimensional crystallography was perturbed by the discovery, first, of incommensurate structures in the 1970s, and then, of quasicrystals in the early 1980s. Although the former did not receive strong reaction because they could be understood as coexistence of two sublattices possessing irrational ratio for their lattice parameters (see Sec. 2.3 and Ref. [10]), the latter have excited the solid-state science community by the presence of rotation axes incompatible, as believed before, with crystalline periodic order [30]. One finds different notations when referring to such atomic arrays (incommensurate or aperiodic structures, quasiperiodic crystals, quasicrystals); nevertheless, all of these structures demonstrate well-defined long-range order confirmed by the existence of sharp Bragg peaks in their diffraction patterns.

Already in early theoretical papers [31–33], the icosahedral quasicrystals were considered as a regular structure periodic in a six-dimensional space. One should accept and remember that the strict conditions allowing only two-, three-, four- and six-fold rotation axes to be compatible with *crystalline* periodicity are totally based on the classical three-dimensional geometry, but in n-dimensional space *quasicrystals* can have any finite subgroup of $O(n)$ as their point group.

Quasicrystals are much more common among materials than they were originally expected when their discovery was first announced in the seminal publication of Ref. [30]. More than 100 different metal alloys have been found to form stable quasicrystalline phases to date [34, 35]. Recent progress in soft and biological matter science and in metamaterials discovered a whole series of new quasiperiodic systems. Among them were micelle-forming dendrimers [36,37], star block copolymers [38], linear terpolymers [39], quasicrystalline arrangements of proteins in two-dimensional viral capsid shells [40], and binary systems of nanoparticles [41]. The discovery of a natural quasicrystal $Al_{63}Cu_{24}Fe_{13}$ [42, 43] has opened a new chapter in mineralogy. Its species was found to have a perfect icosahedral quasicrystalline structure; the corresponding mineral had a degree of structural perfection comparable to the best laboratory specimens [42].

The advanced phenomenological approach allows us to answer, in maximally general form, two fundamental questions still under discussion in the field but essential in further applications of the theory: (i) Is quasicrystalline (quasiperiodic) order a thermodynamically stable state? and (ii) What is the process forming its structure?

7.4.2. *Stability of quasicrystalline structures*

The problem of thermodynamic stability of a state is fundamental in materials science. Indeed, even holding the sample in our hands, we cannot be sure whether we have to apply equilibrium or non-equilibrium approach in order to explain the already studied or predict new properties (including the stability range) of the material, and we cannot decide definitely on its practical applications. Since the discovery of quasicrystals, a number of different mechanisms for the stabilization of quasicrystalline order based on classic Landau theory have been proposed [31–33, 44–46], however, all of the considered models represent certain versions of the weak-crystallization theory. Due to evident mathematical limitations, the aim of such approaches was not to give a detailed quantitative description of real quasicrystalline materials, but only to investigate whether such simple models can have stable quasicrystalline solutions. However, to get definite answers, one has to operate with *structurally stable models*, but this was not the case for early studies. It is worth reminding, to avoid a confusion switching from the stability of a state (thermodynamic stability) to the stability of a model, that we speak on mathematical meaning of "*structural stability*" related to the property of a *polynomial potential.*

The property of structural stability, in mathematical sense, depends on the choice of nonlinearity of the Landau potential (see

Ref. [47]). It follows from the theory of singularities that only in structurally stable models the generic topology of the phase diagrams, phase stability conditions and anomalies predicted for the phase transitions are invariant under small variation of the phenomenological parameters [47], while conclusions of unstable models can be inversed by even very small modifications. An accidental degeneracy, not due to the symmetry requirements, is always related to a structural instability. Allowed by the symmetry, the cubic invariant in the free energy for the liquid–solid phase transition (Eqs. (7.3), (7.6) and (7.12)) requires, for the structural stability of the Landau potential, retaining six-degree terms, at least. The latter limit has never been achieved in the phenomenological models of solidification.

In contrast to model approaches, the advanced phenomenological theory provides all necessary elements allowing to discuss the thermodynamic stability in a general manner. Indeed, minimization of the free energy is a problem of the model-free angular part of the theory, which uses only information on its symmetry (Sec. 1.3.2). The symmetry of the Landau potential is determined by the transformation properties of the order-parameter. The infinite-dimensional order-parameter of the crystallization transformation possesses continuous point group $O(\infty)$ in the order-parameter space ε or σ (Secs. 7.3.1.1 and 7.3.1.2). Minimization Eqs. (7.5) determine the equilibrium nonzero components of the order-parameter $\{\eta^{\mathrm{eq}}_{\langle \boldsymbol{k}j\rangle}\}$, which are vectors in the ε- or σ-space, possessing symmetry of different maximal subgroups of the order-parameter image group $O(\infty)$. Thus, finding out the maximal subgroups of $O(\infty)$, in fact, solves the minimization problem. Figure 7.2 shows icosahedral and dodecahedral arrays of the wave-vectors defining the quasicrystalline structures. The corresponding solutions possessing the I_h symmetry minimize the free energy $F(\eta^{m}_{\boldsymbol{k}_j})$ and ensure global stability of the quasicrystalline states which can be, therefore, the ground state of the system. It becomes evident that the ground state of solid matter is not necessarily lattice periodic: none atoms arraying with "noncrystalline" rotation symmetry is not forbidden.

Thus, the positive answer to the question on the thermodynamic stability of quasicrystalline states is definite, as obtained in the framework of a general model-free theory.

7.4.3. *Formation of quasicrystalline structures: Probability maxima*

Introducing, in Chapter 4, density-wave formalism in order to define the basic functions and parameters of the symmetry-based phenomenological theory of phase transitions, we referred to it as a convenient method widely accepted in crystallography. Indeed, any crystal structure is considered as an interference pattern of probability or mass density waves, while its diffraction pattern is the Fourier image in the reciprocal space. Structure coding in the three-dimensional space has been significantly simplified by tabulating all possible symmetry groups and arrangements of probability maxima in the unit cells, i.e., by introducing 230 space groups and corresponding Wyckoff positions [25]. For multidimensional spaces, the work is not finished yet [48, 49]. Thus, return to the origin of the method appears natural, and the phenomenological model of crystallization offers the regularity in the description of quasiperiodic structures. Moreover, for identifying the corresponding order-parameter, we establish the link between thermodynamics and crystallography of quasicrystals.

The way to solve the quasicrystallographic problem in the framework of the density-wave formalism was first outlined in Refs. [33,50, 51] for icosahedral and fully realized later in Refs. [52, 53] for octagonal MnSiAl and decagonal AlCuCo quasicrystals. Let us extend the procedure of Ref. [52] and describe, in more detail, the formation of an octagonal structure from the isotropic liquid.

In the order-parameter space σ, minimization conditions (7.7) define for the octagonal structure an invariant vector which establishes equations for order-parameter components:

$$\eta_{\boldsymbol{k}_1} = \eta_{\boldsymbol{k}_2} = \cdots = \eta_{\boldsymbol{k}_8} \neq 0, \quad \eta_{\boldsymbol{k}_j} = 0 \; (j \neq 1\text{–}8). \tag{7.23}$$

Reciprocal space vectors $\{\boldsymbol{k}_j\}(j = 1\text{–}8)$ form an octagonal star (Fig. 7.12). The symmetry of the corresponding segregated state is D_{8h} and its structure is formed by two systems of orthogonal density waves (tetragrid $\boldsymbol{k}_1 \perp \boldsymbol{k}_3$ and $\boldsymbol{k}_2 \perp \boldsymbol{k}_4$), resulting, due to their orientation ($\angle \boldsymbol{k}_1 \boldsymbol{k}_2 = \pi/4$), in two-dimensional quasiperiodic lattice (for instance, see Fig. 2.18). Thus, we have a structure aperiodic in the two-dimensional plane space but perfectly periodic in the four-dimensional superspace or, in other terms, the structure has two

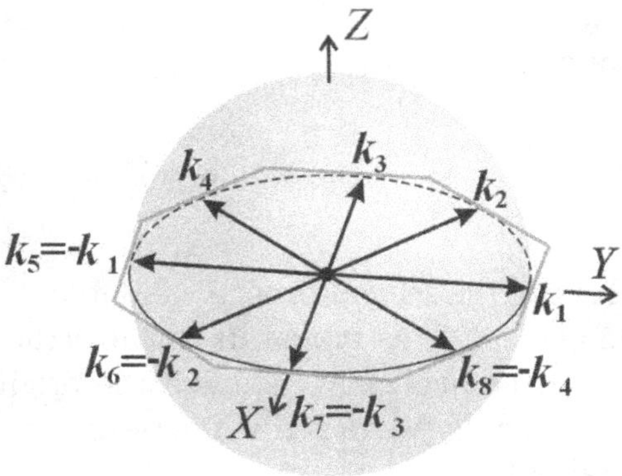

Fig. 7.12. Octagonal reciprocal space vector star $\{\boldsymbol{k}_{1-8}\}$. The basis vectors spanning the superlattice in direct space have the same orientation.

incommensurate length scales in 2D ($a'/a'' = \sqrt{2}$) but a single length scale in 4D ($|\boldsymbol{a}_i| = |\boldsymbol{k}_i^{-1}|$).

Constructing the octagonal density-wave superlattice, we do not need to refer to the full isotropic free energy, it is sufficient to use a local Landau potential and modulation density waves defined by the equilibrium conditions (7.23), which reduce the order-parameter dimensionality from $n = \infty$ to $n = 8$. The periodic deviation of density $\delta\rho(\boldsymbol{r})$ now reads as a Fourier series

$$\delta\rho(\boldsymbol{r}) = \sum_{j=1}^{8} \rho_j \cdot \exp(i\boldsymbol{k}_j\boldsymbol{r}), \tag{7.24}$$

where $\boldsymbol{k}_{5-8} = -\boldsymbol{k}_{1-4}$, and density-wave vectors can be written as $\boldsymbol{k}_j = \boldsymbol{k}_0[\cos(j\pi/4), \sin(j\pi/4)]$.

Each ρ_j is a complex number $\rho_j = |\rho_j| \cdot \exp(i\phi_j^0)$, but since $\rho(\boldsymbol{r})$ is real (being a probability), any ρ_j is equal to its complex conjugate ρ_{j+4}^*. Therefore, $\phi_j^0 = -\phi_{j+4}^0$, and the number of independent phase variables in the octagonal state is $m = 4$.

We concluded in Sec. 7.4.2 that, in order to satisfy the structure stability condition, we have to retain in the energy expansion of the local theory for the octagonal phase terms up to sixth degree, and the Landau free energy is expressed as the following invariant function of ρ_j amplitudes:

$$F(\rho_j, T, P) = a_1 I_1 + b_1 I_2 + a_2 I_1^2 + a_3 I_1^3 + \gamma_{12} I_1^2 I_2, \tag{7.25}$$

where a_n and b_m are phenomenological coefficients dependent on temperature T and pressure P. The non-vanishing invariants are

$$I_1 = \sum_{j=1}^{4} \rho_{\boldsymbol{k}_j} \cdot \rho_{-\boldsymbol{k}_j}, \quad I_2 = \rho_1\rho_2\rho_3\rho_4 + \rho_3\rho_4\rho_7\rho_8 + \rho_2\rho_3\rho_7\rho_8. \quad (7.26)$$

In the octagonal state, the second-order I_1 and fourth-order I_2 invariants are independent on ϕ_j^0 as two of its linear orthogonal combinations correspond to the arbitrary choice of the origin of coordinate system, and the remaining two describe the phason degree of freedom of the quasicrystal [33].

To calculate the coordinates of the quasicrystalline superlattice nodes, we should use the fundamental distinction of the advanced phenomenological theory — the functional form of the order-parameter, i.e., its dependence on local parameters or coordinates (see Chapter 4). Therefore, the amplitudes ρ_j take the form

$$\rho_j(\boldsymbol{r}_n) = \frac{1}{S}\sum_{i=1}^{M} \exp(-\boldsymbol{k}_j \cdot \boldsymbol{r}_i), \quad (7.27)$$

where M is the number of lattice nodes on the considered area S of the plane structure.

To make the free energy (7.25) mathematically treatable, we first consider equilibrium conditions (7.23) for the order-parameter components in the octagonal phase. Equations (7.23) lead to the equality of all $|\rho_j| = \rho_\Delta$. The function $\delta\rho(\boldsymbol{r})$ (7.24) can be transformed to the real form

$$\delta\rho(\boldsymbol{r}) = 2\rho_\Delta \cdot \sum_{j=1}^{4} \cos(\boldsymbol{k}_j \cdot \boldsymbol{r} + \phi_j). \quad (7.28)$$

Second, we parameterize density-wave phases ϕ_j in terms of *hydrodynamic variables*. Generally, the phason variables originate from the extra degrees of freedom involving the choice of phases in the density-wave picture [50]. In such a description, we can identify two hydrodynamic variables: $\boldsymbol{u}$ and $\boldsymbol{v}$. The former, $\boldsymbol{u}$, is the conventional strain displacement variable and the latter, $\boldsymbol{v}$, is the phason strain variable. In the mass/probability density formalism, strains in the quasicrystalline state produce shifts ϕ in the phase of the waves [50, 51]:

$$\phi_j^0 = \boldsymbol{k}_j \boldsymbol{u} + \boldsymbol{k}_j^{\perp} \boldsymbol{v}, \tag{7.29}$$

where wave vectors $\boldsymbol{k}_j = k_0[\cos(j\pi/4), \sin(j\pi/4)]$ and $\boldsymbol{k}_j^{\perp} = k_0[\cos(j3\pi/4), \sin(j3\pi/4)]$. Accordingly, the Landau potential (7.25) is reduced to a simple polynomial

$$F_q = \tilde{a}_1 \rho_\Delta^2 + \tilde{a}_2 \rho_\Delta^4 + \tilde{a}_3 \rho_\Delta^6, \tag{7.30}$$

where new phenomenological coefficients $\tilde{a}_i$ are linear combinations of a_i and b_i from Eq. (7.25).

According to the Birman–Worlock theorem [54, 55], the one-component order-parameter spans a totally symmetric representation in the low-symmetry phase. Thus, the effective order-parameter of the isotropic–octagonal phase transition can be expressed as an average of ρ_j amplitudes:

$$\rho_\Delta(\boldsymbol{r}_i) = \frac{1}{4S} \sum_{j=1}^{4} \sum_{i=1}^{M} \cos(\boldsymbol{k}_j \cdot \boldsymbol{r}_i + \phi_j^0). \tag{7.31}$$

Let us use another distinctive property of the advanced phenomenological theory: the order-parameter is a nonlinear limited function of local parameters (see Chapter 4). Thus, $\rho_\Delta(\boldsymbol{r}_i)$ is limited by its maximal value $|\rho_\Delta| \leq \rho_\Delta^{\max}$, and the functional $F_q[\rho_\Delta(\boldsymbol{r}_i)]$ of Eq. (7.30) becomes identical to the effective free-energy density Eqs. (4.9) and (7.20).

It is worth recalling that the process of crystallization to a quasiperiodic structure, which is periodic in n-dimensions ($n > 3$), is controlled by an order-parameter whose nonlinear functional form is more complex than the same for ordinary three-dimensional symmetry breaking. Introducing two separate groups of variables related to the variation of amplitudes and phases of density modulation waves, we distinguish two problems merged in the latter case. First, in order to segregate components of the parent isotropic mixture, we minimize the free energy by varying the modulation amplitudes. Second, in order to localize the equilibrium positions for the lattice nodes in multidimensional superspace, we minimize free energy by shifting the density waves with respect to each other, i.e., playing with phason variables. Our task becomes easier since we have to minimize the phason contribution to the free energy in the fully segregated state where modulation wave amplitudes reach their maximal value and remain constant.

The maximal value limit $|\rho_\Delta| = \rho_\Delta^{\max}$ of Eq. (7.31) is attained when all maxima ($\delta\rho(\boldsymbol{r}) > \rho_{\text{cf}}$) of Eq. (7.31) coincide with the superlattice nodes, where cut-off parameter ρ_{cf} is defined by the nodes' surface concentration $C_s = M/S$. The global minimum of the free energy, as a function of node coordinates $F_q[\rho_\Delta(\boldsymbol{r}_i)]$, is then attained in the same point $|\rho_\Delta| = \rho_\Delta^{\max}$, and $\boldsymbol{r}_i$ corresponding to the free-energy minimum are equal to those obtained by the maximization of Eq. (7.31). In other words, the free-energy $F_q[\rho_\Delta(\boldsymbol{r}_i)]$ minimization and the effective order-parameter $\rho_\Delta(\boldsymbol{r}_i)$ maximization naturally occur in the stability domain of the fully segregated state (Fig. 7.5) [52].

Free-energy minima and order-parameter maxima are attained simultaneously when the superlattice nodes are located in the highest maxima of function (7.22). To determine their coordinates $\boldsymbol{R} = \boldsymbol{r}_i$, we use the equations

$$\boldsymbol{k}_j\boldsymbol{r}_i + \Delta\phi_j^i = 2jN_j^i, \tag{7.32}$$

where N_j^i are integers, $j = 1 - 4$ and $\Delta\phi_j^i$ are small phase deviations. The latter appear due to the mutual incommensurability of density waves from Eq. (7.28). This is why, even in the highest maxima of $\delta\rho(\boldsymbol{R})$, the phases of individual waves are slightly different from the $2\pi N_j^i$ value. Parametrization of the small phase deviations $\Delta\phi_j^i$ in the form

$$\Delta\phi = \boldsymbol{k}_j{\cdot}\Delta\boldsymbol{r}_i + \boldsymbol{k}_j^{\perp}{\cdot}\boldsymbol{r}_i,$$

allows solving Eq. (7.32) as

$$\boldsymbol{r}_i + \Delta\boldsymbol{r}_i = \sum_{j=1}^{4} N_j^i\boldsymbol{a}_j - \boldsymbol{u},$$

$$\mathbf{r}_i^{\perp} = \sum_{j=1}^{4} N_j^i\boldsymbol{a}_j^{\perp} - \boldsymbol{v}, \tag{7.33}$$

where $\boldsymbol{u}$ and $\boldsymbol{v}$ are homogeneous phonon and phason shifts of the structure, respectively: $\boldsymbol{a}_j = \frac{\pi}{k_0}[\cos(j\pi/4), \sin(j\pi/4)]$ and $\boldsymbol{a}_j^{\perp} = \frac{\pi}{k_0}[\cos(j3\pi/4), \sin(j3\pi/4)], j = 1\text{–}4$. For small deviations $|\Delta \boldsymbol{r}_i| \ll |\boldsymbol{r}_i|$, function $\delta\rho(\boldsymbol{R})$ can be approximated as

$$\delta\rho(\boldsymbol{R}) \approx 2\rho_\Delta \left[4 - \frac{1}{2}\sum_{n=1}^{4}(\Delta\phi_{2n}^{j})^2\right]. \tag{7.34}$$

It yields the maxima for coordinates

$$\boldsymbol{R} = \boldsymbol{r}_j = \sum_{j=1}^{4} N_{2j}^{j}\boldsymbol{a}_j - \boldsymbol{u}, \tag{7.35}$$

and

$$\delta\rho(\boldsymbol{r}_j) \approx 2\rho_\Delta \left[4 - (\boldsymbol{r}_i^{\perp})^2\boldsymbol{k}_j^0\right]. \tag{7.36}$$

It is worth noting that Eqs. (7.33) are identical to the conventional ones in high-dimensional crystallography, which map a plane octagonal superlattice on two-dimensional space. Figure 7.13(a) shows the

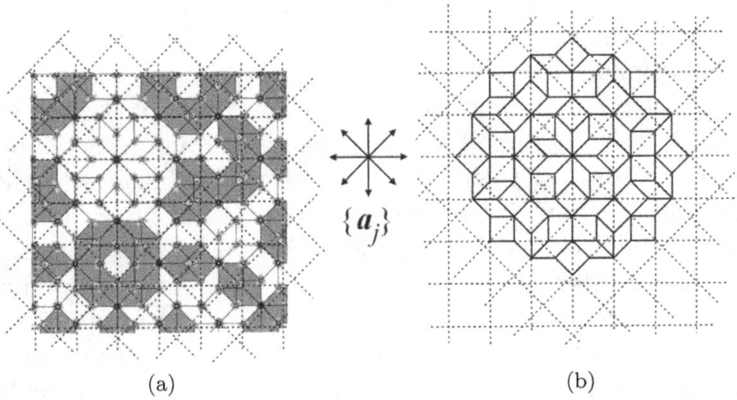

Fig. 7.13. (a) Density function distribution (7.29) above the cut-off value $\delta\rho > 1.5\rho_\Delta$ [52]. Small circles represent the superlattice nodes' (Eqs. (7.33)) projections from the corresponding four-dimensional space. Large gray areas show overlapping tiling elements. (b) A perfect octagonal Ammann tiling pattern with $\sum\phi_j^i = 0$ generated by the dual-grid method [49]. Dotted lines in (a) and (b) show quasiperiodic tetragrid.

octagonal quasiperiodic pattern where the clouds of density distribution $\delta\rho(\boldsymbol{R})$ (Eq. (7.28)) with cut-off limit $\delta\rho > 1.5\rho_\Delta$ are mapped [52]. For reference, the projection of four-dimensional superlattice nodes are shown. One concludes that the deviations $\delta\boldsymbol{r}_j$ (see Eqs. (7.40)) of the nodes' projections from the probability density maxima are negligible.

Figure 7.13 provides us a clear indication of the role of phason strains and the difference between the energy minimization method and the abstract mathematical tiling procedure: mathematical perfection is not exactly the same as the energy gain. Indeed, on Fig. 7.13(b), one can see that the octagonal tiling decoration [49] perfectly matches the reference quasiperiodic tetragrid defined by the vector star of Fig. 7.12. By contrast, the small phase shifts $\Delta\phi_j^i$ minimizing the free energy result [52] in maximization of $\delta\rho(\boldsymbol{r}_j)$ (Eq. (7.43)) in positions localized out of the same regular planes (Fig. 7.13(a)) although the average planes propagate with the same tetragrid period $|\boldsymbol{a}_j| = |\boldsymbol{k}_j^{-1}|$.

7.4.4. *Quasicrystalline structures: From probability maxima to atomic positions*

Two evident ideas make us sort probability maxima as potential nests for atoms in the structure. On the one hand, although we consider four-dimensional space, atoms of the quasicrystal occupy certain positions on the two-dimensional plane and they physically cannot overlap. On the other hand, since we analyze a minimal set of density waves, the full set of their interference maxima belongs to the single Wyckoff-type position in the multidimensional superspace group, and all of them should be occupied with equal probability. Thus, an additional filter should be included in the energy-minimization procedure in order to account for geometrical constraints.

The probability maxima pattern on Fig. 7.13(a) contains two types of nodes: (i) isolated and (ii) coupled maxima corresponding to the overlap of the tiling elementary squares (in gray on Fig. 7.13(a)). The coupled maxima are related by vectors of equal magnitude belonging to the vector star $\{\boldsymbol{S}_j\}$ (Fig. 7.14) whose modulus is $|\boldsymbol{S}_j| = |\boldsymbol{a}_j| \cdot \frac{2-\sqrt{2}}{\sqrt{2}+1}(\boldsymbol{a}_j = \boldsymbol{k}_j^{-1})$. Normally, quasicrystalline structures observed in the experiments are well packed, and simultaneous filling

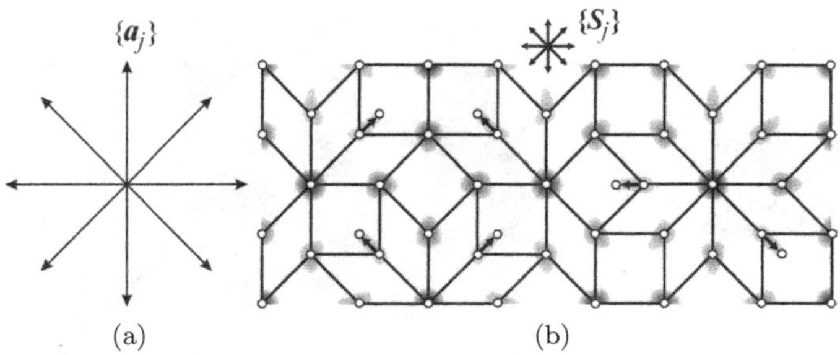

Fig. 7.14. (a) Basis vectors $\boldsymbol{a}_j$ of the tetragrid. (b) Fragment of the density-distribution pattern from Fig. 7.13(a). The vectors $\boldsymbol{S}_j$ point out the blocked positions.

of both the coupled positions is impossible. Thus, the positions corresponding to the lower probability density (tips of the $\boldsymbol{S}_j$-vectors) should be blocked and excluded from consideration. It is worth noting that the conventional projection method [56] gives a superlattice closely similar to the latter which contains only unblocked maxima positions.

Vectors $\boldsymbol{S}_j$ represent not only the geometric characteristics of the position splitting – they correspond, at the fixed phason variable $\boldsymbol{v}$ (Eq. (7.36)), to the single lowest energy phason jumps in the quasiperiodic lattice [52, 53]. The strongest maxima of the function (7.24) are associated, therefore, with the nodes which are more stable with respect to the single phason excitations.

Summarizing, it should be underlined that the advanced phenomenological approach (Chapter 4) considers solidification to the quasiperiodic structures on the model-free basis establishing the symmetry-controlled links between n-dimensional crystal-geometrical procedures and thermodynamics of quasicrystals.

7.5. Formation of Viral Capsid Structures

7.5.1. *Viral capsids*

Crystallization can kill life. Self-developing, which is an inherent process in the living world, is hardly compatible with the rigidity and strict regular organization of a crystal structure. Nevertheless, rigid crystalline-type aggregates can protect, conserve and deliver

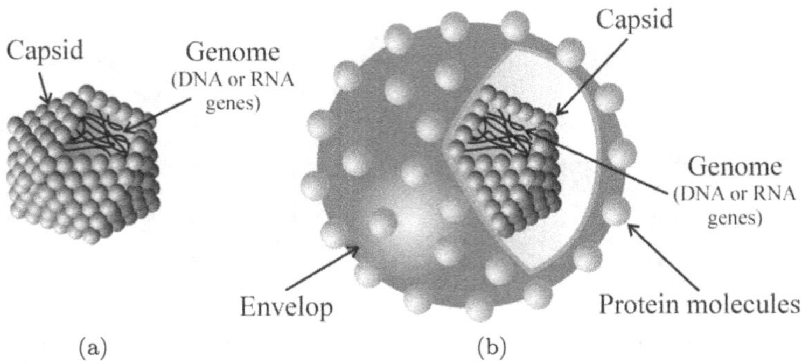

Fig. 7.15. Schematic view of two groups of viruses: (a) non-enveloped or naked, and (b) enveloped viruses. For explanations, see the text.

replication codes from one living host to another. This is the scenario which viruses use to self-protect and self-reproduce.

A virus is an extremely small (~20–100 nm), often disease-causing, agent consisting of a particle (the virion) containing a segment of ribonucleic (RNA) or deoxyribonucleic (DNA) acid (the viral genome) and surrounded by a shield of proteins known as a capsid (Fig. 7.15). The latter protective covering allows the virus to survive between hosts. Unable to replicate without a host cell, viruses are typically not considered living organisms.

Viruses are generally classified into two large groups: (i) non-enveloped (Fig. 7.15(a)) and (ii) enveloped (Fig. 7.15(b)) viruses, depending on the absence or presence of an outer lipid layer. Present in both, the capsid plays a fundamental role in the architecture and the biological function of a virus [57, 58].

Typically, a capsid is self-assembled from several oligomeric structural subunits made of protein molecules (protomers) or from isolated protein molecules. It can be generally described as made from several tens to many hundreds of copies of one or a few different types of capsid protein subunits. In each virus, its oligomeric subunit defines the type of symmetric structure. The basic types are helical and icosahedral. Icosahedral capsids are extremely frequent; they occur in about half of the virus families [58].

Assembly of a viral capsid is a complex process. The helical and icosahedral arrays are the simplest solution to form a regular or symmetric structure. However, in contrast to the periodic or quasiperiodic structures considered in the preceding sections, viruses build a container of defined shape. The capsid is, therefore, considered as a two-dimensional crystal that is limited in size and closed.

The density-wave approach provides a regular method for analyzing the self-assembly process of the capsid structures of viruses with spherical topology and icosahedral symmetry.

7.5.2. *Icosahedral capsids: Crystallization on the spherical surface*

We consider the formation of viral capsids as two-dimensional crystallization of protein molecules – protomers from the isotropic state. What is essential is that the molecular self-organization occurs on the spherical surface which belongs to a bounded three-dimensional object and therefore is closed. The latter property does not modify the density-wave procedures applied in the preceding sections to infinite lattices but only requires replacing the Cartesian coordinate system by the spherical one. The sphere radius R_{sph} is fixed by certain biological limitations linked to the genome's effective size. Thus, Fourier decomposition to plane waves (see, e.g., Eq. (7.24)) must logically be replaced by the decomposition to spherical harmonics. The density-wave theory of the capsid crystallization was developed in the series of publications [59–62], and we follow their logics showing its natural origin from the advanced phenomenological theory.

Localization of density waves only on the sphere surface simplifies Eqs. (7.4) and (7.5), fixing the radial variable $r = R_{\mathrm{sph}}$, then the modulation part of the probability density reads

$$\delta\rho_l(\theta,\phi) = \sum_{m=-l}^{m=l} [\eta_l^m \cdot Y_l^m(\theta,\phi)], \tag{7.37}$$

where l is the harmonic (wave) number and m is its order, η_l^m are the amplitudes of the spherical harmonics Y_l^m, i.e., order-parameter components, polar angle θ and azimuthal ϕ are the remaining spherical coordinate variables. The point symmetry group of the parent spherical surface is $SO(3)$ which contains all continuous rotation axes, mirror planes and inversion center. The majority of "spherical" viruses are indeed icosahedral; simultaneous covering of the surface by *asymmetric* protein molecules and their partial segregation lowers its symmetry to I [52,63]. Thus, even forming a regular icosahedral shape, the capsid possesses no full icosahedral point group as mirror planes and inversion are missing. The loss of the inversion center means, in terms of irreducible representation matrices, that the active representation is not totally symmetric, as in the case of the solidification of metals or liquid crystals, but an *antisymmetric* irreducible representation of the point group $SO(3)$. Two important consequences are: (i) the Landau potential contains only even-degree terms, (ii) only odd-degree harmonics (their parity is $(-1)^l$) form the capsid walls. These facts direct us to the free-energy expansion of Eq. (7.27) and the corresponding phase diagram of Fig. 7.5. One can see, on the diagram, that self-assembling of proteins into capsid shell can occur as not only a strongly first-order crystallization transformation but also, alternatively, can proceed continuously through the intermediate partially ordered state.

The restrictive condition (ii), allowing only the odd l numbers for capsid-forming harmonics, results in the simple selection rule:

$$l = 15 + 6i + 10j, \tag{7.38}$$

where i and j are positive integers or zero [59, 60]. For $l < 45$, the explicit equation for the protein density distribution is obtained by averaging the $Y_l^m(\theta, \phi)$ functions in Eq. (7.37) over the reduced symmetry group I:

$$\delta\rho_l(\theta, \phi) = n_l \cdot \sum_{G=I} \{Y_l^m[\hat{g}(\theta, \phi)]\}. \tag{7.39}$$

Here $\hat{g}(\theta\phi)$ are 60 symmetry operations of the group I, n_l are normalization parameters. Figure 7.16 shows the first four irreducible density functions $\delta\rho_l(\theta, \phi)$ of Eq. (7.39) permitted by the rule (7.38): $l = 15$, 21, 25 and 27. The number of maxima of the density distribution function in Eq. (7.39) is equal to $60N$, where N is the number of different 60-fold positions of the I group. It is worth underlining that density-wave maxima are distributed on the surface in a regular way which possesses local periodicity. The maxima coincide with the general 60-fold position (orbit) of the icosahedral group since asymmetric protein molecules cannot be placed on any rotation symmetry element of the group I.

The density-wave approach demonstrates excellent coincidence of its predictions to the experimentally observed capsid structures.

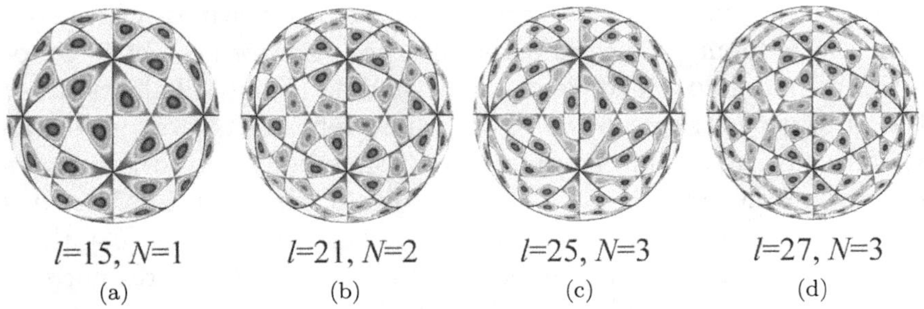

Fig. 7.16. The first four irreducible icosahedral density functions $\delta\rho_l(\theta,\phi)$ (Eq. (7.37)) with the icosahedral symmetry; l is the wave number, N is the number of nonequivalent maxima of the functions. Gray color deepening corresponds to the increase in probability. (Courtesy of O. Konevtsova and S. Rochal.)

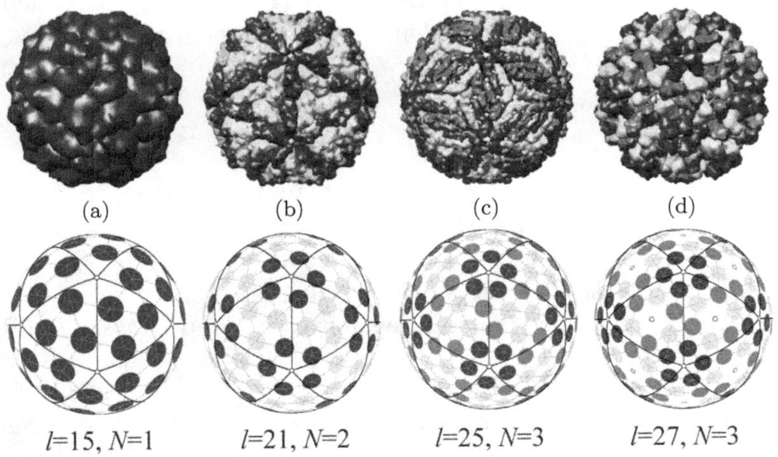

Fig. 7.17. Positions of protein molecules predicted by the models of Refs. [59,60] (lower panels) and compared to the experimental viral capsid structures (upper panels): (a) satellite tobacco mosaic virus, (b) L-A virus, (c) dengue virus and (d) cowpea chlorotic mottle virus [60]. (Courtesy of O. Konevtsova and S. Rochal)

Figure 7.17 gives a few examples comparing the positions obtained by the amplitude maximization method [59–62] to real viral capsid structures.

The symmetry-based density-wave approach is a logical generalization of the pure geometric method [63] actively and successfully used in the models of viral capsid formation starting from the early 1960s. From the theorist's point of view [61], the procedure proposed in Ref. [63] has several problems of arbitrariness which concern the mapping choice and the interpretation of geometric properties of manifolds. The clear understanding of limited character of the pure geometrical approach [63], however, came later, when the statistically significant number of experimentally solved structures had been collected. Though a big number of virus capsid structures are in good agreement with the geometrical scheme of Ref. [63], there is an increasing number of experimentally solved structures not satisfying its empirical geometrical limitations. At the same time, the density-wave theory [59–62], which is a part of the general crystallization theory, successfully describes, in a uniform way, all observed small spherical viruses. Indeed, Fig. 7.17 shows four capsid structures predicted by the density-wave theory and observed experimentally, while two of them, capsids of L-A virus (Fig. 7.17(b)) and dengue virus (Fig. 7.17(c)), are forbidden by the geometrical theory.

References

[1] L. D. Landau, *Zh. Eksp. Teor. Fiziki* **7** (1937) 19; **7** (1937) 627. Transl. in *Collected Papers of L. D. Landau*, ed. D. Ter Haar (Pergamon Press, Oxford, 1965), p. 193; p. 217.

[2] V. I. Marchenko, *Sov. Phys.: JETP* **73** (1991) 759.

[3] D. A. Kirzhnits and Yu. A. Nepomnyaschiy, *Sov. Phys.: JETP* **32** (1970) 1191.

[4] E. I. Kats, V. V. Lebedev and A. R. Muratov, *Phys. Rep.* **228** (1993) 1.

[5] E. Kats, in *Symmetry, Spin Dynamics and the Properties of Nanostructures*, eds. V. Dugaev, A. Wal and J. Barnaś (World Scientific, Singapore, New York, 2015), p. 188.

[6] Wu-Ki Tung, *Group Theory in Physics* (World Scientific, Singapore, 1985).

[7] A. Z. Patashinskii and V. L. Pokrovskii, *Fluctuation Theory of Phase Transitions* (Pergamon Press, Oxford, 1979).

[8] Yu. M. Gufan, *Structural Phase Transitions* (Mir, Moscow, 1983) (in Russian).

[9] P. Toledano and V. Dmitriev, *Reconstructive Phase Transitions in Crystals and Quasicrystals* (World Scientific, Singapore, 1996).

[10] J.-C. Toledano and P. Toledano, *The Landau Theory of Phase Transitions* (World Scientific, Singapore, 1987).

[11] S. Alexander and J. McTague, *Phys. Rev. Lett.* **41** (1978) 702.

[12] V. Dmitriev, *Physica B* **472** (2015) 97.

[13] H. T. Davis, *Introduction to Nonlinear Differential and Integral Equations* (Dover, New York, 1962).
[14] M. H. Vaynberg and V. A. Trenogin, *Theory of Branching of Solutions of Nonlinear Equations* (Noordhoff International Publishing, Leyden, 1974).
[15] D. Errandonea, B. Schwager, R. Ditz *et al.*, *Phys. Rev. B* **63** (2001) 132104.
[16] D. Errandonea, *J. Appl. Phys.* **108** (2010) 033517.
[17] D. Errandonea, *Phys. Rev. B* **87** (2013) 054108.
[18] P. Parisiades, F. Cova and G. Garbarino, *Phys. Rev. B* **100** (2019) 054102.
[19] G. Weck, V. Recoules, J.-A. Queyroux *et al.*, *Phys. Rev. B* **101** (2020) 014107.
[20] *Phase Transitions in Complex Fluids*, eds. P. Toledano and A. M. Figueiredo Neto (World Scientific, Singapore, 1998).
[21] *Self-Assembled Supramolecular Architecture: Lyotropic Liquid Crystals*, eds. N. Garti, P. Somasundaran and R. Mezzenga (Wiley & Sons, Hoboken, New Jersey, 2012).
[22] A. M. Figueiredo Neto and S. R. Salinas, *The Physics of Lyotropic Liquid Crystals* (Oxford University Press, Oxford, 2005).
[23] V. P. Dmitriev and G. R. Ouriques, *Phys. Rev. E* **60** (1999) 1840.
[24] V. P. Dmitriev and G. R. Ouriques, *Eur. Phys. J. B* **12** (1999) 421.
[25] *International Tables for Crystallography*, ed. T. Hahn (Kluwer Academic Press, Dordrecht, 1996), vol. A.
[26] V. P. Dmitriev, P. Toledano, A. M. Figueiredo Neto *et al.*, *Phys. Rev. E* **59** (1999) 771.
[27] P. G. de Gennes, *The Physics of Liquid Crystals* (Clarendon Press, Oxford, 1974).
[28] S. Chandrasekhar, *Liquid Crystals* (Cambridge University Press, Cambridge, 1992).
[29] V. I. Indenbom and E. B. Loginov, *Sov. Phys. Crystallogr.* **26** (1981) 527.
[30] D. Shechtman, I. Blech, D. Gratiasand J. W. Cahn, *Phys. Rev. Lett.* **53** (1984) 1951.
[31] P. Bak, *Phys. Rev. Lett.* **54** (1985) 1517.
[32] P. A. Kalugin, A. Y. Kitaev and L. C. Levitov, *JETP Lett.* **41** (1985) 145.
[33] D. Levine, T. C. Lubensky, S. Ostlund *et al.*, *Phys. Rev. Lett.* **54** (1985) 1520.
[34] A. P. Tsai, *Sci. Technol. Adv. Mater.* **9** (2008) 013008.
[35] W. Steurer, *Z. Kristallogr.* **219** (2004) 391.
[36] X. Zeng, G. Ungar, Y. S. Liu *et al.*, *Nature* **428** (2004) 157.

[37] V. Percec, M. R. Imam, M. Peterca *et al.*, *J. Am. Chem. Soc.* **131** (2009) 1294; **131** (2009) 7662.
[38] A. Takano, W. Kawashima, A. Noro *et al.*, *J. Polym. Sci. Polym. Phys.* **43** (2005) 2427.
[39] S. Lee, M. J. Bluemle and F. S. Bates, *Science* **330** (2010) 349.
[40] O. V. Konevtsova, S. B. Rochal and V. Lorman, *Phys. Rev. Lett.* **108** (2012) 038102.
[41] D. V. Talapin, E. V. Shevchenko, M. I. Bodnarchuk *et al.*, *Nature* **461** (2009) 964.
[42] L. Bindi, P. J. Steinhardt, N. Yao and P. J. Lu, *Science* **324** (2009) 1307.
[43] L. Bindi, *Natural Quasicrystals. The Solar System's Hidden Secrets* (Springer Nature, Switzerland, 2020).
[44] N. D. Mermin and S. M. Troian, *Phys. Rev. Lett.* **54** (1985) 1524.
[45] M. V. Jaric, *Phys. Rev. Lett.* **55** (1985) 607.
[46] L. Gronlund and N. D. Mermin, *Phys. Rev. B* **38** (1988) 3699.
[47] E. I. Kut'in, V. L. Lorman and S. V. Pavlov, *Sov. Phys. Usp.* **36** (1991) 497.
[48] T. Janssen, G. Chapuis and M. de Boissieu, *Aperiodic Crystals: from Modulated Phases to Quasicrystals* (Oxford University Press, Oxford, 2007).
[49] W. Steurer and S. Deloudi, *Crystallography of Quasicrystals* (Springer-Verlag, Berlin-Heidelberg, 2009).
[50] J. E. S. Socolar, T. C. Lubensky and P. J. Steinhardt, *Phys. Rev. B* **34** (1986) 3345.
[51] T. C. Lubensky, J. E. S. Socolar, P. J. Steinhardt *et al.*, *Phys. Rev. Lett.* **57** (1986) 1440.
[52] O. V. Konevtsova, S. B. Rochal and V. Lorman, *Phys. Sol. State* **55** (2013) 1597.
[53] O. V. Konevtsova, S. B. Rochal and V. Lorman, *Phys. Lett. A* **377** (2013) 1215.
[54] J. M. Worlock, in *Structural Phase Transitions and Soft Modes*, eds. E. J. Samuelson, E. Andersen and J. Feder (Universitelsforlage, Oslo, Norway, 1971), p. 339.
[55] J. L. Birman, *Phys. Lett. A* **45** (1973) 197.
[56] C. Janot, *Quasicrystals: A Primer*, second edition (Oxford University Press, Oxford, 1997).
[57] S. J. Flint, L. W. Enquist, V. R. Racaniello and A. M. Skalka, *Principles of Virology: Molecular Biology, Pathogenesis, and Control* (ASM, Washington, 2000).
[58] *Structure and Physics of Viruses: An Integrated Textbook*, ed. M. G. Mateu (Springer, Dordrecht, 2013).

[59] V. L. Lorman and S. B. Rochal, *Phys. Rev. Lett.* **98** (2007) 185502.
[60] V. L. Lorman and S. B. Rochal, *Phys. Rev. B* **77** (2008) 224109.
[61] V. L. Lorman and S. B. Rochal, *Group Theory Methods in Virology: Landau Density Wave Approach.* In *Mathematics for Industry* (Springer, Japan, 2015) **11**, 15.
[62] S. B. Rochal, O. V. Konevtsova, A. E. Myasnikova and V. Lorman, *Nanoscale* **8** (2016) 16977.
[63] D. L. D. Caspar and A. Klug, *Cold Spring Harbor Symp. Quant. Biol.* **27** (1962) 1.

Chapter 8

Phase Transitions in Low-Dimensional Structures

8.1. Introduction

Low-dimensional materials exhibit remarkable distinctive features compared to that of the bulk solids. Physical and chemical properties of a given material are changed after the size of the material is down at least in one of its dimensions. When one or more of the dimensions of a solid are reduced to a nanoscale level (1–100 nm), novel electrical, mechanical, magnetic and optical properties can be introduced [1].

In order to disclose the factors making a low-dimensional material different from its bulk counterpart, it is customary to consider the primary distinctions both microscopic, as confined electron state wave functions giving rise to quantum size effects, and macroscopic, as the high surface-to-volume ratio, for example. The reduction in dimensionality leads, in theory, to peculiar statistical physics, quantum physics and mechanics. Normally, theoreticians consider the effect of dimensionality in terms of short- and long-range interactions. Both short- and long-range interactions can undergo dramatic changes going down from the dimension $D = 3$ to $D = 2$ and $D = 1$, bringing up some unique material properties and advanced functions that cannot be achieved for their counterparts in other dimensions. Due to all these features of the low-dimensional solids, the phase transitions they undergo can be different from those in the bulk samples [2, 3]. However, the transformations occurring in low-dimensional solids exhibit no mechanisms making them an exception from the general picture of phase transitions: external "fields" induced a transition can be distinct, but the intrinsic driving forces remain the same – elastic, phonon, electronic instabilities, etc.

A symmetry-based approach is the most suitable phenomenological formalism to identify the common aspects of phase transitions in different dimensionalities. In this chapter, we consider few notable examples of phase transitions in quasi-1D and 2D systems and apply the phenomenological scheme developed in the preceding chapters, analyzing the structural reconstructions of nanowires, monoatomic layers and bilayers.

8.2. Surface Reconstruction

8.2.1. *Surface crystal geometry*

In our further surface transformation models, we truncate the bulk structure of a perfect crystal and consider the ideal geometry of an atomically clean surface as a parent lattice. Then, we indicate a unit cell corresponding to the new superstructure lattice on a diagram of the ideal bulk-terminated surface simply following the strict symmetry-based procedure outlined in the previous chapters. However, when a surface of a crystalline solid is created as cutting through the bulk, inevitably chemical bonds are truncated in an asymmetric way. As a consequence, the surface structure will be different from its bulk-terminated state. This difference normally is categorized, in terms of surface science, as either a *relaxation* or a *reconstruction* (Fig. 8.1).

Relaxation refers to a collective movement of layer atoms by the same amount and in the same direction which is perpendicular to the surface — the bulk unit cell is preserved at the surface (Fig. 8.1(b)). We focus here, however, on the *reconstruction* process which significantly modifies the two-dimensional structure of the surface layer of a crystal changing its general symmetry, i.e., inducing a phase transition on the surface (Fig. 8.1(c)).

Though the term "reconstruction" is also used to identify the reconstructed surface itself, we refer to it mainly as an active process and a synonym of "phase transition", and we continue using the less common terms in surface science, i.e., phase and surface

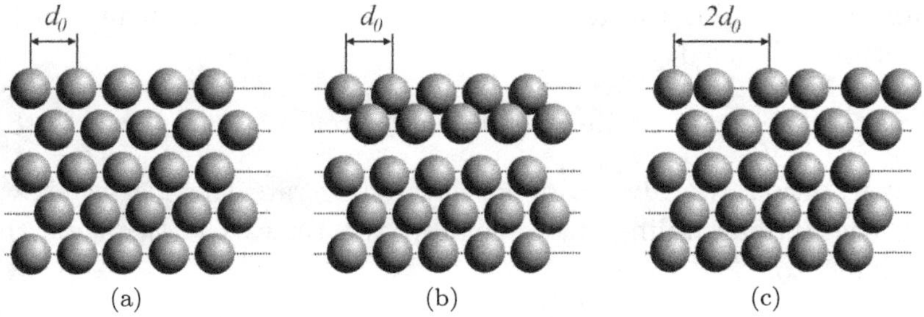

Fig. 8.1. Schematic side view of a crystal when bulk terminated (a), exhibiting surface relaxation (b), and a registry lateral shift – reconstruction (c).

superstructure. Finally, it is worth noting that the reconstruction is observed for a limited number of metal surfaces but is a characteristic feature of surfaces of most semiconductors [2, 3].

In the following sections, our attention is focused on selected examples of phase transitions (surface reconstructions) in topmost layers of the truncated crystals and in monoatomic layers of metals adsorbed on atomic clean surfaces of a single crystal. Substrate surfaces usually provide the template for subsequent deposition and growth of thin films, and we describe the overlayer structures in terms of the underlying structure of the substrate. Surface superstructures are commonly given in *Wood's notation* or *matrix notation*, which link the two-dimensional lattice of the overlayer structure with the ideal lattice plane in the bulk.

8.2.1.1. *Wood's notation*

Wood's notation [4] is the simplest and most frequently used method for describing a surface structure, though it is applicable only for the cases where the two unit cells are of the same symmetry or have closely related symmetries. In this notation, the ratio of the lengths of the basic translation vectors of the superstructure is specified in terms of those of the substrate plane. In addition, the angle of rotation which makes the superstructure lattice be aligned with the basic translation vectors of the substrate is included. Thus, if on a certain Me substrate surface (*hkl*) a superstructure with the basic translations

$$\boldsymbol{d}_1 = m \cdot \boldsymbol{a}_{1s}, \quad \boldsymbol{d}_2 = n \cdot \boldsymbol{a}_{2s}, \tag{8.1}$$

is formed and it is rotated by the angle ϕ, then the surface superstructure is labeled

$$\mathrm{Me}(hkl)(m \times n) - R\phi. \tag{8.2}$$

By convention, we always denote the substrate surface as (1×1) with the understanding that we know the Bravais lattice-type that it belongs to.

8.2.1.2. *Matrix notation*

A more general notation proposed by Park and Madden [5] involves a simple vectorial construction, where one expands the overlayer basis $\{\boldsymbol{d}_i\}$ in terms of substrate basis $\{\boldsymbol{a}_{js}\}$:

$$\begin{aligned} \boldsymbol{d}_1 &= G_{11}\boldsymbol{a}_{1s} + G_{12}\boldsymbol{a}_{2s}, \\ \boldsymbol{d}_2 &= G_{21}\boldsymbol{a}_{1s} + G_{22}\boldsymbol{a}_{2s}, \end{aligned} \tag{8.3}$$

or, in matrix notation,

$$\begin{pmatrix} \boldsymbol{d}_1 \\ \boldsymbol{d}_2 \end{pmatrix} = \begin{pmatrix} G_{11} & G_{12} \\ G_{21} & G_{22} \end{pmatrix} \begin{pmatrix} \boldsymbol{a}_{1s} \\ \boldsymbol{a}_{2s} \end{pmatrix} = \boldsymbol{M} \begin{pmatrix} \boldsymbol{a}_{1s} \\ \boldsymbol{a}_{2s} \end{pmatrix}. \tag{8.4}$$

The determinant of the transformation matrix $\boldsymbol{M}$ ($\det \boldsymbol{M} = G_{11}G_{22} - G_{21}G_{12}$) is given by the ratio of area of the overlayer unit cell to that of the substrate. The superstructure may then conveniently be referred to by the in-line form of the matrix $(G_{11}, G_{12}; G_{21}, G_{22})$. Figure 8.2 shows Wood's and matrix notations for some surface superlattices.

8.2.2. *Layer groups vs space groups*

Layer groups are symmetry groups which describe structures of *three-dimensional* solids possessing the periodicity in only *two dimensions* [6], and it would be relevant if we describe the surface structures by layer groups. A layer group comprehensively characterizes a surface structure by labeling the symmetry group and Wyckoff positions occupied by its atoms. Their smaller number, 80, compared to 230 space groups [7] can be considered as a significant advantage if we restrict our search to solving and characterizing an isolated surface structure and neglect surface structure reconstructions. However, the

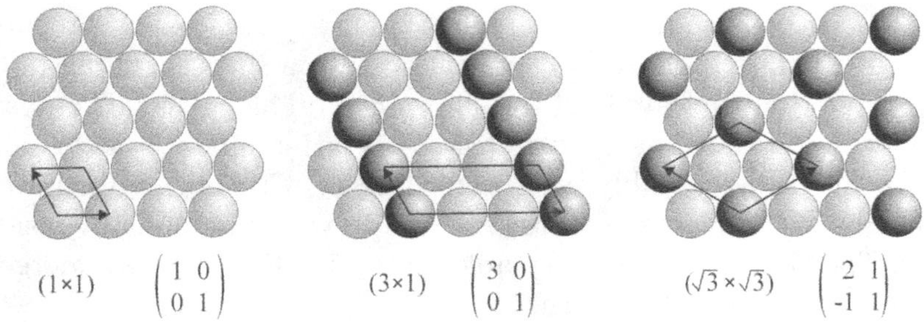

Fig. 8.2. Examples of Wood's and matrix notations at the (111) surface of a face-centered cubic material.

symmetry-based approach to phase transitions, which we developed in the previous chapters, requires much more group-theoretical information and relevant mathematical methods: irreducible representations, image groups and corresponding invariants, order-parameter space symmetry and isotropy subgroups, isotropy manifolds corresponding to stable states and symmetry groups of distorted structures (see Chapter 1). For the 3D space groups, all these are available to the users either in the form of tables [8, 9] or as free online services [10, 11], but it is insufficiently presented for 2D groups. In order to take advantage of such evident completeness, in the following sections we continue using 3D formalism to analyze the surface layer reconstructions even with understanding the redundancy of the method.

8.2.3. *Diamond-type (111) surface reconstruction*

Elemental silicon and germanium are a cornerstone of modern electronics, and their (111) crystal surface is the most studied and well-characterized surface in semiconductor physics. Indeed, it serves as a prototypical substrate for adsorption and growth systems and demonstrates a rich variety of own superstructures [12]. Let us consider, as an introductory example, the reconstruction of the diamond-type (111) topmost bilayer ((1 × 1) phase) to the (2 × 1) superstructure, i.e., the structural phase transition between two classic examples for reconstructed semiconductor surfaces. We will identify the elementary events of a complex scenario of the transformation and show their symmetry-controlled links.

An unreconstructed (1×1) surface for Si(111) can be prepared even at room temperature by annealing with a short laser pulse [12, 13], while simple cleavage of the silicon crystal along the (111) plane produces the (2×1) structure [14]. The latter phase is metastable and converts irreversibly after annealing at 670 K to another well-studied surface superstructure (7×7) [14], which we do not consider here. The (7×7) structure is stable up to 1100 K, where it undergoes transition to the (1×1) structure [15].

8.2.3.1. *Parent structure*

The first step, in the symmetry-based phenomenological approach, should be identification of a proper parent phase. It is logical to follow the scenario worked out in Chapter 6, which considers both two-layer graphite ($n = 2$) and three-layer diamond structures ($n = 3$) as the result of ordering-type phase transitions from a common hexagonal disordered one-layer polytype filled by atoms with concentration $x = 2/3$ (Sec. 6.4). The relevant order-parameters span irreducible representations $\tau_1(\boldsymbol{k}_j)$, where $\boldsymbol{k}_j = \frac{1}{3}(\boldsymbol{b}_1 + \boldsymbol{b}_2) + \frac{1}{n}\boldsymbol{b}_3$. As we focus on the isolated topmost (111) (bi)layer of the diamond crystal, the corresponding parent structure must be one-layer ($n = 1$), and $\boldsymbol{k}_j = \frac{1}{3}(\boldsymbol{b}_1 + \boldsymbol{b}_2) + \boldsymbol{b}_3 = \frac{1}{3}(\boldsymbol{b}_1 + \boldsymbol{b}_2) = \boldsymbol{k}_{13}$. Accordingly, the ordering process ends up at a graphite-like net (Fig. 8.3) with a unit cell tripled compared to the archetype-disordered structure. This new structure of $P6/mmm$ symmetry, containing two atoms per unit cell, becomes a parent phase for the (111) surface reconstructions.

8.2.3.2. *(1 × 1) surface reconstruction*

The (111) diamond-like surface refers, in surface crystallography, to a (1×1) unreconstructed structure. Since we intend to find an appropriate model for its reconstruction to (2×1), i.e., doubling the hexagonal unit cell, two critical points in the hexagonal Brillouin zone should be analyzed: surface M-point, identified by $\boldsymbol{k}_{12} = \frac{1}{2}(\boldsymbol{b}_1 + \boldsymbol{b}_2)$, and central Γ-point ($\boldsymbol{k}_{16} = 0$) (Fig. 6.10(b)) as potentially involved in the transformation scenario. The vector star $\{\boldsymbol{k}_{12}\}$ contains three equivalent vectors: $\boldsymbol{k}_{12}^{(1)} = \frac{1}{2}\boldsymbol{b}_1$, $\boldsymbol{k}_{12}^{(2)} = \frac{1}{2}\boldsymbol{b}_2$ and $\boldsymbol{k}_{12}^{(3)} = \frac{1}{2}(\boldsymbol{b}_1 + \boldsymbol{b}_2)$ [8], but only one is involved in creating a doubled superstructure. Thus, three equivalent translation domains can be

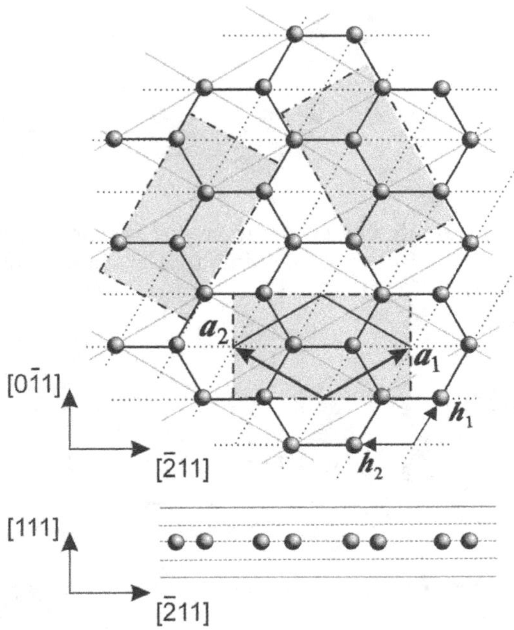

Fig. 8.3. Schematic diagram of the parent hexagonal layer (top and side views). Dotted lines show the archetype lattice of disordered structure (basis vectors $\boldsymbol{h}_1$ and $\boldsymbol{h}_2$). $\boldsymbol{a}_1$ and $\boldsymbol{a}_2$ are the translation vectors of the ordered graphite-like layer structure. Gray rectangles show three symmetry-equivalent translation domains of further (2×1) superstructure (see text).

observed on the (2×1) reconstructed surface corresponding to each of the $\boldsymbol{k}_{12}^{(i)}$ arms. Orientation of the translation domains is shown on Fig. 8.3. To avoid confusion, it should be noted that the superlattice definition ($\boldsymbol{d}_1 = \boldsymbol{a}_1 - \boldsymbol{a}_2, \boldsymbol{d}_2 = \boldsymbol{a}_1 + \boldsymbol{a}_2$), which takes the form (1, −1; 1, 1) (equivalent domains (2, 1; 0, 1) and (1, 2; 1, 0)) in the matrix notation, formally does not correspond to the accepted one in the literature. However, it is crystallographically more correct, while $(2 \times 1) = (2, 0; 0, 1)$ does not satisfy the strict condition established for choosing *Bravais lattices*: a Bravais unit cell must have maximal number of right angles, but orthorhombic (2×1) has a non-rectangular basis.

Figure 8.4 depicts three possible mechanisms which split planar six-membered graphite-like rings. The mechanisms are referred to as puckering (Fig. 8.4(a)) and buckling (Figs. 8.4(b) and 8.4(c)).

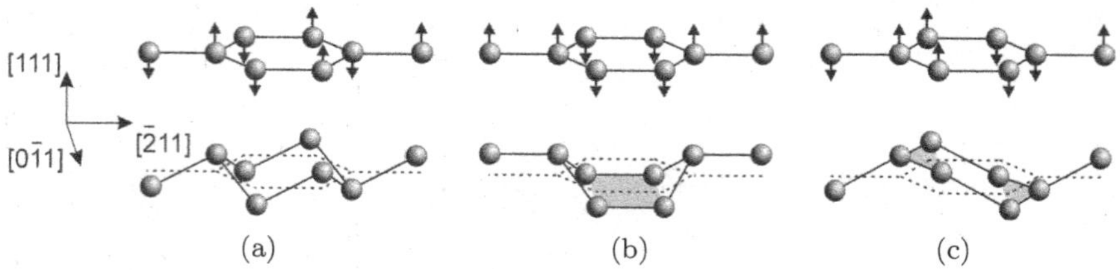

Fig. 8.4. Splitting graphite-like planar layer. (a) Planar ring puckering to the chair configuration. (b) Buckling the ring to the classic boat confirmation. (c) Buckling the layer to C(111) (2 × 1) structure.

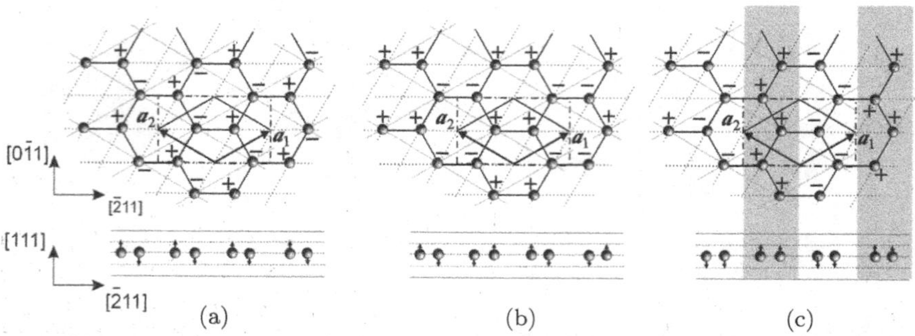

Fig. 8.5. Mechanisms of the graphite-like layer splitting (top and side views). (a) Puckering to a diamond-like bilayer by the $\tau_5(\boldsymbol{k}_{16})$ order-parameter. (b) and (c) Buckled layers ($\tau_4(\boldsymbol{k}_{12})$ and $\tau_5(\boldsymbol{k}_{12})$, correspondingly). Gray bands in (c) show Pandey's π-bonded chains.

First, two crystal distortions create the chair and boat conformation, correspondingly, and are traditionally included in the graphite-diamond transformation scenario. The third mechanism, which is identified by the order-parameter $\tau_5(\boldsymbol{k}_{16})$ (one-dimensional A2a image group in notation of Ref. [9]), puckers a parent planar honeycomb net to the diamond-type (1 × 1) bilayer structure of space symmetry $P\overline{3}m1(Z = 2)$ or $p\overline{3}m1$ (no. 72) in frame of the layer groups (Figs. 8.4(a) and 8.5(a)).

Three order-parameter components of the second mechanism span $\tau_4(\boldsymbol{k}_{12})$ irreducible representation, but a boat-type conformation (space group $Pmmn(Z = 4)$ or layer group $pmmn$ (no. 59)) is irrelevant to the observed (1 × 1) surface reconstruction. On the contrary, the buckling mechanism presented in Figs. 8.4(c) and 8.5(c) and described by the irreducible representation $\tau_5(\boldsymbol{k}_{12})$ (three-dimensional image-group C48a $\sim m\overline{3}m$, see Sec. 1.3.3.5) transforms the planar graphite-like layer to a buckled structure (space group $Pmna(Z = 4)$ or layer group $pmna$ (no. 42)), which can be definitely identified as reconstructed $C(111)(2{\times}1)$. Its untitled zigzag π-bonded chains are a characteristic feature which perfectly corresponds to the latest experimental microscopic structure determination on the low-index diamond surface (see Ref. [16] and Chapter 15 in Ref. [12]).

The π-bonded chain model was proposed by Pandey [17] and is widely accepted as a valid description of the cleaved $(111)(2 \times 1)$ surface structure in Si, Ge and with certain modification, in C [2].

8.2.3.3. *Si(111)(2 × 1) surface reconstruction*

In case of Si(111), (2×1) reconstruction follows the same scenario doubling the orthorhombic unit cell. However, in Si, the two different atoms of the zigzag reside at different heights. The Si atoms in the upper chains are bounded to each other by π-bonds (Fig. 8.5(c)) and are shifted upward and downward making the π-bonded chains slightly buckled [18]. We still use the latter notation "buckled chain", although it would be more correct to discuss the puckering of primary buckled hexagonal net. Indeed, comparing the two structures, Si(111)(2×1) and C(111)(2×1), we find their difference pattern to be identical with the classic puckering picture described in Sec. 8.2.3.2 (Figs. 8.4(a) and 8.5(a)). Thus, we conclude that restructuring from the parent structure to Si(111)(2 × 1) is induced by two independent order-parameters, $\tau_5(\boldsymbol{k}_{12})$ and $\tau_5(\boldsymbol{k}_{16})$, and the distorted structure has $P2/m(\mathrm{Z} = 4)$ symmetry (layer group $p2/m$ (no. 6)), which is a common subgroup of $P\overline{3}m1(2)$ and $Pnma(4)$ space groups.

Figure 8.6 summarizes the elementary reconstruction mechanisms which occur in the upmost (111) (bi)layer of the diamond-like crystal surface. Buckling A- and puckering B-distortions, both of *displacive* type, reconstruct the graphite-like planar layer to "C(111)(2×1)" and "C(111)(1×1)" (or "Si(111)(1×1)") bilayer surface structures, correspondingly. Reconstruction of the C(111)(1 × 1) surface structure to "C(111)(2 × 1)" superstructure requires swapping order-parameters: the transformation switches off $\tau_5(\boldsymbol{k}_{16})$ and activates $\tau_5(\boldsymbol{k}_{12})$. It is worth noting that the direct transformation breaks the group–subgroup relation between the symmetry groups of the structures since $Pmna(Z = 4)$ is not a subgroup of $P\overline{3}m1(Z = 2)$, and the transformation should be classified, therefore, as *reconstructive*, though

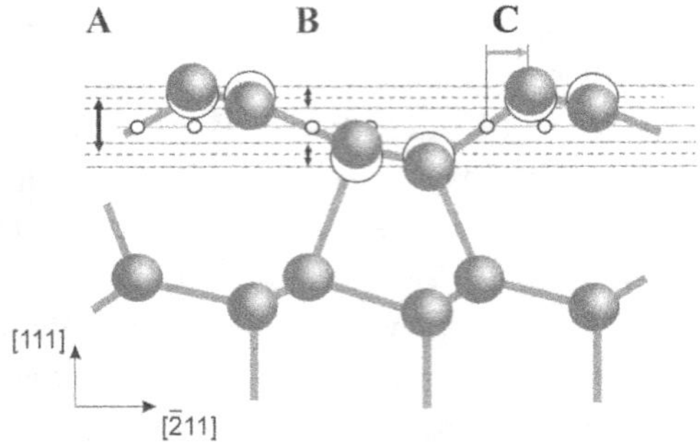

Fig. 8.6. Elementary displacive mechanisms for the (111) surface of a diamond-like crystal (side view). A is buckling with $\tau_5(\boldsymbol{k}_{12})$, B is puckering $\tau_5(\boldsymbol{k}_{16})$, and C is lateral adoptive shift. Small circles are the positions of atoms in the parent graphite-like layer.

both of them are subgroups of the common parent $P6/mmm(2)$. Thus, two mechanisms, A+B, distort "C(111)(1×1)" to Si(111)(2×1) bilayer superstructure.

Last but not least, real uppermost (111) reconstructed surface layers in diamond-like C, Si and Ge crystals are shifted from their ideal positions [12] as shown in Fig. 8.6 (mechanism C). However, such lateral displacement of the entire upmost layer is out of the symmetry-based analysis and is fully controlled by minimization of the adhesion energy.

8.2.3.4. *Positive and negative buckling of the π-bonded chains*

Two different geometric configurations (called isomers) were experimentally found on the Si(111)(2 × 1) reconstructed surface. They can be distinguished in relation to the sign of the π-bonded chain buckling: the positive buckling and the negative buckling [19]. Moreover, it has been demonstrated that two isomers occur and coexist on the surface both at low [20] and room temperature [21]. Density-functional calculations, indeed, predicted similar total energies close to the ground state for differently buckled configurations [22]. So, it was definitely shown, theoretically and experimentally, that the Si(111)(2 × 1) surface belongs to the class of systems where different atomic configurations with equal total energies exist.

It is believed, in the nanotechnology community, that semiconductor systems offering two metastable states with similar energies close

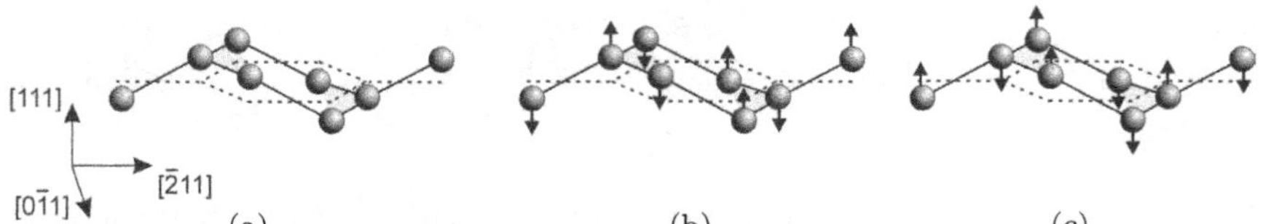

Fig. 8.7. Puckering mechanism ($\tau_5(\boldsymbol{k}_{16})$) applied to the buckled ($\tau_5(\boldsymbol{k}_{12})$) (111) bilayer. (a) C(111)(2 × 1) bilayer surface. Positive (b) and negative (c) puckering of the C(111)(2 × 1) structure resulting, according to Ref. [17], in "positive and negative buckling".

to the ground state are of great interest due to potential applications for nanoscale devices such as rewritable nanoscale memory, for example. In this regard, let us discuss in more detail the phenomenological model of "Si(111)(1 × 1)"–"C(111)(2 × 1)"–"Si(111)(2 × 1)" reconstructions.

The π-bonds buckling appears as the result of an equitranslational *ferroic* "C(111)(2 × 1)"–"Si(111)(2 × 1)" phase transition induced by the primary one-component order-parameter $\tau_5(\boldsymbol{k}_{16})$ (see Sec. 3.1). At the transition, the point group order is reduced by factor two: $[mmm]/[2/m] = 8/4 = 2$. Thus, in accordance with the Curie symmetry principle, two types of domain variants should be stable on the Si(111) surface. They differ in the directions of the puckering displacements (Figs. 8.7(b) and 8.7(c)). Both the buckling isomers, positively buckled and negatively buckled, represent, therefore, the same thermodynamic state (stable or metastable). Domain variants are degenerate in free energy, i.e., they have *equal energies* in any point of the phase space (see Sec. 1.2.2.6).

Switching between domains or producing a monodomain sample requires to apply an external field conjugated (identical in symmetry) to the order-parameter. For example, proper ferroelectrics show, as an order-parameter, spontaneous electric polarization, which can be switched by an electric field; ferromagnets show spontaneous magnetization, which is switchable by a magnetic field.

The puckering mechanism (buckling the chains) possesses the $\tau_5(\boldsymbol{k}_{16})$ symmetry (B_{1g} in spectroscopic notation), and the first relevant non-vanishing multipolar configuration is a *hexadecapole* which relates to g-electron orbitals (quantum number $l = 4$). In terms of interatomic interactions, the same symmetry has a quadrupole–quadrupole interaction. Thus, switching between positively and negatively buckled isomers (Figs. 8.7(b) and 8.7(c)) requires applying a complex tensorial field satisfying the above symmetry conditions.

8.3. Monoatomic Indium Wires on the Si(111) Surface

Although one-dimensional metal conductors largely remain theoretical constructs, a vast number of materials can be viewed as quasi-1D making them attractive due to potential applications in atomic-scale devices created via nanotechnology. Many metallic elements form arrays on the Si(111) surface that are, at least structurally, quasi-1D. In this regard, atomic chain structures on flat and vicinal Si surfaces induced by In adsorbate are the popular model systems. In-induced (4×1) reconstruction on Si(111) is a prototype for an atomic wire-type arrangement of metal atoms on a crystal surface [23]. Arranged in zigzag pattern, the indium atoms induce a *metallic* (4×1) structure [24, 25], which at critical temperature $T_{cr} = 125\,\mathrm{K}$ undergoes a reversible first-order transition to a low-temperature *insulating* (8×2) superstructure (Fig. 8.8) [26–29]. The structural model consists of two In zigzag chains ordered in a (4×1) periodicity that are separated by Si chains resembling the π-bonded chains of the clean Si(111) (2×1) surface (Fig. 8.8(a)). The phase transition slightly rearranges In atoms — a (8×2) surface periodicity is observed (Fig. 8.8(b)) [26, 28]. Let us analyze the structures and their transformations in the frame of the symmetry-based phenomenological approach.

8.3.1. *In/Si(111) archetype structure*

The Si(111)–(4×1)–In surface can be prepared by a two-step growth including (i) *in situ* deposition of In on to the well-ordered

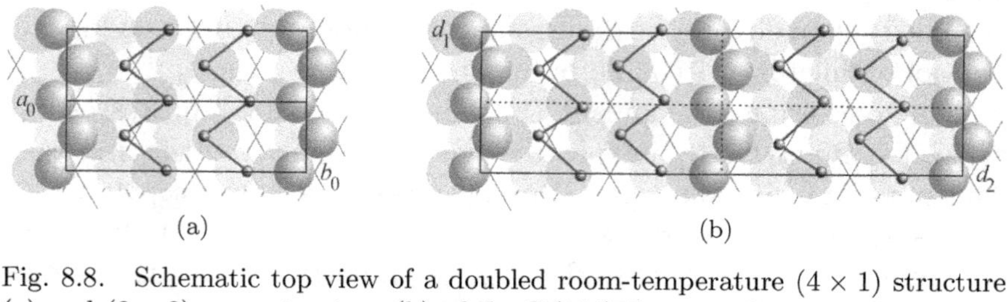

Fig. 8.8. Schematic top view of a doubled room-temperature (4×1) structure (a) and (8×2) superstructure (b) of the Si(111)–In nanowire array. Large gray spheres are Si atoms, and small black spheres are In atoms.

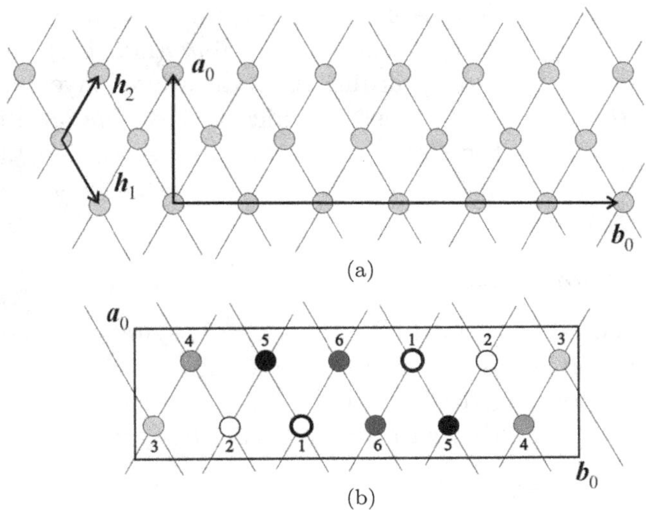

Fig. 8.9. (a) Disordered archetype structure ($\boldsymbol{h}_1 - \boldsymbol{h}_2$). (b) Ordered rectangular superstructure ($\boldsymbol{a}_0 - \boldsymbol{b}_0$).

Si(111) surface at room temperature followed by (ii) slow annealing at ~600 K [30, 31]. Generally, annealing is the fastest way to reach a stable state in the case of a quenched disordered metastable structure. Thus, it is logical to assume that randomly deposited on the crystal surface In atoms are self-assembling via ordering transformation to a certain two-dimensional crystalline form, and start our consideration from this phase transition. However, since this phase (just deposited metal layer) was not characterized experimentally so far, we regard it as a hypothetical archetype structure.

Just like we did in preceding sections, let us establish a unified language and notations for a new class of surface structures. Monoatomic surface superstructures can be regarded as metal-silicon layers on the Si(111) substrate [32]. The layers are slightly buckled but conveniently considered as plane, and consequently, they are described by the two-dimensional plane groups [33]. The lack of widely available group-theoretical information and easy-accessible symmetry tools makes us continue using the three-dimensional space groups, but in order to simplify the identification of the monoatomic surface structures, we provide as well their plane-group references.

We start our analysis from the hexagonal archetype structure shown in Figs. 8.3 and 8.9(a). Two points should be noted. First, due to the (quasi)planar character of the layer structures, we can simplify the description reducing the symmetry group G_0 to non-centrosymmetric $P6mm$ (plane group $p6mm$(no. 17)) instead of $G_0 = P6/mmm$. Second, regarding In–Si disordered layer, we should analyze a *three-component* system which includes metal atoms and vacancies. The archetype lattice is shown in Figs. 8.8 and 8.9 as thin lines.

8.3.2. *Self-organization on the (111) surface*

In our model, hexagonal net of $P6mm(Z = 1)$ symmetry is randomly filled in proportion Si $\sim 1/6$, In $\sim 1/3$, vacancies $\sim 1/2$. A superlattice corresponding to the ordered (4×1) structure is primitive rectangular (Fig. 8.9(b)) with the following basis vectors:

$$\boldsymbol{a}_0 = -\boldsymbol{h}_1 + \boldsymbol{h}_2, \boldsymbol{b}_0 = 6(\boldsymbol{h}_1 + \boldsymbol{h}_2). \tag{8.5}$$

Equations (8.8) define a critical reciprocal space vector $\boldsymbol{k}_6 = \frac{1}{12}(\boldsymbol{b}_1 + \boldsymbol{b}_2)$ which corresponds to the T-point of the hexagonal Brillouin zone (see Fig. 6.10(b)). Ordering occurs on the one-fold Wyckoff position 1(a):(0 0 0), therefore, corresponding six-dimensional permutation representation is induced by the one-dimensional totally symmetric little representation $\hat{\tau}_1$ of the six-arm $\boldsymbol{k}_6$ vector star [8]. According to Eq. (8.5), only two components of the order-parameter are non-zero, and the ordered phase possesses a space symmetry group $Pma2$ (point group $p2mg$(no. 7)). Figure 8.9(b) presents the splitting pattern of initial 1(a) position: In atoms are localized in positions 1 and 5, Si in 4, and the remaining positions are empty.

Irreducible representation $\tau_1(\boldsymbol{k}_6)$ belongs not only to permutation representation but also mechanical (displacive) representation of the archetype structure group. It means that both mechanisms, variation of occupation probability and displacements, are simultaneously involved in the scenario of $P6mm(1) - Pma2(12)$ transformation. Figure 8.10(a) shows atomic displacements corresponding to the $\tau_1(\boldsymbol{k}_6)$ order-parameter condensate $\langle \eta_1 = \eta_2 \neq 0, \eta_3 \div \eta_6 = 0 \rangle$, and Fig. 8.10(b) presents the Si(111)$-(4\times1)-$In structure as a result of combined ordering-displacive-type transformation.

Model-free energy expansion with the $\tau_1(\boldsymbol{k}_6)$ order-parameter components contains odd-degree terms that, according to the Landau condition (see Sec. 2.2), forbid continuous phase transitions in a system. However, the diffusive character of the ordering mechanism

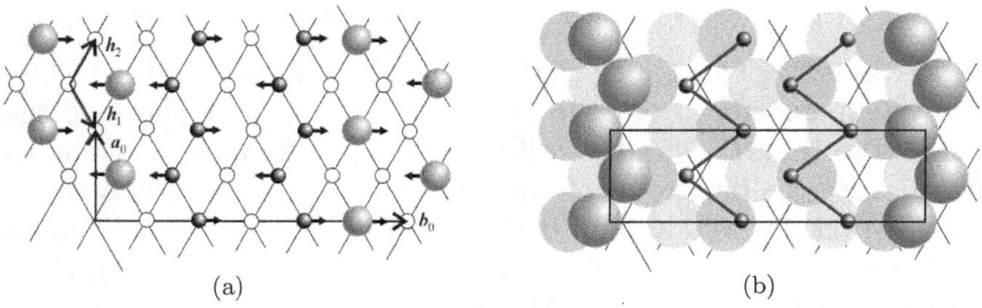

Fig. 8.10. (a) Displacive field corresponding to $\tau_1(\boldsymbol{k}_6)$. (b) Experimentally observed Si(111)–(4 × 1)–In structure [24, 25]. Large gray spheres are Si atoms, and small black spheres are In atoms.

makes it sluggish (pseudo-continuous), and long annealing instead of a jump-like discontinuous behavior is well justified for the (4 × 1) structure preparation process.

8.3.3. *(4 × 1)–(8 × 2) transformation: Hexagon vs trimer formation*

When studying a phase transition – planning an experiment, collecting and analyzing data, experimenters would like to have a set of criteria which allows to unambiguously distinguish between reliable information on the nature and driving forces of a transformation, and the results potentially affected by the conditions of the experiment: its equilibrium or non-equilibrium character, state of the sample, accuracy of the method and measurements, etc. This is an important issue, as actually the most efficient methods controlling the surface structural phase transitions are active methods which excite and steer a system toward the states far from equilibrium (see, e.g., Refs. [34, 35]). The symmetry-based approach, being based on the primary characteristics of the system – its structure and symmetry – provides us with model-free predictions and conclusions on *equilibrium states*, which can serve as selection rules and guidelines for the interpretation of experimental observations. In this regard, the metal–insulator transformation in quasi-1D indium structure on the Si(111) surface is an example case which demonstrates, on the one hand, the predictive capability and, on the other hand, the limitations of the symmetry-based phenomenological theory applied to a low-dimensional system.

8.3.3.1. *Parent phase and primary order-parameters*

It is reasonable, when analyzing a single phase transition in Si(111)–In system, to start from the (4×1) structure as a parent phase without accounting for the self-assembling ordering-type transformation. Symmetry group of the self-assembled structure is $Pma2$ where Wyckoff position 2(c):(0.25, y, 0.75-y) is occupied by Si atoms ($y \approx 0.875$), In1 ($y \approx 0.375$) and In2 ($y \approx 0.71$) (Fig. 8.10(b)). Comparing primitive unit cells of the (8×2) and (4×1) structures, we derive their basis vectors as follows:

$$\boldsymbol{d}_1 = 2\boldsymbol{a}_0, \boldsymbol{d}_2 = 2\boldsymbol{b}_0 \ (V'/V_0 = 4). \tag{8.6}$$

It follows, from Eqs. (8.6), that two symmetry-breaking order-parameters, belonging to X-point ($\boldsymbol{k}_{20} = \boldsymbol{b}_1/2$) and Y-point ($\boldsymbol{k}_{21} = \boldsymbol{b}_2/2$) of the orthorhombic (rectangular) Brillouin zone (Fig. 8.11) form the (8×2) superstructure. The X-order-parameter breaks translational symmetry along zigzag In chains, Y in perpendicular direction. Simultaneous distortion by a reducible order-parameter results in a surface superstructure possessing the space symmetry group Pc (point group pg) [36, 37]. Let us analyze the individual contributions of each order-parameter to the transformation mechanism.

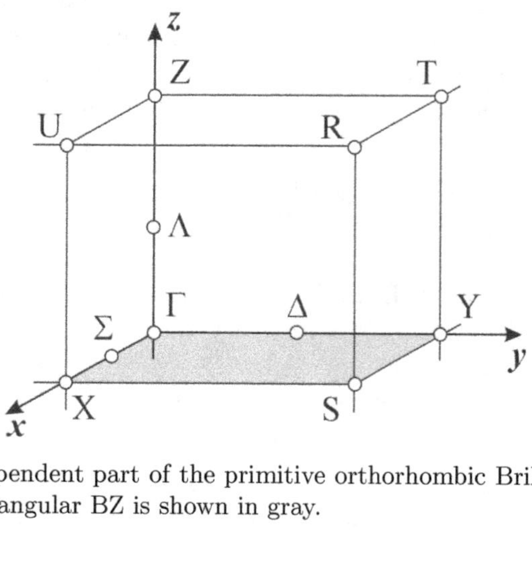

Fig. 8.11. Independent part of the primitive orthorhombic Brillouin zone. Part of the plane rectangular BZ is shown in gray.

8.3.3.2. *X-point order-parameter: Trimer formation in the (4 × 2) superstructure*

Order-parameter components span two-dimensional irreducible representation X_1, the only representation in the X-point ($\boldsymbol{k}_{20} = \boldsymbol{b}_1/2$). Its image group is $I_0 = 4mm$, and the relevant free energy has four minima: I. $\langle x_1 = x_2 = 0\rangle \sim$ parent $Pma2(1)$ phase, and three superstructures of the $(4 \times 2) \sim (\boldsymbol{d}_1 = 2\boldsymbol{a}_0,\ \boldsymbol{d}_2 = \boldsymbol{b}_0)$ type, II. $\langle x_1 \neq 0,\ x_2 = 0\rangle \sim Pm(2)$, III. $\langle x_1 = x_2 \neq 0\rangle \sim P2(2)$, and IV. $\langle x_1 \neq x_2 \neq 0\rangle \sim P1(2)$ (see Sec. 1.3.2.5). Holding a mirror plane in the low-temperature structure (8×2) (symmetry group Pm), we point out the solution no. II. Figure 8.12 shows displacive order-parameter condensate $\langle x_1 \neq 0,\ x_2 = 0\rangle$. It is easy to see that the In atoms form trimers where interatomic distances become shorter than in the parent (4×1) structure and in the neighboring In triangles of the low-symmetry phase.

Selection rules for the mechanisms, due to low orthorhombic symmetry, are not especially restrictive in the X-point of the Brillouin zone. Indeed, existence of a single irreducible representation means that all microscopic mechanisms have the same symmetry (bilinearly coupled, Sec. 2.4.4), and no primary mechanism can be singled out

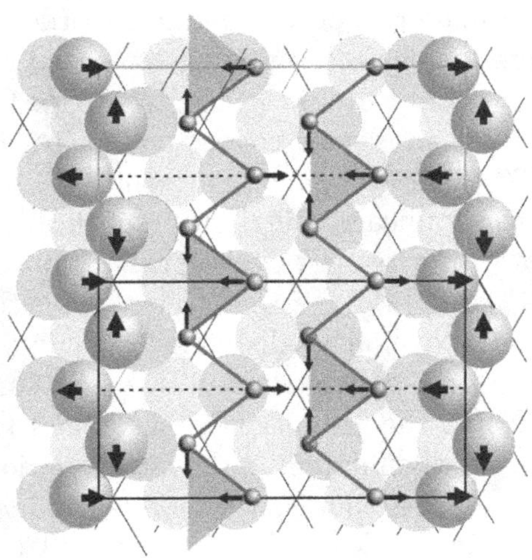

Fig. 8.12. Trimer (gray triangles) configuration in the (4×2) structure.

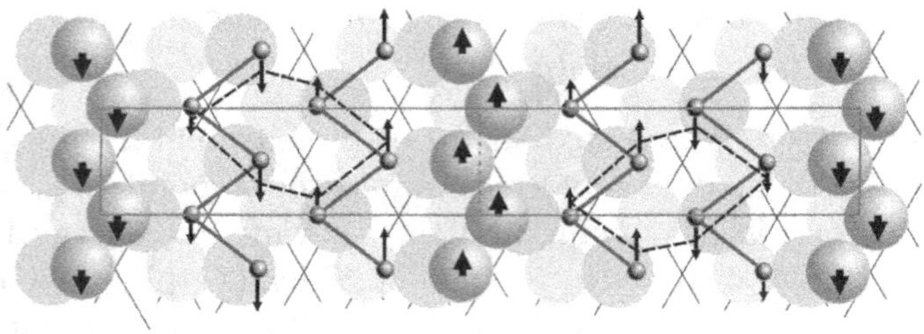

Fig. 8.13. Atomic displacements resulting in (8×1) superstructure. Dashed lines show In hexagons.

only on the symmetry basis. The $(4 \times 1) - (4 \times 2)$ transformation is, therefore, both electron- and phonon-driven, and the corresponding symmetry-breaking distortions occur simultaneously, equally contributing to the total structure distortion.

8.3.3.3. *Y-point order-parameter: Hexagon formation in (8 × 1) superstructure*

Among the four one-dimensional irreducible representations in the Y-point ($\boldsymbol{k}_{21} = \boldsymbol{b}_2/2$), only two, Y_2 and Y_4 (in notations of Ref. [9]), should be considered as candidates to the (4×1)–(8×1) symmetry-breaking order-parameter. Indeed, a reducible order-parameter $X_1 \oplus Y_i (i = 1$–$4)$ lowers the surface symmetry to space group Pm (point group pm) for $i = 1$ and 3, and space group Pc (point group pg) for $i = 2$ and 4, only the latter corresponding to the experimental observation [36, 37]. The difference pattern of (8×2) and (4×1) structures contains significant contribution of the Y_4 displacive basis function that gives priority to Y_4 as a primary order-parameter for both, (4×1)–(8×2) and (4×1)–(8×1), phase transitions. Figure 8.13 depicts Y_4-symmetric displacive distortion which transforms (4×1) structure to the (8×1) superstructure. We can conclude from the pattern that Y_4 favors the formation of hexagonal In configurations in the superstructure when doubling the lattice period perpendicular to the zigzag In chains. Although we confidently identify a Y-point phonon responsible for the destabilization of the (4×1) structure, due to low symmetry of the surface structure, we have again to consider as critical both phonon and electronic degrees of freedom.

8.3.3.4. *Phase diagram and transformation manifestations*

As verification of thermodynamic implications of the mechanisms described above, let us construct the theoretical phase diagram predicted for the surface phases in the Si(111)–In system. The form of the reducible $X_1 \oplus Y_4$ order-parameter expansion is

$$F(x_1, x_2, y) = a_1(x_1^2 + x_2^2) + a_2^{(1)}(x_1^2 + x_2^2)^2 + a_2^{(2)}(x_1^4 + x_2^4) + a_3(x_1^6 + x_2^6) + b_1 y^2 + b_2 y^4 + b_3 y^6 + \delta(x_1^2 + x_2^2)y^2, \quad (8.7)$$

which is identical to the model of Eq. (2.56). Due to the equilibrium conditions $\langle x_1 = x \neq 0,\, x_2 = 0\rangle$ for the order-parameter components in the (4×2) phase, it can be reduced to a simpler effective form

$$\tilde{F}(x, y) = a_1 x^2 + a_2 x^4 + a_3 x^6 + b_1 y^2 + b_2 y^4 + b_3 y^6 + \delta x^2 y^2. \quad (8.8)$$

Equation (8.8) is a symmetric sixth-degree expansion (compared to Eq. (2.55)) and was discussed in detail in Sec. 2.4.2c. In addition to the high-symmetry phase I: $(x = 0, y = 0) \sim p2mg(1) \sim (4 \times 1)$, corresponding equations of state describe three low-symmetry phases:

$$\begin{aligned} &\text{II: } (x \neq 0, y = 0) \sim p2gg(2) \sim (8 \times 1);\\ &\text{III: } (x = 0, y \neq 0) \sim pm(2) \sim (4 \times 2);\\ &\text{IV: } (x \neq 0, y \neq 0) \sim pg(4) \sim (8 \times 2). \end{aligned} \quad (8.9)$$

In order to show the principal topological features of the Si(111)–In phase diagram, we chose Figs. 2.22(f) and 2.22(h) as maximally corresponding to the experimental data (Fig. 8.14).

Inspection of the theoretical phase diagrams of Figs. 2.22 and 8.14 shows that (i) *direct* phase transition (4×1)–$(8 \times 2) \sim$ I–IV is possible only under the condition of negative interaction energy between two destabilizing displacive (electronic)-driven mechanisms ($\delta < 0$ in Eqs. (8.7) and (8.8)), and (ii) this transformation can be *only discontinuous.* The latter conclusion corresponds to the results of accurate experimental study in which a robust hysteresis loop with the width ∼10 K upon slow increase and decrease of the sample temperature at the phase transition was recorded [38].

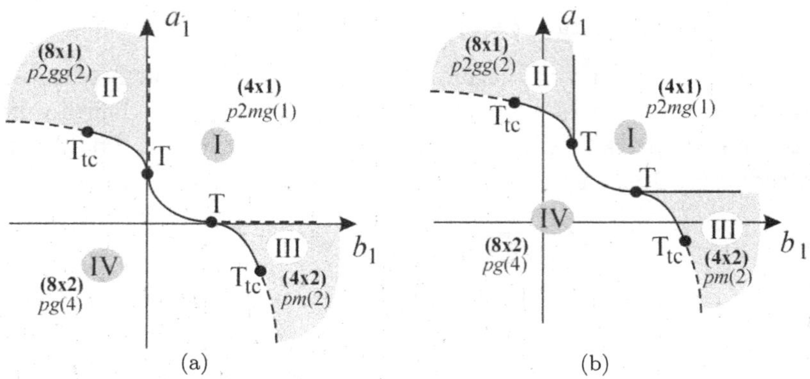

Fig. 8.14. Phase diagram corresponding to the free-energy expansion (8.8) for different relations between phenomenological coefficients: (a) $\delta < 0$, $a_2 > 0$, $b_2 > 0$, $\Delta < 0$; (b) $\delta < 0$, $a_2 < 0$, $b_2 < 0$, $\Delta > 0\,(\Delta = 4a_2b_2 - \delta^2)$. Solid and dashed lines are, respectively, the first- and second-order transition lines. T_{tc} and T points are the tricritical and triple points, respectively.

Equilibrium thermodynamic path (temperature axe t_{eq} on Fig. 8.15(a)) for the (4×1)–(8×2) transformation should be traced regarding the temperature behavior of Raman-active vibrational modes in the low-temperature phase. Figure 8.15(b) shows experimentally observed [39] frequency variation of the shear and rotatory modes, which demonstrate remarkable partial softening when approaching T_{IV-I}. The studied phonons ω_X and ω_Y were reliably assigned to characteristic modes by means of density-functional calculations and symmetry considerations [39, 40]. Comparing their eigenvectors and symmetric coordinates X_1 (Fig. 8.12) and Y_4 (Fig. 8.13) of symmetry-breaking zone-boundary phonons, we conclude on the active role of ω_X and ω_Y in the transformation. Zone-boundary phonons X_1 and Y_4 ($\boldsymbol{k}x \neq 0$) are not Raman-active in the high-symmetry phase but become observable in Raman spectra below T_{I-IV} due to the Brillouin-zone folding ($\boldsymbol{k}_{SM} \to 0$). According to the Birman–Worlock theorem, both of them must be totally symmetric in the low-symmetry phase (see Refs. [54, 55] in Chapter 7). Indeed, the ω_X and ω_Y modes are classified as totally symmetric A' and disappear above the transition temperature $T_{IV-I} = 125\,\mathrm{K}$ [39, 40]. Synchronous softening of ω_X and ω_Y proves simultaneous destabilization of the phase by X_1 and Y_4 order-parameters and allows us

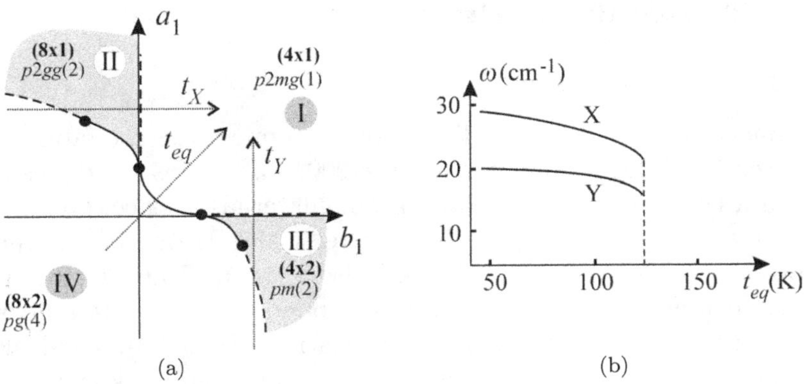

Fig. 8.15. (a) Theoretical phase diagram of the Si(111)–In system (for details, see Fig. 8.14). Dotted lines are the possible thermodynamic paths (see text). (b) Schematic temperature dependence of frequencies of the rotary (X) and shear (Y) modes according to Ref. [39].

to map equilibrium thermodynamic path $t_{\rm eq}$ on the theoretical phase diagram with projections to both the a_1 and b_1 variational coefficients (Fig. 8.15(a)).

On the one hand, there is no reason to assume a competition between the two order-parameters, trying to estimate which mechanism, forming trimers or hexagons, is dominating: both distortions are well synchronized and both occur at the $(4\times 1)-(8\times 2)$ transformation. On the other hand, bi-quadratic coupling of two symmetry-breaking order-parameters offers an interesting option which could be considered in the case of non-equilibrium regime of a surface phase transition. For sufficiently large negative interaction between the two order-parameters ($\delta < 2\sqrt{a_2 b_2}$), variation of one of them, despite the fixed parameters of another one, induces the onset of the second order-parameter. Such a "triggering" scenario is presented in Fig. 8.15(a) by paths $t_{\rm X}$ and $t_{\rm Y}$. Indeed, varying *only coefficient* a_1 (order-parameter $\rm Y_4$, path $t_{\rm Y}$) transforms (4×1) to intermediate structure (4×2) and then to (8×2). The similar applies to the path $t_{\rm X}$ and $\rm X_1$ order-parameters. Thus, a selective pulse perturbation of a single order-parameter is capable of launching a transformation process showing signatures of intermediate phases. Moreover, both transformations on this path can be of second-order, and no energy barrier has to be overcome by the system.

8.4. Twisted Bilayer Graphene

8.4.1. *Graphene*

Graphene was the first two-dimensional atomic structure fabricated by scientists [41,42]. First observed in 2004 [41], it has rapidly established a key role in the domain of nanomaterials science. Graphene is defined as a single layer of graphite, which is one of the main allotropes of carbon (Fig. 8.16; see also Figs. 6.17 and 6.18). It is ideally composed of only carbon atoms bonded together in a hexagonal honeycomb lattice via sp^2 hybridization (Fig. 6.20). The length of the C–C bond is approximately 1.42 Å, and the thickness of the layer is 3.5 Å [43]. In this way, graphite is regarded as being made up of many layers of graphene, with each layer interconnected to each other by weak van der Waals forces (Fig. 6.20).

8.4.2. *Superconductivity in graphene bilayer*

The stacking of a few layers in two-dimensional materials expands the possibilities of tailoring their properties. Besides ordered arrangements, which slightly alter the original single-layer symmetry, a relative rotation between layers can change the behavior of the system dramatically: a small relative twist between the layers induces insulating and superconducting states that were completely absent in a single graphene sheet [44,45]. Such a rotational stacking structure has been widely observed as well in epitaxially grown multilayer graphenes [46–49], and also it can be fabricated by manually aligning single layers using the transfer technique. The most famous twisted

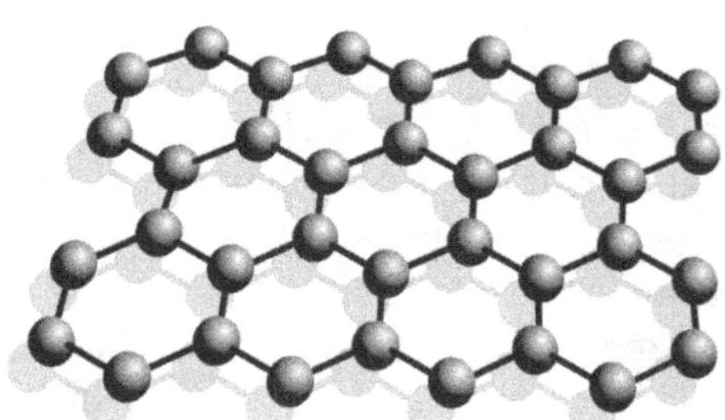

Fig. 8.16. Hexagonal structure of graphene.

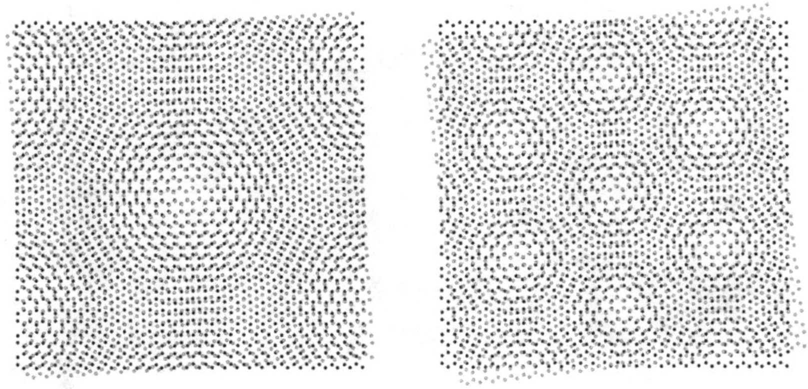

(a) (b)

Fig. 8.17. Moiré patterns of the graphene bilayer twisted by $2\varphi = 3°$ (a) and $2\phi = 6°$ (b).

bilayer is obtained by stacking one layer of graphene on top of another and rotating the two layers by a small twist angle, 2φ (see Fig. 4.14). This causes a long-wavelength moiré pattern (see Sec. 4.2.3.3) to arise between the lattices of two graphene layers (Fig. 8.17). For specific, so-called magic, angles of the order $2\varphi \sim 1°$, strong interlayer hybridization leads to a flat band formation with low energy and narrow bandwidth that is believed to cause superconductivity.

8.4.3. *Low-energy electronic bands*

Graphene has a unique electronic structure, which leads to a number of extraordinary properties not seen in conventional materials. It is common knowledge that the highest valence (π) band and the lowest lying conduction (π^*) band merge at the Fermi level at points K and K′ (the corner of the hexagonal Brillouin zone), known as Dirac points E_{Dirac}, where the energy distance between the two zones is close to zero (Fig. 8.18(a)) [50]. The intersection directly at the intrinsic Fermi level gives rise to graphene's zero-gap semiconducting nature and semi-metallic properties. In the proximity of these intersection points, the electron energy is linearly dependent on the wave vector (Dirac cone, Fig. 8.18(b)). This linear dispersion results in massless excitations described by the Dirac equation rather than the Schrödinger equation.

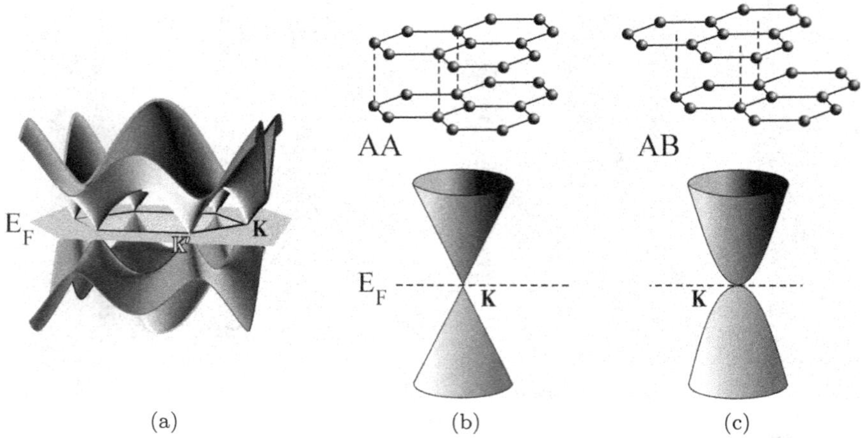

Fig. 8.18. (a) Band structure of single-layer graphene. (b) Magnified Dirac cones in monolayer or AA-stacked bilayer graphene. (c) Low-energy parabolic bands in AB-stacked bilayer graphene.

When two aligned graphene sheets are stacked, these energy bands can be substantially altered by the coupling between the layers. The electronic band structure of monolayer and bilayer graphene has been theoretically worked out by different authors using both a density functional and a tight binding model. It was shown that the energy bands depend on the stacking order: AA-stacked bilayer shows similarity to the monolayer Dirac-type band shape (Fig. 8.18(b)), while Bernal stacking order (AB-stacked, graphite-like) converts the linear bands into pairs of parabolic bands (Fig. 8.18(c)).

The energy spectrum of the bilayer graphene sensitively depends on its rotation angle φ. In a small φ, the interference between incommensurate lattices, manifested by the onset of moiré pattern, significantly influences the low-energy spectrum. However, the superlattice structure, independently of a "parent" stacking order, locally resembles the regular stacking such as AA and AB (Fig. 8.19). It should be mentioned that AB stacking can be considered to be twisted bilayer graphene from AA stacking by 60°.

Theoretical calculations, both in the frame of the tight-binding model [51] and solving a low-energy continuum Dirac model for twisted graphene [52], predicted the occurrence of flat electronic bands. Flat band indicates a type of band structure with constant energy independent of the crystal moment. The key feature of the flat

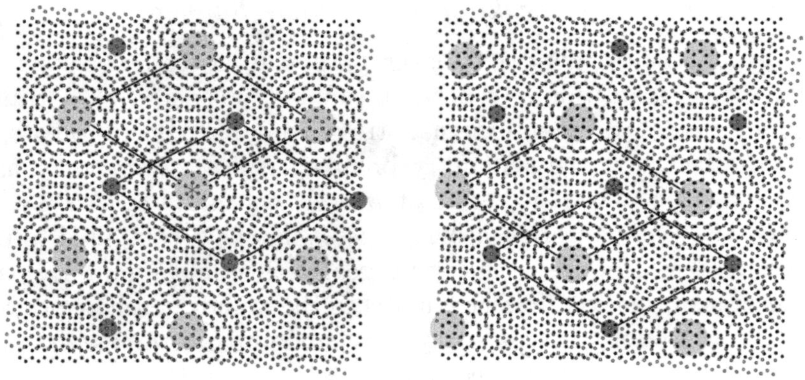

(a) (b)

Fig. 8.19. Twisted bilayer graphene at rotation angle $2\varphi = 5°$: (a) AA parent stacking order, (b) AB parent order. Large and small gray circles mark AA and AB stacking types correspondingly. Solid lines show sublattice unit cells.

band is that the electrons in it have a zero group velocity and infinite effective mass. The group velocity, which is the first derivative of an energy dispersion with respect to crystal momentum, characterizes the mobility of the electrons in the band. In the flat band, the group velocity is zero for all momenta in the Brillouin zone, and this implies that electrons in the flat band are localized. Thus, it becomes easy for the "slow" electrons to see each other and to feel the Coulomb repulsion, which is the so-called *strong correlation phenomenon* and most probably leads the material to become superconductive upon doping, as happens in the high-T_C superconductors. It has been argued that the flat-band system, in general, will enable superconductivity with strongly enhanced T_C value [53]. Indeed, for specific angles of the order $\sim 1°$, where very narrow bands have been seen, twisted bilayer graphene was shown to host a presumably unconventional superconducting state at $T < T_C \sim 1.5\,\mathrm{K}$ [44, 45].

Although the nature of superconductivity in the bilayer system is still being debated, it is clear that the appearance of the flat bands is a fundamental feature, and its emergence at low angles is key. In this regard, let us consider, on the symmetry basis, the mechanism of the rotation symmetry breaking in the bilayer and its generic consequences.

8.4.4. *Symmetry breaking in graphene bilayer*

The symmetry breaking in bilayer graphene, strictly speaking, is not a phase transition. At best, it can be considered as a transformation induced by an external stress field. The conclusion stems from the fact that the symmetry breaking is not *spontaneous*, and it does not occur as the result of *free-energy minimization*, i.e., a twisted layer configuration provides *no energy gain* compared to the untwisted one. However, the crystal-geometrical and symmetry part of the transformation perfectly matches the standard scheme developed in our book. Let us apply this model-free formalism to the bilayer graphene-twisting mechanism.

8.4.4.1. *Parent structures*

Bilayer graphene has two alternative modifications, AA- and AB-stacked layers (Fig. 8.20), which possess different symmetries and different structures. AA-stacked graphene (Fig. 8.20(a)) has a primitive hexagonal unit cell of the $P6/mmm(Z = 2)$ space symmetry (layer group $p6/mmm$), Wyckoff position $(1/3\,2/3\,0,\ 2/3\,1/3\,0)$ occupied (2(c) position in $P6/mmm$ or 2(b) in $p6/mmm$). AB-stacked bilayer (Fig. 8.20(b)) has a space group $P6_3/mmc(Z = 4)$, which logically coincides with the graphite symmetry group, Wyckoff positions 2(b): $(0\,0 \pm 1/4)$ and 2(c): $(1/3\,2/3\,1/4,\ 2/3\,1/3\,3/4)$ occupied. We already mentioned, in Sec. 8.4.3, that the AB stacking can be

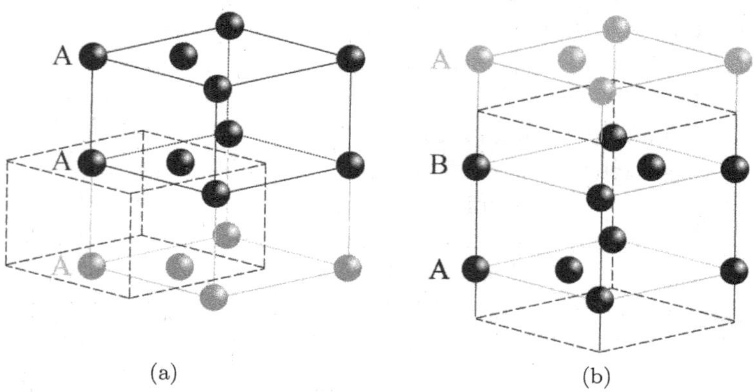

(a) (b)

Fig. 8.20. (a) AA- and (b) AB-stacked graphene bilayers. Dashed lines show a conventional hexagonal unit cell.

considered to be twisted bilayer from AA stacking by 60°. Moreover, a hexagonal moiré pattern in twisted bilayer consists of alternating AA- and AB-stacked regions (see Fig. 8.19). Thus, altering the parent structures in the twisted bilayer is seen as simply choosing a rotation origin point in a *distorted* structure of interest.

8.4.4.2. *Transformation mechanism*

Although, due to non-spontaneous and forced character of the transformation, we skip the stage of the free-energy minimization with respect to the order-parameter, it remains practically useful to define the latter as a crystallogeometric and symmetry-determining parameter. Relative rotation of stacked graphene layers is named, in classic solid mechanics, a torsion, and it is defined as the twisting of an object due to an applied torque. This is a macroscopic deformation, and corresponding symmetry characteristic functions belong to the Brillouin-zone center (Γ-point, $\boldsymbol{k} = 0$). Any rotation can be characterized by an *axial vector* $\hat{\boldsymbol{R}}$, which is a vector product of a radius vector perpendicular to the rotation axis and a body element/point displacement vector. For the twisting lattices we introduced in Sec. 4.2.3.3, more adapted vector product $\hat{\boldsymbol{T}}$ of their basis vector projections (Fig. 8.21):

$$\hat{\boldsymbol{T}} = [\boldsymbol{a}_i \times \boldsymbol{a}_{i\boldsymbol{R}}], \tag{8.10}$$

where $\boldsymbol{a}_i$ and $\boldsymbol{a}_{i\boldsymbol{R}}$ are the basis lattice vectors of untwisted- and twisted-graphene layers correspondingly.

The axial vector $\hat{\boldsymbol{T}}$ defined by Eq. (8.10) spans, in $6/mmm$ point group, a one-dimensional irreducible representation $\mathrm{A_{2g}}$ whose image group A2 (containing only 1 and −1 elements [9]) is, therefore, a symmetry group of the twist order-parameter. Two important conclusions are: (i) the twist order-parameter satisfies the Landau condition (see Sec. 2.2), thus, the corresponding free energy does not contain odd-degree terms, and the transformation (would it be spontaneous) could be continuous with zero enthalpy; (ii) neither phonon [54, 55] nor electron [56] spectra of bilayer graphene contain bands assigned to the $\mathrm{A_{2g}}$ irreducible representation of $6/mmm$. Thus, there is no direct (bilinear) coupling between the twisting mechanism and phonon and electron subsystems of the bilayer graphene.

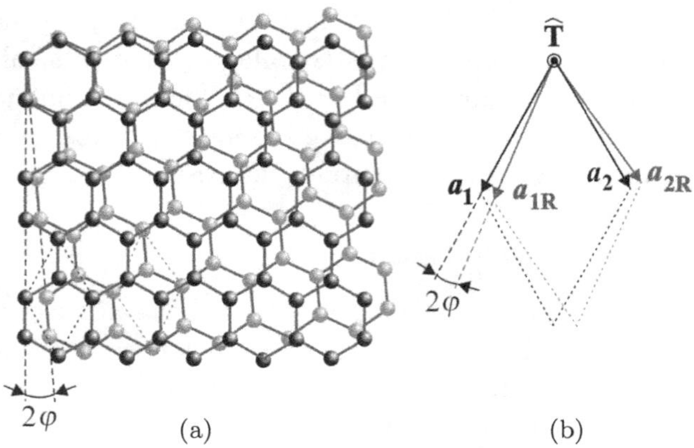

Fig. 8.21. (a) Graphene bilayers twisted by 2φ. Unit cells of graphene are shown by dotted lines. (b) Graphene-twisted unit cells. Axial vector $\hat{\boldsymbol{T}}$ is defined by Eq. (8.10).

8.4.4.3. *Incommensurate and long-periodic structures of bilayer graphene*

Just as infinitesimal twisting of bilayer graphene breaks its rotational and translational symmetry, the $6/m$ point group of twisted bilayer imposes less symmetry constraints than the parent $6/mmm$, however, the transformation is continuous, and accompanied secondary effects, therefore, are weak, one can expect no dramatic modifications in the parameters and properties of the material. By contrast, breaking the translational symmetry of the bilayer at the transition point and, correspondingly, changing the metric of the reciprocal (momentum) space plays a decisive role in establishing its anomalous properties.

Decreasing the translation symmetry of a crystal, i.e., multiplying the lattice parameters, leads, in the reciprocal space, to the folding of its Brillouin zone and energy bands. It is worth recalling that we already considered a case analyzing the effect of the phase transition on phonon dispersion curves in metallic Co (see Sec. 6.3.3 and Figs. 6.10 and 6.11). Moiré pattern observed in the twisted-graphene bilayer represents such a multiplied superlattice with basis parameters

$$|\boldsymbol{L}_1| = |\boldsymbol{L}_2| = \frac{|\boldsymbol{a}_i|}{2\sin\varphi}, \tag{8.11}$$

where $\boldsymbol{a}_i$ are the basis lattice vectors of the graphene structure (Fig. 8.21 and Sec. 4.2.3.3). The superstructure lattice parameters can be commensurate or incommensurate to the reference graphene lattice depending on the rational or irrational value of the sine function of the twist angle φ. For small angles, i.e., in the range of our special interest, the difference between *very long* superlattice period ($L \sim \varphi^{-1} \gg a_i$) and *incommensurate* one has no impact. Thus, at small twist angles, the evolution of the band structure of twisted-bilayer graphene can be viewed as semi-continuous; that is, an infinitesimal change in twist angle does not have a substantial effect on the band structure.

8.4.4.4. *Brillouin-zone folding and energy bands merging*

Two major effects on the energy bands can be predicted as a result of the onset of moiré long-period superstructure: (i) folding of the Brillouin zone merges energy bands close to or at exactly the zone center, and (ii) non-symmetric position of the critical $\boldsymbol{k}$-vector activates gradient invariants in the free energy. Both events should flatten the low-energy bands.

Two typical zone-folding schemas can be realized. Their occurrence depends on the position of the critical vector $\boldsymbol{k}_{\mathrm{OP}}$ which defines the translation symmetry of an order-parameter. Let vector $\boldsymbol{k} = \frac{1}{\alpha}\boldsymbol{a}_i^*$ connect Brillouin-zone center to a special symmetry point on the zone surface, edge or in the corner (general notation is Lifshitz points). An example is a Dirac point in graphene reciprocal space with $\boldsymbol{k}_{\mathrm{D}} = \frac{1}{3}\boldsymbol{a}_i^*$ (K-point). Then, two options are possible: (a) A lattice instability is localized on the Brillouin-zone surface or inside the zone, but it is identified by the vector $\boldsymbol{k}_{\mathrm{OP}} = \frac{1}{2n\alpha}\boldsymbol{a}_i^*$ with even-number denominator ($n = 1, 2, 3, \ldots$). A superstructure basis vector is $\boldsymbol{L}_i = n\boldsymbol{a}_i$, and Brillouin zone folds n times transferring the critical point $\boldsymbol{k}_{\mathrm{OP}}$ to the zone center (Fig. 8.22(a)). (b) A lattice instability is developing at the point $\boldsymbol{k}_{\mathrm{OP}} = \frac{1}{(2n+1)\alpha}\boldsymbol{a}_i^*$. A parent Brillouin-zone boundary point transfers to a position closer to Γ but not exactly at Γ (Fig. 8.22(b)). Its new position is fixed by the vector

$$\boldsymbol{q} = \boldsymbol{k}_{\mathrm{BZB}} = \frac{1}{2}\boldsymbol{a}_i^* \cdot \sin\varphi. \tag{8.12}$$

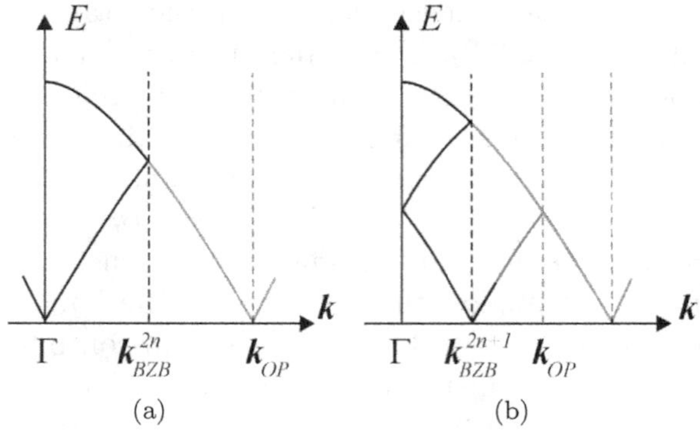

Fig. 8.22. Two schema of Brillouin zone and dispersion curve folding (see text). Dashed lines show parent (gray) and distorted structures (black) show zone boundaries.

Vector $\boldsymbol{q}$ is characteristic of not only the distance from Brillouin-zone center (Γ−K) but also the distances between minima of energy bands (bottoms of the potential wells, K−K′) transferred from the equivalent boundary points (Fig. 8.18(a)). At small twisting angle φ, the inter-minima distances, according to Eq. (8.12), become small ($\boldsymbol{q}(1°) \sim 10^{-3} \cdot \boldsymbol{k}(0°)$), and energy bands significantly overlap, merging and flattening, therefore, their bottoms [57,58]. Summarizing, the zone-folding effect in twisting graphene layers, in reciprocal space, first, *collapses the Brillouin zone* ($|\boldsymbol{q}| \to 0$ when $\varphi \to 0$) and merges the energy bands at Γ-point. With φ increasing, it enlarges a flat bottom part of the energy bands surrounding the superstructure zone center.

8.4.4.5. *Point-symmetry reduction and gradient free-energy terms*

The Brillouin–zone folding is a global effect which affects all, without exception, electron bands and phonon branches independently of their symmetry. Another effect, activation of gradient terms in the bilayer free energy, is more selective. Let us, first, analyze how the twisting bilayer changes the symmetry of the electron energy bands with special attention to the π- and π^*-bands merged at the Dirac point.

Figure 8.23(a) schematically shows $\boldsymbol{k}$-dependence of the p_z-electron energy along the $\Gamma - \mathrm{T} - \mathrm{K}$ line of the hexagonal Brillouin zone and assigns them to the irreducible representations of the point symmetry groups of the corresponding wave-vectors [56]. Twisting reduces the parent group chain $6/mmm{-}mm2{-}\bar{6}m2$ (Fig. 8.23(a)) to $6/m - m - \bar{6}$ (Figs. 8.23(b) and 8.23(c)), and consequently, establishes new assignment of irreducible representations running by the symmetry functions of the bonding (π) and antibonding (π^*) p_z-electrons. Odd- and even-type folding of Brillouin zone (Fig. 8.22) results, in the former case, as well as in case of incommensurate superstructure, to removing energy band degeneracy in the Dirac point ($E'' \rightarrow 2A''$) and to possible emerging of a gap in the energy spectrum.

The torsion order-parameter A_{2g} belongs to the Brillouin Γ-point (Sec. 8.4.4.2), i.e., to a *symmetry point* of the zone (Lifshitz point). A point $\boldsymbol{k}$ in reciprocal space is defined as a symmetry point if there does not exist in the neighborhood of $\boldsymbol{k}$ any point $\boldsymbol{k}'$ which has a point group $G_{\boldsymbol{k}'}$ symmetry equal to or higher than $G_{\boldsymbol{k}}$. A characteristic feature of symmetry/Lifshitz points is that they have numerically fixed coordinates: Γ(0, 0, 0) or K(1/3, 1/3, 0) is a typical example. According to the Lifshitz criterion (see Sec. 2.3), a crystal structure is stable with respect to the transformation to a non-crystalline (incommensurate, aperiodic) form if the order-parameter is located at a symmetry point [59]. However, this is a criterion established *only for spontaneous* symmetry breaking.

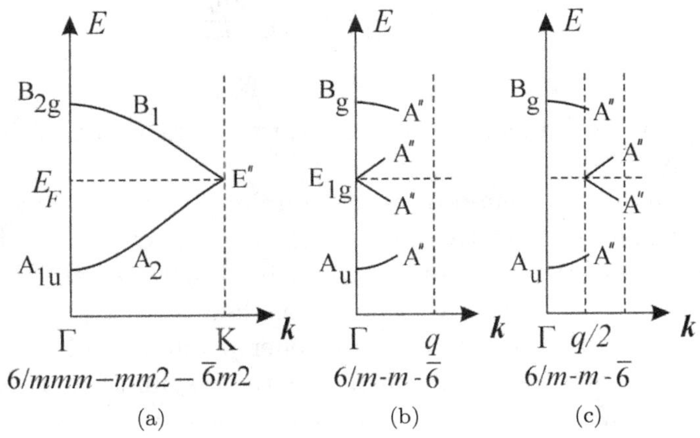

Fig. 8.23. Electron energy dispersion in the hexagonal parent (a), even (b), odd and irrationally folded (c) Brillouin zones (see text). Below, point-group chains of the vectors are depicted.

Since bilayer twisting is a forced distortion, it does not fall under the limitations imposed by the Lifshitz criterion. As a consequence, in order to account for possible or already realized incommensurability, we have to include gradient terms to the bilayer free energy making it a continuous function of the $\boldsymbol{k}$-vector (see Sec. 2.3 and Ref. [59]). It is important that the twisted bilayer graphene structure is a thermodynamically stable state of the material, and its equilibrium free energy is a function of polynomials invariant under operations from the symmetry group of the distorted but not parent phase. Another remark is, due to the violation of the Lifshitz criterion, gradient invariants are allowed to have a form not similar to the Lifshitz invariant $(\eta_i \frac{\partial \eta_j}{\partial x} - \eta_j \frac{\partial \eta_i}{\partial x})$ [59]. Let us construct lowest-degree invariants for the free energy with $\boldsymbol{k}$-vectors inside the Brillouin zone.

Invariant monomials or polynomials in the free energy have to span a totally symmetric representation of the symmetry group. $\boldsymbol{k}$-dependent functions are localized inside the Brillouin zone where relevant points possess the m-point symmetry group. The torsion order-parameter (axial vector) η representation A_{2g} is reduced to totally symmetric A', φ_{pz} functions and gradient $(\partial, \ldots, /\partial x)$ span A″ representation. Thus, the lowest-degree coupling term reads

$$I_C = \left(\varphi_{pz}^* \cdot \frac{\partial \eta}{\partial x} \pm \varphi_{pz} \cdot \frac{\partial \eta^*}{\partial x}\right), \tag{8.13}$$

as it is invariant under operations from the m group: $A''(\varphi_{pz}) \times A''(x) = A'(I_C)$.

The bilinear (pseudo-proper) coupling between the order-parameter spatial derivatives and p_z-electron functions significantly affects the π-electronic bands – the latter shows the same behavior under external parameter variations (see Sec. 2.4.4). The behavior, in turn, reflects the fact that the free-energy minimum in an incommensurate phase does not coincide with a symmetry-determined extremum as in a Lifshitz symmetry point, and one can have a continuous variation of the order-parameter translation symmetry (superstructure lattice parameters), i.e., variation of the $\boldsymbol{k}$-vector without changing the free energy [59].

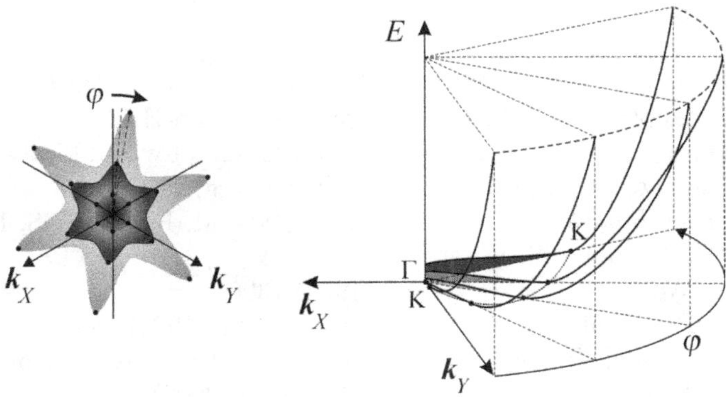

Fig. 8.24. Schematic electron energy band evolution depending on twist angle φ. (a) Top view, and (b) relief sections of the energy band along $\Gamma - \mathrm{T} - \mathrm{K}$ direction.

Thus, the incommensurate distortion can "slide" freely, and electron energy, due to bilinear coupling of Eq. (8.13), follows such a variation without barrier. For electron-energy relief of the twisted graphene bilayer structure, it means flattening of the corresponding energy bands and creating a "valley" along the Γ−T−K line in the hexagonal Brillouin zone (see Fig. 6.10(b)).

Figure 8.24 shows evolution of the merged electron energy band at the zone-center point Γ after Brillouin-zone folding. The process starts from the continuous increase of the central flat part of the integral band.

Then, with twist angle increasing, a low barrier progressively appears at the Γ-point due to the inter-minima distance shortening and band merge decreasing with φ increasing. However, for certain twist angle ranges, the band still can be considered as quasi-flat. Simultaneously, "valleys" are forming along six equivalent Γ−T−K directions. The corresponding lines, along with the entire Brillouin zone, twist by φ when two graphene layers rotate by 2φ angle (see Fig. 4.14).

It is worth underlining that the above scenario is model-free and drives solely from the symmetry of the bilayer and its twisting mechanism without reference to the microscopic details of the bilayer coupling.

References

[1] *Introduction to Nano. Basics to Nanoscience and Nanotechnology*, eds. A. Sengupta and C. K. Sarkar (Springer, Berlin-Heidelberg, 2015).
[2] K. Oura, V. G. Lifshits, A. A. Saranin, A. V. Zotov and M. Katayama, *Surface Science* (Springer, Berlin, Heidelberg, 2003).
[3] *Springer Handbook of Surface Science*, eds. M. Rocca, T. S. Rahman and L. Vattuone (Springer, Cham, 2020).
[4] E. A. Wood, *J. Appl. Phys.* **35** (1964) 1306.
[5] R. L. Park and H. H. Madden, *Surf. Sci.* **11** (1968) 188.
[6] *International Tables for Crystallography*, vol. E, eds. V. Kopský and D. B. Litvin (IUCr and Kluwer, Dordrecht, 2002).
[7] *International Tables for Crystallography*, vol. A, ed. T. Hahn (IUCr and Kluwer, Dordrecht, 1983).
[8] O. V. Kovalev, *Irreducible Representations of the Space Groups* (Gordon and Breach, New York, 1965).
[9] H. T. Stokes and D. M. Hatch, *Isotropy Subgroups of the 230 Crystallographic Space Groups* (World Scientific, Singapore, London, 1988).
[10] *Bilbao Crystallographic Server* (https://www.cryst.ehu.es/).
[11] H. T. Stokes, D. M. Hatch and B. J. Campbell, *ISOTROPY software suite* (https://stokes.byu.edu/iso/isotropy).
[12] *Landolt-Bornstein: Numerical Data and Functional Relationships in Science and Technology – New Series Group III: Condensed Matter*, vol. 45B, eds. G. Chiarotti and P. Chiaradia (Springer, Berlin-Heidelberg, 2017).
[13] D. M. Zehner, J. R. Noonan, H. L. Davis and C. W. White, *J. Vac. Sci. Technol.* **18** (1981) 852.
[14] D. Haneman, J. J. Rownd and M. G. Lagally, *Surf. Sci. Lett.* **224** (1989) 965.
[15] S. Ino, *Jpn. J. Appl. Phys.* **16** (1977) 891.
[16] S. Walter, J. Bernhardt , U. Starke *et al.*, *J. Phys. Condens. Matter.* **14** (2002) 3085.
[17] K. C. Pandey, *Phys. Rev. Lett.* **47** (1981) 1913.
[18] J. E. Northrup and M. L. Cohen, *Phys. Rev. B* **27** (1983) 6553.
[19] S. Nie, R. M. Feenstra, J. Y. Lee and M. H. Kang, *J. Vac. Sci. Technol. A* **22** (2004) 1671.
[20] G. Bussetti, B. Bonanni, S. Cirilli *et al.*, *Phys. Rev. Lett.* **106** (2011) 067601.
[21] K. Loser, M. Wenderoth, T. K. Spaeth *et al.*, *Phys. Rev. B* **86** (2012) 085303.
[22] S.-H. Lee and M.-H. Kang, *Phys. Rev. B* **54** (1996) 1482.
[23] P. C. Snijders and H. H. Weitering, *Rev. Mod. Phys.* **82** (2010) 307.

[24] O. Bunk, G. Falkenberg, J. H. Zeysing *et al.*, *Phys. Rev. B* **59** (1999) 12228.
[25] J.-H. Chao, D.-H. Oh, K. S. Kim and L. Kliman, *Phys. Rev. B* **64** (2001) 235302.
[26] H. W. Yeom, S. Takeda, E. Rotenberg *et al.*, *Phys. Rev. Lett.* **82** (1999) 4898.
[27] T. Kanagawa, R. Hobara, I. Matsuda *et al.*, *Phys. Rev. Lett.* **91** (2003) 036805.
[28] T. Tanikawe, I. Matsuda, T. Kanagawa and S. Hasegawa, *Phys. Rev. Lett.* **93** (2004) 016801.
[29] Y. J. Sun, S. Agario, S. Souma *et al.*, *Phys. Rev. B* **77** (2008) 125115.
[30] V. G. Lifshits, A. A. Saranin and A. V. Zotov, *Surface Phases on Silicon: Preparation, Structure, and Properties* (Wiley, Chichester, 1994).
[31] A. A. Saranin, A. V. Zotov, K. V. Ignatovich *et al.*, *Phys. Rev. B* **56** (1997) 1017.
[32] *Monatomic Two-Dimensional Layers. Modern Experimental Approaches for Structure, Properties, and Industrial Use*, ed. I. Matsuda (Elsevier, 2019); ch. 1.
[33] *International Tables for Crystallography*, ed. T. Hahn (Kluwer Academic Press, Dordrecht, 1996); vol. A, Part 6.
[34] T. Frigge, B. Hafke, T. Witte *et al.*, *Nature* **544** (2017) 207.
[35] J. G. Horstmann, H. Bockmann, B. Wit *et al.*, *Nature* **582** (2020) 232.
[36] C. Gonzalez, J. Ortega and F. Flores, *New J. Phys.* **7** (2005) 100.
[37] C. Gonzalez, F. Flores and J. Ortega, *Phys. Rev. Lett.* **96** (2006) 136101.
[38] F. Klasing, T. Frigge, B. Hafke *et al.*, *Phys. Rev. B* **89** (2014) 121107.
[39] E. Jeckelmann, S. Sanna, W. G. Schmidt *et al.*, *Phys. Rev. B* **93** (2016) 241407.
[40] E. Speiser, N. Esser, S. Wippermann and W. G. Schmidt, *Phys. Rev. B* **94** (2016) 075417.
[41] K. S. Novoselov, A. K. Geim, S. V. Morozov *et al.*, *Science* **306** (2004) 666.
[42] A. K. Geim and K. S. Novoselov, *Nat. Mater.* **6** (2007) 183.
[43] B. Partoens and F. M. Peeters, *Phys. Rev. B* **74** (2006) 075404.
[44] Y. Cao, V. Fatemi, S. Fang *et al.*, *Nature* **556** (2018) 43.
[45] Y. Cao, V. Fatemi, A. Demir *et al.*, *Nature* **556** (2018) 80.
[46] C. Berger, Z. Song, X. Li *et al.*, *Science* **312** (2006) 1191.
[47] J. Hass, R. Feng, J. E. Millan-Otoya *et al.*, *Phys. Rev. B* **75** (2007) 214109.
[48] G. Li, A. Luican, J. M. B. Lopes dos Santos *et al.*, *Nat. Phys.* **6** (2009) 109.

[49] D. L. Miller, K. D. Kubista, G. M. Rutter *et al.*, *Phys. Rev. B* **81**, (2010) 125427.
[50] A. H. Castro Neto, F. Guinea, N. M. R. Peres *et al.*, *Rev. Mod. Phys.* **81** (2009) 109.
[51] E. Suárez Morell, J. D. Correa, P. Vargas *et al.*, *Phys. Rev. B* **82** (2010) 121407.
[52] R. Bistritzer and A. H. MacDonald, *Proc. Nat. Acad. Sci. USA* **108** (20011) 12233.
[53] N. B. Kopnin, T. T. Heikkilä and G. E. Volovik, *Phys. Rev. B* **83** (2011) 220503.
[54] N. Mounet and N. Marzari, *Phys. Rev. B* **71** (2005) 205214.
[55] S. Piscanec, M. Lazzeri, F. Mauri and A. C. Ferrari, *Eur. Phys. J. Special Topics* **148** (2007) 159.
[56] E. Kogan and V. U. Nazarov, *Phys. Rev. B* **85** (2012) 115418.
[57] J. Wang, S. Deng, Z. Liu, Z. Liu *et al.*, *Natl. Sci. Rev.* **2** (2015) 22.
[58] T. T. Jia, M. M. Zheng, X. Y. Fan *et al.*, *Sci. Rep.* **6** (2016) 18869.
[59] J. C. Toledano and P. Toledano, *The Landau Theory of Phase Transitions* (World Scientific, Singapore, 1987).

Index

M

N

O

P

CPSIA information can be obtained
at www.ICGtesting.com
Printed in the USA
JSHW010724170223
37688JS00002B/15

9 781800 612914